KEW

The History of
the Royal Botanic Gardens

The Palm House after its latest restoration in 1984–90, and ready for restocking with plants.

KEW

The History of
the Royal Botanic Gardens

RAY DESMOND

Foreword by
Professor Sir Ghillean Prance

THE HARVILL PRESS
with
THE ROYAL BOTANIC GARDENS, KEW

First published in 1995 by
The Harvill Press
84 Thornhill Road
London N1 1RD

1 3 5 7 9 8 6 4 2

End papers show an aerial view of the Royal Botanic Gardens, Kew
in May, 1992 © Hunting Aerofilms 1992
The ornament on the title page is a detail of the Victoria Gate; that at the top
of each chapter is taken from the Palm House

A CIP catalogue record for this title is
available from the British Library

ISBN 1 86046 076 3

Set in Linotron Caslon Old Face No. 2
by Rowland Phototypesetting Ltd,
Bury St Edmunds, Suffolk

Printed and bound in Great Britain by Clays Ltd, St Ives plc

The Random House Group Limited supports The Forest Stewardship
Council® (FSC®), the leading international forest-certification organisation.
Our books carrying the FSC label are printed on FSC®-certified paper.
FSC is the only forest-certification scheme supported by the leading
environmental organisations, including Greenpeace. Our
paper procurement policy can be found at
www.randomhouse.co.uk/environment

To
Margaret and Craig
Josephine and Hannah

CONTENTS

FOREWORD

The Royal Botanic Gardens, Kew has had a long and rich history and has played a most important rôle in the life of the Nation. Since Princess Augusta, Princess of Wales and mother of King George III, installed a small nine acre garden at Kew in 1759, the gardens have constantly developed and progressed to their current prominence in the world of botanic gardens. As we read in this fascinating account of Kew's history we see that it has not been without its ups and downs, arguments with royalty, politicians and local residents, and a constant battle to obtain adequate funding to carry out its mission. However, we also see the progressing of a most dynamic institution that has constantly built on its past history rather than reject it. It is vital that an institution of the stature of Kew acknowledges its history and develops from it, and so this new, up-to-date, and thoroughly researched history will not only be intriguing reading for the public, it will also be of great use to our staff as we plan the future.

The vast number of prominent people who have interacted with Kew over the years gives many glimpses into the recent history of our nation and not just of the Royal Botanic Gardens, Kew. We read about well-known scientists such as Charles Darwin, prime ministers, members of the Royal Family, architects, landscape gardeners, engineers; but of greatest importance is the long series of staff gardeners, scientists and directors, who have dedicated their lives to making Kew what it is today.

The author of this definitive history of Kew, Ray Desmond, is highly qualified to write about Kew. Nobody learns more about the institution where they work than the librarian. A former librarian of Kew and the author of several other books about the history of botany and botanists, Ray Desmond is an expert in his field. The thoroughness of his research for this book is seen by the extent of the notes on each chapter and by the broad coverage of the text. I only wish that this volume had been available to me when I became Director, because it is essential to know the history of the institution which one is leading. I thank Mr Desmond for making such an important contribution to the life of Kew. It will be of great importance as we lead this great botanic garden into the twenty-first century.

GHILLEAN T PRANCE
Director
MARCH 1995

LIST OF COLOUR PLATES

INTRODUCTION

Most visitors come to Kew to admire its floral displays and to relax in pleasant
surroundings, perhaps unaware that it is primarily a scientific institution dedicated to research in taxonomy, anatomy, cytogenetics, biochemistry and conservation. It is also custodian of a historic landscape which has evolved over
several centuries. Charles Bridgeman introduced innovative features to Queen
Caroline's garden, Princess Augusta's gardeners implemented her late husband's plans for his Kew estate, 'Capability' Brown indulged his skills in
naturalism, while W.A. Nesfield pursued a preference for geometric display.
Kew today is a palimpsest of garden styles, of piecemeal development and
improvisation, of pragmatic solutions to needs as they arise – finding space,
for instance, for a heathery, or a rock garden, or a rose pergola, or a pond
for aquatic plants. Fragments of earlier layouts still survive. All the undulations
are man-made, the oldest probably being a mound in the south-west corner
of the grounds, a podium, perhaps, for a garden temple in Richmond Gardens;
the latest has been formed during the making of the Secluded Garden. The
Pond in front of the Palm House is a remnant of a lake conceived by Frederick
Prince of Wales.

The River Thames imposed its presence on several schemes: the Duke of
Ormonde placed a summerhouse on its banks; Queen Caroline extended the
celebrated riverside terrace; and 'Capability' Brown integrated it as he did with
any convenient sheet of water. Sir William Hooker reluctantly obscured a
view of the river when he made a plantation of trees to hide the industrial
development in the Brentford docks. The stretch of the Thames from Hampton
Court to Chiswick – embracing Richmond Gardens and the garden at Kew –
has been called the cradle of the English Landscape movement. In 1994 the
Thames Landscape Strategy was launched with Government support and the
backing of the Countryside Commission, English Heritage, the Royal Fine
Art Commission and local authorities. Kew is sympathetic to its aims to preserve
what has survived of the natural and man-made landscape on this part of the
Thames, to restore some of its lost vistas, and to develop its potential for
recreation and tourism.

"Imperial Kew by Thames glittering side" was how Erasmus Darwin hailed
the Royal Gardens in his poem, *The Botanic Garden* (1789–91). He had in
mind its exotic flora, culled from Britain's colonies. Sir Joseph Banks, the two
Hookers and Sir William Thiselton-Dyer, however, viewed the Gardens they

"The river now proudly flows between the spreading lawns of Sion, and the Royal Gardens of Richmond; which, together, form a scene of superior grandeur and beauty." J.J. Boydell: *History of the Principal Rivers of Great Britain*, vol.2, 1796, 30. Both gardens had been landscaped by 'Capability' Brown.

controlled as an instrument of imperial endeavour, as a legitimate means of developing the natural resources of the Empire. Banks, for instance, supervised the transfer of the breadfruit from the Pacific to the West Indies. He treated colonial botanic gardens as useful centres for the acquisition and propagation of economic crops. The Hookers encouraged the development of these overseas gardens. Kew's involvement in the transfer of cinchona and rubber from South America to India and elsewhere has been well-documented, but comparatively little has been written about the Colonial Office's dependency on Kew's expertise in botanical and horticultural matters.

While an advocate of such commercial enterprise, Sir Joseph Hooker was well aware of some of the disastrous consequences of indiscriminate exploitation. A letter from Downing Street to the Governor of Ceylon in June 1873 conveyed his alarm about the destruction of the forests there by coffee planters.

> As however I observe that you dwell rather upon the loss of valuable timber than upon the injury to climate caused by extensive clearings, I take the opportunity afforded to me by Dr Hooker's letter of directing your particular attention to the very serious consequences to the climate of the island which may result from the further destruction of the forests, and of strongly impressing upon you the importance of effective steps being at once taken to preserve the timber as far as practicable on lands over which the Crown has control. ("Ceylon forests" 1873–1900, f.2. Kew archives).

The Thames between Syon House and Kew Gardens still retains much of its rural charm.

The deterioration of the environment has worsened since Sir Joseph Hooker was Director, and Kew has now made a firm commitment to do whatever it can "to reduce the rate of destruction of the world's plant species and their habitats" (to quote the mission statement published in Kew report for 1987– 90). Kew's initiatives include the compilation of regional floras – the first step in any systematic conservation exercise, the formation of a Seed Bank, and participation in international conservation programmes.

Kew's evolution from the two royal gardens into the world's premier botanic garden has not been smooth or uneventful. Retrenchment or undesirable directives from government ministers have from time to time threatened its existence or its character. It suffered from severe underfunding during the 1830s when its future was in the balance. Sir William Hooker, ordered by one of his First Commissioners to plant an excessive number of flower-beds, feared that Kew's role as a botanic garden would be subordinated to its attractions as a public park. Mr Gladstone had to intervene in Sir Joseph Hooker's bitter confrontation with A.S. Ayrton who did his best to diminish Kew's status as a scientific institution. Official policy had to reconcile the needs of recreation and research, of aesthetic considerations and scientific display.

My narrative, which encompasses the broader perspectives of Kew's overseas activities, the fulfilment of scientific and educational objectives, and the pattern of its evolving landscape, is also intended to serve as a work of reference on most aspects of Kew's history. I know from my own experience how varied, and sometimes how searching, are the questions the public ask. For instance, when were admission charges first imposed? When was a local constabulary formed? What is Kew's largest tree? When was the Temple of Bellona moved to its present site? When did Kew last use the Wardian case? What is the significance of the dated small stones scattered about the grounds? In order not to overload the general text with a density of facts, it seemed to me that the best solution would be to relegate this kind of information to a series of appendices. The longest appendix is a chronology. Besides being a repository of facts and contemporary comment, it will also, I hope, provide a useful overview or summary of the important occurrences in Kew's long history.

ACKNOWLEDGEMENTS

My research could not have been accomplished without the facilities of the Kew Library and the co-operation of its staff. In particular I wish to thank Miss Sylvia FitzGerald, John Flanagan, Mrs Cheryl Piggott, Miss Marilyn Ward and Miss Barbara Lowry. I am also grateful for the occasional help I have received from Paul Davies, David Field, Jim Keesing, Michael Maunder, John Simmons and Bernard Verdcourt. Media Resources, and in particular Andrew McRobb, expedited my numerous photographic requests, and Mrs Ann Lucas and her typists efficiently and patiently deciphered my handwriting. I am also grateful for the support this project has received from the Director, Professor Sir Ghillean T. Prance, and the Keeper of the Herbarium, Professor G.Ll. Lucas.

I am much indebted to the Royal Library and Royal Archives at Windsor, and to the Secretary and Keeper of the Records at the Duchy of Cornwall office for allowing me to consult their collections. Other institutions whose resources were made readily available to me include the Bodleian Library; the Botany Library of the Natural History Museum; the British Library (especially its Manuscripts Department and the Map Library); Buckinghamshire Record Office; Glynn Vivian Art Gallery, Swansea; Gray Herbarium Library at Harvard University; Guildhall Library, London; Heinz Gallery of the Royal Institute of British Architects; Huntington Library, California; Minet Library; National Library of Scotland; Nottingham University Library; Public Record Office; Richmond Local Studies Library; Sir John Soane's Museum; Surrey Record Office.

Among individuals who have assisted me I must mention my friend of many years, the late Leslie Paton; Mrs K. Cooke; Edward Diestelkamp; Miss Gina Douglas; Miss Olwen Hedley and John Cloake, who both generously gave me the benefit of years of research; and, lastly, this book's designer and my friend, Miss Vera Brice.

CHAPTER ONE

A Queen's Garden

*Queen Caroline commissions Charles
Bridgeman to landscape the grounds of
Richmond Lodge — William Kent adds garden
buildings — the Hermitage — Merlin's Cave —
public comment*

THE NEW FASHION in landscape gardening blended the Georgian villages bordering the Thames from Richmond to London into a succession of "beautiful buildings, charming gardens and rich habitations of gentlemen of quality", making the river, in Daniel Defoe's opinion, infinitely superior to the Seine, the Danube and the Po. It was England's equivalent to the country villas that lined the Brenta in Veneto in northern Italy, and the stretch of the river between sprawling Hampton Court and diminutive Kew Palace especially attracted Defoe's approval. In this rural retreat, extolled as the "matchless Vale of Thames" by Richmond's poet, James Thomson, the Court and aristocracy, statesmen, city merchants, men of letters and painters created gardens with mounts and summerhouses to contemplate the meandering river and the contours of Richmond Hill.

Alexander Pope, in Twickenham with "A river at my garden's end", believed that "No scenes of paradise, no happy bowers, [are] equal to those on the banks of the Thames." He made a *camera obscura* of his grotto on whose walls he could view "the objects of the river, hills, woods, and boats". His neighbour, Horace Walpole, thought his own "prospect is as delightful as possible, commanding the river, the town and Richmond Park". Mrs Pye in her survey of Twickenham's notable houses, published about 1760, confessed "to a repetition of the words view, prospect, etc. till I am tired of them myself".

One of the riverside dwellings commended by Defoe in the early 1720s was Richmond Lodge, the summer residence of the Prince and Princess of Wales, situated in the Old Park of former Richmond Palace (now the Old Deer Park) with its views to Isleworth and the Duke of Northumberland's Syon House.[1] Possibly once occupied by Cardinal Wolsey after his fall from royal favour, it became a hunting lodge for James I; William III enlarged it and his gardener, George London, created a tree-lined avenue to the Thames. The property was leased in 1707 to James Butler, Duke of Ormonde, who partially rebuilt the house and extended the grounds by renting additional land lying north towards Kew, improvements which impressed John Macky when he visited Ormonde Lodge in 1714.

> A perfect Trianon, everything in it and about it answerable to the grandeur and magnificence of its great master . . . There is a fine avenue that runs from the front of the house to the Town of Richmond at a half a mile distance one way [known as Wild Chestnut Walk], and from the other front to the Riverside, both [avenues] inclosed with balustrades of iron. The gardens are very spacious and well kept. There is a fine terrace towards the River. But above all the woods cut out into walks, with plenty of birds singing in it, make it a most delicious habitation . . .[2]

After the unsuccessful rebellion of the Jacobites in 1715, the Duke of Ormonde, being one of their supporters, was forced into exile, forfeiting all his estates, including Ormonde Lodge.

The red-brick house, its arcadian situation and perhaps the potential of its large garden, attracted the attention of the Prince and Princess of Wales who, banished from St James's Palace by a hostile George I in 1717, moved into it in 1718, renaming it Richmond Lodge. Caroline, Princess of Brandenburg-Anspach, had married George Augustus, the only son of George I in 1705. During her youth in Germany she had watched the formation of a Le Nôtre-style garden at the Charlottenburg Palace in Berlin, and knew intimately the baroque garden at Herrenhausen in Hanover. Thus she had acquired a fondness for gardening not shared by her husband, who preferred hunting and studying the genealogy of Europe's princely families. Soon after their arrival in England in 1714, when they were created Prince and Princess of Wales, they incurred the displeasure of George I, an obstinate man with few intimate friends. It has been said of Princess Caroline that she shrewdly exercised "the virtues of compromise and conciliation"; after their enthronement in 1727 she indulged her desire for power through the King, sometimes by deferring to his wishes, often by deception and by a mutually beneficial alliance with the Prime Minister, Sir Robert Walpole. For relaxation she enjoyed theological speculation and gardening. We learn from a letter that Alexander Pope wrote to his friend, Lord Bathurst, on 13 September 1719 that he had been invited by the Princess to a meeting to discuss alterations to her garden at Richmond Lodge. He described the consultation with typically cynical amusement.

> Several Criticks were of several opinions: One declar'd he would not have too much Art in it, for my notion (said he) of gardening is, that it is only sweeping Nature; Another told them that Gravel walks were not of a good taste, for all of the finest abroad were of loose sand: A third advis'd peremptorily there should not be one Lyme-tree in the whole plantation; a fourth made the same exclusive clause extend to Horse-chestnuts, which he affirmed not to be Trees, but Weeds; Dutch Elms were condemn'd by a fifth; and thus about half the Trees were

Engraving of Richmond
Lodge. John Lawrence: *A
New System of Agriculture*
1726, frontispiece. George
London created the avenue
leading to the Thames
during the 1690s, and
possibly the line of trees
from the front of the
house.

proscribed, contrary to the Paradise of God's own planting, which is expressly said to be planted with *all trees*. There were some who cou'd not bear Ever-greens, and call'd them Never-greens; some, who were angry at them only when cut into shapes, and gave the modern Gard'ners the name of Ever-greens Taylors; some who had no dislike to Cones and Cubes, but wou'd have 'em cut in Forest-trees; and some who were in a passion against any thing in shape, even against clipt hedges, which they call'd green walls.[3]

Pope, now in his thirtieth year, had just leased a riverside villa with five acres of land at Twickenham where he was to spend the rest of his life obsessively shaping and planting his garden. On 22 March 1725 he proudly announced to the Earl of Oxford that he had "just turfed a little Bridgmannick Theatre myself. It was done by a detachment of His workmen from the Prince's [i.e. Prince of Wales at Richmond Lodge]."[4] This is the first indication of the participation of the landscape gardener, Charles Bridgeman, in the development of the Richmond Lodge estate. Confirmation that it was undergoing transformation appears in a letter that César de Saussure wrote to his family in France in June 1726. "The Princess takes a great interest in the gardens, which are spacious and she has greatly embellished them."[5] A frontispiece engraving in a book published in the same year and dedicated to the Princess of Wales, shows two solid rectangles of woodland flanking the riverside avenue with just a glimpse of formal planting.[6]

After the coronation in 1727, Parliament approved a generous Civil List for the monarchy including £100,000 a year for Queen Caroline, about double that ever received by any previous Queen of England. In addition, Somerset House and Richmond Lodge were bestowed upon her, the latter possibly intended as a dower house should she survive her husband. By 1729 Caroline had extended her estate by leasing several parcels of land, including the Dutch House and three adjacent houses near Kew Green. The Bowling Green in Richmond was also incorporated, and houses around the Green and in the Kew Foot Road were acquired and demolished to make new entrances into her grounds, which now extended over 400 acres from Richmond Green tapering to Kew Green in the north, bounded by the Thames on the west and Love Lane (which corresponded roughly to the line of Holly Walk in Kew Gardens today) on the east. The subsequent landscaping, although carried out with professional assistance, interpreted her constant objective of "helping nature, not losing it in art".[7]

Although Queen Caroline's horticultural education had been influenced by German precepts of formality, she was nevertheless receptive to current innovations and changes in English garden styles. English seventeenth-century gardens had reflected continental fashions: Italian waterworks, French grand formal vistas and Dutch intricate flower-beds and topiary. The early eighteenth-century garden was an intermediate stage, a landscape reflecting contemporary philosophical, literary, aesthetic and even political thought. Lord Shaftesbury, deploring "the formal Mockery of Princely Gardens", advocated "things of a natural kind". Joseph Addison also promoted the concept of the landscape garden, convinced that "there is something more bold and masterly in the rough, careless strokes of Nature, than in the nice touches and embellishments of Art". Alexander Pope, denouncing the excesses of clipped hedges, urged a return to the "amiable simplicity of unadorned nature". He was one of the first to make the connection between landscape painting and the disposition of elements – buildings as well as planting – in a garden. In early eighteenth-century gardens there was a conscious effort to mix the formal and the natural, to create variety and irregularity. Boundary walls were often concealed as ha-has to extend the visual limits of the landscape. These gardens invited exploration, introduced carefully-contrived surprises, and evoked a calculated reaction to a medley of garden buildings. In order to experience the desired impact visitors were recommended to follow a prescribed route; such was the procedure at Richmond Lodge, one of the major showpieces during this transitional phase.

The early English landscape garden was nurtured by a group of professional designers and architects among whom Charles Bridgeman and William Kent were leading practitioners. Both worked for Queen Caroline at Richmond Lodge. Little appears to be known of Bridgeman's youth, but it is possible that he started his professional career in the Brompton nursery of George London and Henry Wise. In 1709 he drew a plan of the grounds of Blenheim Palace for Wise, who had been appointed Royal Gardener by Queen Anne seven years earlier. In 1724 he, together with Alexander Pope and Lord Bathurst, designed a garden at Marble Hill in Twickenham for Mrs Henrietta Howard, George II's mistress. As partner of Wise in 1726, he was automatically elevated to "one of HM's Principal Gardiners". When Wise retired two years later, Bridgeman succeeded him as Royal Gardener, perhaps with the covert support of Queen Caroline. His responsibilities for the royal gardens excluded Richmond Lodge which was subject to a separate contract. He began work there by the mid 1720s and possibly earlier. The Public Record Office has a plan of Richmond Lodge, Gardens and Park which is unsigned and undated but since it refers to the Prince and Princess of Wales it has to be before June 1727, the occasion of their accession.[8] Probably by Bridgeman, it shows the house and gardens occupying about 380 acres; eleven acres around the Lodge were laid out in formal plots; 22 acres of woodland with sinuous paths lay to the north of the riverside avenue but much of the woodland recorded on Rocque's plan of the estate in 1734 is still virgin fields; there is, for instance, no Great Oval, Grass Plot or Amphitheatre. Bridgeman retained the two principal avenues from the house and the Duke of Ormonde's riverside Terrace. His progress can be observed on a series of plans of Richmond Lodge by the Huguenot surveyor and cartographer John Rocque.[9] Lord Egmont noticed some changes in August 1730: "several new walks made through the Park and gardens. One of them is a mile long, reaching from Richmond town to Sir Charles Ayre's [i.e. Eyre's] house on Kew Green which the Queen bought at his death".[10] In 1733 a mount near Bridgeman's canal was raised to view the Thames and countryside.[11] A few years later another mount was sited

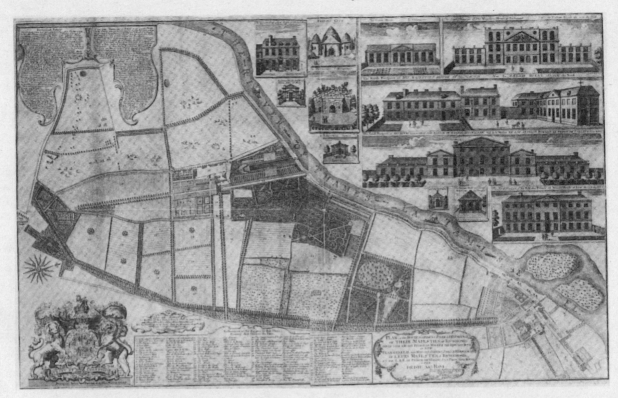

near the Duck Pond and Merlin's Cave. The extension of the riverside Terrace with a retaining wall was started in 1729 and finished in 1734 when it reached Lady Clinton's house at Kew. This grassed Terrace, lined by a row of elms on the garden side, attracted fashionable crowds, especially on Sunday evenings in summer.

Visitors beginning their tour of the grounds from the Richmond Green entrance were directed to the neat brick dairy overlooking the canal, where they paused to admire its display of delicate china bowls and jars. A path through a grove of trees led to the Tuscan temple on a mount with fine views beyond the garden. Retracing their steps to the dairy, sightseers would cross the riverside avenue into a wood (probably part of the original planting by the Duke of Ormonde) to the Queen's Pavilion. Here some paths led to the Duke's summerhouse on the Terrace; others plunged into the wood where an exploration of its labyrinthine walks with closely clipped hedges would eventually disclose the first of the garden's two extraordinary buildings – Merlin's Cave, which faced the large rectangular Duck Pond and was flanked by a tree-covered mount. The Forest Oval, 500 feet in diameter, and the Hermitage were reached after a short walk. Adjacent was an amphitheatre and a diagonal wilderness. This itinerary may give the impression that the estate was entirely wooded; much of it, however, comprised a vast expanse of meadows and cornfields.[12] When visitors reached the diagonal wilderness they had a choice of a ten-minute walk to Kew Green or of returning to Richmond Green by the much longer Forest Walk which ran parallel to Love Lane.

When this *Plan of Richmond Gardens* was produced by John Rocque in 1734, much of Bridgeman's landscaping had been done. The only known illustration of the Queen's House at Kew is at top left. This had been purchased by Queen Caroline in 1728. The White House and its modest garden are in the bottom right-hand corner.

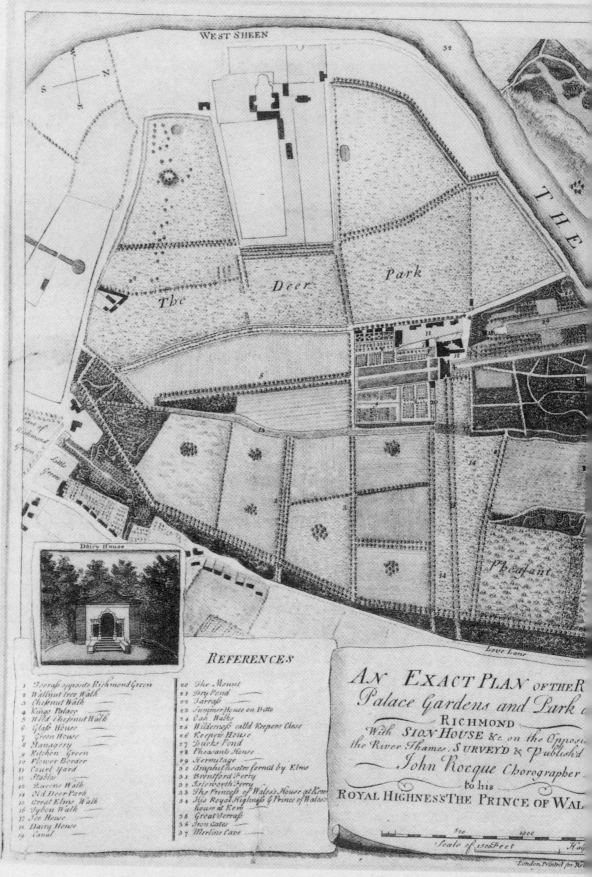

John Rocque: *Plan of Richmond Gardens*, 1754. Features of particular interest: 19 – long canal; 20 – riverside mount; 15 – Great Elm Walk from the Lodge to Love Lane; 37 – Merlin's Cave near bottom left-hand side of Duck Pond; 29 – Hermitage above Diagonal Wilderness and Amphitheatre

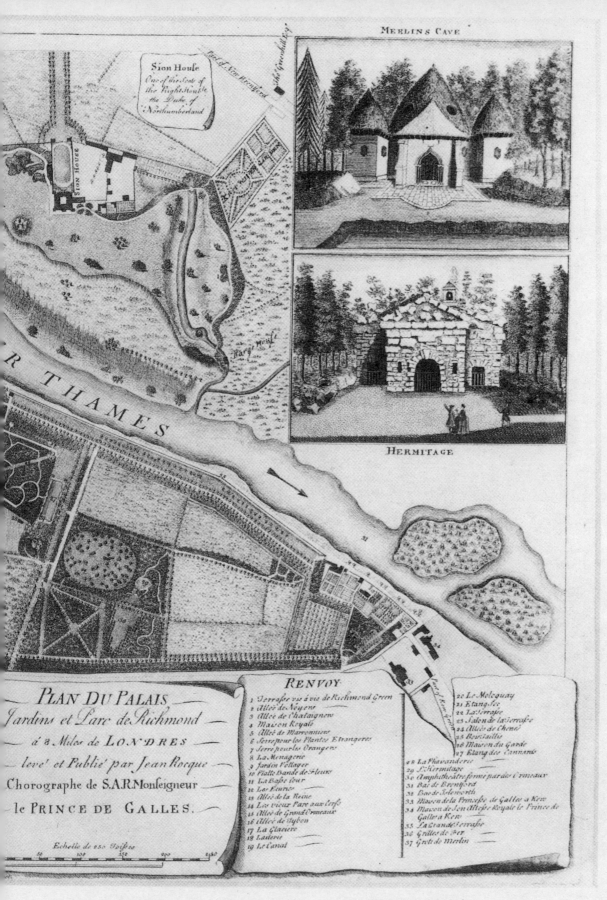

(30). The Pheasant Ground, also known as Broom Fields, belonged to Lady Elizabeth St André before it was added to Richmond Gardens in the 1740s.

THE SUMMER-HOUSE ON THE TERRASS.—

THE DUKE'S SUMMER-HOUSE.

THE DAIRY.

THE QUEENS PAVILION.

Illustrations from *A Description of the Royal Gardens at Richmond*, 1740s.

a. Duke of Ormonde's summerhouse on the Terrace with large windows for viewing the river. The Isleworth Ferry Gate now stands approximately on its site.

b. Duke's summerhouse near G. London's walk to the river.

c. William Kent's Dairy at the head of the canal. It had stucco walls and was "furnished suitable to a royal dairy, the utensils for the milk being of the most beautiful china".

d. William Kent's Queen's Pavilion, c. 1730. It had "a beautiful chimney piece" and displayed a model of a new palace proposed for Richmond Gardens.

Richmond Lodge demonstrated Bridgeman's protean style: a straight canal, a rectangular pond, serpentine walks, mounts, woods and open fields. A contemporary writer approved its "agreeable wildness and a pleasing irregularity", and Horace Walpole applauded Bridgeman's innovation of cultivated fields juxtaposed with a dense planting of trees suggesting miniature forests. Bridgeman was an enthusiastic arboriculturist; at Richmond he planned walks of chestnut or walnut. The Great Elm Walk proclaimed his preference for that tree; he generously clothed the New Mount near Merlin's Cave with young elms and used them to good effect in forming a large Amphitheatre.

Visitors to Richmond Gardens were usually unanimous in their praise but there were those who voiced dissent. Batty Langley, a respected landscape designer of Twickenham, considered the canal "too narrow for its length" and its plantation of trees "stiff and regular"; he also lamented the absence of shade-giving trees on the Terrace. Lord Egmont looked in vain for waterworks. The *Daily Gazette* for 18 September 1735 reported a savage attack in *The Craftsman* which ridiculed "a great Number of close Alleys, and clip't Hedges, without any Variety or Prospect".

It was, however, some of Kent's architectural contributions rather than Bridgeman's landscaping which attracted most comment and frequently censure. William Kent, who had studied painting and buildings in Italy, at first practised as an artist and architect, before establishing a reputation as an interior decorator and landscape designer. Hailed by Horace Walpole as the "father of modern gardening", he adroitly interpreted Alexander Pope's dictum that "all gardening is landscape painting". His first commissions in landscaping were not undertaken until the late 1720s, when Bridgeman and other pioneers had already introduced informality in gardens.

OPPOSITE: Temple on the Mount in Richmond Gardens, by Bernard Lens the Younger (1680–1740), enameller and miniature painter to George II. This Mount was created in 1733, and Kent designed the domed Tuscan Temple with a circular altar in the centre. Traces of this Mount can still be seen in the south west corner of Kew Gardens. *London Borough of Richmond upon Thames*

BELOW: Engraving of drawing by P. Brookes of Richmond Gardens with Syon House on the opposite bank of the Thames. John Boydell: *Collection of One Hundred Views in England and Wales* 1770, plate 28. The Temple on the Mount is in the distance. The roof of the Duke of Ormonde's summerhouse appears above the Terrace's brick wall. Queen Caroline extended this famous Terrace to Kew, 1729–34.

Marco Ricci. Walk from Richmond Lodge to Richmond Green. J. Badeslade and J. Rocque: *Vitruvius Britannicus, volume the fourth* 1739.

Kent may have been presented to Queen Caroline by the Prime Minister, Sir Robert Walpole, who had employed him at Houghton, or by his patron, Lord Burlington. In October 1729 the Board of Works sanctioned the building of a pavilion in the woods at Richmond Lodge. The Queen's Pavilion – its subsequent name – was, so far as is known, the first of a number of garden buildings designed by Kent; a dairy, a domed temple with Tuscan columns, and a summerhouse followed, but his Hermitage and Merlin's Cave received the most attention and established the garden's reputation and notoriety.

The assortment of buildings, monuments and statues which adorned so many Georgian estates declared not only their owners' reverence for antiquity and their classical scholarship but also their political beliefs. Lord Cobham at Stowe in Buckinghamshire erected a number of these architectural manifestoes; there was a Gothic temple dedicated "To the liberty of our ancestors", and a Temple of Friendship commemorating a group of dissident Whigs committed to Frederick, Prince of Wales, whose bust it housed. The busts of Alfred the Great and of Edward the Black Prince, who personified British patriotism, occupied niches in the Temple of British Worthies. Frederick later defiantly acknowledged his alliance with the Whig opposition by finding space for these two 'patriots' in his own garden pavilion at Carlton House.

However, the message conveyed by Richmond's Hermitage was philosophical rather than political. The Board of Works approved its erection in November 1730; completed under the supervision of the Clerk of Works, Henry Flitcroft, in 1731, it lay on a site about 300 yards north east of Merlin's Cave.[13] Approached across a circular lawn, it announced itself as a ruin of rough-hewn stones, crudely laid together to form a triple-arched façade with a central pedimented bay. It jutted out of a hill created by Bridgeman, about 30 or 40 feet high, around which a path led to the top clothed in a ragged clump of trees.[14] "Very Gothic," reported *The Craftsman* approvingly, "being a Heap of Stones thrown into a very artful Disorder, and curiously embellished with Moss and Shrubs, to represent rude Nature."[15] This seemingly ideal accommodation for a hermit was belied by the classical order and comfort of its interior. A central octagon, just over 12 feet in diameter, flanked on either side by a 'cell', was surmounted by a dome with a small, square lantern. Another cell beyond this octagon contained an altar with a bust of Robert Boyle radiating the golden rays of the sun. Busts of Isaac Newton, Samuel Clarke, John Locke and William Wollaston filled niches in the octagon itself. It was, in truth, a pantheon honouring Newtonian science and latitudinarian theology. Newton, through the discovery of the laws of the universe, and Locke, through

Marco Ricci. Walk from Richmond Lodge to the ferry. To the left of the Temple on the Mount is a bridge over a culvert from Bridgeman's canal. J. Badeslade and J. Rocque: *Vitruvius Britannicus, volume the fourth*, 1739. The plates were engraved by J.B.C. Chatelain.

To Her most Excellent Majesty QUEEN CAROLINE This View of the HERMITAGE in the Royal Garden at Richmond, And of the Heads of ye Honble. Robt. Boyle Esq. Jn. Locke Esq. Sr. Isaac Newton Willm. Wollaston Esqr. & of ye Revd. Dr. Saml. Clarke Done after the Marble Bustos placed therein, is most Humbly Dedicated.

rational debate, sought confirmation of Christian theism. Clarke was a regular contributor to Queen Caroline's 'seminars' on metaphysics and theology, and Wollaston, one of his enthusiastic supporters, wrote a bestseller, *The Religion of Nature delineated* (1724). Boyle, a chemist and a founder of the Royal Society, had also participated in theological disputation. The Queen, who so easily could have assembled a galaxy of German worthies, by shrewdly confining herself to British intellectuals identified herself with the island's cultural heritage. According to the *Grub Street Journal* for 30 August 1733, a bust of Bacon was also destined for the Hermitage, but whether it was ever installed remains uncertain. The Italian sculptor Giovanni Guelfi, brought to England by Lord Burlington, executed the bust of Clarke. It has been suggested that he made terracotta models of the others which John Michael Rysbrack copied in marble.[16]

OPPOSITE: William Kent's Hermitage, 1735, stood somewhere near the site of the present Azalea Garden. It was backed by a mound planted with trees, and encircled by a path. A bell hung in the small turret.

The Hermitage was extravagantly praised in the *Gentleman's Magazine*, which during 1732 and 1733 sponsored competitions for the best poems acclaiming "Her Majesty's Grotto at Richmond". "Every man, and every boy, is writing verses on the Royal Hermitage," mocked Alexander Pope. Surprisingly, it escaped 'Capability' Brown's obliteration of Caroline's garden during the 1760s and 1770s. Fragments of it survived until about the mid-nineteenth century, when it was confusingly known also as Merlin's Cave and the Grotto. John Smith, gardener, foreman and then Curator at Kew from 1822 to 1864, told Sir Joseph Hooker that "What is called Merlin's Cave is really the stone of the Hermitage and the underground cave was done by sons of George III assisted by a bricklayer whom I knew."[17] When a tree was

Interior of the Hermitage. John Vardy: *Some Designs of Mr Inigo Jones and Mr William Kent*, 1744, plate 33. Two symmetrical wings flanked an octagon with an altar on which rested the bust of Robert Boyle; other busts filled the niches.

Kent's Merlin's Cave.
*Merlin: or, The British
Inchanter and King Arthur,
the British Worthy,* 1736.
In the foreground is the
Duck Pond which it
overlooked.

taken out of the Beech Clump near the present Azalea Garden in 1983 a large
shaped stone was discovered beneath it. It may have come from the Hermitage,
which had stood somewhere in the vicinity.

William Wrighte's *Grotesque Architecture* (1767) offered a range of hermit-
ages in a choice of styles. Occasionally such buildings had a resident recluse,
hired in exchange for free accommodation, food and modest wages, but usually
dummy figures, suitably attired, sufficed. Merlin's Cave in Richmond Gardens
not only displayed dummies but also had a live attendant.

This Cave, another of Kent's creations, appeared on the south side at the
east end of the rectangular Duck Pond in 1735.[18] Charles Bridgeman raised
a mount nearby which, over the next three years, he grassed and covered with
more than 130 English elms.[19] The extraordinary edifice comprised a central
pavilion flanked by octagonal wings, all surmounted by three conical thatched
roofs. An ogee-arched entrance and buttresses established its Gothic pedigree.
Fanlights in the larger central cone illuminated a large circular room supported
by four wooden pillars. The side chambers were furnished with rustic bookcases
filled with vellum-bound books.

In an apse with three Gothic arches a tableau of six life-size figures had
been fashioned by Mrs Salmon, the proprietor of a London waxworks: a
youthful Merlin seated "in a musing posture" at a table loaded with books on
magic and mathematical instruments; he and his secretary were joined by Henry
VII's consort and Queen Elizabeth. The two remaining figures invited

Interior of Merlin's Cave. John Vardy: *Some Designs of Mr Inigo Jones and Mr William Kent*, 1744, plate 32. It had octagonal wings and a central chamber. Wooden pillars supported the ceiling and the bookcases were filled with white vellum-bound volumes.

The central chamber of Merlin's Cave contained a tableau of six life-size wax figures. [Edmund Curll]: *The Rarities of Richmond*, 1736.

speculation: one may have represented Minerva, Britannia, Bradamante or Britomart from Spenser's *Faerie Queene*; the other may have been Mother Shipton, or Melissa an obscure prophetess, or perhaps a nurse to Queen Elizabeth, or Britomart. The iconography was not at all clear. A page of the Backstairs had posed for Merlin and a Grenadier Guard for his secretary; it was rumoured that Lady Suffolk from nearby Marble Hill had stood in for Henry VII's queen; a Mrs Poyntz and a Miss Paget were willing models and a tradesman's wife in Richmond was portrayed as the nurse.

Stephen Duck, a rustic poet, was Merlin's Cave's custodian. The son of an agricultural labourer in the Vale of Pewsey in Wiltshire, and largely self-taught, he wrote verses which came to the attention of a lady-in-waiting to the Queen. Caroline graciously rewarded his modest efforts with an annual allowance of 30 guineas and a small house in Richmond. His luck held: an appointment of Yeoman of the Guard, Keeper of Duck Island in St James's Park and marriage to the Queen's housekeeper at Kew followed. For several years he enjoyed the sinecure of Keeper of the Queen's Library in Merlin's Cave, with his wife as "the necessary woman". When royal patronage ceased with the Queen's death in 1737, he sought a career in the Church – chaplain to a Regiment of Dragoon Guards, then preacher at Kew and finally a living at Byfleet in Surrey. Perhaps a loss of status and three marriages contributed to his suicide in 1756.

While Stephen Duck attended to his duties in Merlin's Cave, his wife attempted to explain to puzzled sightseers the significance of its display. Horace Walpole ridiculed it as "an unintelligible puppet show". Nor were other observations always complimentary. *Fog's Weekly Journal* for 6 December 1735 concluded rather ambiguously that "it is Hieorglyphical [*sic*], Emblematical, Typical and Symbolical, conveying lessons of Policy to Princes and Ministers of State". A putative link between the prophecies of Merlin and the Hanoverian succession might have suggested this mythical person to the Queen. Judith Colton offers several plausible explanations including a suggestion that the Cave was Royal Richmond's riposte to the Whiggism of Stowe.[20] *The Craftsman*, that outspoken, spirited journal of the Opposition, likened Merlin's Cave to "an old Haystack, thatch'd over".[21] When Caroline complained to George II about such attacks, he replied unsympathetically "I am very glad of it . . . you deserve to be abused for such childish silly stuff."[22]

Such notoriety encouraged a rash of imitations. Several inns in London featured a Merlin's Cave; the Crown coffee house in King Street offered "Merlin in Miniature; or, a lively Representation of Merlin in his Cave, as in the Royal Gardens at Richmond, being a New and Entertaining Piece of Moving Machinery, such as never before appeared in Publick". The edifice was demolished in 1766 during 'Capability' Brown's relandscaping of Richmond Gardens[23], but its reputation endured; the Ordnance Survey evidently believed the building had survived since its 25 inch map of Kew Gardens in 1865 identified a huddle of stones a few hundred yards north west of 'Mossy Hill' as being Merlin's Cave. This was, in fact, the remains of the Stone House, reputedly built by some of the sons of George III. The Ordnance Survey corrected this error in its 1894/96 map by placing the site of Merlin's Cave somewhat imprecisely in the vicinity of Sir William Hooker's Lake west of the Temperate House.

When Queen Caroline died in 1737 she left debts amounting to £20,000, due in part to her extravagance in improving the grounds at Kensington Palace and especially at Richmond, her favourite residence. Although she had her detractors, among them Horace Walpole who contended that she "made great pretensions to learning & taste, with not much of the former and none of the latter", she never lacked admirers. When the Duke of Leeds visited the Duchess of Orléans's gardens at Baquolet near Paris in 1733 he judged them "extremely pretty" but decided they "did not come up to her Majesty's at Richmond". The naturalist and antiquarian, Daines Barrington, commended her improvements at Kensington and Richmond whose gardens were laid out "upon a larger scale and in better taste than we have any instances of before that period".[24]

After the death of Charles Bridgeman on 19 July 1738, responsibility for the Royal Gardens was shared among several gardeners. In August, A.S. Milward was appointed to St James's and Kensington, G. Lowe to Hampton Court and J. Kent to Windsor and Newmarket. At the same time the Treasury requested the Board of Works to survey Richmond Gardens, "distinguishing in the admeasurement the ground to be kept in fine order from the grounds requiring no such keeping".[25] The result of this survey was submitted in October and on 21 December 1738 Thomas Greening, father and son, were engaged as chief gardeners at Richmond. Thomas Greening senior had been a nurseryman in Brentford, Middlesex and, during the 1720s, gardener to the Duke of Newcastle at Claremont in Surrey. Their contract[26] required 79 acres to be kept "in fine order"; these included the principal features: the Terrace, the walks in the woods, the hill near Merlin's Cave, and the long Forest Walk from Richmond to Kew. The Terrace Walk in the Deer Park, being little used and therefore requiring less maintenance, was paid at a lower rate. Grass was to be mown, gravel paths swept, flower borders dug, hedges clipped, fruit trees pruned and the two ice-houses refilled. The generous salaries royal gardeners received were expected to cover the provision of all garden implements, the wages of subordinates, the purchase of manure and of fuel for the hothouses. In 1751 Kensington and St James's passed to the control of Thomas Greening senior; when he died in 1757 his son John succeeded him there and at Richmond. John Haverfield, Princess Augusta's gardener at Kew, had taken over Richmond Gardens by May 1762.

After Queen Caroline's death, Richmond Gardens passed to George II who continued to visit the property every Saturday during the summer although little was now spent on the estate. In 1749 a thousand feet of walls, 12 feet high, were constructed for espalier fruit trees and in 1752 a hothouse for strawberries was built. It would be George II's grandson who ordered a radical reshaping of the grounds, destroying Bridgeman's layout and removing Kent's buildings.

CHAPTER TWO

Prince Frederick at Kew

The Prince leases Kew House – altered by
William Kent – Carlton House garden –
House of Confucius – improvements to the
garden planned

WHEN THE ELECTOR OF HANOVER succeeded to the English throne in 1714 as George I, he brought with him his son, George Augustus, and daughter-in-law, Caroline, who left their son Frederick, then seven years old, at the Electoral Palace at Herrenhausen. According to Chancellor King's diary, it was the wish of Frederick's parents that he should eventually become Elector of Hanover while his younger brother, William, should follow them on the English throne. Even at this early age Frederick was excluded from his parents' affections. George I persistently snubbed George Augustus who, in turn, treated his eldest with the same contempt. Strained relations were to exist between George III and the Prince Regent. This family trait survived in Queen Victoria's aversion, so she told a certain statesman, to being in the same room as her son and heir, the future Edward VII. It must be said, of course, that this adversarial stance adopted by the sovereign to the heir apparent was not just a peculiarity of the Hanoverian dynasty; the ruling families of Russia and Prussia indulged in a like antipathy.

In 1728, George II yielded reluctantly to his ministers' advice that his eldest son should assume his rightful place at the English court. Frederick was summoned to London and, denied a separate establishment, lodged in St James's Palace where his parents could keep an eye on his activities. When he was made Prince of Wales in January 1729, his father decided that his allowance should be less than a third of what he himself had enjoyed in the same position. George II clearly suspected his son to be as desirous of power as he had been when Prince of Wales. While Frederick's modest, friendly mien captivated Londoners, it alarmed his parents. Always apprehensive that his popularity threatened the King's authority, the Queen's dislike lapsed into pathological hatred. "My dear first born," she told Lord Hervey, "is the greatest ass, and the greatest liar, and the greatest *canaille*, and the greatest beast, in the whole world, and . . . I most heartily wish he was out of it." She maintained this enmity even on her deathbed, stubbornly refusing to see her son.

When this ill-concealed hostility gained some public sympathy for Frederick, he was emboldened to demand not only his own establishment but also permission to marry. George II, persuaded by his Prime Minister, reluctantly agreed to the marriage, choosing from among the few available Protestant princesses Augusta of Saxe-Gotha. The wedding took place in St James's Palace on 8 May 1736, the modesty and docility of the 17-year-old bride convincing George and Caroline that she posed no threat to them. She was ever cautious and discreet in her demeanour – "Princess Prudence", Horace Walpole dubbed her; these qualities she effectively exercised in her role as her husband's hostess. Frederick's circle of friends and acquaintances, composed mainly of politicians in opposition to the Government, and writers and artists, were entertained at one of his London houses or at Cliveden overlooking the Thames.

Frederick displayed a dilettante's interest in art, literature and science, especially astronomy. He collected the works of sixteenth- and seventeenth-century Italian and Flemish masters; his appointment of Philip Mercier as his principal painter indicated a preference for French rococo and a rejection of classicism; he accepted the patronage of St Martin's Lane Academy whose members included Gainsborough and Hogarth. He cultivated the acquaintance of his near neighbour, the poet Alexander Pope, presenting him with some marble urns for his garden. A competent cellist, PLATE 1 he sometimes entertained servants and officials in Kensington Palace, "singing French and Italian songs to his own playing for an hour or two together". His musical taste led him into conflict with his father, who admired Handel, while he perversely supported Buononcini.

With this mutual antagonism pervading all facets of their relationship, it was perhaps unexpected that in 1731 Frederick leased a house within a mile or so of Richmond Lodge and adjacent to the Dutch House (now Kew Palace) which Queen Caroline had leased about 1728. Could it have been an act of defiance? Frederick was certainly capable of such irrational behaviour. Lord Egmont says it was, in fact, a gesture of reconciliation. He had it on good authority that the Prince of Wales "put himself to an inconvenient expense to purchase his house at Kew, that he might be near his Majesty when at Richmond".[1]

The house, a timber-framed property, stood a very short distance south of the Dutch House. Frederick had acquired its contents including a series of seventeenth-century historical portraits, some by Van Dyck, in March 1731 from Lady Elizabeth St André, daughter of the 2nd Earl of Essex; about six months later he leased the house itself, although he had moved in during Christmas 1730.[2] Some 80 years earlier Richard Bennet, son of the Lord Mayor of London, had owned it; through his daughter's marriage it had passed to Sir Henry Capel, subsequently Lord Capel of Tewkesbury and Lord Deputy of Ireland.

Gardening had been in the Capel blood; Capel's brother, the Earl of Essex, had created a notable garden at Cassiobury Park in Hertfordshire, and his sister, Mary, Duchess of Beaufort, gardened enthusiastically at Badminton in Gloucestershire. Capel's friend, the diarist John Evelyn, after a visit to Kew in August 1678 praised its orchards and when he returned five years later oranges and myrtles filled two houses. He noted the tall "palisades of reeds" shading the oranges in tubs out in the open during the summer. When John Gibson toured London gardens in 1691, Capel's garden was on his itinerary.[3] He admired two mastic trees (*Pistacia lentiscus*), "said to be the best in England"; four white-striped hollies and six laurustinus were also singled out for commendation. A terrace walk, bordered by grass with a rue hedge on one side and a row of dwarf trees on the other, topiaried yew, flower-beds and fruit trees were all duly appreciated. A few years later Lord Capel died; his widow continued to enjoy the garden until her death in 1721 when her grandniece, Lady Elizabeth Capel, daughter of the 2nd Earl of Essex, inherited the property. Her husband Samuel Molyneux, George II's Secretary, installed a telescope in the house; in 1725 his friend, James Bradley, while staying there discovered the aberration of light and the nutation of the earth's axis.[4] Following Molyneux's death in 1728, Lady Elizabeth found another husband in the notorious Nathaniel St André in 1730, a marriage which met with royal disapproval compelling the unfortunate couple to retire in disgrace to the country.

With the Prince of Wales as its new occupant, plans were put in hand to make it a suitable royal residence. William Kent was appointed architect in August 1732[5] but the Prince of Wales's household accounts indicate that alterations had begun in 1731. It would appear that the core of the original house was retained – Fanny Burney visiting it in 1786 observed that it had been repaired rather than rebuilt. "When the house was taken down in 1802, it was found to have been originally built of red brick, worked in ornamental grooves and patterns: over this, wooden planks had been fastened on which a smooth coating of stucco had been laid."[6] It was, however, enlarged by the addition of one- and two-storey wings, employing some of the builders and craftsmen engaged on projects at Richmond Lodge. An imposing pediment dominated the garden façade, and William Kent compensated for the restraint of its Palladian exterior in the opulence of the rooms. He designed the great staircase,

Frederick, Prince of Wales (1707–1751), by J.B. Van Loo.

White House, also known as Kew House, drawn by Joshua Kirby. William Chambers: *Plans . . . of the Gardens . . . at Kew*, 1763 (also includes floor plans). Capel's timber house had been transformed by Kent into a Palladian mansion for Prince Frederick.

chimney pieces, furniture and the painted ceilings; Isaac Mansfield decorated the gallery with plasterwork of flowers, foliage and feathers; the walls were hung with Capel heirlooms – tapestries and paintings. The dwelling (which because of its white stucco rendering was known as the White House) was ready to receive Frederick's bride in 1736. Unlike his other residences, it served primarily as a rural retreat for the Prince's family.

As he needed a more prestigious home for receptions and functions, Frederick acquired Carlton House above The Mall in June 1732. It had been built for Lord Carleton in 1709 and inherited in 1725 by his nephew Lord Burlington, who presented it to his mother in 1732. Frederick, existing on an inadequate allowance from his father, borrowed money for its purchase but never lacked eager creditors who saw the heir to the throne as a good investment. Some alterations were made to the house, supervised by William Kent, whose main contribution was improving the 12-acre estate.

Kent had already helped to landscape the grounds at Chiswick House for Lord Burlington who may have recommended him as a garden designer to the Prince of Wales. Kent built a neo-Palladian temple to terminate a vista stretching the whole length of the garden at Carlton House. Busts by Michael Rysbrack of Alfred and the Black Prince were duly installed and *The Craftsman* approved this overt political statement.

> . . . his royal Highness the Prince of Wales has order'd a fine statue of King Alfred to be made for his Gardens in Pallmall, with a Latin Inscription; in which it is particularly said, that this Prince was the Founder of the Liberties and Commonwealth of England . . . his Royal Highness hath likewise order'd another statue to be set up there, in Memory of the famous Prince of Wales, commonly call'd the black Prince; in the Inscription upon which he declares his intention of making that amiable Prince the Pattern of his own Conduct.[7]

Sir Thomas Robinson, an amateur architect, wrote approvingly to his father-in-law, Lord Carlisle, of the remodelling of the gardens, which Kent completed in two years.

> There is a new taste in gardening just arisen, which has been practised with so great success at the Prince's garden in Town, that a general alteration of some of the most considerable gardens in the Kingdom is begun, after Mr Kent's notion of gardening, viz., to lay them out, and work without either level or line. By this means I really think the 12 acres the Prince's garden consists of, is more diversified and of greater variety than anything of that compass I ever saw; and this method of gardening is the more agreeable, as when finished, it has the appearance of beautiful nature, and without being told, one would imagine art had no part in the finishing, and is, according to what one hears of the Chinese, entirely after their models for works of this nature, where they never plant straight lines or make regular designs.[8]

TOP: Carlton House
garden, designed for
Prince Frederick by Kent.
Detail of a drawing and
engraving by W.
Woollett, 1760. The
garden statues or terms are
probably the ones now at
Kew, brought there by
Princess Augusta.

The only evidence of
Kent's involvement with
the Prince's garden at Kew
is this garden shelter. W.
Chambers: *Plans . . . of
the Gardens . . . at Kew*,
1763.

Although it is frequently asserted that William Kent redesigned the gardens
at Kew for Frederick, no documentary evidence confirming this statement has
yet been found. The only horticultural link between Kent and Kew is an
engraving of a garden seat or shelter he designed, reproduced in Sir William
Chambers's *Plans . . . of the Gardens and Buildings at Kew* (1763). Princess
Augusta burnt her husband's papers on his death to conceal, presumably,
evidence of his political allegiances, and in so doing may have destroyed valuable data on Frederick's gardening activities at Kew. For information about
his improvements there we have to rely on his household accounts in the Duchy
of Cornwall Office, together with reports and observations from knowledgeable
people like Horace Walpole, George Vertue and Peter Collinson. A cautionary
word: it is not always absolutely clear whether work itemised in the household
accounts refers to Kew or another property belonging to the Prince of Wales.

The garden of the White House appears on John Rocque's map of 1734
and subsequent editions. On the north side of the house two central grass plots
are flanked by counterbalancing courtyards with a narrow strip – probably the
kitchen garden – on the western edge; on the south side two rectangular areas
of shrubs thrust into a lawn, altogether ten acres[9] – hardly a Kentian garden!
The estate extended south through a succession of hedged fields with evocative
names – Warren Fields, Lime Tree Close, Great and Little Sycamore Closes,
Orchard Field, Oat Close, Dean Leas, and Canary Close[10]. The only contemporary comment on this garden, so far discovered, is provided by Rocque on
his 1734 map:

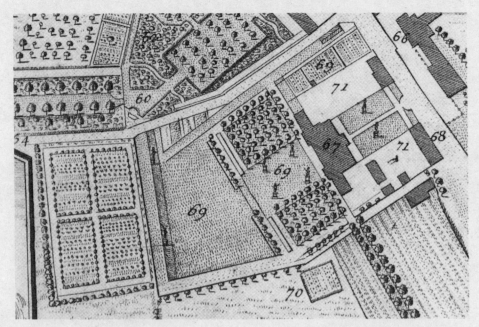

Grounds of the White House (enlargement of John Rocque's *Plan of Richmond Gardens*, 1748). Key to the figures: 60 – part of the garden of Queen's House; 64 – Love Lane; 66 – Dutch House, then occupied by Princess Royal; 67 – White House; 68 – stables; 69 – gardens (note the statues); 70 – melon ground; 71 – courtyard.

Kew Palace [i.e. White House] the seat of his Royal Highness ye Prince of Wales, a compleat place very beautifull in its situation. Gardens, etc. all laid out at his Highnesses expense and affords a delightfull Prospect of ye River and the opposite Country.

Frederick retained the services of Lady Elizabeth's gardener, John Dillman, who had come from Dillinesburg in Germany, eventually becoming a British citizen. From 1732 Dillman also maintained the Prince's garden at Carlton House and an acre plot at Leicester House; in 1740 Durdans near Epsom was added to his responsibilities. He retired in 1753 and died about seven years later. One of Dillman's assistants at Kew and Leicester House was John Abercrombie, now remembered as the author of many horticultural books.

In 1734/5 eight terms or statues were carved for Frederick: two by Joseph Pickford, two by Peter Rubens and four by Peter Scheemakers[11]. The household accounts fail to make it clear whether they were destined for Kew. Terms similar to the ones now in the Queen's Garden at Kew are depicted in Woollett's engraving of the garden at Carlton House and it may be that they were later transferred to Kew. Some engravings of the 1760s show a term standing near the Ruined Arch at Kew; one can be seen on the north side of the Arch in Chambers's *Plans . . . of the Gardens and Buildings at Kew* (1763); now decapitated, it is part of the heap of carved stone at the base of the Ruined Arch. John Smith who knew Kew from 1822 remembered that "for many years [the terms] were made to appear part of what is called the Ruined Arch. Sir William [Hooker] had them taken up, and for a time they were stood against the wall of King William's Temple. Afterwards they were taken to the Office Yard and cleaned with the view of being put up in some part of the Garden but

nothing was done . . ."[12] On 24 February 1886 the Director at Kew, William Thiselton-Dyer, informed the Board of Works that his Office Yard had five terms "removed many years ago from the Pleasure Grounds".[13] He recommended resiting them in the curved recess of the yew hedge at the south end of the Palm House. There they stayed until their transfer to the Queen's Garden in the 1960s.

Possibly a stone plinth with cherub heads also has associations with Frederick, but this is pure speculation. Michael Rysbrack has been proposed as the sculptor but it could be the work of one of several stone masons the Prince employed. For many years it stood in front of the Temple of Bellona and is now to be found in the Temperate House.

Kew and Frederick's other estates were stocked by Dillman with plants purchased from London and local nurseries such as Robert Furber of Kensington and Richard Butt of Kew Green. One bill for trees and shrubs delivered by Richard Butt from September 1734 to March 1735 came to the considerable sum of £938.[14] With the purchase of Cliveden and Durdans it is unlikely that Frederick spent much time at Kew. After his mother's death in 1737, relations with his father showed some improvement and in the late 1740s he turned his attention once more to Kew.

Sir William Irby (later Lord Boston), Princess Augusta's Lord Chamberlain, told Lord Bute that it had been the Prince of Wales's intention "to put himself to very considerable expence at Kew".[15] During 1749 and 1750 Frederick purchased or leased another 42½ acres; by 1751 32 additional acres had been laid out and planted. In 1750 Horace Walpole confirmed the start of "great works in the garden".[16] George Bubb Dodington (later Lord Melcombe), political adviser to the Prince of Wales, noted petulantly in his diary for 27 February 1750: "Work'd in the new walk [at Kew]" and in the following day's entry: "All of us, men, women and children worked at the same place – cold dinner."[17]

The antiquary and engraver, George Vertue, the most informative source on Kew's gardening projects, entered the Prince's service in 1748, specifically to produce a catalogue of the dispersed picture collection of Charles I. When Vertue visited Kew on 12 October 1750, Frederick

> was directing the plantations of trees [and] exotics with the workmen – adviseing & assisting where wee were receivd gratiously and freely walking and attend the Prince from place to place – for 2 or three hours, seeing his plantations, told his contrivances, designs of his improvemnts in his Gardens, water works, canal, etc. great numbers of people labouring there, his new Chinesia Summer hous, painted in their stile & ornaments The story of Confusius & his doctrines, etc.[18]

Vertue later reported that Frederick "was planting about his Gardens also many curious & forain trees [and] exotics".[19] The ledgers of Frederick's Treasurer itemises £216 for the purchase of trees and shrubs for Kew in May 1750.[20]

In 1751 Vertue was commissioned to collect or make drawings of ancient and modern philosophers. It was Frederick's wish "to make an aquaduct thro his Gardens at Kew and the earth thrown up was to make a mount which he intended to adorn with the statues or busts of all these philosophers and to represent the Mount Parnassus".[21] Among Vertue's papers in the British Library[22] there is a list of 42 eminent historical figures under the headings 'Inside', 'Outside' and 'Scenes' – their suggested distribution as statues inside and outside a building on the mount. Early in 1751 Joseph Goupy submitted to Frederick "A Drawing of a Grecian Pavillion or Building to be upon the Mount".[23] Classical writers and philosophers were paired by Vertue with those he perceived to be their modern counterparts: Archimedes with Newton, Aeschylus with Shakespeare, Horace with Pope, Vitruvius with Inigo Jones, and so on. Frederick, familiar with the garden buildings at Stowe, was clearly influenced by Cobham's Temple of British Worthies; indeed Lord Cobham who had recently died found a place in Frederick's Pantheon of great men.

In June 1753 Robert Greening replaced John Dillman who had been instructed by Princess Augusta to implement Frederick's 'Plan' for his Kew estate. Greening's contract (now with Sir George Lee's papers) stipulates that he was to plant the Mount and to continue dredging the lake started by Dillman. It is not inconceivable that this lake was Vertue's "canal" or "aquaduct" and that the present Mound near the Pond was Frederick's "Mount Parnassus" formed from "the earth thrown up" in dredging it.

His Mount Parnassus, with its artistic, philosophical and political reson-

OPPOSITE:
Five terms flanking the semi-circular hedge in the present Queen's Garden behind Kew Palace; probably the ones formerly at Carlton House. The pedestal of another term lies at the base of the Ruined Arch. *Photograph by A.McRobb.*

Large pedestal, now in the Temperate House. It has been attributed to Michael Rysbrack, who was one of a number of sculptors employed by Prince Frederick. *Photograph by A. McRobb.*

ances, would have aptly complemented Kew's House of Confucius, dedicated to the teaching of the Chinese philosopher and moralist, then much admired in the West. Sir William Chambers in his *Plans . . . of the Gardens and Buildings at Kew* (1763) attributed this octagonal two-storey structure to Joseph Goupy, PLATE 2 although the engraving illustrating it in his book credited it to Chambers. John Harris[24] argues that the House of Confucius was in fact by Chambers – a youthful commission undertaken when he met Frederick during the summer of 1749. Chambers, Harris believes, now preferred to disown its rococo exuberance which conflicted with the scholarly record of Chinese architecture in his *Designs of Chinese Buildings* (1757). But Chambers's name on the engraved plate could have been a printer's error or a misleading reference to the bridge he designed for the House of Confucius. Joseph Goupy, who was drawing master, 'cabinet painter' and art adviser to the Prince of Wales, in March 1750/1 submitted a bill for several designs including a 'Chinese Arch'[25] which was subsequently built.[26] It is therefore not unreasonable to suppose that he also designed the House of Confucius which was erected at Kew in 1749.

Another confusion concerning Frederick's gardening endeavours at Kew relates to a very large hothouse, (then called a "stove") which, it is frequently stated, he intended building at Kew. The evidence for this is a letter from Thomas Knowlton to Richard Richardson which says that "the Prince of Wale[s] is now about preperations for bulding a stove 300 feet in length for plants & not pines . . ." Unfortunately Dawson Turner who edited Richard Richardson's letters for publication[27] had transcribed the date of this letter as "November 13th 1750" instead of "1758". The hothouse in question is, of course, the Great Stove designed for Princess Augusta by Chambers; the "Prince of Wales" is the future George III.

Thirteen statues of Greek and Roman deities by the Renaissance sculptor, Pietro Francavilla, were ordered from Florence in 1750 for the royal gardens. They reached their destination a year after Frederick's death and, still in their packing cases, languished in a shed at Kew for some years. Transferred to Hampton Court, they eventually reached Windsor Castle, where four of them still stand on the East Terrace; another is in America; Apollo and Zephyr, formerly in the possession of the Royal Horticultural Society, came to Kew, were placed in the portico of King William's Temple, then briefly displayed in the Orangery before being transferred to the Victoria and Albert Museum. It is regrettable that their departure left Kew with few tangible links with Frederick.[28]

Frederick never sited his Italian statues, possibly intended for his Mount Parnassus, because on 20 March 1751, still in his early forties, he died unexpectedly. Just two weeks earlier he had been supervising his gardeners at Kew, caught a cold from a soaking in a sudden storm, and a few days later retired to Leicester House when pleurisy was diagnosed. He was blistered and bled and, when believed to be out of danger, suffered a spasm of coughing and died. A post-mortem established that death had been caused by "the sudden breaking of a large abscess under the sternum bone, where it had been gathering for two or three years past". It was suggested that this abscess might have been

OPPOSITE: Two of 13 statues by Pietro Francavilla (1548–1615), purchased by Frederick in 1750: A. Apollo, B. Zephyr. "I shall be curious to know what the Princess of Wales does with them . . . I hear that they are still destined for Kew, though I should have thought Kensington a better place." (5 May 1752. Sir Horace Mann to Horace Walpole).

One of the brass locks in Kew Palace with Prince Frederick's initials and badge. These brass fittings were probably transferred from the White House at the time of its demolition in 1802. Items of Georgian woodwork may have been transferred at the same time.

the result of a blow from a cricket ball while playing with his sons at Cliveden some years earlier. William Chambers, then studying in Rome, quickly prepared several sketches of a neo-classical mausoleum for the late Prince of Wales, unfortunately never built.

The hostility of his father, the loathing of his mother and Lord Hervey's malicious reportage all helped to foster Frederick's reputation as a weak and feckless man. That he was impetuous, indiscreet and profligate not even his friends denied; but they also recognised his charm, his wit and generosity. His death was genuinely lamented by the gardening fraternity in England. Peter Collinson told John Bartram in North America that "Gardening and planting have lost their best friend and encourager; for the Prince had delighted in that rational amusement a long while: but lately, he had a laudable and princely ambition to excel all others." The sentiment that was shared by Dr John Mitchell who declared that "Planting and Botany in England would be the poorer for his passing."[29]

CHAPTER THREE

Princess Augusta's "earthly Paradise"

The Princess continues Frederick's improvements – is advised by Lord Bute – Greening, Haverfield and Aiton are employed as gardeners – John Hill and his Hortus Kewensis *– Orangery – the Great Stove – trees transferred from Duke of Argyll's garden*

GEORGE II never harboured any lasting resentment towards his daughter-in-law, now the Dowager Princess of Wales; in fact as Regent she once deputised for him during one of his visits to Hanover. A reticent woman, she never sought popularity. Lord Waldegrave judged her to be a woman of moderate intelligence with few ideas of her own. She survived her husband for more than twenty years, content to live a retired life in London at Carlton House and Leicester House or at the White House in rural Surrey where she dedicated herself and her wealth to the improvement of its garden. Her household accounts for the early 1750s presage the intense horticultural activity of the next decade: regular maintenance of the garden, the gradual extension of cultivated areas and the erection of buildings. In June 1752 her head gardener, John Dillman, was contracted "to compleat all that part of the Garden at Kew that is not yet finished in the manner proposed by the Plan and to keep all that is now finished, which together is computed at 35 acres."[1] Throughout all these initiatives she was guided by friends of her late husband – especially Lord Bute and the Reverend Stephen Hales, both competent botanists.

An entry in her household accounts for 1752 records a consultation with Stephen Hales. "By an order of Her Royal Highness & The Reverend Doctor Hales, the men's time at Marsh-gate [Richmond], opening the springs & work belonging to it". Hales was perpetual curate at Teddington, about four miles from Kew, for 51 years. Famous as the author of *Vegetable Staticks* (1727) and a Fellow of the Royal Society, his investigations in plant physiology had implications for the ventilation of greenhouses and the pruning of trees and attracted the attention of the Prince of Wales. As a grateful acknowledgement of his cordial relations with her husband, Princess Augusta appointed him her chaplain in 1751 and Clerk to the Closet. On his death she had a monument by the King's sculptor, John Wilton, erected in Westminster Abbey.

She depended on John Stuart, 3rd Earl of Bute, for advice on landscaping, the acquisition of plants and the recruitment of staff. Born in Edinburgh, educated in England, Bute lived for five years in modest circumstances on the family estate on the Isle of Bute where he planted trees and studied botany, agriculture and architecture. During this enforced retirement, he kept in touch with Peter Collinson and others who shared similar interests in botany and horticulture. The naming of 'Steuartia' (*Stuartia*), an American shrub which first bloomed in England in 1742, in honour of Bute was a well-deserved tribute. Urged by his uncle, the 3rd Duke of Argyll, to improve his prospects by boldly plunging into London society, he lived for a while in fashionable Twickenham before moving to a more prestigious address at Caen Wood House (Kenwood) in Hampstead. A chance meeting with the Prince of Wales at the Egham races led to the social advancement he sought, and a mutual enthusiasm for the arts gained him in September 1750 the influential post of Lord of the Bedchamber in the Prince's household. Although he lost the appointment when Frederick died, he established a relationship with his widow with remarkable success if contemporary gossip is to be believed. None could surpass Horace Walpole in malicious rumour. When Frederick wished to be alone with a woman, confided Walpole, "he used to bid the Princess walk with Lord Bute. As soon as the Prince was dead, they walked more and more in his memory."

Augusta, Princess of Wales (1719–1772), by J.B. Van Loo.

Princess Augusta consulted Bute and by appointing him Groom of the Stole – a kind of household steward – his position as a trusted family friend was confirmed. He consolidated his position in 1755 when he accepted the responsibility of 'finishing tutor' to George, Prince of Wales, teaching him morality and statecraft and inculcating a love of art, literature and science. The future George III, then a shy and insecure young man, readily accepted Bute as his mentor, ever obedient to his instructions, constantly seeking his approval. "I am young and unexperienced," he confessed to Bute, "and want advice."

In 1763, after his wife came into her inheritance, Bute purchased Luton Hoo in Bedfordshire with over 4,000 acres which were landscaped by 'Capability' Brown with trees supplied by Haverfield from Kew. A decade later Bute acquired Highcliffe looking towards the Isle of Wight. Again Brown was called in to transform the grounds. All these properties – Caen Wood, Luton Hoo and Highcliffe – benefited from Bute's horticultural expertise, especially Highcliffe, where he filled a large conservatory with exotics. A.B. Lambert who visited this garden in 1791 "was astonished at the profusion of fine plants, & at the magnificence of the whole apparatus for the accommodation of them".[2]

John Stuart, 3rd Earl of Bute (1713–1792), by Allan Ramsay.

Bute's gardens provided him with abundant material for the botanical observations which he recorded "in a curiously crabbed hand of extraordinary neatness". These researches culminated in *Botanical tables, containing the different families of British plants distinguished by a few obvious parts of fructification rang'd in a synoptical method*. This costly work which questioned Linnaeus's classificatory aims proposed an entirely artificial arrangement of the plant kingdom. The recipients of the 12 copies in nine volumes, published in 1784, included Catherine, Empress of Russia, Sir Joseph Banks and, naturally, Queen Charlotte, who graciously acknowledged the gift in a letter from the White House at Kew.

> I can never hesitate a moment of taking under my protection any work Lord Bute recommends; but particularly so in this instance, as I know the work on British plants to be the produce of Lord Bute's studies.[3]

Johann Sebastian Müller PLATE 7 who contributed 566 plates to this rare publication was one of a number of flower painters engaged by Bute. Plant portraits were also executed for him by the country's doyen of botanical artists, Georg Dionysius Ehret; when Ehret died the young Simon Taylor who emulated his master's style drew Kew's flora for Bute. The sale of Lord Bute's botanical library in 1794 attracted many potential bidders but the choicest items were undoubtedly his collection of flower paintings: examples from the brushes of Ehret, George Edwards, Ann Lee, Margaret Meen, Maria Merian, and Chinese and Indian artists; there were also 684 drawings by Simon Taylor "from plants chiefly at Kew Gardens".

Bute's political activities brought him universal opprobrium but his dedication to botany and horticulture earned him praise; "few persons cherished the study of nature more ardently, or cultivated it so deeply," said Sir James Edward Smith. Peter Collinson, whose garden at Mill Hill in north London boasted one of the finest collections of foreign plants, was another admirer. "You desire to know our botanical people," he wrote to Linnaeus. "The first in rank is the Right Honourable the Earl of Bute." North America's leading plant collector, John Bartram, was assured that Bute "is the only great man that encourages ingenious men in planting botanic rarities". Books were dedicated to him: by William Curtis in the first volume of his *Flora Londinensis* (1777) in gratitude for financial assistance in producing it; by Samuel Pullein in *An Essay towards a Method of Preserving the Seeds of Plants* . . . (1759); by the Swiss botanist Albrecht von Haller in *Bibliotheca Botanica* (1771); and by Richard Weston in *Flora Anglicana* (1775). The plant genera *Stuartia* and *Butea* commemorate him. Yet he was never elected a Fellow of the Royal Society, and when the Linnean Society was formed in 1788, his admission to membership was not recommended by its Treasurer because of his "dubious politics".

Lord Bute is mainly remembered today as an unpopular minister of state, as an undesirable influence on George III, while his pivotal role in raising Princess Augusta's garden to one of the most remarkable in Georgian England has been forgotten, overshadowed by the architectural contributions of Sir William Chambers. But it was not so in his day. Chambers's book on Kew in 1763 acknowledged "the great botanical learning of him who is the principal manager, and the assiduity with which all curious productions are collected from every part of the globe, without any regard to expence". Horace Walpole

Aquatint of F.I. Mannskirsch's view of the White House. Robert Greening's estimate of the work to be done in June 1753 included "sinking two parts of the field which is to be added to the lawn, so as to give a view of the water [lake] from the house". This Great Lawn of 41 acres attracted much favourable comment. Greening's contract in 1758 included the provision of "a sufficient flock of sheep to feed the lawn".

put it more succinctly: "Lord Bute had the disposition of the ground [at Kew]."[4]

Although Bute is seldom mentioned in Princess Augusta's household accounts or in the papers of her Treasurer, Sir George Lee, he was nevertheless the garden's mentor – the "director", in Chambers's words, "equally skilled in cultivating the earth, and in the politer arts".[5]

In 1752 a new greenhouse supplemented the existing one; the following March the bricklayer, Solomon Brown, was paid for building a terrace; a reed hedge now enclosed the Indian pheasants. From June 1753, John Dillman, who also tended the gardens at Carlton House and Leicester House, had his duties at Kew confined to a six-acre kitchen garden, the melon ground and orangery. Robert Greening now replaced Dillman in Kew's pleasure grounds; an unusual stipulation was that he should look after Princess Augusta's herd of nine cows. For these duties he received 300 guineas a year and the use of 54 acres of farmland on the estate. Greening, whose father, Thomas, was head gardener at Richmond Gardens[6], had experience as a landscape designer. He had planned a garden at Wimpole Hall in Cambridgeshire about 1752[7]. During his first year at Kew he continued to dredge the six-acre lake with its three-acre island, planted the adjacent mound and levelled the field beyond the lawn to give a better view of the lake from the White House.[8]

In 1754 the path that led to Goupy's Chinese Arch was gravelled. In the household accounts for 1761–2 (f.53) payment is made to John Greening as executor of the late Robert Greening who, in 1754, had supervised the erection of "a large Chinese temple on one column with a neat Chinese chair to go round the same" and "a small bell temple neatly painted with ornaments fix'd up in the flower garden". This is the only known reference of the existence of these two intriguing garden buildings. The Reverend John Mulso, who preached a charity sermon at Kew one Sunday in 1754, walked in Princess Augusta's garden afterwards. "It is a beautiful country retirement, but not royal," he informed the Reverend Gilbert White. "The gardens will be handsome when finished. The lawn is grand, & ye only grand thing there."[9]

By 1755 110 acres had been walled or fenced at Kew. In 1756 John Dillman resigned from Princess Augusta's service, having received a legacy of £300 from Lady Elizabeth St André. Jonathan Hood replaced him at Carlton House and Leicester House and Greening took over the kitchen garden, melon ground and orangery at Kew.

An estimate for converting the nursery at Kew into a wilderness "according to a plan by Robert Greening" was submitted in 1757. It would occupy a ten-acre site and require 50,000 plants,[10] and Greening sought Bute's advice and authorisation for the tree-planting.[11] With the renewal of Greening's contract in March 1757 the extent of Princess Augusta's pleasure grounds was defined as that "contain'd between the house & the hedge which goes from the little mount in Love Lane [probably the mound on which the Mosque was subsequently built] to the Chinese Arch by the road on the other or opposite side of the gardens".[12] The terms of Greening's contract were extended early in 1758 to include the management of "a sufficient flock of sheep to feed the lawn".

When Robert Greening died in March 1758, his brother John at Richmond Gardens, 'Capability' Brown and Thomas Burnes applied for his vacant post. But it was John Haverfield, a surveyor at Twickenham and known to Lord Bute, who became Greening's successor in April 1758.[13] He was responsible for "ordering and cultivating" the gardens, and supervising the work of labourers; "cultivating the Physic Garden" appears for the first time in the household accounts for July to December 1759.

John Haverfield and his son, also John, were jointly in charge of Richmond Gardens from 1762 when John Greening left. When Haverfield senior died in 1784, William Aiton followed him at Kew and John Haverfield junior assumed sole charge of Richmond Gardens until 1795, when he resigned, the result of changes in the terms of his contract which reduced his margin of profit. He then went into private practice as a landscape designer, one of his clients being Sir John Soane for whom he remodelled the gardens of his residence at Pitshanger Manor in Ealing. Soane also consulted him about projects in which he was involved; they made frequent visits together to Tyringham Hall in Buckinghamshire where Soane designed a lodge and a bridge for William Praed. Between 1804 and 1812 Haverfield made extensive alterations to the grounds at Walsingham Abbey in Norfolk. Though he was never on Princess Augusta's establishment, one nevertheless may wonder whether he influenced in any way the planting at Kew during his father's superintendency. Thomas Haverfield, another of Haverfield senior's sons, was employed at Kew before taking charge of the gardens at Hampton Court in 1785. Haverfield House which still stands on the east side of Kew Green was home for several generations of the family.

It was John Haverfield senior who, according to his grandson, recruited William Aiton,[14] destined to be an important figure in the early history of Kew. There seems to be some confusion about Aiton's place of birth. *The Dictionary of National Biography* and other sources favour a village near Hamilton in Lanarkshire. Sheriff William Aiton in his *Inquiry into the Origins of the Aiton Family in Scotland* (1830) confidently asserts he was born in Boghall, Carnwath, also in Lanarkshire, where his father, John Aiton, was a farmer. John Aiton moved to Woodhall at Calderbank near Airdrie as farm manager on the Shawfield estate. There William and his two brothers trained as gardeners. Like so many other Scottish gardeners, William Aiton in 1754 sought employment in England with, it has been claimed, a letter of introduction to Philip Miller of the Chelsea Physic Garden who had a reputation of looking favourably on fellow Scots. For much of the eighteenth century the Chelsea Physic Garden functioned as the country's principal centre for the receipt and despatch to other gardens of exotics from abroad. At the peak of its reputation there was no better place for a young gardener to extend his knowledge and experience. Since Aiton's name does not appear in the surviving archives of the Society of Apothecaries who administered the Chelsea Physic Garden, he must have been one of the anonymous 'Apprentices' frequently mentioned in the Garden minutes.

All printed records of Kew agree that Aiton arrived at Kew in 1759 but the first mention of his name in Princess Augusta's accounts is in volume 47: "To William Aiton for cultivating and keeping in order our Physick Garden there for the same time", that is, January to March 1763. The most likely explanation is that he had been included in the expenses allocated to Haverfield for "men's labour". By the mid 1760s a division of responsibilities emerged; Haverfield supervised the pleasure grounds, kitchen garden, melon ground and orangery while Aiton looked after the physic garden. Their status can be gauged from their annual salaries: in 1770 Haverfield got £700 and Aiton £120.

Detail of W. Woollett's view of the White House, the Lake, Temples of Bellona, Pan and Eolus and House of Confucius.

William Aiton (1731–1793). Originally attributed to J. Zoffany, this portrait is now believed to be by George Engleheart (1752–1839), miniature painter to George III. In Engleheart's fee-book there is an entry that 'Mr Acton' [Aiton] was a sitter during 1786–7. The plant in Aiton's hand is *Aitonia*, named after him by C.P.P. Thunberg.

RIGHT:
Portrait of John Hill (1714–1785) by Coates. J. Thornton: *New Illustration of the Sexual System of Carolus von Linnaeus*, 1807. Thornton described Hill in this engraving as the "First Superintendent of the Royal Gardens at Kew", and included a vignette of the White House.

No one knows what salary that intriguing character John Hill received for the work he did at Kew or whether he was paid at all. No one as yet has identified his role there – a casual assistant, perhaps, a botanical adviser or, most improbably, 'Superintendent' as Robert Thornton extravagantly described him. A native of Peterborough, apprenticed to a London apothecary, possibly an itinerant actor for a few years, Hill first found congenial employment collecting plants for the Duke of Richmond and Lord Petre. A few minor parts on the stage at the Haymarket and Covent Garden and the chance to write the libretto of an opera gave him a taste for the theatre that he never lost. While uncertain where his future lay, he wisely retained his partnership in the shop of a London apothecary. Writing appealed to him: his two novels and two farces were failures but he made a name as a gossip columnist on the *London Advertiser and Literary Gazette*. Under the *nom de plume* of 'The Inspector' he daily entertained or informed his readers with topics as disparate as female emancipation, air pollution and contemporary fashion. Endowed with confidence, curiosity and phenomenal energy, he was willing to tackle any subject, to undertake any commission. But science in its broadest sense attracted him most: microscopy, astronomy, geology, zoology, botany and medicine. He hoped ultimately to be admitted to the exclusive membership of the Royal Society; when his application was rejected, he responded characteristically by denouncing the Society in a couple of books brought out in quick succession. He antagonised its Fellows by calling them "cockelshell merchants", "medal scrapers" and "butterfly hunters". They responded by accusing him of employing ghost-writers to produce his books. They were solidly united in their condemnation: "a mean designing undermining fellow", raged William Arderon of Norwich. "A man of great pretensions but of little credit in botany and physic," sneered Sir James Edward Smith. During this vituperative exchange with the scientific community, Hill may have received some satisfaction from the award of MD by Glasgow University in 1750; selling herbal remedies such as 'essence of water-dock' and 'tincture of bardana' proved a profitable venture. He used to get his supply of herbs from the Chelsea Physic Garden until his excessive collecting there led to his expulsion.

The urge to write never for one moment waned. His first ambitiously conceived botanical work appeared in weekly parts between January 1756 and January 1757. In *The British Herbal: an History of Plants and Trees, Natives of Britain, cultivated for Use, or raised for Beauty*, as it was called, Hill questioned Linnaeus's taxonomy where he thought it in error, but always "with decency and good manners" as Peter Collinson assured the touchy Swedish naturalist. Erasmus Darwin accused Hill of reusing discarded engraved plates, recutting them with a careless disregard to accuracy.

Hill's *Eden: or, a Complete Body of Gardening* appeared almost simultaneously with the *British Herbal* in 60 weekly parts. When the entire work was issued as a folio volume in November 1757, it was dedicated to Lord Bute. Professor George Rousseau, the author of a forthcoming biography of John Hill, believes that Hill first made Bute's acquaintance during the latter part of 1757. At last he had found a supporter who was to become an influential patron and a collaborator. The Duke of Northumberland, one of Hill's few allies, probably brought the two men together. His protégé made such a favourable impression that by the summer of the following year Hill was advising Bute's uncle, the Duke of Argyll, about his garden at Whitton in Twickenham.

In 1758 Hill undertook at the earnest behest of Bute the implementation of a 'great Plan', a decision that was to have devastating consequences for him and his wife. Only a man with Hill's impetuous acceptance of any challenge would have succumbed to Bute's blandishment. The 'great Plan' revealed itself as the compilation of a truly artificial classification of the plant kingdom, rejecting any 'natural' concept on the premise that such an ideal was unattainable. According to Hill's widow, it was to be "the most voluminous, magnificent and costly work that ever man attempted". Furthermore Bute had promised to bear the cost of its publication and to indemnify Hill against any personal loss. The first volume of *The Vegetable System* appeared in 1759 and when it reached the 26th volume 16 years later, 26,000 figures had been drawn for 1,600 plates, and 7,000 footnotes elucidated the text.

A grateful Lord Bute rewarded Hill for this commitment to predictable years of drudgery. It so happened that in 1758 Hill had recommended the establishment of a botanical garden at Kensington Palace in a slim pamphlet entitled *An Idea of a Botanical Garden in England: with Lectures on the Science*. In February 1761 Thomas Worsley, Surveyor of the Office of Works, informed Bute that he had "notified according to your lordship's order to Mr Greening that he is no longer to have the care of the Kensington Gardens".[15]

Perhaps it should come as no surprise that Hill got Greening's post! "Dr Hill . . . is one of the first men preferred in the new reign," wrote an envious Horace Walpole; "he is made gardener of Kensington, a place worth two thousand pounds a year."[16]

Through the intervention of Lord Bute, Hill also found employment in Princess Augusta's garden, confirmation of which was supplied by Lady Hill some years after her husband's death.

> The first employment that Lord Bute proposed to Sir John Hill . . . was the disposing and superintending a part of the Princess of Wales's Garden, at Kew, destined for Botany; which was to contain all the plants known on earth. In order to do this, he formed a correspondence with men of distinguished learning everywhere; receiving and giving seeds . . . His attendance was likewise at least once a week required at Kew.[17]

Hill himself gratefully acknowledged the "royal patronage and protection" he received from Princess Augusta[18] but nothing, so far, has come to light which clarifies his duties. Robert Thornton's elevation of Hill to "First Superintendent of the Royal Gardens at Kew" is not substantiated by any evidence. In a letter to the King of France's herbarium on 29 December 1772, Hill called himself 'L'Directeur du Jardin Royal de Kew' but the man's self-aggrandisement makes this appellation suspect. His Kew connections are tantalisingly perceived in incidental remarks and brief comments. Charles Knight wrote that many of the "curious plants" in the "Physic or Exotic Garden" had been collected "with great diligence and judgment, by the late Dr Hill".[19] His obituary in the *Annual Register* for 1775 claimed that "he obtained the patronage of the Earl of Bute, through whose interest he acquired the management of the Royal Gardens at Kew, with a handsome salary" – surely a confusion with his appointment at Kensington Palace since there is no mention of any salary paid to Hill in Augusta's household accounts. Perhaps the cursory notice of his death in the *Middlesex Journal* for 25 November 1775 is nearer the truth when it described him as "Botanist to the Royal Gardens at Kew".

But Hill's involvement at Kew – indeed the entire redevelopment there – had been put in jeopardy by Princess Augusta's problems in 1759 with the leasehold of the White House. When her late husband had leased the property from Lady Elizabeth St André, the 4th Earl of Essex, then a minor, had agreed to sell it to Frederick on his inheriting the estate. This event came about in 1759 but now the Earl reneged on his promise. "The Princess was piqued," reported Horace Walpole, "& was upon the point of leaving it [i.e. Kew]."[20] Fortunately she was able to negotiate a renewal of the lease for a period of 21 years[21] but after her eldest son's accession in 1760 Lord Essex was not reappointed a chamberlain to the King "and the reason is supposed to be that he has not behaved very well about this garden".[22]

On 13 November 1758 Thomas Knowlton, gardener to Lord Burlington at Londesborough in Yorkshire informed Richard Richardson, a physician with a notable garden at North Biesley, also in Yorkshire, that

the Prince of Wale[s i.e., the future George III] is now about preperations for bulding a stove 300 feet in length for plants & not pines [i.e.
pineapples] & my Ld Bute has already seatled a correspondence in asia
africa america europe & everywhere he can as to be shure my Ld is ye
most knowing of any in this kindome by much of any in it: such is his
great abilitys there in; & he is ye person as has prompted ye young prince;
& from such what may not be exspected & next Spring it will rise &
grow apace as all glasse & frams [frames] will be ready.[23]

The Great Stove, as it was called, one of a number of hothouses William
Chambers designed, did not reach the dimensions reported by Knowlton. But
it was then one of the largest in the country, 114 feet long, consisting of a
central bark stove 60 feet in length with a dry stove at either end, each of 25
feet; the overall height being 20 feet. The dry stoves heated by flues beneath
the floor and in the back wall were furnished with stepped stands on which
tender plants such as "succulent plants which are impatient of moisture" (Philip
Miller) were placed. The bark stove, also heated by flues, had a tan pit along
its length, 10 feet wide and 3 feet 6 inches deep. A Dutch invention, bark
stoves reached England early in the eighteenth century. Oak bark, ground to
a powder, and used in tanning leather, when mixed with elm sawdust produced
a fermenting heat into which potted plants were plunged. Temperatures of 30
degrees centigrade could be attained but there was no means of regulating this
moist heat. What species should be allocated to the dry stove or bark stove
was a matter of trial and error, and many exotics succumbed in the process.

The Reverend Stephen Hales designed the flues for the Great Stove and
suggested technical improvements.

> And as there will be several partitions in the green-house, I have
> proposed to have the glass of one of the rooms covered with shutters
> in winter, to keep the cold out, which will make a perpetual spring
> and summer, with an incessant succession of pure warm air. What a
> scene is here opened for improvements in green-house vegetation![24]

The only surviving plant house, designed by Chambers at Kew, is the
Orangery, elegantly classical in its proportions. Records indicate it was started
in 1757 but Chambers said it was built in 1761. It is made of brick covered
with the architect's special brand of stucco which, as he anticipated, has proved
extremely durable. Two furnaces in a shed behind the building heated flues
beneath the floor. Sir John Parnell who saw it in 1769 said it was "filled
completely, chiefly with oranges which bear extremely well and large". However, the experience of a succession of gardeners all confirms that its tiled roof,
inadequate light and insufficient heat made it an unsatisfactory building for
growing citrus fruit. The armorial bearings in the centre of its façade are those
of Frederick and Augusta.[25]

The Great Stove stood in the original botanic garden with the Orangery
just outside. East of the Orangery stretched an arboretum of about five acres

of hardy trees and shrubs; south of the Great Stove lay the Physic or Exotic Garden of roughly four acres. John Smith, the source of much of Kew's early history, stated that the Physic Garden displayed herbaceous plants on a one-acre site in long rows, arranged in Linnean classified order, with a path between each two rows. Each genus was identified by a named label and every species by a numbered label which could be identified in a manuscript record of herbaceous plants.[26] Elsewhere he mentions a collection of grasses in circular beds and British and alpine plants enclosed by low clipped hedges of lilac and hornbeam, but these probably occurred at a later stage of the Physic Garden's development. From a calculation of entries in John Hill's *Hortus Kewensis* (1768) it would appear that the herbaceous collection exceeded 2,700 species.

One of the Botanic Garden's generous donors was John Ellis, an amateur naturalist with an obsessive interest in the survival of plants on long sea voyages. When he was about to be appointed Provincial Agent in London for West Florida in 1763 he assured the President of the Board of Trade and Plantations that "All the pay I demand from the Province is to be in rare plants and seeds for the Royal Garden at Kew and your Lordships."[27] In 1764 he asked the British post at Canton to send him tea seeds embedded in wax, a method he favoured for preserving their viability. Apparently some of them survived since J.C. Lettsom in his *History of the Tea Tree* (1772) mentioned that "the largest tea plant in this kingdom is, I believe, at Kew; it was presented to that royal seminary by John Ellis, who raised it from seed". Ellis boasted to the Governor of New York that he had "introduced many rare & valuable plants into the Royal Gardens at Kew".[28] Aiton's competence as a gardener undoubtedly encouraged Ellis's generosity. "I

John Hill: *Twenty-five New Plants, rais'd in the Royal Garden at Kew*, 1773, plate 5: *Carthamus laevis* (now *Stokesia laevis*), drawn by Hill.

J. Hill: *Hortus Kewensis*, 1768. One of 20 engravings of plants, drawn by Hill, in what was the first published catalogue of plants cultivated at Kew.

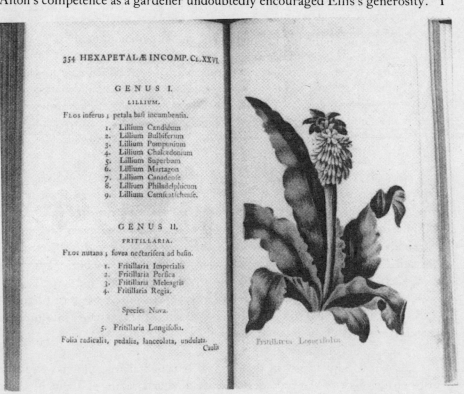

have sent Mr Aiton, her Royal Highness the Princess Dowager of Wales's Botanic Gard'ner at Kew, a parcel of seeds and dont doubt but he will raise them; as he is a perfect master of his business."[29]

Directed by Lord Bute and Hill and cosseted by the skills of Aiton, Kew's reputation grew. Thomas Knowlton spoke for his generation of gardeners in recognising it as having "one of the best collections in the kingdom if not in the world",[30] a comment prompted perhaps by the recent arrival at Kew of a collection of choice trees from the late Duke of Argyll's estate.

Archibald Campbell, Lord Islay and later 3rd Duke of Argyll, was Lord Bute's maternal uncle and his guardian after the death of Bute's father. Horace Walpole spoke approvingly of him as "one of the first encouragers of planting in England, most of the curious exotics which have been familiarized to this climate being introduced by him". Work on creating his estate at Whitton from waste land on Hounslow Heath started in 1725. A formal canal led from his house into an extensive plantation of trees and shrubs, many of them new introductions to this country. The Cedars of Lebanon, raised from seed, were a particular favourite and he took pride in having a representative collection of North American conifers. When he died in 1761, Lord Bute moved as many as could be safely transplanted to Kew in the spring of 1762.[31] Today's survivors of that transfer may include the magnificent *Ginkgo biloba* close to the Orangery and the *Robinia pseudoacacia* and *Sophora japonica*, both sadly near the end of their lives and only kept from collapse by metal bands and struts.

When Dr John Hope inspected Kew in September 1766, he saw a Guilandina (*Gymnocladus dioica*) and a Kermes oak (*Quercus coccifera*) among the newly introduced trees from the Whitton estate. King's Botanist for Scotland and Professor of Botany at Edinburgh, Hope had come south to see for himself

W. Woollett's view of the canal and Gothic Tower in the gardens of the 3rd Duke of Argyll. The Duke can be seen in the foreground on the left showing some guests around his grounds. After his death, some of his choicest trees came to Kew (*see* M. Symes, A. Hodges and J. Harvey: 'The plantings at Whitton'. *Garden History*, vol. 14, no. 2, pp 138–72).

what plants were being cultivated in English gardens and to acquire as much as he could to take back with him. He conscientiously listed their names, occasionally adding concise comments on their size, shape and flowers. Aiton was closely interrogated about his methods which were duly recorded: "Campanulas should all stand by themselves during winter in a frame among sawdust"; cuttings of *Mesembryanthemum* were planted in a mixture of soil and sea sand; Gardenias "require a strong loam & much water and much air"; "trees in the lawn protected from sheep by stakes placed in an inverted cone;" the flues of the Great Stove provided the glass frames bordering it with sufficient heat to flower African bulbs and tender annuals.[32] Hope never mentioned Smeaton's water engine, hidden in a thicket near the House of Confucius. The engineer, John Smeaton, better known for his building of Eddystone lighthouse, designed this Archimedes screw engine to raise water from a 12-foot-deep well for the lake, ponds and flower-beds.

Lord Bute had good reason to brag to the Governor of Georgia that "the Exotic Garden at Kew is by far the richest in Europe . . . getting plants and seeds from every corner of the habitable world".[33] Nor did Kew overlook the more accessible resources of the country's nurseries and individual gardens. The invoice of just one carrier, Thomas Layton, discloses that from February 1768 to March 1769 he delivered to Kew seven consignments from the nurseryman James Gordon, five from the Chelsea Physic Garden, four from John Hill, and one each from Peter Collinson and John Ellis. During the same period Aiton received plants from private gardens in Cambridgeshire, Dorset, Lancashire, Oxfordshire, Yorkshire and Wales.

Foremost among the British suppliers of foreign plants were James Gordon of the Mile End Nursery and Lee and Kennedy of the Vineyard Nursery, Hammersmith. Magnolias, Carolina poplars, *Ceanothus* and white brooms were part of one consignment to Kew in 1769 from Lee and Kennedy. James Gordon was renowned for his skill in raising kalmias, rhododendrons and azaleas from seed, and doubtless specimens of these also reached Kew. He is reputed to have made £500 over four years in the marketing of gardenias alone. In 1753 he introduced *Sophora japonica* from France where it had been introduced from China in 1747, and it is conceivable that Kew's ancient specimen came – perhaps via the Duke of Argyll – from his stock. Other nursery suppliers included John Busch of Hackney, William Malcolm of Kennington and William Watson of Islington. Kew also bought from one of its former gardeners, John Cree, who collected plants in Carolina during the 1760s before establishing a nursery at Addlestone in Surrey. One of his invoices in 1768 lists the sale to Kew of *Clematis*, *Eupatorium*, *Vaccinium*, *Aster*, *Convolvulus*, *Agave*, *Rudbeckia*, *Hypericum* and a specimen of the "new toothach tree" (*Zanthoxylum americanum*).

With perhaps a touch of hyperbole, Peter Collinson rated Kew as "the Paradise of our world, where all plants are found, that money or interest can procure. When I am there, I am transported with the novelty and variety; and don't know which to admire first or most."[34] He told everyone that Kew's transformation into a mecca for discerning gardeners was due to Lord Bute's

"great knowledge in the science of botany". Bute's protégé, John Hill, published the garden's first catalogue in 1768. Entitled *Hortus Kewensis*, it lists 3,400 species: 2,712 herbaceous plants, 488 hardy and 200 tender trees and shrubs. Hill adopted Linnaeus's binomials but rejected his classification for herbaceous plants, preferring his own system – '*methodo florali nova dispositas*' as the book's subtitle puts it; trees and shrubs were arranged in alphabetical order. Twenty indifferent plates, hand-coloured in some copies, reproduce his drawings. *Hortus Kewensis* was re-issued in a so-called second edition in 1769. Hill illustrated 25 more Kew plants as appendices in his *Vegetable System*;[35] 20 were re-issued in his 1773 edition of *Eden* and all appeared again in *Twenty-Five New Plants, rais'd in the Royal Garden at Kew* (1773). Sixteen of these plants had a North American provenance; the rest came from the West Indies, China, South Africa, Italy and Portugal. The *Hortus Kewensis* and these engravings remind us that during the 1760s and 1770s Kew was well on the way to becoming one of the richest repositories of the world's flora.

The maintenance and replenishment of the ice-house usually came within the routine duties of the gardener. John Haverfield's accounts itemise his expenditure on the one at Kew, which has now been restored and made accessible to visitors. John Evelyn refers to "conservatories of ice and snow, and other hortulan refreshments", but it is not until the eighteenth century that ice-houses became a normal accessory to large households, a source of ice during the summer and a cold store for perishable food. Crushed ice packed around with straw filled a deep brick-lined shaft surmounted by a domed roof for additional insulation. Its entrance was usually at an acute angle to exclude light and secured by close-fitting doors. Covered with soil and shaded by trees, the ice-house was a familiar grassy undulation on many estates. The ice was obtained from nearby pools – in Kew's case, the large lake. In 1765 Haverfield employed a man and a boy to clean the ice-house; another person made ice-skimmers; four horses and carts and a gang of labourers took three days to fill the shaft; 24 bundles of straw were used to insulate the walls. The work was never popular with the men who could obviously only collect ice during extreme cold. Kew's labourers were accordingly bribed with liberal quantities of beer.

CHAPTER FOUR

Sir William Chambers
contributes

His appointment as architect – opposed to
'Capability' Brown's concept of landscaping –
his buildings – public reaction to Kew

ON 6 APRIL 1757 a relatively unknown architect announced 'Proposals For
Publishing by Subscription, Designs of Villas, Temples, Gates, Doors and
Chimney Pieces'. William Chambers was a young man in a hurry to establish
a reputation in his chosen profession. The following month his *Designs of
Chinese Buildings, Furniture, Dresses, Machines and Utensils*, dedicated to
George, Prince of Wales, appeared. From the little we know of Chambers's
youth he never lacked self-confidence, ambition and drive, and so it comes as
no surprise that he soon made the acquaintance of Lord Bute, one of the most
influential men in the country. The precise circumstances of their meeting are
not known but Chambers's friend, Robert Wood, wished him well: "I heartily
congratulate you upon the complim[ent] paid you by the Prince [of Wales]
& upon Lord Bute's friendship."[1] According to Thomas Hardwick, a former
pupil of Chambers, when John Carr of York was asked by Bute to recommend
someone to instruct the Prince of Wales in architecture, Chambers was his
nomination. Since, however, Hardwick's brief biography of Chambers contains numerous errors his recollection cannot be accepted without reservation.[2]
Chambers's acceptance as a suitable tutor for the Prince was followed that
summer of 1757 by his appointment as architect to Princess Augusta at her
properties in London and Buckinghamshire and at Kew in Surrey. The Dowager Princess would have been favourably inclined towards Chambers, remem-
bering, no doubt, his drawings of a projected mausoleum for her late husband.
Horace Walpole noted that Bute "had carried Chambers the architect thither
[i.e. to Kew]"[3] and Chambers signalled his indebtedness to his lordship by
dedicating the first two editions of his *Treatise on Civil Architecture* (1759,
1768) to him, a work which had emerged from his tutorials with the Prince.

Chambers makes his first appearance in Princess Augusta's household
accounts during the period Lady Day to Michaelmas 1757 when he certified
the payment of work carried out at Kew. "The prince employs me three
mornings in a week to teach him architecture," he tells a correspondent; "the
building [and] other decorations at Kew fill up the remaining time. The
princess has the rest of the week which is scarcely sufficient as she is for ever
adding new embellishments at Kew [in] all [of] which I direct the execution

[and] measure the work."[4] Other commissions supplemented his annual remuneration of £100 from the Princess and £50 from the Prince.

Chambers's progress by the age of thirty-four was impressive in the light of his background. Born in 1723 in Gothenburg in Sweden where his parents were merchants of Scottish extraction, and educated in England, he returned to Sweden when he was sixteen. He entered the service of the Swedish East India Company, making a voyage to Bengal in 1740–2, and two voyages to China in 1743–5 and 1748–9. He left the Company's employ on his return from the last voyage, determined to become an architect. With that objective in mind, he went to Paris where he attended lectures in architecture at J.F. Blondel's Ecole des Arts for a year before completing his studies in Rome. Towards the end of 1755 he arrived in London, obtained modest accommodation in Covent Garden and, while waiting for clients, converted his notes and sketches of Chinese buildings and artifacts into his first book. His knowledge of oriental buildings found tangible expression at Kew as did his five years' study of Rome's classical architecture. In engaging Chambers, Princess Augusta may have hoped to fulfil Prince Frederick's vision of an exotic Kew. In this task Chambers was assisted by the English translation of *A Plan of Civil and Historical Architecture* (1730) by the Austrian imperial architect, J.B. Fischer von Erlach, who discussed and illustrated Islamic and Chinese buildings as well as traditional classical edifices.

When *Designs of Chinese Buildings* appeared in 1757 the vogue for chinoiserie, that is the absorption of Chinese or more often pseudo-Chinese decoration into European art, was at its peak. The establishment of trading posts in China by Portugal and Holland during the sixteenth century had significantly improved commercial links with Europe. Fragile porcelain which hitherto could not be easily transported overland by traditional trade routes was now being shipped to Europe in great quantities. Soon opportunist Chinese craftsmen were manufacturing patterned porcelain specifically for the European market. When supply could not match demand French and Dutch factories obligingly produced their own versions of blue and white pottery. And similarly when demand for oriental lacquerwork outstripped supply, European imitators quickly went to work. Chinoiserie was in the mainstream of the evolution of rococo, the abandonment of formal restraint for frivolous decoration. The engravings in *An Embassy . . . to China* (1665 and subsequent editions) by Jan Nieuhof, a member of the Dutch mission to the Manchu Emperor a decade earlier, encouraged this fashion. De Halde's *Description géographique, historique de l'empire de la Chine* (1735) revealed few new aspects of Chinese life and culture but it, too, served as a useful compendium for artists, craftsmen and architects.

The patrons of buildings which reflected the prevailing taste for the exotic were, as one would expect, royal households and fashion-conscious members of the gentry. Louis XIV had ordered a Trianon de Porcelaine to be erected in the park at Versailles in 1670 for his favourite concubine. The first Chinese feature to adorn any European garden, this single-storey building, clad with blue-and-white *faience* tiles, imitated the porcelain which mistakenly was believed to cover the Nanking Pagoda.

An awareness of China did not intrude on the English landscape until the late 1730s, however, and then only diffidently, making a greater show of decoration than of architectural form. A modest rectangular hut stood in a pond on wooden piles in the grounds of Stowe in 1738, an incongruous neighbour to substantially built classical temples. Its painted exterior panels depicted bouquets of flowers and scenes of Chinese life; more paintings inside confirmed its oriental derivation. In the late 1740s Shugborough Park near Stafford not only had a Chinese house modelled on a dwelling in Canton but also a hexagonal six-storey wooden pagoda. About the same time Benjamin Hyett's garden at Marybone House, Gloucester, sprouted a diminutive pagoda, a mere four storeys high. A Chinese pavilion, its curved roof hung with bells, stood on the river's edge of the Earl of Radnor's garden at Twickenham. The Earl's neighbour, Horace Walpole, who assured Horace Mann in 1750 that "Gothic or Chinese [buildings] . . . give a whimsical air of novelty that is very pleasing," a few years later expressed doubts about the proliferation of oriental structures – "the progenitors of a very numerous race all over the Kingdom". Some of these buildings owed their inception to the pattern books of Chinese designs which were more concerned with decorative detail than with technically accurate construction. The earliest reliable engravings of Chinese architecture appeared in *Designs of Chinese Buildings* by Chambers, who had made measured drawings during his stay in Canton. He may have borrowed from pattern books, been guilty of invention and, having only visited Canton, been ignorant of regional variations in style, but *Designs of Chinese Buildings* became an approved text, especially in continental Europe. Perhaps of greater interest to some of his English audience was the essay "Of the art of laying out gardens" which he added to the book. He had visited several Chinese gardens, supplementing his observations, so he tells us, with information from "a celebrated Chinese painter". Both the *Gentleman's Magazine* and the *Annual Register* reprinted this provocative essay which was also included in Thomas Percy's *Miscellaneous Pieces relating to the Chinese* (1762). Sir William Temple in his *On the Gardens of Epicurus* (1685) had promoted an image of Chinese gardens which extolled carefully-contrived irregularity in their layout. Chambers, too, praised such "beautiful irregularities". He commended their "winding passages cut in the groves, to the different points of view, each of which is marked by a seat, a building, or some other object. The perfection of their gardens consists in the number, beauty, and diversity of these scenes."

Chambers's *Dissertation on Oriental Gardening* (1772) expanded the essential details of this essay with help from Thomas Whately's *Observations on Modern Gardening* (1770). Chinese gardeners, contended Chambers, sought an emotional response through contours, an imaginative display of vegetation and, if required, through the disposition of buildings, statuary and fountains. Garden displays, varied throughout the year, projected "the pleasing, the terrible and the surprising". In actual fact, Chambers was voicing his own theories dressed in the fiction of a Chinese intermediary, a literary device used by Montesquieu, Goldsmith and other writers. He rejected what he perceived to be bland and insipid landscapes of lawns, water and a limited range of plants.

PLATE 1 "The Music Party". Frederick, Prince of Wales with his sister Ann at the harpsichord, Caroline playing the mandora and Amelia with a book on her lap, but listening attentively.

This charming conversation piece was executed by Frederick's court painter, Philip Mercier (1689–1760) about 1733. Another version has the Banqueting House at Hampton Court as its setting.

Its portrayal of familial harmony is misleading since the three Royal princesses shared their parents' antipathy to Frederick, especially Princess Ann who, at that time, was not on speaking terms with her brother. Frederick is practising on the violoncello. Lord Hervey, who also disliked him, sardonically observed that "Nero was not fonder of his harp than the Prince of his violoncello".

In the background of this picture is the Dutch House (now Kew Palace), leased to Queen Caroline. The road from Kew Green separated it from the entrance to the White House where the quartet was sitting. *National Portrait Gallery*

PLATE 2 1763 plan of Princess Augusta's garden at Kew. 1 White House; 2 Orangery; 3 Temple of the Sun in the Arboretum; 4 Great Stove; 5 Physic and Exotic Gardens; 6 Flower Garden; 7 Menagerie; 8 Temple of Bellona (subsequently moved); 9 Temple of Pan; 10 Temple of Eolus; 11 Smeaton's Water Pump; 12 House of Confucius; 13 Lake and Island; 14 Theatre of Augusta; 15 Temple of Victory; 16 Ruined Arch;

PLATE 3 House of Confucius, designed by Joseph Goupy for Frederick, Prince of Wales, about 1749. Drawing by Charles Edward Papendiek, assistant to Sir John Soane, 1818–24. *Sir John Soane's Museum*

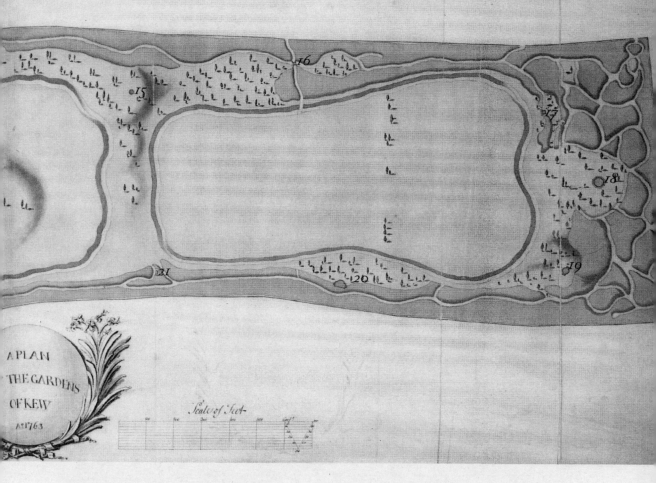

17 Alhambra; 18 Pagoda; 19 Mosque; 20 Gothic Cathedral; 21 Gallery of Antiques; 22 Temple of Arethusa? (subsequently moved); 23 Palladian Bridge; 24 Temple of Solitude?; 25 Great Lawn. Below the bottom boundary of the plan is Love Lane, a public right of way. *British Library*

PLATE 4 Alhambra, designed by Sir William Chambers, and built in 1758. It consists of a single room with a painted ceiling. *Sir John Soane's Museum*

PLATE 5 *Protea neriifolia*. Watercolour by Francis Bauer (1758–1840). This particular species of *Protea* was discovered in South Africa in 1597, but most of them were introduced to Europe in the late 18th and early 19th centuries, many of them by Kew's collector in the Cape, Francis Masson. Twenty three species were listed in the *Hortus Kewensis* (1810–13) but towards the end of the 19th century they were seldom grown. Sir Joseph Hooker blamed their disappearance on "improved systems of heating houses" and "incessant watering". *Natural History Museum, London*

Since the principal perpetrator of this style was 'Capability' Brown, his book can be seen as a veiled attack on him. Chambers, who did not object to natural landscaping, disliked compositions which suggested "chance [rather] than design; and a stranger is often at a loss to know whether he is walking in a meadow, or in a pleasure ground". He denounced the obliteration of mature gardens "to make room for a little grass, and a few American weeds".

It is conceivable that personal antipathy to some extent motivated Chambers's onslaught on Brown. Hardwick wrote that Chambers considered "Brown an intruder on an art in which neither his talents nor his education could entitle him to any respect".[5] It was even hinted that Chambers had resented the loss to Brown of a commission to build Lord Clive's house at Claremont. Horace Walpole saw the *Dissertation* as a "wild revenge against Brown". *The Monthly Review* for August 1772, commenting on Chambers's dislike of Brown's relandscaping of Richmond Gardens, identified the *Dissertation* as an attempt "to avert his royal majestie's attachment from the plan on which his garden at Richmond has been improved". The press, too, questioned his accuracy and objectivity. That his descriptions were "highly entertaining" was a dubious compliment, but the indictment that held him up to ridicule came from an anonymous poem, *An Heroic Epistle to Sir William Chambers* (1773). It denounced the Tory establishment whom Chambers, as a senior servant of the Crown, represented. "I laughed till I cried and the oftener I read it the better I liked it," wrote Walpole who had discreetly encouraged the author, the Reverend William Mason, to write it. Mason, a devotee of Kent and Brown, effectively discredited the waning fashion for Chinese features in English gardens. It went through ten reprints in 1773 alone whereas the *Dissertation* enjoyed only a second edition in which Chambers corrected some misconceptions about the book and added "An explanatory discourse by Tan Chet-Qua". French and German translations introduced the *Dissertation* to a wider audience in Europe where it became an influential guide to '*le jardin anglo-chinois*'.

When Chambers entered Princess Augusta's employ, he was confronted with two existing Chinese features at Kew: the House of Confucius and a Chinese arch, both probably designed by Goupy. Chambers's first assignment in 1757 was a bridge for which Joseph Wilton carved a 'River God's Head' to support the House of Confucius when it was moved the following year, most likely from the island, to the eastern extremity of the lake. Chambers ordered essential repairs to be done to this fragile structure: a hole to be blocked, the canvas to be tacked, and a sash to be mended. It survived in a dilapidated state until 1844 when the Office of Works sanctioned its demolition, accepting W. Lang's tender for the purchase of the materials of both the House of Confucius and the Temple of Pan.[6] It may have been reassembled, since Edwin Simpson's *History of Kew* (1849) noted that it then stood "in a meadow near Richmond Bridge".

About 1760 Chambers designed a Chinese-style Aviary of considerable depth at one end of the Flower Garden to house Princess Augusta's collection of small native and foreign birds. A short walk from this garden led to the Menagerie, an oval enclosure lined by cages or pens for exotic pheasants and bigger birds. A large pool occupying most of the space, a refuge for goldfish

and the more delicate waterfowl, featured a wooden octagonal pavilion "in imitation of a Chinese open Ting" on a small island.[7] A ting, that is a pavilion consisting of a minimal roof with supporting pillars, traditionally provided unobstructed views and shelter from light showers. By about 1785 the Menagerie had been converted into "an agreeable lawn"[8] and the pheasants presumably transferred to the New Menagerie pens adjacent to the Queen's Cottage.

The Pagoda, one of the few survivors of Chambers's Chinese phase, now stands like an uneasy exclamation mark terminating W.A. Nesfield's vista. It was not the first to intrude upon the English landscape; it has already been noted that Shugborough flaunted an oriental tower nearly a decade before Chambers began building one at Kew. In July 1761 Horace Walpole reported its construction[9] which was completed within months. Chambers's own drawing of a pagoda in Canton[10] would have influenced his design which reflected orthodox classicism and fashionable rococo as much as it did authentic Chinese architecture. At the time it offered the most accurate copy of a Chinese building to be seen in Europe. It soars 163 feet, divided into ten octagonal storeys, each successive one diminishing one foot in diameter and one foot between storeys. In his 1763 folio Chambers included plans of its ground floor and two storeys, an elevation, a cross-section and a folding plate which shows some minor revision to the original design. The 80 sinuous creatures, looking like coy seahorses, each with a bell projecting from its mouth, were replaced by aggressive dragons, wings poised, ready to launch themselves into the air, but without bells. If the colours are accurate in the faded watercolour drawing now in the Royal Institute of British Architects, the dragons were gilded (Chambers said they were "covered with a kind of thin glass of various

Thomas Sandby's view of the oval-shaped Menagerie with pens or large cages for birds around its perimeter. Chambers also designed the octagonal pavilion in the centre of the pool. The Temple of Bellona was then a near neighbour. The Menagerie had been converted into a lawn by about the mid-1780s. W. Chambers: *Plans . . . of the Gardens . . . at Kew*, 1763.

A replica of Chambers's pavilion was placed in the lake in the grounds of Osterley House in 1987.

colours"), the Chippendale-like railings around each balcony painted blue and red, and red applied to the underside of each roof. It is difficult to imagine roofs "covered with plates of varnished iron of different colours". Coloured glazed tiles would surely have been easier to fix. We can, however, be certain about the finial or terminal pole which glowed with several layers of gilt paint. But no amount of paint could ever disguise the ponderous solidity of the tower.

Now deprived of its colour and its dragons, it still remains an impressive building. Daniel Solander who saw it during the summer of 1761 shared public apprehension about its stability.

All thought that a building so much out of proportion should have fallen down before it was finished & no one believed it wd stand the

Theatre of Augusta,
c. 1820. Chambers
designed this
semi-circular Corinthian
colonnade as an open-air
theatre in 1760. *Sir John
Soane's Museum.*

terrible thunderstorms & tempests which we experienced there a month
ago, & which were more severe than had ever been remembered. The
Pagoda is now considered a masterpiece of art which in fact has stood
these shocks so well.[11]

Chambers boasted that there was "not the least crack or fracture in the whole
structure" which was built of "very hard bricks".[12] The quality of the work
of his bricklayer, Solomon Brown, was severely tested in 1941 when several
bombs fell near the Pagoda without damaging it. Unfortunately other materials
proved not to be as durable as the bricks. The roofs were the first to require
attention; in 1784 a coppersmith and a tiler were engaged to slate them. An
Office of Works minute for 1824 recorded that the "Pagoda in Kew Gardens
is to be kept up".[13] By 1915 the finial which had been renewed in 1825 had
been replaced about six times.[14]

The Pagoda again needed repairs when Sir William Hooker became Director in 1841. With Decimus Burton's appointment as architect in 1844, Hooker
sought his advice. In August 1844 Burton submitted a couple of sketches of
his colour scheme – yellow roofs, white gallery railings and a reddish pink
ground storey – and a proposal for curved roofs from which bells but no
dragons would hang. When Hooker took over the Pleasure Grounds in which
the Pagoda stood from the aged W.T. Aiton in 1845, he assured his father-in-
law that "my first object is to restore the Pagoda (under Mr Burton's directions)
& in ornament and painting make it like the very beautiful one now exhibiting
in the Panorama of Nankin in Leicester Square".[15] He discussed the restitution
of the building's lost features with the Commissioners of Woods and Forests
while it was still clad in scaffolding for workmen replacing the finial and
painting the woodwork.

A   B

Two versions of the Pagoda; B being the one built. The bottom storey is 26 feet in diameter and 18 feet high; each successive storey diminishes by one foot in both diameter and height. Chambers said he based his pagoda on an engraving in his *Designs of Chinese Buildings*, but it has been suggested that his inspiration was the Porcelain Tower in J. Nieuhof's *An Embassy . . . to . . . China*. W. Chambers: *Plans . . . of the Gardens . . . at Kew*, 1763.

The restoration of the dragons which had disappeared many years earlier became a desirable aim. They do not appear in John Buxton's painting done at Kew in June 1813 for Sir John Soane, and may have been removed when the Pagoda was tiled in 1784. Hooker said that having been made of wood, they soon rotted and were consequently removed.[16] W.T. Aiton who had managed Kew since 1793 must have been his informant. It has also been suggested that these wooden dragons had been painted with "highly-coloured enamels".[17] Burton, who supported Hooker's recommendations, not only wanted the ornamental features reinstated but also the slate roofs replaced with copper or cement and painted in the original colour; even the brickwork was to be coated with colour.[18] His estimate of £3,500 proved predictably unacceptable to both the Commissioners and the Office of Works. Hooker tried again in 1856 and, as recently as 1979, the matter of replicas of the dragons was considered but the Pagoda still awaits the restoration it richly deserves.

Its legacy has been a rash of pagodas in European parks: at Potsdam, Chanteloup, Scoonenberg, Montbéliard, Munich, Oranienbaum and Tsarskoie Selo. At home George IV's Fishing Temple at Virginia Water swarmed with a plague of dragons. John Nash's pagoda erected in St James's Park for victory celebrations in 1814 was spectacularly incinerated during a firework display. Of the few that have survived, the most delightful is the cast iron pagoda fountain built by Robert Abraham at Alton Towers about 1827. Marooned on an island in a still pool, it compensates for its smallness – only three storeys high – with an elegance and frivolity that the staid Kew Pagoda sadly lacks.

Chambers's Pagoda was appropriately flanked by two equally exotic buildings: the Alhambra and the Mosque. Frederick had intended erecting a Moorish building at Kew. An inscription in ink on a drawing, once owned by Lord Bute and now at the Royal Institute of British Architects, states: "Plan & Elevation of a building in the old Moorish taste"; added to this is a pencilled note: "this I drew in 1750 for the Prince, a model was made of it & it was built in 1758." It is a two-storey structure with five-bay open loggias on both floors. John Harris identified the artist as the Swiss painter, Johann Heinrich Müntz, an attribution confirmed by an annotation in Horace Walpole's hand in a copy of Archibald Robinson's *A Topographic Survey of the Great Road from London to Bath and Bristol* (1792)[19]: "The Turkish building [at Kew] was borrow'd by H.I. Muntz from a drawing by Richd Bentley Esq. in the collection at Strawberry Hill". A draughtsman as well as a distinguished scholar, Richard Bentley had lived abroad, partly in France, where he may first have

met Müntz, then an engineer in the French army. Later he introduced Müntz to Walpole. Chambers's Alhambra PLATE 4, erected in 1758, and indisputably influenced by this 1750 drawing was, in John Harris's words, "a more fanciful rococo Gothic design laced with eastern motifs". Like the Mosque on the western side of the Pagoda it was boldly — one might even say garishly — painted, its brilliant red columns emphasising its vertical lines.

The Mosque, built three years later, also owed its conception to another's

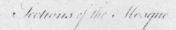

Interior of the Mosque with stucco palm-trees in the octagonal saloon and its domed ceiling painted by Richard Wilson. W. Chambers: *Plans . . . of the Gardens . . . at Kew,* 1763.

drawings despite Chambers's assertion that he had "endeavoured to collect the principal particularities of the Turkish Architecture".[20] Plates in Fischer von Erlach's rather romantic interpretation of the world's building styles[21] have been suggested as his main source of reference. The Mosque stood on a small mound, probably the 'little mount' mentioned in Robert Greening's contract of 1757. A large central dome between two smaller ones, a slender minaret at both ends and a florid Arabic text from the Qu'ran proclaimed an Islamic temple but the interior displayed light-hearted rococo. The two small rooms painted a delicate rose colour contrasted with the dazzling yellow of the main saloon. Eight green stucco palm trees supported the central dome on which Richard Wilson painted drifting clouds in a summer sky. This piece of whimsy failed to impress Walpole: "anybody might have done them," he wrote dismissively.

To counterbalance the exoticism of the East, Chambers also introduced into the Princess's garden familiar styles – dignified classical structures designed with the confident assurance acquired from years of study in Italy. He used the Ionic order for the Temples of Arethusa, Peace and Victory and the Gallery of Antiques, Doric for the Temples of Bellona and Pan, Corinthian for the Temple of the Sun and the Theatre of Augusta, and Composite for the Temple of Eolus.

His Ruined Arch was not just a folly; it served also as a functional bridge over a path for sheep and cattle to be brought from the Kew Road to the enclosed pastures within the garden. Georgian sensibility appreciated ruins as picturesque objects and as reminders of the passage of time which, in Gibbon's sonorous prose, "spares neither man nor the proudest of his works". Chambers in his *Dissertation on Oriental Gardening* alleged that the Chinese adopted a similar view of garden ruins: "to indicate the debility, the disappointments and the dissolution of humanity; which fill the mind with melancholy and incline it to serious reflections". Those landowners fortunate enough to own genuine ruins incorporated them into the design of their gardens; Studley Royal, for instance, effectively absorbed the sprawling remains of Fountains Abbey. Sanderson Miller, an expert in contrived decay, provided ruins to order – a sham castle, a crumbling tower, a solitary Gothic arch or whatever his client wanted. Batty Langley published a selection of designs of ruins for his clients. Gothic was usually preferred but a classical form was chosen by Chambers who knew intimately the impressive ruins of Rome. He conceived "a Roman antiquity, built of brick, with incrustation of stone"[22], a "triumphal arch" with three openings, two of which he subsequently converted into rooms. A generous clothing of ivy and other rampant vegetation with a scattering of stone fragments at the base, "seemingly fallen from the buildings", added to the appearance of deterioration. In 1864 the two side arches were reopened,[23] and in 1932, when the structure had truly become a ruin and a danger to the public, it required partial restoration.[24]

No fashionable Georgian garden was denied a specimen of Gothic architecture, genuine or fake. Alfred's Hall at Cirencester Park, the Gothic Temple at Stowe and the Ruined Castle at Hagley afforded evidence of a vigorous

North side of the Ruined Arch, by Joshua Kirby. The two side rooms are now open arches. Note also the term – one of those carved for Frederick? – the remains of which still lie at the base of the Arch. W. Chambers: *Plans . . . of the Gardens . . . at Kew,* 1763.

Ruined Arch, by Richard Wilson. Presumably a preliminary sketch for his oil painting which was exhibited simply as the 'Ruin' at the Royal Academy in 1762. Until 1949 this painting was thought to be of a garden feature at the Villa Borghese. *Glynn Vivian Art Gallery, Swansea.*

Gothic Revival movement in this country; Kew had "one of the richest, and most delicately ornamented, of the gothic garden buildings of the century".[25] But Kew's Gothic Cathedral, as it was called, was the creation not of Chambers but of Johann Heinrich Müntz. Bentley, who first knew him in France, met him again in Jersey in 1755. He introduced him to Horace Walpole in whose employ he remained for four years until his dismissal following an affair with one of the servants at Strawberry Hill. Having demonstrated his skill in the

Gothic Cathedral by
Johann Heinrich Müntz,
briefly an architect of the
Gothic revival in
England. Built of wood, it
measured about 50 feet
across, and was concealed
in a grove.
W. Chambers: *Plans . . .
of the Gardens . . . at
Kew*, 1763.

Plan & Elevation of the Gothic Cathedral.

Gothic mode with a charming garden house for Thomas Hudson in Twickenham in 1757, he was invited to contribute a building to Kew. In 1759 his
Gothic Cathedral, built of wood and painted plaster, added another style to
the garden's eclectic architecture.

Inherently fragile, it was repaired in 1764 and probably required frequent
attention before its demolition in 1807.[26] Routine maintenance took up a great
deal of Chambers's time – patching roofs, seeing to defective guttering, reglazing, repointing of brickwork and repainting. It should be remembered that
these temples and other garden features were erected quickly – the Palladian
bridge to the island in the lake and the Temple of Victory were reputed to
have been built in a night – and were constructed of flimsy materials, sometimes

no more than lath and plaster. Count Kielmansegge discovered for himself the truth of Horace Walpole's remark that "they are all of wood."[27]

> All these buildings consist of wood only, but are so cleverly covered with plaster, and painted in oil colours, that you swear they were solid buildings of quarry stone, unless by knocking them you discovered the truth from the sound . . . I can well believe that these wooden structures will not last long.[28]

Garden buildings were often insubstantial constructions but Kielmansegge believed that since Princess Augusta had only leased the White House she was unwilling to go to the expense of building in durable stone. An exception, of course, had to be the Pagoda whose height and proportions demanded more permanent materials. Of the 20 or so buildings designed by Chambers only the Pagoda, the Temples of Arethusa and Bellona (both reconstructed when later moved) and the Temple of Eolus (rebuilt by Decimus Burton), the Ruined Arch (partly rebuilt), the Orangery and a few garden shelters survive.[29]

Chambers has been credited with landscaping Kew but nowhere in his *Plans . . . of the Gardens and Buildings at Kew* (1763) did he claim that he had designed it. Indeed he attributed it by implication to Lord Bute:

> Originally the land was one continued dead flat: the soil was in general barren, and without either weed or water. With so many disadvantages it was not easy to produce anything even tolerable in gardening but princely munificence, guided by a director, equally skilled in cultivating the earth, and in the politer arts, overcame all difficulties. What was once a desart [sic] is now an Eden. The judgement with which art has been employed, to supply the defects of nature, and to cover its deformities, hath very justly gained universal admiration, and reflects uncommon lustre on the refined taste of the noble contriver; as the vast sums which have been expended to bring this arduous undertaking to perfection, do infinite honour to the generosity and benevolence of the illustrious possessor.[30]

The "illustrious possessor" was, naturally, Princess Augusta and the "director" and "noble contriver" Lord Bute. Modesty was never one of Chambers's failings; in his book he claimed credit for most of the buildings and it is inconceivable that he would not also have admitted to the layout of the grounds had he been their designer. Certain features had already been formed before his arrival at Kew. John Dillman had managed "the piece of water . . . with the island being 6 acres water & above 3 acres island".[31] With surplus soil dredged from the lake he had created the Mound where the Temple of Eolus stands. Dillman's successor, Robert Greening, levelled the wide expanse of lawn from the White House to the lake and made the wilderness at the southern extremity of the garden. The barrier of trees skirting Love Lane and Kew Road was there as much for privacy as for any aesthetic reason; Greening may have planted this screen, taking the gardens at Pope's villa and Carlton House

The Alhambra, Pagoda and Mosque, by William Marlow. A section of the sunken fence or ha-ha confining livestock to a field can be seen.

as his model. If it can be assumed that the principles of garden design Chambers ascribed to the Chinese were, in fact, his own opinions then he must surely have disapproved of Kew's central core of open space. "Neither do they [i.e. Chinese gardeners] ever carry a walk around the extremities of a piece of ground, and leave the middle entirely open, as it is too often done among us."[32] This is precisely what happened at Kew. That perceptive gardener, Peter Collinson, always recognised Bute as the arbiter of Kew's development. "When it would be suitable," he wrote to Bute, "I shall with pleasure see your wonderful operations & the improvements those gardens have receiv'd from your great skill in every branch of science."[33] Walpole attributed many of the garden's gentle undulations to Bute. "Being on a flat, Lord Bute raised hillocs [sic] to diversify the ground."[34]

The disposition of walks, the peripheral belts of trees with serpentine paths, a stretch of water and acres of grassland do not add up to a trend-setting landscape. What, however, gave Kew its distinction, its unique position among contemporary gardens were its renowned plant collections and its extraordinary assembly of buildings, designed and subtly disposed by Chambers. He utilised existing mounds as viewing platforms for the Temples of Eolus and Victory; he terminated the southern vista with the eye-catching trinity of the Pagoda, Alhambra and Mosque; he moved the House of Confucius to a site which offered better views; and everywhere he enticed the visitor with glimpses of garden buildings among the trees. They illustrated his axiom that "Variety is always delightful, and novelty, attended with nothing inconsistent or disagreeable, sometimes takes the place of beauty."[35]

Some of the paintings of this landscape executed by Joshua Kirby, William Marlow and Thomas Sandby were engraved to complement plans of all the buildings, published in 1763 by command of George III who gave £800 towards its production.[36] Chambers dedicated this impressive iconography of a royal garden to Princess Augusta whose munificence and constant support had made the transformation of Kew possible. This was a gracious tribute to his employer; a greater debt, however, was owed to Lord Bute whose patronage he acknowledged by presenting him with all the original drawings for the book. "I beg pardon for taking this liberty, and am with the warmest sentiments of gratitude for the many great and generous favors you have conferr'd on me."[37]

Kew's enhanced reputation inevitably attracted many visitors – knowledgeable gardeners and casual sightseers. Like neighbouring Richmond Gardens, Kew had a designated route for viewing the grounds PLATE 2. It began in the arboretum to the east of the Orangery where the Temple of the Sun, modelled on a ruin at Baalbek, stood in the centre of a small lawn. Only a few yards south of it, but partly concealed by a line of trees, stood the Great Stove facing "the prodigious variety of curious plants" in the Physic or Exotic Garden, all within an enclosure of trees and shrubs. A baroque entrance arch to the Flower Garden within another screen of trees led to a floral display, a Chinese aviary and a well-stocked fishpond. Beyond lay the wooden lattice fence of the oval Menagerie or pheasant pens where a Chinese garden pavilion

View by William Marlow of the Lake, Orangery, Temples of Bellona and Eolus, House of Confucius and of the Palladian Bridge which was erected at night "to surprise the Princess" next day (H. Walpole). W. Chambers: *Plans . . . of the Gardens . . . at Kew*, 1763.

Theatre of Augusta,
Temple of Victory and
Palladian Bridge, sketched
from the perimeter walk by
William Gilpin in 1765.
Bodleian Library.

dominated a large ornamental pool. The classical dome of the Temple of
Bellona rose rather incongruously above the encircling trees. The Temple of
Pan terminated a short alley to the left of the exit from the Menagerie. Ahead
the path skirted the Mound surmounted by the Temple of Eolus with its
revolving seat, passed Smeaton's water engine, and approached the House of
Confucius at the lake's edge. From the upper of its two furnished rooms the
Temple of Arethusa could be seen on the far side of the lake. The woodland path
continued to hug the garden's eastern boundary. A garden shelter, designed by
William Kent for Prince Frederick's garden, was one of a number of seats
installed along its length. An open space broke the continuous belt of trees
with an alternative path leading to the lakeside and, on a slight incline, to the
semi-circular colonnade of the Theatre of Augusta where open-air perform-
ances were sometimes staged. The Temple of Victory, commemorating the
battle of Minden, stood clear of the trees on a mound which offered a view
north across the lake to the White House or south towards Richmond Hill.
Princess Augusta or Queen Charlotte looking at the Temple of Victory from
the White House is supposed to have remarked on its likeness to a mushroom,
and thereafter for many years it was known as the Mushroom Temple.
Returning to the path, visitors passed under the Ruined Arch and admired a
statue of a muse, before approaching the Pagoda and its attendant buildings,
the Alhambra and the Mosque. One of the best views anywhere in the neighbourhood could be had from the top storey of the Pagoda in a room thoughtfully
furnished for the comfort and relaxation of those who had toiled to the top.
Cedars of Lebanon planted around its base had become so thick a century later
that Joseph Hooker said that the view of the lawn from the upper storeys had
been completely obscured by them. The Wilderness behind the Pagoda formed
the southern boundary. By this time most newcomers to the garden, realising
that the itinerary presented a sequence of discrete compositions, would have
accepted the Gothic Cathedral without a shock of surprise. Some would miss

the Gallery of Antiques, richly decorated with bas reliefs and statues, almost hidden in a thicket. The Temple of Arethusa faced the island with its wooden bridge, and the Temple of Solitude and possibly the Temple of Peace ended the circular tour which terminated to the west of the White House.

Public opinion was divided on the merits or suitability of all these buildings but there was unanimous approval of the uncluttered acres stretching from the White House to the Wilderness – the "noble lawn" as far as the lake and beyond that the two pastures confining sheep and cattle within ha-has – giving spectacular views to all parts of the estate. The Reverend William Gilpin awarded Kew, with a few reservations, his accolade of approval in 1765.[38] He found no fault with the perimeter walk but recommended occasional breaks in the trees to open up vistas to the lawn and lake. While sympathetic to the difficulties in contouring a completely flat terrain, he disapproved of the little mounds thrown up near the Temple of Victory (Sir John Parnell, on the other hand, rather liked them). He judged the Pagoda "a whimsical object" which, with its companions, the Alhambra and the Mosque, presented "a disagreeable regularity". Count Karl von Zinzendorf, who visited Kew a few years later, thought the group had been placed too close together. The London Magazine for August 1774 dismissed the Pagoda as "the puerile effort of an overgrown boy". Some people found difficulty in accepting the cultural diversity represented by the garden's architecture. "Amazing absurdity," noted Sir John Parnell in his journal. "Where different nations are thus introduced into an improvement, they shou'd at least be hid from one another, by a hill, wood or clump of trees."[39]

Notwithstanding Horace Walpole's belief that "there is little invention or taste shown" at Kew, it attracted the attention of artists like Richard Wilson PLATE 11 whose paintings of the Pagoda, Palladian bridge and the Ruined Arch rejoiced in its exotic contrasts. Wilson's canvases presumably met with the approval of Chambers who tried unsuccessfully to persuade George III to purchase them.

After his visit to Kew in 1765, the French academician, P.J. Groseley, wrote that the "gardens which the princess dowager of Wales has lately laid out at Kew, in the neighbourhood of Richmond, unite all that the English taste has been capable of producing, most magnificent and most variegated".[40] Kew inspired the creation or influenced the design of a number of continental gardens. A French engineer and geographer, George-Louis Le Rouge, who published a vast pictorial record of gardens in England, France and Germany,[41] pirated most of his engravings including many of Kew. Kew and Stowe – Kew's rival in the number of garden buildings – provided copy for the two temples Mme du Barry built at Louveciennes. Continental pagodas, however, such as, for instance, the one at Chanteloup, owed little to Chambers's design.

In 1763 Queen Charlotte requested Lord Bute to send a plan of Kew's garden to the Duke of Mecklenburg who had already acquired a copy of Chambers's book. The Landgraf Friedrich II and his English wife, a daughter of George II, erected a copy of the Mosque and another Kew building in their rococo garden at Schloss Bellevue. The Mosque reappeared in the garden at

Ruined Arch and Temple of Victory, by F.M. Piper. *K. Akademien för de fria konsterna, Stockholm.*

Schwetzingen in Baden-Württemberg. The Swedish architect, Fredrik Magnus Piper, who toured English gardens in 1772 and again in 1780, sketching some of the buildings at Kew, was the first to introduce English concepts of landscape design to his own country.

Towards the end of her life Princess Augusta only visited Kew briefly on Tuesdays and Saturdays, when she had breakfast with her eldest son and Queen Charlotte. She had now retired to a secluded existence at Carlton House. With commendable fortitude she bore the pain of the cancerous growth that eventually killed her on 8 February 1772, aged 52. Horace Walpole reckoned she had spent between £30,000 and £40,000 on improving Kew and concurred with gossip that she had squandered her private income on Lord Bute. So unpopular had she become, due to this friendship with Bute and the influence they reputedly exerted on the King, that at her funeral "the mob huzzaed for joy and treated her memory with much disrespect". Oliver Goldsmith was a lone voice praising Augusta. In a couple of days he wrote *Threnodia Augustalia* which was performed in Mrs Cornely's Great Room in Soho Square on 20 February 1772. His verses speak of Kew as

> . . . *a scene beyond Elysium blest;*
> *Where sculptur'd elegance and native grace,*
> *Unite to stamp the beauties of the place;*
> *While sweetly blending, still one seen,*
> *The wavy lawn, the sloping green;*
> *While novelty, with cautious cunning,*
> *Through ev'ry maze of fancy running,*
> *From China borrows aids to deck the scene* . . .

In contrast, several hundred lampoons and caricatures appeared ridiculing Princess Augusta and Lord Bute and denouncing their illicit relationship. Without doubt they were close friends, Augusta admiring Bute's erudition and respecting his advice, but that they were lovers is surely unlikely. All the evidence suggests that both exercised marital fidelity. Within a year of Augusta's death, Bute bought a villa and began a garden at Christchurch which he assiduously cultivated during the last six years of his life. He died at the age of 78 on 10 March 1792.

Bute's collaborator at Kew, John Hill, died on 22 November 1775, shortly after the publication of the 26th volume of his *Vegetable System*. His prodigious outpouring of books had been publicly recognised the previous year with the award of the Order of Vasa by King Gustavus III of Sweden; with royal permission he now pompously adopted the style of an English knight. Hill had beggared himself in publishing the *Vegetable System*. Possibly because its production became so protracted, Bute had lost interest in it and withdrawn financial support. Hill's widow tried in vain to claim from Bute outstanding expenses allegedly due to her late husband.

William Chambers was also honoured by the King of Sweden, who made him a Chevalier of the Order of the Polar Star in 1770. His award, too, was recognised as equivalent to an English knighthood. He remained Princess Augusta's architect until her death. George III briefly retained his services to make some alteration to the White House which he and his family used as a summer residence. Chambers had at last reached the pinnacle of his profession: a prosperous private practice enabled him to lease in 1765 the villa and estate at Whitton of the late Duke of Argyll, Lord Bute's uncle. In 1769 he was appointed Comptroller of the Works and upon the reorganisation of the department in 1782 promoted to Surveyor-General and Comptroller. After Kew, he had few garden commissions: a Chinese temple at Ansley in Warwickshire in 1767, the reconstruction of a Chinese temple at Amesbury Hall in Wiltshire in 1772 and designs for a garden at Svartsjö Castle in Sweden in 1774, unfortunately never implemented.

CHAPTER FIVE

The Royal Family in residence

*George III engages 'Capability' Brown to
relandscape Richmond Gardens — move to the
White House and Kew Palace — domestic life
— Queen Charlotte's interest in botany —
Queen's Cottage — Richmond Gardens and
Kew Garden merged — Castellated Palace —
King's illness*

ON 25 OCTOBER 1760 the eldest son of Frederick Prince of Wales
ascended the throne as George III. Two days later he appointed his mentor
and former tutor, his "Dearest Friend", Lord Bute, a Privy Councillor. As
now more than ever he needed Bute's guidance, he succeeded in May 1762 in
obtaining his appointment as First Lord of the Treasury, a post which carried
the responsibility and authority of Prime Minister.

It was Bute who arranged George III's marriage in September 1761 to
Princess Sophie Charlotte of Mecklenburg-Strelitz, a small state in northern
Germany. "She is not a beauty," reported his emissary Colonel David Graeme,
but "she is amiable". Like Princess Augusta she was only seventeen at the time
of her betrothal; neither then nor later did she show any interest in the affairs of
state — she was, in Lord Chesterfield's approving judgement, "an unmeddling
Queen". A loyal wife, constantly attentive to the welfare of her 15 children,
she preferred domesticity and the seclusion that the royal residences at Richmond, Kew and Windsor assured. From 1764 Richmond Lodge became the
summer retreat of the royal family where they relaxed in Queen Caroline's
garden — soon to disappear — and drank coffee in the Duke of Ormonde's
summerhouse on the Terrace. They retired discreetly to the White House at
Kew when the public were admitted to their garden on Sundays.

Following a survey of the grounds of Richmond Lodge ordered in November 1761, the Haverfields took charge of its maintenance, soon to be disrupted
by the appearance of Lancelot 'Capability' Brown. Brown, already a landscape
designer of some repute, sought a royal appointment to further his career. He
had failed in his bid to be Princess Augusta's gardener when Robert Greening
died in March 1758; he was unlucky in his attempt to secure the vacant post
at Kensington Palace despite the support of several peers of the realm. Success
at last came in July 1764 when he was made Master Gardener at Hampton

Court. Now a royal gardener, George III commissioned him to relandscape Richmond Gardens and the adjacent hamlet of West Sheen. Brown had perfected a formula which he imposed ruthlessly, sometimes mechanically, on all his landscapes: undulating lawns right up to the house, judiciously-placed clumps of trees, and a lake, often achieved by damming a stream. The result, complained Chambers, differed "very little from common fields, so clearly is vulgar nature copied in most of them". An opponent of inflexible formality, his schemes obliterated straight avenues and destroyed intrusive buildings. His proposals for Richmond conformed to this practice;[1] Bridgeman's layout and most of Kent's buildings yielded to mandatory grass and acres of woodland linked by meandering walks. His plan approved, he recruited Michael Milliken, a gardener at Chatsworth, to be his foreman at Richmond. In a letter of 27 January 1765, Milliken tells his wife that he had "a great many men at work and am much confin'd. The King & Quen come 2 or 3 days a week here and talkes as free, and I think is as bent on the work as ever the Duke of Devonshire was." Later the same year the Clerk of Works at Richmond, Joshua Kirby, queried whether "Mr Haverfield should be paid his whole allowance for taking care of Richmond Gardens, notwithstanding his having nothing to do for the present with that part of the Garden where the alterations are making".[2]

The garden buildings were the first casualties. Even celebrated Merlin's Cave did not escape; lead salvaged from it was purchased in 1766 by Mr Devall, a plumber.[3] The only survivor was the Hermitage, left as an inoffen-

W. Woollett's view of the Lake, the Temple of Arethusa (on its original site) and the Orangery. The extraordinary Swan Boat was made on the occasion of the Prince of Wales's 17th birthday in 1755. Designed by John Rich, the manager of Covent Garden, its neck and head reached 18 feet, and the boat could hold 10 people. It was christened the *Augusta*.

sive ruin. The 18 houses constituting West Sheen were razed to provide additional pasture for the King's cattle. For Brown's contemporaries the greatest loss was the Terrace, which had extended almost the whole length of Richmond Gardens, buttressed by a brick wall and separated from the River Thames by a public road. After the erection of Kew Bridge in 1759, this road had fallen into disuse and by an Act of Parliament in 1766 the King was allowed to close it provided the towpath, essential for river traffic, remained. Brown followed the example of Richard Owen Cambridge at Cambridge Park on the Twickenham side of the river where walls and terraces were supplanted by lawns descending close to the water's edge. The Terrace, "a boast of the country"[4] and "admiration of all foreigners",[5] gave way to sweeping lawns that brought Richmond Gardens into greater intimacy with the river.[6] Uvedale Price, a devotee of the picturesque, applauded this transformation. So did the agronomist Arthur Young: "the terrass and the grounds about it are now converted PLATE 10 into waving lawn that hangs to the river in a most beautiful manner".[7] Those, however, who had enjoyed strolling along this riverside walk never forgave Brown.

> Richmond gardens now declare the hand that spoilt them; nor is there
> a person who can recollect the beauty of the lengthened terrace, but
> censures the innovator – Mr Capability Brown.[8]

When Sir John Parnell saw Richmond Gardens in 1769 they were still in the process of being "laid out partly in the old and partly in the new taste". And when he described the ground as being "dead flat" we realise that what we now call Rhododendron Dell had not yet been excavated. It is the only surviving evidence of Brown's handiwork at Kew. Created as a Hollow Walk in 1773, it was then planted with laurels.[9]

There were those like Chambers and the anonymous contributor to the *Middlesex Journal* who saw only destruction and desolation as the consquences of Brown's work.

> . . . but when I reflected that he had destroyed that Terrace which Queen Carolina [sic] made at great expence, and pulled down her Merlin's Cave, overturned her Hermitage, filled up her pond, removed her dairy, and drove the plough through her paddock, I own I grieved . . .[10]

Portrait of George III (1738–1820), by Allan Ramsay.

Others, more receptive to innovation and change in gardening fashions, welcomed what they judged were positive improvements. Arthur Young admired Brown's manipulation of tree cover in groves, small clumps and single specimens, and the manner in which the river was constantly brought into view. It was a "mild agreeable landscape, which seems created by the hand of unpresuming taste".

Some 20 years after Brown had left Richmond, J.J. Boydell's assessment of his mature landscape reflected current opinion. Brown, he wrote

> broke the avenues; rooted up the long line of dressed hedges; gave the woods a natural shape; unveiled extensive lawns; destroyed . . . Merlin and his cave; dilapidated every tasteless building; formed plantations, which are now grown into effect and beauty; and, conducting a gravel path around the whole, gradually displayed the varying scenery of this charming domain.[11]

'Capability' Brown's landscaping formed part of George III's scheme to replace Richmond Lodge with a palace fit for royal occupation. The Dodsleys were not alone in thinking the Lodge "unsuitable to the dignity of a King of England".[12] Not long after George II's accession in 1727, Sir Edward Lovett Pearce had designed a Palladian palace for him but his plans and those of William Kent were shelved, due perhaps to royal disapproval or lack of funds. Kent's design, executed about 1735, had got no further than a pearwood model which Count Kielmansegge saw displayed in the Queen's Pavilion in Richmond Gardens in 1761.[13] Repairs and minor alterations continued to be made to Richmond Lodge until the reign of George III.

Queen Charlotte (1744–1818). Studio of Allan Ramsay. *National Portrait Gallery.*

Kielmansegge noted in his diary that George III intended building early in 1762 a new palace in Richmond Gardens, probably near the Terrace. However economy dictated its deferment to 1765 when Chambers submitted plans of a massive rectangular building with pavilion towers and imposing steps leading up to a Corinthian portico (its site is shown on 'Capability' Brown's 1764 plan of Richmond Gardens). But since £28,000 had been spent on purchasing Buckingham House, the Privy Purse could not at the time sustain the cost of another royal residence.

One of Chambers's later designs for a palace in Richmond Gardens. Quinlan Terry made use of it in his riverside scheme in Richmond. *Royal Institute of British Architects.*

Nevertheless George III had no intention of abandoning his pet project, especially since Richmond Lodge could no longer accommodate his growing family, and so in 1769 Chambers produced yet another set of plans. The site chosen, about 60 yards north of the Lodge, coincided almost exactly with the spot where Queen Caroline's dairy had stood. Foundations were laid in the summer of 1770 and the building had acquired its high-vaulted basement before all work ceased on it three years later. Mrs Papendiek, Assistant Keeper of the Wardrobe in the Royal Household, tells us that the King reluctantly gave up the project when Richmond Parish refused to sell him a plot of land essential for the "elegance and convenience" of the royal estate. As she is not always reliable it is more likely that financial factors again decided the matter. In 1775 the long-suffering Chambers informed Thomas Worsley, an architectural friend, that he had completed yet another scheme for the elusive palace – a central block with pavilions linked by colonnades, probably a modification of the 1769 design. Nothing happened, perhaps because George III's building obsessions were about to be transferred to Windsor.

A pencil sketch by George III of Syon House seen from Richmond Gardens. Joseph Goupy, who designed a number of garden features for Prince Frederick, taught George III landscape drawing, but the King preferred architectural drawing; his version of the Temple of Victory is in the Royal Library. *Reproduced by gracious permission of Her Majesty The Queen.*

The royal preoccupation with building was deplored by Peter Collinson. "I wish the King had any taste in flowers or plants; but as he has none, there are no hopes of encouragement from him, for his talent is architecture."[14] Both Chambers and John Joshua Kirby, an authority on perspective, had instructed him; he made meticulous studies of the classical orders and copied the works of William Kent, Colen Campbell and Chambers. The succession of plans for Richmond's new palace were all subjected to his critical scrutiny. Kirby's *Perspective of Architecture* (1761) includes His Majesty's version of a colonnaded Palladian house. Even a few of Kew's garden buildings were credited to him: "some were designed by King George 3ᵈ," noted Horace Walpole enigmatically.[15] Several contemporaries believed the Temple of Victory had been designed by him: "*temple de la victoire qui est du dessein du roi d'un fort mauvais goût*".[16] But his drawing of the Temple of Victory, now in the Royal collection at Windsor, differs in detail from the illustration in Chambers's 1763 book on Kew: the order is Doric, not Ionic, the frieze is not the same and no windows are installed. Perhaps Chambers adapted the King's design but claimed, as he did for most of the temples, the authorship of this building. Chambers stated that the Temple of the Sun had been "begun and finished under my inspection in the year 1761"[17] – but Sir John Parnell, obviously misinformed, reported after his visit to Kew in 1763 that this particular temple had been "built by the present King from a design of his own".

The death of Princess Augusta provided an alternative home for the royal family. "We are going to change our habitation this summer exchanging Richmond for Kew, there our home will be for the best and greater solitude," Queen Charlotte told her brother, Prince Charles of Mecklenburg-Strelitz on 20 March 1772. Two months later he learned that "we have arrived at Kew". Soon after, Richmond Lodge was demolished.[18]

Chambers as Comptroller of the King's Works was commanded to plan alterations to Princess Augusta's White House, now called Kew House by the royal family. One wing was scheduled to become a library and the other possibly a music room while the rebuilding of the riverside façade was contemplated. Nothing in the archives confirms that these conversions were ever executed, but it is unlikely that any substantial extensions were made; when the Reverend D. Lysons saw Kew House some 20 years later he dismissed it as "small, and calculated merely for occasional retirement". We know, however, of one addition – a brick clock tower close to the eastern wing. It can be seen in a drawing by Bishop Fisher of Salisbury, tutor to Prince Edward from 1780 and Chaplain-in-Ordinary to the King.[19] The clock itself had been made at the Observatory in the Old Deer Park at Richmond by John Smith in 1777. The tower, which survived the demolition of Kew House, was in a ruinous state when George Papendiek drew it about 1820. In 1849 the clock was installed in Queen Victoria's residence at Osborne House on the Isle of Wight.

The move to the former White House never solved the royal family's need for extra accommodation. So the Prince of Wales and Prince Frederick were lodged in the Dutch House; adjacent 'Queen Caroline's House' served as a

Drawing of the White House by John Fisher (1748–1829), tutor to Prince Edward in 1780, and later Bishop of Salisbury. A competent artist, he taught drawing to Princess Elizabeth. The clock tower was added by Chambers in 1772–3. *Leslie Paton collection: London Borough of Richmond upon Thames.*

Clock tower of the White House. Lithograph by George Papendiek, *c.* 1820. The clock, now at Osborne House, was installed at Kew about 1777.

nursery for Prince Ernest, and the Princes William and Edward moved into the west wing of Cambridge Cottage, once the home of Lord Bute. The six princesses had bedrooms in Kew House and every morning all the children were reunited with their parents for breakfast.

The Dutch House – Kew Palace as it is known today – went through a

Dutch House or Kew Palace where two of the sons of George III – the Prince of Wales and Prince Frederick – lived.

confusing permutation of names: the Princes' House, the Prince of Wales's House, Royal Nursery, the Old Red House and the Old Palace. It and a neighbouring house (subsequently known as Queen Caroline's House) and nearly 70 acres of land had been leased by Queen Caroline between 1728 and 1729 from Sir Richard Levett.[20] Sir Richard's grandson, Levett Blackborne of Lincoln's Inn, expressed willingness in 1772 to sell the estate for only £20,000, "so little do I desire to profit myself, by encroaching on their Majesties' love and attachment to Richmond". The freehold of the property comprising several houses, stables, coach-houses and outhouses was conveyed to Queen Charlotte in July 1781.

Undercroft beneath Kew Palace, possibly contemporary with the Palace, or part of an earlier dwelling (but not the Dairy House as some authorities suggest).

The proximity of Kew hamlet to the Tudor Richmond Palace on the river had always attracted courtiers and public figures as residents. Robert Dudley, created Earl of Leicester by Queen Elizabeth, owned the Dairy House which, on his death, may have been acquired by the lawyer, Sir Hugh Portman. Sir John Puckering, Speaker of the House of Commons and Lord Keeper of the Great Seal, occasionally entertained the Queen at his home in Kew. All these dwellings have disappeared with the exception of the Dutch House which may have been built on the site of one of them – its rib-vaulted cellars and brick well could be the remains of an earlier structure. It is possible that Samuel Fortrey, a London merchant of Flemish descent, destroyed the earlier house when he built the Dutch House in 1631; the date of its erection and the initials of its first owners S.C.F. (Samuel and Catherine Fortrey) are cut into the brick above the entrance.

The Dutch House, which could have acquired this name from Fortrey's

W. Woollett's view of the Great Lawn seen from the White House with the Temple of Victory and the Pagoda in the distance. "The only beauty of this garden is the first view of the lawn, with the Pagoda at the end, which, though the termination hath not the effect of conclusion . . . gives a dignity and an extent to the gardens very unexpected." *London Magazine*, August 1774, p. 360.

connections with Flanders, is an outstanding example of what Sir John Summerson called the 'Artisan Mannerist style'. It represents a fashion briefly favoured by those who eschewed both Tudor and Inigo Jones's Palladian. Cromwell House at Highgate, Swakeleys at Ickenham and Broome Park in Kent are survivors of the genre, which favoured brick and esteemed prominent gables. Kew's crow-stepped gables are replicated in a number of East Anglian manor houses. Its Flemish bond brickwork, that is, headers and stretchers alternating in every course, is a very early example of the technique. The Dutch House also pioneered moulded and carved brick as a decorative feature – the Ionic and Corinthian capitals to the pilasters on the first and second storeys were adroitly shaped by hand. Unfortunately the Doric pilasters flanking the doorway on the ground floor were removed when a clumsy and inappropriate wooden porch was installed early in the nineteenth century. The replacement of the original mullion and transom windows by sashes was another desecration. The functional simplicity of the interior suited the modest living of the royal family. His Majesty's breakfast- and dining-rooms were located on the ground floor; the first floor had separate bedrooms for George III and Queen Charlotte, who also had a drawing-room and a boudoir. The princesses retired at night to small bedrooms in the attic.

George III was the first British monarch to distinguish between court and home; St James's Palace served for state occasions and official receptions; Buckingham Palace provided privacy, but after 1775 domestic life centred on

Kew and Windsor. He lived at Kew for about three months of each year, beginning about the middle of May, arriving on alternate Tuesdays and staying until Friday. Unlike his grandfather and great-grandfather, George III's personal life was exemplary – no extramarital affairs enlivened royal gossip. Moral rectitude, prudent economy and well-regulated routine governed his home life. He dispensed with formality at Richmond and Kew. "The Royal Family are here always in so very refined a way, that they live as the simplest country gentlefolk," wrote Fanny Burney, a Keeper of the Robes. "The King has not even an equerry with him, nor the Queen any lady to attend her when she goes her airings." They rose early; he perused official papers before they breakfasted separately, the King drinking tea while the Queen preferred coffee. The King never ate again until dinner at four: a plain dish such as "roast mutton, lamb, veal, beef and fowls, generally cold, with sallads [sic] is the diet of the royal family; made dishes are never touched. The King's beverage is wine considerably diluted with water."[21] For exercise he rode; for relaxation he played chess or backgammon; the Queen sat at the card table or embroidered with her daughters. By eleven o'clock everyone was in bed.

After breakfast with their parents, the children were collected by their tutor or governess. Practical work formed part of the education of the older

From the Temple of Victory looking north to the White House, by Hendrik de Court (1742–1810). Chambers's Ionic temple commemorated the battle of Minden on 1 August 1759. It was in a ruinous state when Sir William Hooker arrived in 1841, and was replaced by the flagstaff. *Reproduced by gracious permission of Her Majesty The Queen.*

Looking from the Temple of Victory towards the Ruined Arch, Alhambra, Pagoda and Mosque. The fence confined cattle to part of the bottom field. Anonymous drawing. *Reproduced by gracious permission of Her Majesty The Queen.*

boys – gardening and agricultural tasks or building a small stone house, the ruins of which survived at Kew until the 1880s. Once a week they accompanied their parents to Richmond Gardens where they played with their pets and inspected any additions to the exotic fauna in the New Menagerie.

Queen Caroline had kept a menagerie at Richmond for "deer and wild beasts";[22] tigers were still housed in the "Royal Paddock Garden" in 1740.[23] The first record we have of George III's New Menagerie (so called possibly to distinguish it from the remnants of Queen Caroline's) appears on Thomas Richardson's survey of the Royal Manor of Richmond in 1771. Situated in a

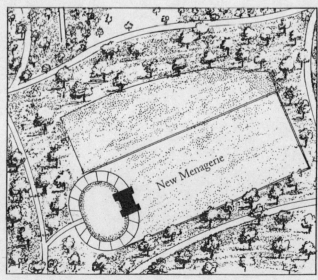

New Menagerie

The building in the circle of pheasant pens is the predecessor and core of the Queen's Cottage. Reproduced from Burrell's and Richardson's plan of Richmond, 1771.

clearing in the woods in the grounds of Richmond Lodge, it consisted of a paddock of about three acres with pheasant pens forming an oval ring of which a small building was part – a residence for the keeper of the menagerie perhaps? Whatever its function, it eventually became the Queen's Cottage. Besides birds with colourful plumage, this royal zoo displayed cattle from Algeria and India and "a hog like a porcupine in skin, with navel on back". Its most prized inhabitants were the kangaroos. In December 1793 David Dundas tells Sir Joseph Banks that one of them had been "if I may be allowed the expression, for some time in the state of parturition. I have taken the liberty to inform you that I learnt yesterday that the man who has the care of it had perceived the head of the young one appearing out of its pouch so long as 30th October."[24] By the time the collection was dispersed ten years or so later the population of kangaroos had reached nearly 20.

The Queen's Cottage survives today as a charming example of cottage orné – small villas simulating rustic prettiness with thatched roofs, ornamental gables and bay windows. Some of the earliest examples had been erected on the Badminton estate during the 1750s; they enjoyed a vogue from the late eighteenth century until the Regency – John Nash composed an entire village of these bijou residences at Blaize Hamlet near Bristol. No-one knows who designed the Georgian cottage at Kew. A similar vernacular style can be seen in the model village at Milton Abbas in Dorset, built by 'Capability' Brown from designs by William Chambers. No evidence exists for their involvement in the Queen's Cottage which the *London Magazine* for 1774 attributed to Queen Charlotte.[25] Her contribution probably went no further than ideas for

The original single-storey building extended as far as the two slightly projecting doors. The lighter-coloured brick clearly indicates the enlargement of the Queen's Cottage with wings and an upper floor. Photograph by T. Harwood.

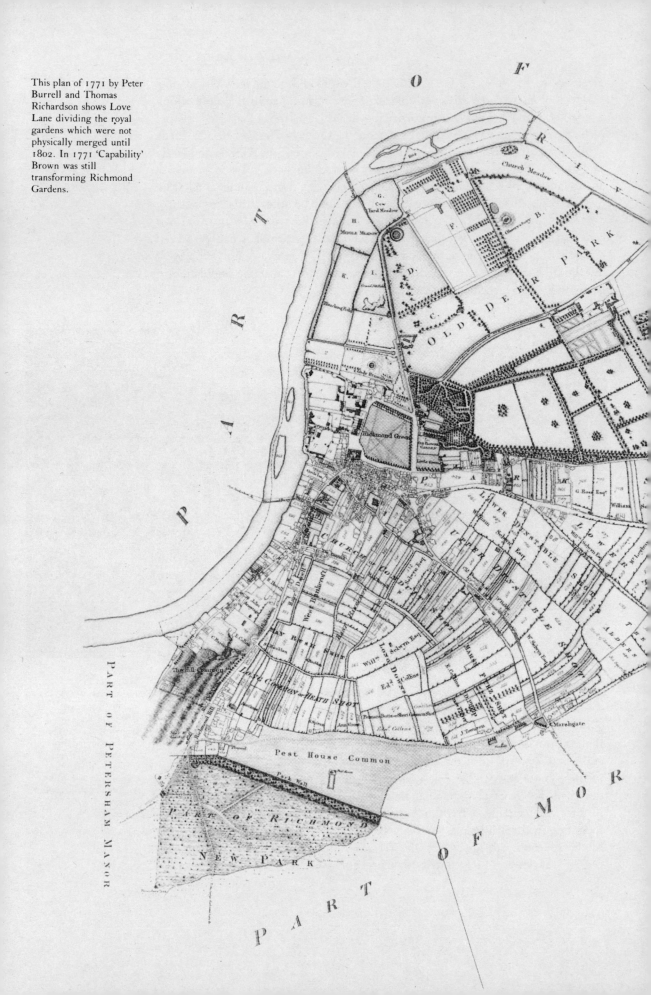

This plan of 1771 by Peter Burrell and Thomas Richardson shows Love Lane dividing the royal gardens which were not physically merged until 1802. In 1771 'Capability' Brown was still transforming Richmond Gardens.

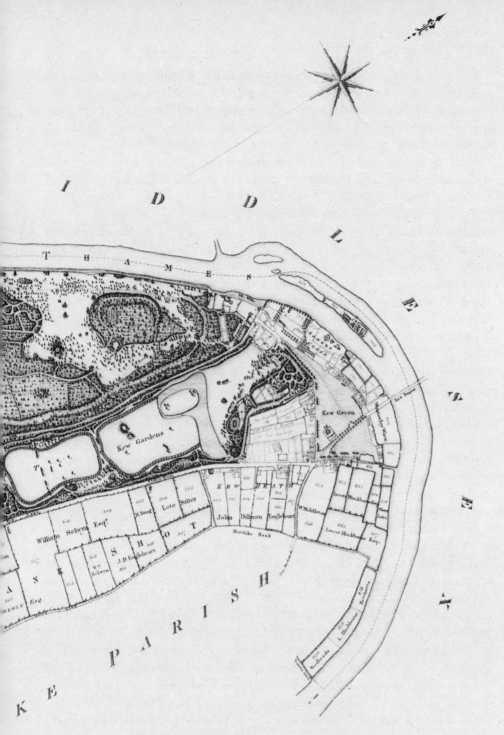

PLAN

of the Royal Manor of

RICHMOND,

otherwise

WEST SHEEN,

in the County of SURRY;

in GRANT to

HER MAJESTY.

Taken under the Direction of

PETER BURRELL ESQ:

His Majesty's Survr Genl

in 1771. by Thos Richardson in

York Street, Cavendish Square.

the interior décor. Built of red brick, with Tudor pretensions, it started as a single-storey structure – the two projecting doors on either side match the configuration of the structure linked to the pheasant pens on Richardson's 1771 map. It is not known when the first floor clad in decorative timber framing was added. Queen Charlotte, writing to her brother, Prince Charles of Mecklenburg-Strelitz, on 20 March 1772, mentions that her cottage had four or five rooms, which suggests to Olwen Hedley, the author of the excellent biography of Queen Charlotte, that the additional floor had already been built. The cottage served solely as a picturesque shelter for the royal family, a place for snacks and the occasional meal. The one large room on the ground floor displayed Hogarth prints on its walls;[26] these were removed during the 1890s[27] and reinstated in 1978. A curving staircase leads to the bower-like picnic room with its chinoiserie resonances PLATE 18; painted convolvulus and nasturtium climb the walls and the door mouldings and pelmets are composed of bamboo motifs. The same floor houses a china closet and a discreetly concealed commode.

Queen Charlotte's third daughter, Princess Elizabeth, whose artistic talents added floral garlands to wall panels in Frogmore House, may have executed the climbing plants in the picnic room. The *St James's Chronicle* announced in 1805 that "Kew Cottage in Kew Gardens has undergone considerable alteration and improvement under the direction of the Princess Elizabeth . . . The Cottage likewise has been furnished. The outside of the building stands in great need of being made to correspond with the inside. It was completed during the late stay of the Royal Family there."[28] We do not know the nature of these improvements but John Charlton stated that they involved adding another floor. In December 1806 William Townsend Aiton was instructed to convert the kangaroo paddock into a flower garden[29] and the inmates of the menagerie were dispersed.

When Queen Anne's Garden House in the grounds of Windsor Castle became vacant in 1775, Queen Charlotte acquired it and Chambers converted it into a castellated mansion. There the royal pair lived until they moved into the new private apartments in the Castle in 1804. George III, who welcomed the greater informality that Windsor allowed, soon spent more time there than at Kew. In May 1790 the Queen purchased a cottage about half a mile south of the Castle as a repository for her botanical collections and library. Two years later it was demolished when the acquisition of adjacent Frogmore enabled her to enlarge the grounds. The architect, James Wyatt, transformed Frogmore into a colonnaded 'Trianon' where the Queen and her daughters botanised, painted and embroidered. Mary Moser's profusion of painted blossoms on the ceiling and walls of Frogmore House and Princess Elizabeth's more restrained floral panels proclaimed the family's enthusiasm for botany.

The garden at Frogmore was laid out according to orthodox principles of picturesque planning. Wyatt added a Gothic Ruin and Princess Elizabeth is said to have designed a thatched Hermitage, a barn or garden ballroom and an octagonal Temple of Solitude. Aiton from Kew supervised the erection of a greenhouse. Prince Augustus was informed by his mother in 1791 that the "plants from Kew [were] beginning to cut a figour [sic] in the greenhouse

POINCIANA *pulcherrima, aculeis geminis. Linn' sp. pl. p.544.*

Sydney Parkinson pinx 1767. Kew

PLATE 6 *Caesalpinia pulcherrima.* Watercolour on vellum by Sydney Parkinson (*c.*1745–71), 1767. Joseph Banks, just returned from Newfoundland, who met Parkinson at the Vineyard Nursery, Hammersmith, persuaded him to draw plants both there and at Kew, of which this is one of twelve paintings now in the Natural History Museum. Banks engaged Parkinson as his botanical artist on the *Endeavour* voyage. The *Hortus Kewensis* (1789) records this plant as being in cultivation at Kew.

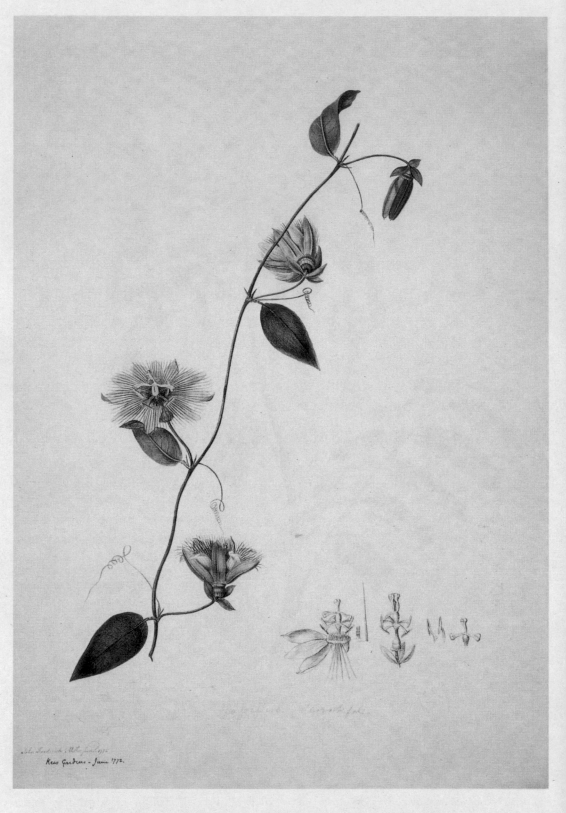

PLATE 7 *Passiflora serratifolia*. Watercolour by John Miller (Johann Sebastian Müller, 1715–*c*.1790). Drawn at Kew in June 1772. Miller had been Joseph Banks's "articled draughtsman" on his trip to Iceland; on his return he drew plants at Kew and other gardens for Banks. *Natural History Museum, London*

Stanhopea insignis Bauer.

PLATE 8 *Stanhopea insignis*. Watercolour by Francis Bauer (1758–1840) who was resident artist at Kew from 1790 to 1840. This Brazilian orchid reached Kew about 1826. The *Hortus Kewensis* (1810–13) listed 115 species of orchids of which 84 were exotics. At that time gardeners were uncertain about the cultivation of epiphytic orchids. Kew placed them on a bed of "loose turfy soil interspersed with small portions of stems of trees". (*Gardeners' Chronicle* 31 January 1885, 143). *Natural History Museum, London*

Tab. 12.

J.ᵉ Sowerby delt 1787.

Limodorum Tankervillia vol. 3 pag. 302.

PLATE 9 *Phaius tankervilleae.* Watercolour by James Sowerby (1757–1822), 1787. Introduced to this country in 1778, it was one of the first tropical orchids to flower in the glasshouses at Kew where it is still in cultivation. This drawing is engraved on plate 12 of vol. 3 of the *Hortus Kewensis* (1789). *Natural History Museum, London*

. . . My chief plants are to be natives of England & all such foreign ones as will thrive in our soil."[30] Sir Joseph Banks was relieved to inform a friend that "the Queen's new garden at Frogmore near Windsor will be elegant in the extreme & would rival Kew if H[er] Majesty did not forbid all thoughts of introducing anything there till it has been at Kew".[31]

The Queen also tended a small flower garden at Buckingham Palace but Frogmore gave her more scope and the incentive to create a "little Paradise", her "*Paradis Terrestre*". Her gardening activities stimulated an interest in the structure of flowers. The Queen "studies botany intensively & really reads with perseverance elementary books", reported a gratified Banks.[32] With some diffidence she told Lord Bute, who had presented her with a copy of his *Botanical Tables*, that she was about to compile a "Herbal from impressions on black paper",[33] inspired no doubt by the diligent Mrs Delany PLATE 15 who had patiently assembled plant portraits in coloured paper against a black background. But the Queen proposed mounting the plants themselves rather than paper images.

Within a few months of declaring her intentions to Lord Bute, she received a ready-made herbarium from the King. The Reverend John Lightfoot, Chaplain to the Dowager Duchess of Portland, had corresponded with many naturalists, had joined friends in botanical excursions, had published a *Flora Scotica* (1777), and had in the course of these activities amassed a collection of dried plants, mainly British. Following his early death in 1788, George III paid 100 guineas for his herbarium which was housed in 24 mahogany cabinets at Frogmore. (Sir) James Edward Smith, President of the newly-formed Linnean Society of London, who rearranged this herbarium, subsequently gave "regular lectures (or rather conversations) in Zoology & Botany" to the Queen and the Princesses. They all sat around a table, Smith speaking from his notes which Queen Charlotte earnestly copied and then read aloud at the conclusion of each session to check her accuracy. Smith's botanist friend, Thomas J. Woodward, was very concerned about a possible breach of propriety during his zoological lectures.

> . . . you have doubtless found out a proper method of decently conveying your instructions in that branch to her Majesty & the four Princesses, & have not shocked their delicacy by any dissertations on the generation of animals, nor hazarded any opinions upon the physical causes which occasion the boar to be so long, & the bull so short a time in performing the same natural functions.[34]

Smith's zoological revelations apparently never excited the same level of enthusiasm as did his botanical tuition. The Queen continued to add to her collection of flower books, Princess Elizabeth resolved to excel as a botanist, and one royal holiday at Weymouth was agreeably passed in collecting, pressing and mounting marine plants.

Francis Bauer at Kew and the artist, Margaret Meen, gave lessons in flower painting to the Queen, the Princess Royal and Princess Elizabeth who meticulously coloured Bauer's *Erica* engravings, copied plates in botanical works, and tested the capabilities of pencil, pen-and-ink, crayon, watercolour and oils, usually in the intimacy of a family group. Robert Thornton ingratiat-ingly praised their industry in his *New Illustration of the Sexual System of Linnaeus* (1799–1807):

> There is not a plant in the Gardens at Kew (which contain all the choicest productions of the habitable globe) but has been either drawn by her gracious Majesty, or some of the Princesses, with a grace and skill which reflect on these personages the highest honour.

Other authors, too, dedicated their books to the Queen: Margaret Meen her *Exotic Plants from the Royal Gardens at Kew* (1790), Mary Lawrance her *Collection of Roses from Nature* (1796–9) and Charles Abbot his *Flora Bed-fordiensis* (1798) which hailed her as "the first female Botanist in the wide circle of the British Dominions". When the extraordinary Bird of Paradise flower was introduced to Kew from the Cape in 1773, Banks and Aiton agreed it to be most fitting to name it *Strelitzia reginae* after the Queen's natal home of Mecklenburg-Strelitz.

The monarch's preoccupation with Windsor never meant that Kew was forgotten or neglected. When the Revd Thomas Methold, residing on the south side of Kew Green, died in the 1780s, George III purchased his garden of over three acres, adding it to the royal garden. The leasehold of the Levett Blackborne estate, including the Dutch House, was acquired in 1781 and this consolidation of the Kew estate continued over the next 20 years. The death of the 4th Earl of Essex in 1799 provided the long-awaited opportunity to procure the freehold of Kew House and its grounds. While this was being negotiated, Kew's botanical garden passed to the Establishment of the Royal Household at Michaelmas 1801 and, like other royal gardens, became the responsibility of the Lord Steward. After the King had inherited his mother's Kew estate in 1772, he sought its union with neighbouring Richmond Gardens. The *London Gazette* for 5 October 1773 reported that "The workmen have begun to pull down the pailing [sic] in Richmond Gardens, which divide them from Kew, and a grand walk is to be formed in the spot where the present division stands." A paragraph in the *Middlesex Journal*, 21–23 April 1774, noted that "Upwards of four hundred labourers are now at work in H. Majesty's Gardens at Richmond & Kew – The old walls which divided the two gardens are pulled down and the ground levelling." Such announcements were actually premature since the authority of Parliament and the consent of local parishes were necessary before a public right of way could be closed. Love Lane, which separated the two gardens PLATE 12, ran from West Sheen Lane near Richmond Green, proceeded north along a line either on or parallel to the present Holly Walk, and terminated at the Horse Ferry west of Kew Palace. A wooden bridge spanning it linked the two gardens. An Act of Parliament in 1785 authorising its closure compensated Richmond and Kew parishes by making the bridle road from Kew Bridge (the present Kew Road) suitable for vehicular traffic. It was not, however, until 1802 that the boundary walls were eventually demolished and the two gardens merged into one.[35]

Now the freeholder of Kew House, George III perversely decided to pull it down, having resolved to replace it with an imposing edifice looking across the river to Brentford. A few small items such as the brass door plates, engraved with the letters 'F.P.' and the ostrich feathers of Frederick, Prince of Wales, were transferred to the Dutch House, and by 1803 Kew House, except for the separate kitchen wing and porter's lodge, had disappeared. Now converted into flats for garden staff, the plain pedimented building still retains its eighteenth-century kitchens.

George III had stubbornly persisted in his goal of a new palace at Kew. Joseph Farington noted in his diary for 12 January 1794 that "Wyatt lately shewed Hodges a set of designs for a Palace to be built at Kew which the King has a serious intention of doing." The architect James Wyatt had been engaged by Queen Charlotte in 1792 to make alterations and additions to Frogmore House. Farington learned from Benjamin West that through the Queen's influence Wyatt had obtained the post of Surveyor-General and Comptroller in 1796 following the death of Chambers. Already a master of neo-classical architecture, Wyatt now added neo-Gothic to his repertoire. He had expressed the quintessence of ecclesiastical Gothic at Fonthill Abbey in Wiltshire; the Kew commission undertaken in a private capacity offered him the opportunity to indulge in the symmetry of medieval castles.

Its foundations were dug on a site where once Queen Caroline's House, Lady Charlotte Finch's residence and the Ferry-house had stood.[36] This low-lying spot of some seven acres, a few hundred yards from the present Kew Palace, faced "the smoky and dusky town of Brentford, one of the most detestable places in the vicinity of London". "I never thought I should have adopted Gothic instead of Grecian architecture", George III ruefully confessed to the Duchess of Württemberg, "but the bad taste of the last forty years has so entirely corrupted the professors of the latter I have taken to the former from thinking Wyatt perfect in that style, of which my house will I trust be a good example."[37] PLATE 13 Work started in 1801 and very quickly a central keep, four storeys high, surrounded by turrets and towers and a high curtain wall, all extravagantly crenellated, confronted the bemused citizens of Brentford. Wyatt adopted his brother Samuel's invention of tubular, hollow cast iron units to support this massive structure; cast iron – an early use of this material in building – also fabricated the window tracery. The King accepted the fact that a shortage of skilled workmen caused some of the protracted delays in its erection but he also suspected "a certain want of diligence in Wyatt". By 1800 £40,000 had already been allocated to the project but when the costs had soared to £100,000 in 1806 the King had lost interest, perhaps to the relief of the Queen who had always preferred the unassuming and intimate Dutch House. Benjamin West believed that "Wyatt had ruined the King, as he had done Mr Beckford of Fonthill. The King's private purse had been exhausted by him in building at Kew and in other places."[38] It had been almost completely roofed by 1811 when costs had reached an astronomical £500,000. But now the King's complete mental collapse had removed the need to continue and no one regretted the decision to abandon it. Someone likened it to the Bastille. Sir Nathaniel

Castellated Palace from Kew Palace lawn. Anonymous, 1810s-20s. The small building to the right is an aviary. The Queen's House and other dwellings were demolished to provide a site for the Castellated Palace.

Wraxall was equally dismissive: "a most singular monument of eccentricity and expense . . . an image of distempered reason".[39]

With dry rot spreading through the empty shell of the Castellated Palace, as many called it,[40] George IV ordered its demolition in 1827. John Nash removed oak flooring, skirting, stone paving, the principal staircase and two other staircases for his conversion of Buckingham House[41], (Sir) Jeffry Wyat-ville appropriated windows, doors and the screen at the foot of the grand staircase, while William T. Aiton installed one of the small Gothic windows in his residence on the corner of Kew Green;[42] what remained was sold by public auction. After an unfortunate accident in December 1827 when several workmen dismantling it were killed, explosives were used to complete its destruction.

Escalating costs apart, the King's health had ultimately decided the fate of the Castellated Palace. Progressive blindness and eventually insanity forced him to withdraw to the seclusion of Windsor Castle. George III had first manifested disturbing signs of erratic behaviour in 1765. Disconcerted by this unexpected attack, he prudently suggested that Parliament should make provision for a Regency in case "it should please God to put a period to my life, whilst my successor is of tender years". His second attack, this time much more severe, occurred in the autumn of 1788. His family decided to move him from Windsor to the greater privacy of Kew. With no signs of improvement, his doctors consulted the Reverend Dr Francis Willis, the pro-prietor of an asylum in Lincolnshire, well-known for the successful treatment of mental illness. Willis, who had assured the Parliamentary Committee in

December 1788 that he had "great hopes of His Majesty's recovery", subjected his patient to a savage regime of purges, bleeding, blistering, the application of leeches, and frequent restraint in a straitjacket. Domination of his patient appeared to be the doctor's method and the King soon feared his oppressor. Colonel Robert Fulke Greville, George III's equerry, recounted an episode at Kew on 11 February 1789.

> H. My's second walk was in the Exotic Garden where he saw Mr Eaton [Aiton] his botanic gardener – who was talking to Dr Willis. The King overheard his promise to make up a basket of exotic plants for the doctor one of these days; & on hearing this, He added, "Get another basket Eaton at the same time, & pack up the doctor in it, and send him off at the same time."[43]

There were intervals of rational behaviour when George III was aware of his surroundings and capable of lucid conversation.

> We came to the Exotic Garden, and to the several hothouses in it, where He was much amused & talked of various plants with recollection – After this throughout the remaining part of the walk He conversed on various subjects & much like himself – He talked of the purchase of Kew for £20,000 & renewed the former plans of laying Kew & Richmond Gardens together.[44]

A Regency Bill which had been passed by the Commons was being read in the House of Lords when the King suddenly recovered. Church services were held on 23 April 1789 to give thanks for his deliverance "from the severe illness with which he hath been afflicted".

But in 1801 after 12 years' remission, George III succumbed to another major attack. A room was hastily prepared for him in Kew House, then in the process of being stripped of its furnishings in readiness for its demolition. Willis was again his tormentor. Once more the King recovered, but with a recurrence in 1804 he was again confined to Kew, this time in the Dutch House. It would not be surprising if Kew now symbolised imprisonment and suffering for him. He made his last visit to the gardens on 9 January 1806 when Princess Elizabeth had planned a meal in the Queen's Cottage, but rain unfortunately prevented this family gathering in a much-loved retreat. In 1810, now a septuagenarian and almost completely blind, permanent madness descended upon him. The Prince of Wales, empowered by Parliament, became Regent on 5 February 1811.

The doctors Ida Macalpine and Richard Hunter, having studied contemporary medical records and eyewitness accounts of George III's behaviour, concluded in 1968 that he was not mad in the usual sense but a victim of a rare inherited metabolic disorder which their researches revealed had also afflicted other members of the royal houses of Stuart, Hanover and Prussia. Namier shrewdly assessed him as a "much maligned ruler", a man whose eccentricities, we now learn, derived not from a weak and volatile personality, but from an inherited disease, a 'royal malady'.

Queen Charlotte seldom visited Kew after her husband's final collapse. The double wedding at the Dutch House of the Duke of Kent to Victoria, Dowager Princess of Leiningen, and of the Duke of Clarence to Princess Adelaide of Saxe-Meiningen marked the last formal occasion. An improvised altar covered with crimson cloth was placed before the fireplace in the Queen's drawing-room where the ceremony was performed on 11 July 1818. Four months later, on 17 November 1818, the Queen died in the Dutch House, sitting upright in an armchair, the Prince Regent and her family in attendance. Three days after her funeral, the establishment in this old palace was dispersed, leaving only the housekeeper and a few retainers. Christie's auctioned most of her personal effects in a series of sales beginning on 7 May 1819. Her library, excluding books with personal associations with the royal household, realised nearly £4,000. The Lightfoot herbarium was purchased by Robert Brown for 50 guineas.[45]

The Times for 21 December 1818 reported that "the Prince Regent's intention [is] to make Kew Palace one of his principal summer residences. It is certain very extensive alterations are about to be made there." These "alterations" included its demolition and a new dwelling to be built a little to the east of it! After his accession as George IV in January 1820, the Dutch House was reprieved although certain outbuildings disappeared. With the dismantling of the Castellated Palace, the name 'Kew Palace' referred unambiguously to this surviving royal building.

One of the windows salvaged from the Castellated Palace and inserted in Descanso House in Kew Gardens. *Photograph by A. McRobb.*

CHAPTER SIX

Sir Joseph Banks takes charge

*Banks's early life — the Endeavour voyage —
friendship with George III — involvement with
Kew — formation of a botanic garden — foreign
visitors — increase in plants arriving at Kew
— W.T. Aiton succeeds his father as head
gardener*

GEORGE III HAD SECURED the post of Prime Minister for his friend, Lord Bute, who, while wishing to retain the privileges of royal patronage, lacked the stamina for high public office and political conflict. When the King became disenchanted with Bute, who resigned in 1763, and William Pitt was invited to form a government, their now fragile relationship abruptly ended. Some years after his father's death, the Duke of York recalled George III emphatically stating that the last time he had any communication with Bute was when they had accidentally met at Kew in 1766. Three years later Bute went abroad for his health for several years. He now no longer had any influence at Kew, which even Princess Augusta seldom visited. Her death in 1772 precipitated a change of direction for this newly-landscaped and well-stocked garden. Within a decade it was to acquire an international reputation through the intervention of Joseph Banks, the arbiter of Kew's destiny for almost half a century.

When Banks, the son of a rich landowner at Revesby Abbey in Lincolnshire, came into his inheritance with estates in Derbyshire, Staffordshire and in Wales, it yielded him an annual income of £6,000, placing him in the top stratum of the nation's wealthiest people. He had been educated — inadequately by all accounts — at Harrow and Eton where, quite independently, he developed a taste for natural history and, guided by Gerard's *Herball*, began the formation of a herbarium. A portrait of Banks, reputedly painted about this time, presciently shows him seated at a table, his hand lightly resting on an open volume of flower drawings and a globe of the world at his side. When he entered Christ Church, Oxford as a gentleman-commoner, the Sherardian chair of Botany was held by Humphrey Sibthorp who never published a scientific paper and delivered only one public lecture during the 35 years of his professorship. Sibthorp therefore did not object when Banks paid a young botanist at Cambridge to present a short summer season of botanical lectures at Oxford — an early instance of his initiative and enterprise.

Sir Joseph Banks (1743–1820), by Thomas Phillips in 1809, in the presidential chair of the Royal Society.

Banks left Oxford without a degree, plunged into London society, assiduously cultivated the friendship of scientists, and was elected a Fellow of the Royal Society at the early age of 23. He was then a supernumerary on board a naval frigate bound for Newfoundland on fisheries protection duties. During this trip he collected plants PLATE 14, birds, insects and rocks, observed the habits of Esquimos, and triumphantly added an Indian scalp to his miscellaneous collection of specimens and artifacts. He returned to London in 1767 with a reputation for boldness and tenacity in his natural history pursuits.

The successful voyage whetted his appetite to see more of the world, and when the Government and the Royal Society joined forces to equip and despatch a ship to the Pacific to observe the transit of the planet Venus in 1769, Banks's opportunity arose. The ship's captain, James Cook, also had secret instructions to find *Terra Australis Incognita* in the southern ocean. Prompted by Banks, the Royal Society requested that he be allowed to join Cook's ship, HMS *Endeavour*, for "the advancement of useful knowledge". Banks, now a man of substance since his father's death in 1761, paid his own passage and that of his eight companions, who included the Swedish botanist Daniel Solander, Herman Spöring – another Swede – as secretary and draughtsman, Alexander Buchan as topographical artist, and Sydney Parkinson as scientific artist. Though Banks was not the first to enrol a botanist on a voyage of exploration – the French expedition under Louis-Antoine de Bougainville which sailed for the Pacific in 1766 included the botanist Philibert Commerson on board – his preparations were perhaps more thorough, as John Ellis reported to Linnaeus.

> No people ever went to sea better fitted out for the purpose of Natural History, nor more elegantly. They have got a fine library of Natural History; they have all sorts of machines for catching and preserving insects; all kinds of nets, trawls, drags and hooks for coral fishing; they have even a curious contrivance of a telescope by which, put into the water, you can see the bottom to a great depth, where it is clear. They have many cases of bottles with ground stoppers of several sizes to preserve animals in spirits. They have the several sorts of salts to surround the seeds; and wax, both bees-wax and that of the *Myrica* . . . They have two painters and draughtsmen, several volunteers who have a tolerable notion of Natural History, in short Solander assured me this expedition would cost Mr Banks ten thousand pounds.[1]

Banks and his assistants wasted no time getting down to their allotted duties after the *Endeavour* left Plymouth on 25 August 1768: their nets trawled marine life, birds that alighted on the ship's rigging were shot and examined, and plants were collected while the ship anchored off Madeira, Rio de Janeiro and Tierra del Fuego.

Their destination, Tahiti in the Pacific, was reached in April 1769, where a base for astronomical observations was established and solemnly named Point Venus. Considering the seductive charm of the Polynesian women, nothing could have been more appropriate, but of course the name was never intended to commemorate these unexpected delights. Even Banks was not immune. He confided to his journal his appreciation of one "very pretty girl with a fire in her eyes".

His journal reveals not only his appetite for experience but also his eager curiosity about the customs of these islanders: the ritual of a wrestling match, the painful tattooing of the buttocks of a girl of twelve, or a funeral ceremony during which Banks allowed himself to be stripped and blackened with charcoal and water. He quickly mastered enough of their language – the only foreign tongue he ever learned – to serve as Captain Cook's interpreter.

When Alexander Buchan died soon after the ship reached Tahiti, Parkinson took on the additional burden of drawing landscape, people and artifacts. With so much material to record, he could often do no more than make a cursory sketch of a flower, adding detailed colour notes to facilitate the execution of a finished drawing later on. Before his death from malaria and dysentery, Parkinson had produced 269 finished flower paintings and nearly 700 sketches. On his return to England, Banks had copper engravings made of over 700 of these drawings for publication but they were never commercially printed until nearly two centuries later.

With the completion of astronomical observations, the *Endeavour* superficially explored the coasts of New Zealand and eastern Australia before turning towards home, reaching England in July 1771 – a voyage of just over a thousand days but one that had cost the lives of 38 of the crew, including five of Banks's entourage. Banks had good reason to be well satisfied with his collections; he had, for instance, dried specimens of some 3,600 species of plants, of which 1,400 were new to science. He and his collections of seeds and plants, bottles of pickled animals, curios and portfolios of drawings temporarily eclipsed the achievements of Captain Cook's seamanship. He was fêted by society and had an audience with George III at St James's Palace. The three years Banks spent on the *Endeavour* profoundly influenced his future perception of exploration, scientific voyages, transoceanic transfer of plants, and even imperial expansion. The *Endeavour* enterprise with its complement of scientists and artists set a standard for future voyages to emulate.

Within a few months of its return, a second voyage was being planned, this time with two ships: HMS *Resolution* and *Adventure*. Banks automatically assumed he would be joining it and, in anticipation, nominated a party which included Solander, four artists, two secretaries, six servants and two horn players. Cook, not unreasonably, complained about the size of this party and the drastic modifications the ship would require to accommodate it. The public adulation Banks had been enjoying had clearly gone to his head and when Cook objected, he petulantly withdrew from the venture and went to Iceland instead.

Iceland followed by a brief visit to Holland marked the end of his foreign travel. He now turned to the sorting of his collections, the supervising of the engraved versions of Parkinson's drawings, and the management of his estates. At the same time he was gradually being drawn into the affairs of the Royal Gardens at Kew.

Banks never tired of telling his correspondents that botany had always been his "favourite pursuit". When he petitioned the East India Company on behalf of the Botanic Garden in Calcutta in 1806, he reminded its Court of Directors that:

> I am bold in pleading this cause because it is in favor of a science to which I have devoted the best part of my life & for which I have voluntarily exchanged my prospects [of] Parliamentary consequence & possibly of high office . . .[2]

During his youth he had botanised in England and Wales, had studied Linnean methodology, and was seriously considering a visit to the great Swedish naturalist when the *Endeavour* voyage intervened. He tried to keep up to date with new schemes for classifying plants and, in later life, considered Jussieu's natural orders to be superior to Linnaeus's artificial arrangement. The collections he made in Newfoundland and his accompanying notes confirm his botanical competence. He wrote little himself on botany, preferring to leave such matters to the qualified botanists he employed – Daniel Solander, Jonas Dryander and Robert Brown – who also organised his large herbarium and his predominantly botanical library which their employer made freely available to any *bona fide* researcher. He was a collector rather than an investigative botanist, a promoter of other people's researches. His contemporaries nevertheless respected his botanical knowledge. "Great as the loss of Linnaeus must certainly be to science," wrote Johann Reinhold Forster, "it will not be severely felt whilst we have so enlightened botanists as Mr Banks and Dr Solander."[3] When Banks and his assistant formally sought permission to visit Lord Bute's Luton estate they were unreservedly welcomed.

> Can there be a spot dedicated to vegetables, & shut up from the inspection of the first Patron of Botany. The truth is Ld Bute has long wishd for the pleasure of seeing Mr Banks & Doctor Solander at that place, when he might profit by their superior knowledge in his favourite science.[4]

Princess Augusta's garden at Kew, renowned for its exotics, inevitably attracted the attention of Banks. The obituary of William Aiton in the *Gentleman's Magazine* for 1793 noted that the young Banks had made his acquaintance in 1764. Three years later Banks commissioned Sydney Parkinson to draw some of Kew's flowers PLATE 6. Kew provided an appropriate venue for the formal presentation of Banks and Solander to George III on August 10 1771. The *Westminster Journal* for 24–31 August reported that

> Dr Solander and Mr Banks have the honour of frequently waiting on his Majesty at Richmond, who is in a course of examining their whole collection of drawings of plants and views of the country.

On his return from Iceland in 1772 Banks gave some of the ship's ballast of lava to the Chelsea Physic Garden and to Kew where it provided the structure of a moss garden.

It has been impossible to establish when Banks actually embarked on his advisory role at Kew, but the favourable impression he had made on the King prepared the way for his involvement after the death of Princess Augusta. Although it was Princess Augusta's physician, Sir John Pringle, shortly to be elected President of the Royal Society, who had persuaded George III in 1772 to despatch a plant collector to the Cape of Good Hope, most likely the idea had come from Banks who had spent a month there on the return voyage of the *Endeavour*. By 1773 Banks had created an unofficial role for himself at Kew: he modestly described himself to the Spanish ambassador as operating "a kind of superintendence over his Royal Botanic Gardens".[5] William Townsend Aiton, responsible for the routine management of Kew, never questioned his authority or his instructions. "This establishment is placed under the direction of Sir Joseph Banks," he admitted to a correspondent in 1801.

Banks's authority came directly from the King, whom he would often meet at Kew on Saturdays, no doubt discreetly making suggestions, deferentially soliciting his monarch's approval. The diary of the King's equerry, Colonel Greville, frequently notes their encounters.

> The King however continued his walk with Sir Joseph Banks about three hours. They first visited the Exotic Garden, thence walked through Richmond Gardens where the sheep were looked at, thence they proceeded across the London Road [now Kew Road] to Marsh Gate where H. My has lately made a farm yard & erected farm offices.[6]

Banks confirms such assignations in a letter to one of his correspondents.

> I was sent for on Saturday as usual, and attended in the Gardens and Farm for three hours, during all which time he gave his orders as usual, and talked to me on a variety of subjects . . .[7]

The friendship that matured between monarch and subject sprang from a common concern about agriculture and rural affairs, an interest that had earned the king the soubriquet 'Farmer George'. On one occasion when he and Colonel Greville were inspecting the Royal Wiltshire flock at Marsh Gate Farm they considered the feasibility of improving the inferior wool of British native sheep. The equerry recommended cross-breeding with the superior Spanish Merino sheep, cautioning secrecy in obtaining the animals since the Spanish authorities might view the experiments as a threat to their lucrative export trade in Merino wool. When Banks was consulted, he impulsively offered his services. The King approved. "Sir Joseph Banks is just the man. Tell him from me that I thank him & that his assistance will be most welcome." George III had complete confidence in a man who bred sheep on his Lincolnshire estates, was an active member of the Lincolnshire Wool Committee, and was experimenting with the cross-breeding of sheep at his home at Spring Grove. Furthermore he had access to a network of useful contacts abroad. He missed seeing the King to discuss his plans, and George III's concern when he learnt of this abortive visit to Kew shows the respect and regard he had for Banks:

> The King is much hurt he was not apprized on Tuesday last that Sir Joseph Banks was at Kew (indeed he never heard of it till he received his note this day) as he would have found time to have seen him.[8]

Banks used British merchants in Lisbon to organise the smuggling of Merinos into Portugal. The Spanish flock, which ultimately grazed and multiplied at Windsor and Kew, was publicly auctioned on 15 August 1804 in the field below Kew's Pagoda. At this public sale Captain John Macarthur purchased seven rams and three ewes for his estate near Parramatta in New South Wales; those that survived the sea voyage joined the descendants of his Cape Merinos. This was the genesis, it could be claimed, of Australia's wool trade.

The procurement of Merino sheep was a temporary distraction from Banks's resolve to expand Kew's botanical functions. The phrase, 'His Majesty's Botanic Garden' competed with 'Royal Garden' in his correspondence. Francis Masson, no doubt following Banks's lead, published a paper in 1776 in the *Philosophical Transactions of the Royal Society* on collecting in South Africa "undertaken for the discovery of new plants towards the improvement of the Royal Botanical Gardens at Kew".[9]

European botanical gardens began as physic gardens attached to schools of medicine where plants with healing properties were grown for demonstration and study. The Orto Botanico in Pisa founded in 1543 is probably the first of such gardens, a priority challenged, however, by Padua, a near contemporary. Similar gardens followed at other centres of learning: Zürich (1561), Lyons (1564), Rome (1566), Bologna (1567) and Montpellier (1593). Their function changed during the seventeenth century into repositories for samplings of the flora of Europe's colonies. Britain, a latecomer in this fashionable trend, established its first botanical garden in 1621 at Oxford. Edinburgh, with a progressive medical school, followed in 1670, and in 1673 the Society of Apothecaries founded a garden at Chelsea for the use of its members. Richard Bradley, who held the chair of botany at Cambridge from 1724 until his death in 1732, desperately wanted one for his university. He called attention to the utility of such a garden in the preface of his *Survey of the Ancient Husbandry and Gardening* (1725). In addition to an obvious educational function, he saw it emulating the garden at Amsterdam where economic crops were grown for the Dutch colonies.

> When I . . . reflect upon the state of our American plantations, and our extensive trade, I can see no reason but that we may render them more advantageous than they are at present, by sending to them many plants of use, which will grow freely there, and may be collected and prepared for them, in such a Garden as I speak of.

This vision of botanical gardens participating in the commercial life of the nation came to fruition with Banks's transformation of Kew into a centre for the global transfer of plants through its collectors and links with colonial botanical gardens. Unlike the Jardin du Roi in Paris, which had its own research facilities, Kew concentrated on practical horticulture, being privileged to call upon Banks's herbarium and botanists for the identification of new species. In order to establish Kew's superiority to gardens at home and abroad, Banks insisted that "as many of the new plants as possible should make their first appearance at the Royal Gardens".[10] It became a matter of personal pride to outstrip rival botanical gardens in Vienna and Paris. To this end he was prepared to sacrifice his own collecting instincts.

> I have long ago given up the collection of plants in order that I might
> be the better able to promote the King's wishes to make the Royal
> Botanic Gardens at Kew as respectable as possible, so that everything
> that comes to me, from whatever quarter it may be, is instantly sent
> to Kew . . .[11]

In pursuance of this objective, the random acquisition of plants which had
prevailed during Princess Augusta's time needed to be replaced by purposive
collecting in selected regions of the world. Banks also required the support of
a reliable official at Kew. John Haverfield's assistant, William Aiton, responsible for the small physic or exotic garden there since 1759, had gradually
gained the approval and respect of the country's gardening fraternity. John
Fothergill judged him "a very ingenious, sensible, honest man". With the
decline of Chelsea Physic Garden's reputation, John Ellis observed that "everything curious was sent to Mr Aiton, the Princess of Wales' gardener at Kew".[12]
The smooth and swift conversion of Kew into a dynamic botanic garden, its
glasshouses filled with the floral discoveries of its collectors, owed much to
harmonious working relations between Banks and Aiton. One of Banks's directives in 1773 required the replacement of all numbered plant tags by labels
giving their names. The same year witnessed a large tree and shrub planting
programme – nearly 800 species, the majority of North American origin.[13]

Captain Cook returned from his second circumnavigation in July 1775
with only a token gathering of seeds for Kew; indeed the Royal Gardens
received little living material from any of the three voyages. James Bruce
during his exploration of Abyssinia found time to collect seed for Kew. Francis
Masson returned in 1774 with a bounty of plants

> unknown at that time to the Botanical Gardens of Europe. By means of
> these, Kew Gardens has in great measure attained to that acknowledg'd
> superiority which it now holds over every other similar establishment
> in Europe, some of which as Trianon, Paris, Upsala, etc. till lately
> vyed with each other for pre-eminence without admitting even a compe-
> tition from any English garden.[14]

Visitors to Kew admired Masson's discriminating eye: proteas, crassulas,
mesembryanthemums, ericas, pelargoniums, ixias and gladioli formed the bulk
of his consignments. Two years later he sailed to Madeira and passed the next
six years there and in the Canaries, the Azores and the West Indies, tirelessly
collecting on behalf of Kew.

In August 1777, Jonas Dryander, a student from Uppsala in Sweden,
made the first of numerous visits to the Royal Gardens. With the death of his
fellow countryman, Daniel Solander, in 1782 he became Banks's botanist-
librarian with the responsibility of identifying plants in cultivation at Kew.
Another Swedish botanist, Carl Thunberg, had come to Kew in 1778 "to see
the beautiful gardens . . . abounding with living plants and under the care of
Mr Aiton".

When John Haverfield, in his ninetieth year, died in 1784, William Aiton automatically assumed responsibility for the Kitchen Garden and the Pleasure Grounds, at last the sole superintendent of the entire Kew estate.

In the same year that saw this change in management, the French naturalist, B. Faujas de Saint Fond called at Kew,[15] crediting its "present state of perfection" entirely to Aiton. The pervasive orderliness, presentation and good maintenance of the small Botanic Garden had made Kew, in his opinion, superior to all other gardens he had seen. He applauded its consummate taste in the mixing of foreign trees and shrubs. Particularly memorable for him was the Venus flytrap (*Dionaea muscipula*) from Carolina flourishing at Kew (it had died at the Jardin des Plantes in Paris), the heavily perfumed *Magnolia grandiflora*, and the moss carpeting the Icelandic lava donated by Banks. His enthusiasm was shared by another French visitor, Madame Roland, for whom Kew was "the most interesting that I have ever seen; the most skilful art cannot be better disguised; everything breathes nature and freedom; everything is grand, noble and graceful." It seems that the Gallic nation had taken Kew to its heart. "*La collection de Kew est la plus belle qui existe,*" was the verdict of D. de Courset in 1789.[16]

Unquestionably the most distinguished Frenchman to be admitted to the Royal Gardens during the 1780s was Charles-Louis L'Heritier de Brutelle, a magistrate and aristocrat who embraced the revolutionary cause in France and was to lose his life during those turbulent years. Banks, who met this amateur botanist in 1783, characteristically allowed him unimpeded access to his herbarium and library during his prolonged stay in London in 1786 and 1787. L'Heritier was the one responsible for recognising the talent of an unknown artist from Luxembourg, Pierre Joseph Redouté, when he came to the Jardin du Roi in Paris, instructing him in the discipline of botanical illustration and engaging him to illustrate his *Stirpes Novae* (1785–1791). L'Heritier brought Redouté to London in 1787 to paint plants in notable gardens in the capital and home counties, including Kew. Redouté and the artist James Sowerby, also recruited for the project, depicted flowers in bloom there, in a number of the plates made for the ensuing publication, *Sertum Anglicum* (1788).

More than 40 of the specimens described in *Sertum Anglicum* had been collected by Francis Masson, at that time Kew's main source of new plants, but there were many other contributors for whom collecting was a pleasant distraction from official duties or business. Banks asked diplomats, army and navy officers, merchants and missionaries to remember Kew's needs during their travels abroad. Sir John Murray on active service with the army in India despatched several packages of seeds; Daniel Corneille, Governor of St Helena, sampled the island's flora for Kew. Hinton East, the Receiver-General in Jamaica, made several large donations:

> I rejoice exceedingly in hearing your account of the Epidendrons and
> am not a little proud of having been the hand to introduce them into
> the King's Garden.[17]

Voyages and expeditions often had a royal gardener on their complement. In 1785 the Government asked Banks to nominate a naturalist for HMS *Nautilus*, about to seek a site on the coast of West Africa suitable for a convict settlement. Banks instructed Aiton to select a gardener with some botanical knowledge.

> I wish you would look out without delay as I understand the matter is pressing & acquaint Dryander with your progress. The opportunity is not to be lost; we shall get [dried] specimens for me & seeds for Kew Garden . . . We may have future opportunities so that I lay much stress on this . . . For God's sake be active & do not let such an opportunity slip you.[18]

Kew picked Anton Hove, a Polish-born gardener, who had come to Kew for experience.

When *Nautilus* failed to find a suitable location for a penal colony, the Government reconsidered Banks's recommendation to a committee of the House of Commons in 1779 that Botany Bay in eastern Australia might be suitable provided European livestock and crops were introduced. Eleven ships, known as the First Fleet, under the command of Captain Arthur Phillip, left Spithead in May 1787. They carried administrators, marines and convicts destined to found Britain's first settlement in the Antipodes. Governor Phillip, whose jurisdiction covered the whole of eastern Australia and neighbouring Pacific islands, had no botanist or gardener to advise him, and consequently believed Botany Bay to be devoid of any useful vegetation. When HMS *Guardian* sailed in mid-1789, its quarterdeck carried a greenhouse filled with agricultural crops for the new colony. Hopes that its return to England would bring a harvest of Australian flowers for the King's garden at Kew were dashed when the ship was wrecked on its outward voyage. Two young Kew gardeners, James Smith and George Austin, were lost along with many of the crew. However, as the ships of the First Fleet arrived back in England during 1789 some seeds were delivered to Banks and the Hammersmith nurseryman, James Lee.

> We have this summer had vast additions to our exotic gardens principaly from Botany Bay, the plants of which will stand in our greenhouses.[19]

It would be another decade, nevertheless, before British gardeners could gauge the diversity of the Australian vegetation through the collections of Caley, Good, Brown, Cunningham and others.

In the meanwhile other countries were yielding coveted novelties. A specimen of *Fuchsia coccinea* (*F. macrostemma?*) from South America was presented to Kew in 1788. According to Sir James Everard Home, "Sir Joseph Banks carried the original fuchsia into the Garden on his head, not choosing to trust it to any other person."[20] In 1789 Banks at last succeeded in obtaining a specimen of the Chinese tree paeony, brought back by a surgeon from Canton.

The first *Hydrangea hortensia* (*H. macrophylla?*), also from the Far East, was on display at Kew early in 1789.

Such splendid flowers were bedded in a small lean-to house, a mere 12 feet in length and 6 feet wide. Kew's expanding plant population now urgently needed more glasshouse accommodation. A new lean-to house, 110 feet long, 17 feet wide and 14 feet high, heated by flues, was erected in 1788 for Masson's South African collections. Here potted plants stood on stair stages and when six feet tall were transferred to the Orangery. For many years Cape bulbs were grown in glazed frames in front of hothouses, heated by their flues, and protected by shutters during severe winters.

Optimism and confidence permeated Banks's letter to Professor Thunberg at Uppsala. "Of late his Majesty & the Queen have paid more attention to the Botanic Garden than they formerly did."[21] In 1794 the King purchased the Revd Thomas Methold's house and garden on Kew Green in which some new glasshouses would be built. It was also during the 1790s, according to the Revd T.T. Haverfield, that George III, desiring more "arable land", had the greater part of the lake filled in.[22] 1790 was an auspicious year for Kew. Francis Bauer became its first resident botanical artist. It was also the year in which plans were completed for a naval survey of the west coast of North America under the command of Captain George Vancouver. Archibald Menzies, nominated as Vancouver's surgeon and naturalist, belonged to that band of men whom Banks infected with his own enthusiasm for exploration and discovery.

> From the first moment I had the honour of being your correspondent,
> I found within me a particular desire of traversing unknown regions
> in quest of my favourite pursuit.[23]

Banks had great hopes of this voyage.

Many unsolicited gifts of plants from officials of the East India Company reached Kew including in 1791, a consignment grown in the Company's garden on St Helena. The Company's Court of Directors, ever grateful for advice freely given by Banks, became Kew's most generous benefactor. Individuals like Sir John Murray sent Indian seeds; officials like William Roxburgh, a surgeon and botanist in the Madras Presidency, were reminded to be alert to the needs of His Majesty's Garden. After Roxburgh transferred to Calcutta in 1793 as Superintendent of the newly-formed botanical garden there, the links he established with Kew endured through a succession of Superintendents/ Directors until India's independence in 1947.

Banks had good reason to be well satisfied with Kew's prosperity and renown.

> Kew Gardens proceeds with increased vigor. The additions of plants
> lately receivd are indeed very interesting. We have 3 Magnolias from
> China, only one of which was before known among us & that only from
> Kaempfer's *Icones*. Epidendrons blossom away daily; E. vanilla is as high

as the glass & will soon produce flowers. Ferns are propagated from seed [sic] of the West Indies so that the Garden must soon overflow.[24]

The flow of foreign material was increasing. In 1792 Lord Macartney headed a diplomatic and commercial mission to China, financed by the East India Company which had a post at Canton. His secretary, Sir George Staunton, anticipated some botanical forays with the help of two gardeners and guidance from a report which Banks readily provided – 'Hints on the subject of gardening suggested to the gentlemen who attend the Embassy to China'. Needless to say, Banks appended a list of plants he desired for Kew. When the mission reached Java in March 1793, Sir George sent Banks a sturdy nutmeg tree and some mangosteens; six pots of plants from Canton followed in December. Bulbs were received from Adam Afzelius, botanist to the Sierra

The entry in the 'Kew record book, 1793–1809' listing plants brought to Kew by HMS *Providence* in 1793. *Kew Library.*

Leone Company; from South Africa more seeds from the indefatigable Francis Masson. Returning from Australia, Governor Phillip handed over 82 tubs and boxes of plants. Parcels were beginning to arrive from Archibald Menzies on Captain Vancouver's ship.

The most memorable delivery of 1793 – nearly 800 pots representing what was at that time the largest single despatch ever made to Kew – filled HMS *Providence* when it docked at Deptford in August. After the unsuccessful attempt of HMS *Bounty* to transport breadfruit trees from the Pacific to the West Indies, a second voyage was planned in 1791 with two ships, HMS *Providence* and *Assistant*, again under the command of Captain Bligh, and again masterminded by Banks. Two gardeners tended the plants during the sea journey to the West Indies. The acquisition of plants for Kew was a secondary objective of the expedition, the bulk of the South Pacific collections being destined for West Indian planters. Kew was to receive whatever could be spared, and Bligh recorded that this amounted to 1,283 plants from Tahiti, Tasmania, New Guinea, Timor, St Vincent and Jamaica. Many apparently died subsequently since only 147 species are listed in the second edition of the *Hortus Kewensis*.

William Aiton never had the pleasure of inspecting HMS *Providence*'s cargo. He had died on 2 February 1793 after 33 years' service, for much of the time 'Gardener to His Majesty', and admired for his horticultural competence. In a letter to Banks, barely two years earlier, Hinton East had remarked that "Aiton has infinite merit in his management." The Marchioness of Rockingham praised "the plainness of his manners, without a grain of that pomposity one might have expected". Carl Thunberg named *Aitonia capensis* in his honour. Banks and Jonas Dryander were among the pallbearers at his funeral at Kew, where he was buried.

William Townsend Aiton (1766–1849), probably by L. Poyot.

William Townsend Aiton, who had worked under his father since 1782, was an obvious successor, and the King's Apothecary, David Dundas, interceded with Banks on his behalf. Banks supported his candidature, confident that the son needed only more experience to be a dependable superintendent. W.T. Aiton had been educated at a private school in Chiswick and at Mr Smith's Academy in Camberwell; while apprenticed to his father, private commissions gave him the opportunity to design a number of villa gardens.

In 1795, only two years after being promoted to his father's post, Aiton took over Richmond Gardens when John Haverfield junior resigned. At last the two contiguous royal estates were under one management, a step nearer to their eventual physical amalgamation. The renewal of his contract in 1796[25], itemising his duties, provides some useful acreage figures and an insight into routine maintenance. The Kitchen Garden, for instance, which supplied produce for the royal table covered 12 acres. The Great Lawn of 41 acres south of Kew House was cut with scythes and swept every fortnight – and more frequently in summer – and fed by grazing sheep. Watering was done by a 'water engine' pulled by two horses.

Aiton had a fluctuating work force of eight to twelve gardeners with an experienced man as foreman. A particularly good foreman, James Donn, became Curator at the Cambridge Botanic Garden with Banks's recommendation. He left in 1794 with his advocate's best wishes.

> I am sure you may also rely on Mr Aiton's goodwill to you and that
> he has it in his power to send what plants he pleases to Cambridge.[26]

When William Forsyth, royal gardener at Kensington and St James's Palaces, died in 1804, Banks failed to persuade the King to appoint Donn. The post went instead to W.T. Aiton, who thereupon surrendered his share of the Windsor garden contract to his brother John, thus fulfilling George III's desire "that both brothers will now be able to live comfortably".[27] The post of foreman at Kew was usually a stepping-stone to preferment: the foreman, William Kerr left for China in 1804 and his successor, William McNab, took charge of the Royal Botanic Garden in Edinburgh in 1810.

George Noe, a German gardener who had come to Kew to complete his horticultural training and had followed James Donn as foreman, escorted a large consignment of plants to Russia in 1795. One reason for his selection was his ability to converse in German with the recipient of this gift, the Grand Duchess Maria Feodorovna, daughter-in-law of the Empress Catherine. The hold of a small cargo boat was adapted to accommodate more than 300 potted plants. *Strelitzia reginae* was among the 226 species chosen for their beauty or rarity; indeed the majority came from Masson's Cape collections, ericas predominating. In July George Noe sailed for St Petersburg with precise instructions from Banks on the care of the plants in his charge. Since he was expected to instruct Russian gardeners on Kew's gardening practices, he took scaled drawings of the Cape glasshouse and another house for tropical plants. He returned in December with a gift of seeds.

The Royal Gardens were now in sight of achieving Banks's goal of superiority, both in the number and quality of its plants.

> . . . our King at Kew & the Emperor of China at Jehol solace them-
> selves under the shade of many of the same trees & admire the elegance
> of many of the same flowers in their respective gardens,

a grateful Banks wrote to Sir George Staunton in January 1796. Staunton's Chinese plants were largely dried specimens but he did add distinction to British gardens with his *Rosa bracteata*, known as 'Macartney's Rose', and the plume poppy, *Macleaya cordata*.

When HMS *Royal Admiral* returned from Calcutta in May 1796 it brought from the Botanic Garden there the largest presentation of plants since that of HMS *Providence* in 1793. Two years earlier the same ship had taken Christopher Smith, one of the two gardeners who had sailed with Captain Bligh, to Calcutta where he had been appointed botanical gardener or nurseryman. With him went an assortment of European fruit trees and crops for trial cultivation in India. Peter Good, a Kew gardener, looked after the tubs of plants on both voyages.

In 1797, Francis Masson, restless after nine years in South Africa, got his wish for further service abroad by being sent to North America.

"Who knows but that England may revive in New South Wales when it has sunk in Europe."[28] So wrote Banks encouragingly – and perceptively – to its despondent Governor, John Hunter, in March 1797. The small community of convicts and their guards at Port Jackson (now Sydney) had barely survived appalling deprivation.

> Whenever prosperity returns, I shall solicit the King to establish a botanist with you. The plants we have received, which are now tolerably numerous, make a most elegant addition to the gardens. I trust, good sir, that when you make your excursions, or when you send parties into new districts, you will not forget that Kew Garden is the first in Europe, and that its Royal Master and Mistress never fail to receive personal satisfaction from every plant introduced there from foreign parts when it comes to perfection.

With Treasury's authority in June 1798 to despatch crops and fruit trees to Port Jackson, Banks found a young kitchen gardener, George Suttor, eager to settle in Australia, and set him to work propagating at Kew. Eighteen boxes of plants grown for their medicinal or culinary uses were in due course loaded into a portable greenhouse or plant cabin on HMS *Porpoise*. Despite Suttor's constant attention, few had survived when Australia was reached in November 1800 after a seven-month voyage.

Banks discovered another young man willing to go to Australia. After a crash course in elementary botany and plant identification at Kew, George Caley was employed by Banks personally and began his collecting career in New South Wales in April 1800.

On 18 July 1801, the year in which Kew came, like other royal gardens, under the authority of the Lord Steward,[29] HMS *Investigator* left Spithead on a scientific survey of New Holland. During the late 1790s, concerned that nothing of commercial value had yet been discovered in Australia, Banks had conferred with the Admiralty about mounting an expedition. His proposal for intensive exploration received a sympathetic response, and the support both of Governor Hunter and Matthew Flinders, a seaman who had navigated the coast south of Port Jackson and sailed around Van Dieman's Land (Tasmania). Flinders was appointed captain of the *Investigator* which was well-equipped for its task, and had on board a complement of scientists and artists: Robert Brown as naturalist; Peter Good from Kew as his assistant and gardener; an astronomer; an experienced miner as mineralogist; and natural history and landscape painters. Banks planned every detail: he drafted the duties of the scientific personnel, selected their equipment, provided a library of works of reference which included every available map of Australia and the published journals of previous voyages, and shipped a prefabricated plant cabin for assembly and use at Port Jackson. The collection of plants for Kew fell to Peter Good, who conscientiously forwarded seeds on any ship returning to England. The *Hortus Kewensis* credits 116 Australian species to him.

The flora of the Cape had been searched and now it was the turn of Australia. South America was still sparsely represented in Kew's hothouses. With Francis Masson in mind, Banks had vainly sought permission of the Spanish ambassador in 1796 to station a man at Buenos Aires "to collect seeds & living plants for the use of the Royal Gardens at Kew & the Queen's elegant retreat at Frogmore".[30] Spain was still having problems with rebellious colonists in 1802 when Banks raised the matter with the Foreign Secretary. He hoped that describing his collector as "a young gardener who has had no education beyond his line of life"[31] would allay any Spanish suspicions about his motives. He even volunteered to have all plants and seeds routed via Spain so they could be inspected. As Masson was now in North America, William Kerr would have taken his place but diplomatic overtures failed and Banks decided to send him to China instead.

He secured the cooperation of the East India Company who offered not only a free passage for Kew's collector on one of their East Indiamen, but also generously waived his accommodation expenses while he was stationed at their base in Canton. Banks assured Kerr that if he were "diligent, attentive & frugal", this appointment promised a good chance of "a better station in life than your former prospects permitted you to expect",[32] and in 1803 he set off.

In the same year Kew had an unexpected haul – 100 tubs of Guiana plants taken from the French ship *L'Union* by British privateers. Most died – unfortunately common in those days of experimental glasshouse management – but at least eleven plants survived to find a place in the *Hortus Kewensis*.

As seeds gathered on the *Investigator* expedition continued to arrive at Kew, Banks thanked Robert Brown in Australia.

> They are all sown in Kew Gardens & much hopes built on the success of them which we expect will create a new epoch in the prosperity of that magnificent establishment by the introduction of so large a number of new plants as will certainly be obtained from them.[33]

Among the first lot of Chinese plants from Kerr, inspected by Banks and Aiton in August 1804, was the vigorous shrub *Kerria japonica*, which has so well adapted itself to the English climate. It appropriately commemorates Kerr's horticultural successes, which would give British gardens such favourites as *Rosa banksiae* and *Lilium lancifolium*. The latter, a trumpet lily with its mass of pink to red-orange blossoms, grew so well with Aiton that he propagated and distributed more than 10,000 bulbs between 1804 and 1812.

Few ships left Calcutta without a chest of plants or a bag of seeds from William Roxburgh. In Europe, the royal gardens at Schönbrunn were a regular source of supply, and purchases were often made from London's leading nurserymen – Hugh Ronalds, Richard Williams, George Loddiges, Joseph Knight, Lee and Kennedy and John Fraser. John Lyon's spectacular sale of North American plants in 1806 filled many gaps at Kew. Some of Aiton's purchases were intended for gardens abroad. Kerr received hardy fruit trees and shrubs such as apple, pear, peach, plum, cherry and apricot; others went to replenish colonists' gardens in Australia. Kew reciprocated the generosity of overseas donors, especially botanical gardens in the West Indies, Mauritius and India.

A new glasshouse in 1803 provided additional accommodation for the steady influx of plants from Australia and the Cape, but as the pomologist, Thomas Andrew Knight, tactfully pointed out to Banks, "the houses at Kew are not in the best possible form".[34] More practically, in 1801 Kew grew fruit and vegetables to the value of £1,891 for Windsor Castle and St James's Palace. In 1806 Aiton requested funds to rebuild a dilapidated pinery which supplied fruit for the King. Aiton's skills in forcing figs supplied the royal tables with more than 200 baskets in 1810. Perhaps with this market gardening activity in mind, he presumably felt justified in asking the Board of Green Cloth in November 1807 for an increase of £100 in Kew's annual vote. The annual expenditure on Kew in the financial year 1801/2 – £1,240 – had only risen to £1,402 by 1811/12.[35] During 1800–20 more was being spent on individual gardens at Windsor, Kensington Palace and Hampton Court than on Kew. It should be remembered, however, that Kew's botanical activities were confined to an insignificant portion of the estate. In the Pleasure Grounds sheep and cattle grazed and pheasants were raised; in 1813, 26 acres stretching south from Stafford Walk to the Pagoda were sown with cereal crops.[36]

During the summer of 1810 Banks visited Windsor Castle for his last consultation with George III, now blind, his health deteriorating and his delusions soon to return. In February 1811 the Prince of Wales took the oaths of office as Regent; by the end of the year the King had withdrawn into a world of his own. The absence of the King's authority does not appear to have adversely affected Kew, now controlled by the Lord Steward's department. Banks's reputation and his influential friends largely compensated for the lack of direct royal support and intervention. The first part of a new edition of the *Hortus Kewensis* appeared in 1810 and further volumes were printed until its completion in 1813.

The post of Superintendent of a new botanical garden in Ceylon in 1810 rewarded William Kerr's diligence as a collector in the Far East. Sir Joseph Banks anticipated receiving from him an

> abundance of mountain plants, such as will thrive in our green houses if not in the open air. In truth our hothouses overflow at present to[o] much with intertropical plants that I scarce wish for additions of that kind.[37]

The war had made Kerr the last collector to go abroad. With Napoleon's abdication in 1814 and the prospect of a lasting peace, Aiton, anxious to resume collecting, informed Banks that he had gardeners well-qualified and willing to serve in any part of the world. Banks replied promptly and positively:

> . . . the connection I have been permitted to form with the Royal Gardens at Kew is among those most grateful to my feelings, and I beg you to be assured that as long as I shall be permitted to continue it I shall cherish and improve it to the best of my power.[38]

He was determined to re-establish Kew's supremacy, judging the imperial gardens in Vienna to be its only serious competitor. Provided the Prince Regent approved, he preferred to concentrate on the Cape and Australia since their flora did not require hothouses for their cultivation. Moreover he felt certain he could rely upon the cooperation of the governors of these territories. The Treasury, accepting the case presented to them, agreed to finance two collectors who were to sail on the first available warship to Rio de Janeiro to await transport to their respective destinations.

The Kew gardeners, Allan Cunningham and James Bowie, spent nearly two years in Brazil, collecting in the neighbourhood of Rio and as far distant as São Paulo, adding bromeliads and orchids to the Royal Gardens' display of South American plants. In 1816 they separated, Bowie bound for the Cape and Cunningham for Australia.

Lord Macartney's frustrated attempt in 1792 to establish cordial relations with China did not deter a second mission led by Lord Amherst in 1816. Dr Clarke Abel, who combined the duties of the party's medical officer and naturalist, was briefed by Sir Joseph Banks on the plants that his assistant, the Kew gardener James Hooper, should collect, but again British diplomacy failed to appease Chinese xenophobia. The ship was wrecked off the coast of Sumatra on its return voyage and Hooper's collections were lost.

In March 1816 the Kew gardener, David Lockhart, joined Captain J.K. Tuckey's exploration of the Congo river as assistant to the botanist, Christen Smith. Fever killed Smith, along with most of the crew, but Lockhart survived to deliver the late botanist's dried specimens to Banks.

William Kerr served barely two years as Superintendent at the Botanic Garden in Ceylon before he died; Alexander Moon, "a smart young man trained at Kew", sailed in July 1816 to fill the vacant post. Soon seeds were being shipped to Kew from the latest addition to Britain's expanding empire.

The infirmities of old age and severe attacks of gout which confined Banks to a wheelchair never affected his dedication to Kew's interests. When he learnt in 1817 that the French were fitting a ship to explore the north and north-west coasts of Australia, he instructed Allan Cunningham to join a small British cutter about to undertake a hydrographic survey of the same coastline.

> . . . this will give you an opportunity of collecting plants, which could by no other means be obtained; & of enriching the Royal Gardens at Kew with plants which otherwise would have been added to the Royal Gardens at Paris, & have tended to render their collection superior to ours.[39]

In 1818 he warmly congratulated Cunningham and Bowie on the results of their collecting. Their Brazilian plants were flourishing and bulbs from the Cape were in bloom.

> When the seeds you have now sent home shall flower, I think Kew will be in as high a state of beauty & scientific excellence as it ever was

when Masson sent home the beautiful & curious novelties as they then were from the Cape.[40]

Francis Masson had been the first and perhaps the best of all Kew's collectors, and it was to see a plant he had introduced in 1775 that brought Banks to the Royal Gardens in 1819 for the last time. The occasion was the very first cone produced by the African cycad, *Encephalartos altensteinii*.

Many of the exotics that came to Kew had been presented to Banks as a grateful acknowledgement of assistance he had rendered or, more often, as an expression of friendship. It was through his recommendation to the East India Company that Nathaniel Wallich succeeded William Roxburgh as Superintendent of the Calcutta Botanic Garden in 1817. Wallich showed his appreciation by showering his backer with choice specimens from his Botanic Garden and new species from Nepal. "I confess I was not aware[of] the increased energies which have been this year exercised in favor of my favourite establishment, the Royal Gardens at Kew," Banks wrote in thanking Wallich. "Scarce a ship has arrivd that was not chargd with some valuable present from you . . . Kew Gardens already feels sensibly the effect of your kind attentions."[41]

CHAPTER SEVEN

Publicising Kew

Daniel Solander — Jonas Dryander — Robert Brown — Hortus Kewensis — Francis Bauer

THE ROYAL GARDEN'S DISPLAY of trees and ornamentals was unrivalled in Britain — only the private gardens of John Fothergill at Upton in West Ham and of William Pitcairn in Islington offered any serious competition. Now a printed catalogue of all its species seemed desirable as a means of publicising Kew's uniqueness and, with the addition of brief cultivation notes, of providing useful data for gardeners. Its genesis can be traced back to 1773 — although at that time publication was not considered — when William Aiton, presumably at Banks's behest, began compiling a descriptive plant list.[1] Adopting the same title Hill had used for his enumeration of the stock in Princess Augusta's garden — *Hortus Kewensis* — Aiton published his three-volume catalogue in 1789. His casual acknowledgement of "the assistance of men more learned than himself" hinted the extent of his indebtedness to Daniel Solander and Jonas Dryander.

Daniel Solander, a former pupil of the Swedish naturalist Linnaeus, arrived in England in May 1760 with the intention of staying no more than a year. But the rapport he established with the capital's naturalists and his popularity with English society changed his mind. A post of assistant in the British Museum in 1763, followed by a fellowship of the Royal Society a year later, were further inducements to stay. It was about this time that he met the young Banks who, impressed by his botanical expertise, sought his advice before his trip to Newfoundland. A firm friendship grew out of this relationship, with Solander becoming Banks's companion and naturalist on the *Endeavour* voyage. Shortly after their return the *Gazetteer and New Daily Advertiser* for 4 September 1771 reported that "Dr Solander has presented the Princess Dowager of Wales with several curious exotic plants for Her Royal Highness's gardens at Kew," specimens, perhaps, from the recent voyage. He resumed his duties at the British Museum, becoming Keeper in 1773, evidently not a particularly onerous post since he also looked after Banks's museum and library in Soho Square. Aiton, with a meagre knowledge of botany, always consulted Solander whenever he had plants for identification. The letters of the younger Linnaeus, written during his visit to England in the summer of 1781, disclose that he worked with Banks, Solander and Dryander examining Kew's cultivated plants for a new catalogue. Solander's unexpected death in May 1782 might have led to its abandonment had not Dryander been on the spot and available. B. Faujas de Saint Fond reported in 1784 that the book was still in progress.

Jonas Dryander, Solander's fellow countryman, who had also studied at Uppsala, was introduced to Banks soon after his arrival in London in July 1777. Banks engaged him as a botanical assistant in 1778 and, on Solander's death, made him his curator and librarian. Since Solander identified plants by comparing them with dried specimens, Aiton had to take his queries to the Soho Square herbarium. Dryander, on the other hand, preferred the living plant and dealt with Aiton's new acquisitions during his frequent visits to Kew. When Dryander took over Solander's manuscript notes he expected to finish the catalogue within about four years. A sense of urgency entered the proceedings when it appeared that the French botanist, L'Heritier, was stealing a march on Kew with the serial publication of his *Stirpes Novae*. Banks, annoyed by L'Heritier's apparent abuse of his hospitality, complained to James Edward Smith.

> We have thought of getting forward Aiton's catalogue of the Garden this winter for publication next summer. That will a little molest L'Heritier if he means to use any of our plants without acknowledging them to have come from England.[2]

W. Aiton: *Hortus Kewensis*, 1789. The Linnean family and genus names are given at the head of each page with specific names in the margins. Every entry has a concise Latin description, the publication in which the plant is cited, an English name, country of origin, date of introduction and introducer (when known), and flowering period.

Banks may have harboured doubts about L'Heritier's integrity but Dryander enjoyed cordial relations with him, submitting proof sheets of the *Hortus Kewensis* in 1788 for comment while L'Heritier responded by freely giving Dryander details and some proofs of his *Sertum Anglicum*. But such exchanges of unpublished matter were not uncommon at that time. Dryander toiled on – "plodding" according to the impatient Treasurer of the Linnean Society – and at last the catalogue appeared between August and October in 1789, its three volumes illustrated by 13 engravings of drawings by Francis Bauer, G.D. Ehret, J.F. Miller, F.P. Nodder and J. Sowerby PLATE 9. George III had ordered 100 copies in advance of publication and within two years the edition had sold out.

The title-page proclaimed William Aiton as author but Banks's two assistants had provided the taxonomic expertise. "It was Solander who reduced our garden plants to order," wrote Sir James Edward Smith, "and laid the foundation of the *Hortus Kewensis* of his friend Aiton."[3] Dryander contributed most of the third volume and edited the entire text. Aiton added the dates of introduction of plants to British gardens, and distinguished between trees, shrubs and herbaceous plants, whether they were annual or perennial, hardy or tender, and indicated their flowering periods. Although it purported to be a record of plants that had been cultivated at Kew, Jonathan Stokes stated that it included a few specimens grown in the Chelsea Physic Garden, Fothergill's collection at Upton and other London gardens.[4]

The work was well reviewed by contemporary periodicals, applauded by botanists and used by the Schönbrunn gardens in Vienna as a desideratum list for exchanges between the two institutions. The Lewisham nurseryman, John Wilmott, extracted all the names for an *Alphabetical Enumeration of the Plants contained in the Hortus Kewensis with additions from Dr Coyte's Botanic Garden and Hortus Cantabrigiensis* (1798).

With no diminution in the number of new species added to the Royal Gardens, Banks soon contemplated a new and enlarged edition. The first intimation of this intention appeared in the preface to *Delineations of Exotick Plants at Kew* (1796–1803). At Banks's suggestion, William Townsend Aiton had proposed this revision to Dryander, offering an equal share in any profits from sales.[5] Dryander's agreement secured, the labour of compilation and searching botanical literature began. From 1808, Richard Cunningham, one of Aiton's gardeners at Kensington Palace, took charge of the clerical chores – arranging plant diagnoses, establishing dates of introduction, and assisting in proof reading.

Dryander's death in October 1810 only temporarily delayed progress. Robert Brown was not only Dryander's replacement at Soho Square but also his successor as editor of the *Hortus Kewensis*. Brown had been naturalist on HMS *Investigator* and since his return to England in 1805 had been sorting and identifying the collections made during the voyage. Dryander had got nearly half way through the third volume when Brown took over without much enthusiasm for a work whose format had already been decided. With his last contribution finishing on page 228 of the final volume, Brown had dealt with Asclepiadeae, Cruciferae, Orchidaceae, Proteaceae and genera of other families.

The second edition of Hortus Kewensis appeared in five volumes during 1810 to 1813 in a print-run of 1,250 copies. W.T. Aiton, who acknowledged much more generously than had his father the assistance of his collaborators, had greater claim to authorship. There is no doubt about his participation as surviving records at Kew confirm: his corrections to the first edition, for instance, his painstaking checking of dates of introduction, and his overall supervision of the book's production. The following year gardeners were offered a cheaper and more portable version, an *Epitome*, with an appendix adding another 300 species and a selection of fruits and vegetables cultivated at Kew.[6] Banks wanted the *Hortus Kewensis* to be an accurate inventory of Kew's stock of plants. When James Bowie arrived in South Africa in 1817, one of his tasks was to find replacements for Masson plants that had died since their introduction at Kew. Banks reminded him that "their names stand in the *Hortus Kewensis*. It is therefore most important to the credit of the Garden to replace the plants & make the Garden correspond with the catalogue."[7]

A comparison of both editions reveals the strengths and weaknesses of the collections and the fluctuations in regional representation resulting from the activities of Kew's collectors. Hill's record of 3,400 species in 1768 had increased to 5,600 by 1789. One would expect a botanical garden to give priority to the flora of its own country and Kew was no exception: just over 2,100 British species are recorded, followed by 1,400 from continental Europe. The flora of temperate North America was well established in Britain by the mid-eighteenth century and nearly 700 species could be seen in the Royal Garden. The availability of glasshouses inevitably governed the population of tender plants. The South African flora exceeded 700 species, a third being Francis Masson introductions. Masson's forays in Madeira and the Canary Islands harvested nearly 100 specimens with token offerings from Tenerife and the Azores. A mere 50 or so plants gave an inadequate conception of the resources of India's jungles and plains. Inaccessible China supplied about 50 flowers, Japan, even more reclusive, fewer than 10. The few seeds that had survived the long sea voyages from Australia and New Zealand often failed to germinate; the explosion in the antipodean flora began with the receipt of Peter Good's collections on HMS *Investigator* from 1801 and with George Caley's discoveries.

Dominant newcomers in the 1810–13 edition of the *Hortus Kewensis* are Australia (*c.* 300 species), South America (*c.* 260 sp.), Siberia (*c.* 220 sp.) and China (*c.* 120 sp.). The British and European floras together amount to a third of the total of 11,000 species in this second edition. Francis Masson's score reaches nearly 1,000 from the Cape (183 species of *Erica*, 175 of *Mesembryanthemum*, 102 of *Pelargonium*, 57 of *Oxalis* and 42 of *Stapelia* were his biggest hauls). William Kerr's Chinese contributions were a mere 15 but others were probably casualties on the sea voyage or lost in cultivation at Kew. Of over 1,000 plants unloaded from HMS *Providence* in 1793, more than 140 new species survived in the Royal Gardens. Some of the capital's nurserymen were credited with a number of introductions: James Gordon, Lee and Kennedy, Loddiges, William Rollisson, and Richard Williams from whom Kew almost certainly obtained some of their stock.

The usefulness of these two editions of the *Hortus Kewensis* today, apart from their historical insight into Kew's horticultural growth, lies in their dating introductions and naming the introducers of garden plants.[8]

A work illustrating some of the choice flowers listed in the *Hortus Kewensis* now became Banks's next priority. Kew's principal rival, the Royal Gardens at Vienna, had selected 300 of its ornamental and economic plants to feature in *Hortus Botanicus Vindobonensis* (1770–6). Hill, in a very modest way, had attempted such a publication with *Twenty-five New Plants rais'd in the Royal Garden at Kew* (1773). Rather than a finite book, however, Banks opted for a periodical, undoubtedly influenced by the *Botanical Magazine*, a monthly illustrated journal, launched by William Curtis in February 1787. The appearance in January 1789 of the first part of L'Heritier's *Sertum Anglicum (an English Wreath, or Rare Plants which are cultivated in the Gardens around London especially in the Royal Gardens at Kew observed from the Year 1786 to the Year 1787)* prodded him into action. Although most certainly peeved that Kew's treasures should first appear in a foreign publication, Banks nevertheless congratulated L'Heritier, requesting a list of the plants he intended to figure in future issues so that Kew could avoid duplication in its projected periodical.

Kew needed ready access to the brush and pencil of a botanical artist to illustrate this periodical. But, quite apart from this consideration, Banks had always believed that a botanical garden of any standing ought to have on its staff a flower painter to record new plants as they bloomed or fruited.

The opportunity to recruit a competent flower painter presented itself when a young Austrian artist, Franz Andreas Bauer, accompanying Joseph F. von Jacquin on a European tour, reached England in 1789. Franz (Francis in its anglicised version) and his brother Ferdinand had been trained as artists at the Schönbrunn Imperial Gardens under its director, Nikolaus J. von Jacquin. "I believe it is high time I brought him [i.e. Francis Bauer] away from here, otherwise he will be filched from me. Banks has his eye very much on him," the young Jacquin warned his father on 3 November 1789. Sir Everard Home, who was to become an intimate friend of Bauer, recalled years later how on the day before Bauer was to return home, he dined with Banks who tempted him to stay with an attractive salary which he himself promised to pay.[9] And so in 1790 Francis Bauer, a victim of Banks's persuasive charm, set up house in Kew where for the next half-century he subjected his talent to the discipline of meticulous botanical portraiture. In the meantime Jacquin, resigned to his loss, had found another promising if unusual pupil – a former cook and monk!

Banks was not alone in believing Kew deserved a serial record of its flowers. Margaret Meen, a native of Bungay in Suffolk, who now taught flower and insect painting in London, published in 1790 the first number of *Exotic Plants from the Royal Gardens at Kew* with every intention of issuing two numbers a year. It sold at 16 shillings if the plates were coloured and 12 shillings plain, but foundered after only two issues, probably the consequence of high

Francis Bauer (1758–1840). Artist unknown. *Kew Library.*

Ixora coccinea, a South
Asian plant, drawn by
Margaret Meen in her
*Exotic Plants from the Royal
Gardens at Kew*, 1790.

production costs. Although a competent artist who exhibited at the Royal Academy and the Watercolour Society, Miss Meen fell far short of the high stan-
dards consistently maintained by Bauer.

Kew's periodical was eagerly awaited. In 1791 Banks assured the Dutch
scientist, Martijn van Marum, that it would appear *"en nombres d'une manière
splendide"*.[10] Two years later Adam Afzelius learnt that the first number was
imminent; in 1795 some of the engraved plates were presented to the Grand
Duchess Maria Feodorovna. Despite Banks's fears, expressed in a letter to
Nikolaus J. von Jacquin in May 1796, that the worsening of the war might

Erica grandiflora, introduced by Francis Masson from South Africa in 1785. Drawing by Francis Bauer in his *Delineations of Exotick Plants cultivated in the Royal Garden at Kew*, 1796–1803.

delay its publication, the first number was published by W.T. Aiton in the same year. *Delineations of Exotick Plants cultivated in the Royal Garden at Kew* depicted ten of Francis Masson's Cape heaths. Its anonymous preface (by Banks, according to Sir Everard Home[11]) promised annual issues provided it was favourably received. The scientific precision of Bauer's work in Banks's opinion rendered any text superfluous, each drawing being "intended to answer of itself any question a botanist can wish to ask", but descriptions were promised in the next edition of the *Hortus Kewensis*. Another ten ericas were illustrated

in the second number in 1797 and ten more in 1803.[12] With that third number Kew's periodical on which Banks had set such great hopes ceased publication. Bauer wrote that its failure could be attributed to its conception being too ambitious, to its being sold at a loss, and to the death of its principal engraver, Daniel MacKenzie.[13] The thirty ericas engraved in this short-lived periodical superbly demonstrate Bauer's fastidious attention to detail, their minute and critical characters drawn with the deftness and sensitivity of a miniaturist.

The title-page of this work describes Francis Bauer as 'Botanick Painter to His Majesty', an inaccurate appellation since his salary came from Sir Joseph Banks. It is true he taught flower painting to Princess Elizabeth, receiving for his pains a mere half-stick of Indian ink. As previously mentioned, Queen Charlotte and Princess Augusta coloured some of his erica plates which were sold in the sale of the Queen's possessions after her death. Bauer may also have instructed Thomas Christopher Hofland, a soldier in the King's Own Company at Kew and a self-taught artist. Hofland's work had impressed George III, who directed him to draw plants at Kew and then proposed him for the post of topographical painter on HMS *Investigator*, an offer Hofland regretfully declined when his mother objected.[14] He is now remembered as the engraver of his wife's drawings of Lord Blandford's house and gardens at Whiteknights

Fragment of plate-printed linen and cotton, manufactured in 1766 by John Collins of Woolmers, Hertfordshire. The Pagoda and some other chinoiserie buildings were copied from Chambers's 1763 book on Kew. *Victoria and Albert Museum.*

near Reading and of some topographical prints.[15] Bauer's influence can be discerned in the work of William Hooker (1779–1832), his only truly talented pupil, who contributed over 100 plates to R.A. Salisbury's *Paradisus Londinensis* (1805–8), and recorded fruit varieties for the Horticultural Society of London.

Banks boasted that Francis Bauer and his more adventurous brother, Ferdinand, were "the most skilful painters of natural history in this Kingdom, & in my poor opinion are not equald [sic] in any part of Europe" PLATES 5 and 8.[16] Posterity would not quarrel with that judgement. His technical skills were flawless, and his botanical knowledge enabled him to recognise significant morphological characteristics, which he drew with infinite patience. He examined and dissected plants with the aid of microscopes, of which he had assembled 16 by the time of his death. Vegetable structure, pollen grains, human and animal organs and hair were submitted for his analysis. When T.A. Knight sent Banks some leaves of his pear trees afflicted with a fungal disease, Bauer "subjected them to his double microscope". The hand-coloured lithographs in his *Strelitzia depicta* (1818) miraculously preserve the freshness and fragility of his pure wash drawings. He illustrated John Lindley's *Illustrations of Orchidaceous Plants* (1830–8) and towards the end of his life, with failing eyesight, drew ferns for Sir William Hooker's *Genera filicum* (1838–42). Banks's will endowed Bauer with an annuity of £300 for as long as he continued to work at the Royal Gardens. The artist remained at his post until his death in 1840 but his drawings of Kew's plants are now among the treasures of the Natural History Museum in London.[17]

Oval dish from a table and dessert service consisting of 952 pieces, made by Josiah Wedgwood for the Empress Catherine II of Russia in 1773–4. Each piece was decorated with a different British town or country seat taken from contemporary engravings. Kew was represented on 22 pieces.

CHAPTER EIGHT

Collectors and collecting

Recruitment of plant collectors – qualities
required – their duties – transportation
problems – botanic gardens in the colonies –
Banks as an imperialist – his services to Kew

SIR JOSEPH BANKS, the initiator of Kew's collecting programme, naturally took a personal interest in the recruitment of its collectors. Preference was given to royal gardeners whose performance and botanical knowledge satisfied the Aitons. When Banks wanted a collector for HMS *Nautilus*'s survey of the west coast of Africa he had in mind a gardener

> active & healthy, able to write a good hand & willing to write down
> such observations as he may make. Whenever he lands he should have
> a little idea of botany & be well acquainted with the manner of gathering
> & drying specimens and be able to give some idea of the soil, whether
> sandy, loamy, clayey, boggy, etc., etc.[1]

Aiton recommended a young Pole, Anton Hove. There were no other candidates but Aiton's choice met with the approval of the Secretary of the Royal Society who found him well-educated, a fluent linguist, experienced as a medical practitioner, and "bold, active & animated with the most laudable ambition of being distinguished".[2] This paragon outshone Kew's normal intake of gardeners who wanted only to widen their horticultural experience before seeking employment with the landed gentry. Any showing an aptitude for botany, as well as a lively curiosity about the diverse range of plants grown at Kew, were viewed as potential collectors. When George Caley, a young man from Manchester with enthusiasm and little else, wanted to be a collector in New South Wales, Banks advised him to serve an apprenticeship at the Chelsea Physic Garden and at Kew where he would have an opportunity to familiarise himself with the Australian flora.

> How can you be useful to your employers as a botanical traveller to
> send home species of plants from thence [Australia], till you have made
> yourself acquainted with those already in England, I do not know. We
> have now several hundreds of such, and to send them again would be
> idle and useless.[3]

Banks valued personal integrity and loyalty to Kew and to himself almost as much as professional skills. Allan Cunningham and James Bowie learned that they owed their

> good fortune to be selected from among the very great number of excellent young men who have been educated at the Royal Gardens of Kew under the eye of their worthy director Mr Aiton, not so much from any superiority you possess of many others in botany or horticulture as from a firm persuasion in Mr Aiton that you do both excel in the virtues of honesty, sobriety, diligence, activity, humility and civility, and that you will not for a moment lose sight of these essential qualifications which above all others ensure to a traveller respectability among strangers and assistance from those in high life who have the power either of giving or of withholding it.[4]

The "mildness of your temper" and "ready obedience" were what secured James Hooper's appointment. Banks, appreciating Hove's "good sense & daring character" displayed during his time on *Nautilus*, confidently chose him for a covert collecting operation in India. Banks regarded a Scottish education with particular favour, because it inculcated "the habits of industry, attention & frugality". The collectors Archibald Menzies, William Kerr and Peter Good all came from north of the border and Allan Cunningham was of Scottish extraction.

Banks demanded total dedication and for that reason preferred bachelors unencumbered by family responsibilities. He did not conceal his irritation when George Caley who eventually got to Australia wanted to marry a colonist. "I did not take him to beget a family in New South Wales. I fear if he is not more active than is compatible with a married life I must get rid of him."[5] Caley revealed himself to be irascible, tactless, impetuous and jealous, but Banks perversely chose to ignore these flaws, possibly impressed by his enthusiasm, ability and persistence in adversity.

He usually looked for humility in his collectors, an awareness of their proper place in society, a willingness to be "directed by their instructions, [and] not to take upon themselves the character of gentlemen, but to establish themselves in point of board and lodgings as servants ought to do".[6] It was predictable that outspoken George Caley would antagonise practically everyone in the colony. When the Governor complained of his behaviour, Banks sympathised:

> Had he been born a gentleman, he would have been shot long ago in a duel. As it is, I have borne with much more than even you have done, under the conviction that he acted under strong, though mistaken, feelings of a mind honest and upright . . .[7]

It was the absence of discernible status that sometimes brought collectors into misunderstanding and conflict with colonial officials and settlers. Even diffident

Allan Cunningham was accused of arrogance by Governor Macquarie, who denounced him as an "unbred [sic], illiterate man whose only pretensions to personal attention from me arose from the opinion you have entertained of his usefulness in the line of his profession".[8]

The pay these men received did nothing to enhance their social standing. David Nelson on HMS *Bounty* got only £50 a year with a £25 kit allowance and free messing while on board ship. Anton Hove in India was paid £60 a year, another £50 for kit and £50 for his passage. Banks, unaware of Hove's losses in several robberies, judged his submission of expenses as "most unjustifiably enormous" and threatened his instant return if he did not keep within prescribed financial limits. Francis Masson was more handsomely treated with an annual expense allowance of up to £200 and another £100 a year payable on his return. William Kerr was cautioned that he could not be "too frugal or too scrupulous" with his £100 annual allowance as everything but clothes would be provided by officials of the East India Company in Canton:

> . . . the less you indulge yourself in the supply of unnecessary wants during the time of your pilgrimage in foreign parts, the more reserve you will make for future indulgence, when the enjoyment of your friends & a residence in your own country will double every pleasure that money can purchase for you.[9]

John Livingstone, the Company's surgeon at Canton, believed Kerr's modest salary lost him the respect of the Chinese with whom he had to deal. "I have not the slightest doubt but his failure is to be attributed, chiefly, to the necessity he was under of associating with inferior persons, from his deficiency of means to support himself more respectably."[10] When Banks obtained the Treasury's consent to resume plant collecting in 1814, he also succeeded in getting a collector's salary raised from £100 to £180. Cunningham and Bowie who received this revised rate of pay were counselled by Banks to observe "frugality" so that they would return home "before the afternoon of life has closed with a fair prospect of enjoying the evening in ease, comfort and respectability".[11]

No collector set out on his travels without a thorough briefing from Banks. William Kerr's first mission was treated as an "apprenticeship" under the tutelage of the superintendent of the East India Company's post at Canton. Archibald Menzies's medical training and previous collecting experience encouraged Banks to demand a great deal of him on Captain Vancouver's voyage: when surveying new lands for instance, he had to assess their suitability for settlers. The fauna also came within Menzies's remit and he was expected to make observations on the social life and crafts of natives. Keeping a journal was mandatory for all, and while Banks welcomed letters, fuller accounts had to be submitted to Aiton. Notes on habitats were to accompany the despatch of plants. No doubt Kew's propagators derived some benefit from James Bowie's terse comments on some bulbs he sent them in 1817: "generally in gritty or stiff soil", "black soil, moist situations", "sandy soil", "should be planted very deep". Dried specimens of the living plants collected were appreciated since their size and shape might suggest to Kew's gardeners appropriate methods of cultivation.

Although collectors were sometimes instructed to seek specific flowers, the more frequent search for "useful, curious or beautiful" plants gave them considerable latitude. However, Banks judged it prudent to remind Anton Hove not

> to neglect those [plants] which are small or unsightly as it is just as likely that qualities useful to physic or manufacture and singularity of structure interesting to the botanist should be found in the minute & ugly as in the conspicuous & beautiful productions of nature.[12]

With Kew's hothouses full to overflowing, hardy or semi-hardy plants were preferred to those which needed special treatment.

The unauthorised disposal of plants to others, especially nurserymen, was a cardinal sin. The contracts of Cunningham and Bowie included this awesome warning:

Francis Masson (1741–1805). Artist unknown of this only portrait of Kew's first plant collector, once belonging to his friend, the nurseryman James Lee of Hammersmith. *Linnean Society.*

> should any new plant sent by you to Kew appear in any other garden, an enquiry will be immediately set to find out in what manner it was procured & if it proved to have been obtained from you in any circuitous manner whatever, your having parted with it will be deemed a breach of the fidelity you unquestionably owe to your employers.[13]

Banks jealously guarded Kew's reputation of having one of the finest collections, many plants flowering there for the first time in Britain.

Despite distance and delays in communication Kew tried, not always successfully, to control the movements of its representatives abroad. Banks reprimanded Masson for undertaking two unsanctioned journeys into the interior of the Cape. "What I recommend is a fixed residence during the ripening season at any place where plants are abundant."[14] Kerr also incurred Banks's displeasure for choosing to visit the Philippines when marauding pirates prevented his collecting near Macao on the Chinese coast. Banks feared that the Philippine flora would prove more difficult than that of China to grow at Kew.

Such prohibitions were disregarded by Caley, who made several attempts to find a way through the seemingly impenetrable Blue Mountains west of Sydney. Cunningham's pioneering explorations, undertaken after Banks's death, opened up a route to the lush pastures of the Darling Downs. He christened prominent landmarks that he discovered with the names of friends: Mount Aiton and Mount Brown, Dryander's Head and Good's Peak. He himself is similarly commemorated on Australian maps by Cunningham's Gap, Mount Cunningham and Mount Allan.

Many collectors were, of necessity, explorers. They often travelled without accurate maps, through country lacking roads and with few habitations. Their allowances denied them the luxury of a well-equipped expedition. Cunningham had a horse, occasionally a cart, and convict assistants on his shorter journeys, but even for his longest and most ambitious expedition he could only afford eleven pack horses and absolutely essential navigational instruments. The loyal service rendered by some of his convicts eventually earned them their freedom. Masson moved through South Africa on horseback with his gear stowed in an ox-waggon; years later Bowie travelled in similar fashion but protected by a Boer escort. In Brazil, Bowie and his companion, Cunningham, had to be content with mules and a couple of negro labourers, and were forced to arm themselves when their request for a soldier was refused by the Portuguese authorities, even though they offered to pay him. Several collectors were the victims of violent robbery. Attacked several times by dissident tribesmen in India, Anton Hove lost all his personal possessions and specimens, even suffering the indignity of being stripped. Escaped convicts assaulted and robbed Cunningham on Philip Island near the penal colony of Norfolk Island. After French forces had invaded Grenada in 1778, Masson was forcibly recruited into a hastily-formed militia and subsequently taken prisoner. Disaster struck again when he lost all his equipment and collections during a hurricane on St Lucia two years later. Tropical disease or physical exhaustion took the lives of Nelson on Timor, Good and Cunningham in Australia, Kerr in Ceylon and possibly Masson in Canada.

Crayon drawing of Allan Cunningham (1791– 1839) by Sir Daniel Macnee. One of the last of Sir Joseph Banks's collectors, he was in Australia when his patron died. *Kew Library*.

Bringing the flora of distant lands to Europe depended not only on competent collectors but on reliable transport facilities as well. A voyage from the Far East on a slow East Indiaman could take as long as six months, and during the Napoleonic wars when ships travelled in convoy, protracted voyages were unavoidable. Few plants survived the extreme variations in climate, the effect of salt sea spray, and the absence or inadequacy of care and attention.

Difficulties arose soon after the collector had harvested his plants. They were usually kept in a healthy state by replanting them in a nursery garden set aside for that purpose. Masson had such a garden a few miles outside Cape Town. When a ship became available they would be potted and placed in special containers. Masson discovered that the few British vessels calling at the Cape on their way home were reluctant to accept bulky cargoes of living plants. "I have many new succulent plants. Some very curious stapelias, euphorbias, etc., but how I shall get them home, God only knows," he complained on one occasion. Kerr was sensibly advised by William Lance in Canton to pack his plants in small portable cases to ensure that sailors would not refuse them because of their weight. Hove was humiliated by being denied a passage when the captain feared that a former Indian governor on board might object to the ship's deck being cluttered with tubs of plants.

So alarming was the casualty rate among these migrant plants that Banks became personally involved in methods for ensuring their well-being. Guidelines for packing them before loading them on ships recommended the abundant use of damp moss in casks cut down to a manageable size with drain holes in the bottom. Aiton favoured moss, sharp sand or open baskets for bulbs but he had no particular preference: "try different methods until you are fully satisfied of the better means of sending home your precious treasures," he encouragingly

The type of ox-waggon that plant collectors would have used in the Cape during the late 18th and early 19th centuries. W.J. Burchell: *Travels in Interior of Southern Africa*, vol.1, 1822, p. 148.

counselled Cunningham.[15] Many ingenious though futile methods were tried in order to prolong the viability of seeds: boxes, bottles, pots, and tins, sometimes hermetically sealed, were used as containers; seeds were bedded in sand, brown sugar and other unpromising substances, even coated with wax and resin. Banks suggested simply hanging them in muslin bags suspended in baskets. Sir John Murray in India usually divided a batch of seeds between several ships, hoping that at least one lot would miraculously survive.

Any Kew gardener escorting a cargo of plants at sea had precise instructions to observe especially regarding watering, light and ventilation. Rain water was preferable to stale cask water, and fresh water was to be used to wash off salt deposited on leaves by sea spray. Plants needed some sort of protective barrier against dogs, cats and the miscellaneous livestock usually found on board ship. Drastic measures were employed to drive out rodents.

> A boat with green boughs should be laid alongside with a gangway of green boughs laid from the hold to her, and a drum kept going below in the vessel for one or more nights; and as poison will constantly be used to destroy them and cockroaches, the crew must not complain if some of them who may die in the ceiling make an unpleasant smell.[16]

The daily routine of tending plants always took precedence over the gardener's personal comfort. David Nelson, the collector on HMS *Bounty*, was solemnly reminded that:

> One day or even one hour's negligence may at any period be the means of destroying all the trees and plants which may have been collected, and from such a cause the whole of the undertaking will prove not only useless to the public, but also to yourself. I therefore cannot too strongly recommend it to you to guard yourself against all temptations of idleness or liquor.[17]

HMS *Bounty*, formerly the *Bethia*, had been converted and equipped to

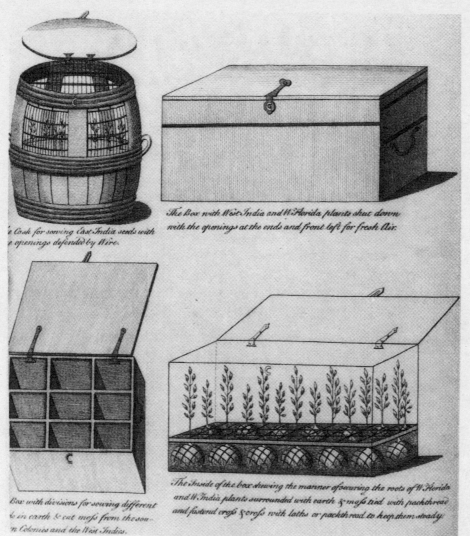

The Box with West-India and W.Florida plants shut down with the openings at the ends and front left for fresh Air.

e Cask for sowing East-India seeds with e openings defended by Wire.

Box with divisions for sowing different
e in earth & cut mofs from the sou-
n Colonies and the West-Indies.

The Inside of the box shewing the manner of securing the roots of W.Florida and W.India plants surrounded with earth & mofs tied with packthread and fastened crofs & crofs with laths or packthread to keep them steady.

Examples of containers in use in the late 18th century to transport plants. The failure rate was, not surprisingly, high. J. Ellis: *Directions for bringing over Seeds and Plants from the East Indies and other distant Countries,* 1770.

collect breadfruit trees on the Pacific islands for West Indian plantations. The great cabin was enlarged and a false floor with many holes for pots was inserted; two skylights supplemented the light from the stern windows; gratings ensured a supply of fresh air and a stove provided heat during cold weather. Below this false floor, the deck, lined with lead, drained surplus precious water into a reservoir. According to Captain Bligh, all the breadfruit were thriving at the time of the infamous mutiny.

Captain Vancouver on HMS *Discovery* never shared Bligh's pride in the health of the plants in his custody. The tension between him and his surgeon/naturalist, Archibald Menzies, may have been generated by the antagonism already existing between the captain and Banks. A rift occurred when Menzies refused to hand over his journal, which Vancouver probably assumed contained unfavourable comments about his command. Menzies complained to Banks about the plant cabin on the quarter deck being invaded by "goats-dogs-cats-pigeons-poultry, etc., etc." and of water impregnated with tar and turpentine

Kew's specimen of *Araucaria araucana*, drawn in the early 1830s when it was 12 feet tall. Some seedlings of this Chilean tree had been presented by Archibald Menzies to Sir Joseph Banks in 1795. One was planted just north of the ice-house at Kew where it was known as 'Sir Joseph Banks's pine'. A.B. Lambert: *Description of Genus Pinus*, 1832, plate 56.

dripping from the rigging on to his plants.[18] When he protested to Vancouver about the loss of plants through the transfer to nautical duties of the servant who had looked after them, Menzies was placed under arrest. "I can now only show the dead stumps of many that were alive and in a flourishing state when we crossed the Equator for the last time."[19]

Despite exhortation from Banks, ships' crews, understandably reluctant to be part-time gardeners, left plants unprotected during storms and neglected to ensure they received adequate watering. So Kew willingly rewarded any seaman prepared to be responsible for their care. A convict gardener in Australia, delegated to that duty, was promised his freedom should he succeed in

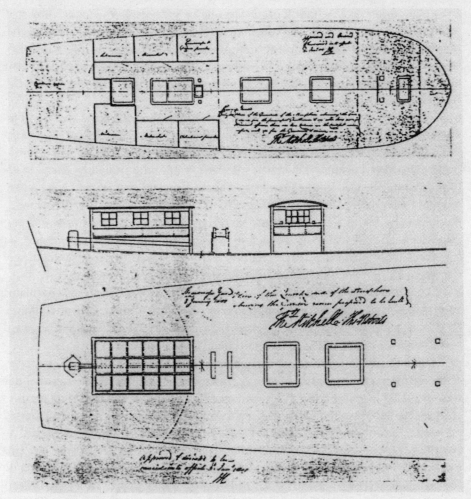

A plant cabin, favoured by Sir Joseph Banks, on the quarter deck of HMS *Investigator*. R.A. Austin: *Matthew Flinders on the Victorian Coast*, 1794, p. 17.

bringing the plants in his charge safely to England. HMS *Porpoise*, on its maiden voyage to Australia in 1800, had the exceptional good fortune to have three gardeners, including George Caley and George Suttor, to take care of the plants stowed on board. Banks reminded the captain of HMS *Guardian* when it returned from Australia with plants for the Royal Gardens that "if any other plants except those intended for the King be taken on board, no water shall be issued to them until the King's plants shall have had their full allowance."[20]

Whether or not they were looked after by an attentive member of the crew, Banks firmly believed that plants stood a better chance of survival bedded in boxes of soil placed in a small glasshouse, variously known as a 'plant cabin', 'garden hutch' or 'botanic conservatory'. He designed some of these miniature glasshouses himself, even chalking out their position on the quarter deck. Since their weight, when completely filled, could exceed three tons, seamen were concerned that they could affect the stability of the ship. The *Guardian* in 1789/90 had the distinction of being the first to carry one, crammed with fruit trees for settlers in Australia. When the ship was holed by an iceberg, however, the plant cabin was the first item of cargo to be jettisoned. Vancouver objected strongly to his ship being encumbered with a plant cabin. HMS *Investigator* stowed away a prefabricated plant cabin for assembly and use during Flinders's survey of the Australian coast. They became familiar features on East Indiamen and some naval ships on the Far East route during the first decade of the nineteenth century.

William Kerr at Canton used these glasshouses with variable success. The almost total loss of his consignment on the *Winchelsea* in 1806 finally convinced Banks that a gardener should be in attendance on board ship. Kerr was accordingly instructed to engage a Chinese gardener who could profitably learn something of Kew's horticultural practices while in England before returning home with the next cargo of plants. When Au Hey arrived at the Royal Gardens, Aiton lodged him in a room near the Aviary while a plant cabin was being fitted on an East Indiaman bound for Canton. Au Hey, who had insisted on a present for his services, received a silver watch, and departed without any regret on the part of the Kew staff. The experiment was not repeated.

St Helena in the South Atlantic offered a convalescent home for any plants suffering during the leisurely voyages from Australia, China and India. There they were planted and after they recovered (many did not) they were repotted and loaded on the next available ship bound for England.

The hazards of these voyages, however, did not cease when ships finally docked. After many months at sea, crews were in no hurry to unload their cargoes and customs formalities prolonged the delays. Unnecessary casualties occurred in a collection of plants from Cayenne in 1803 when they lay neglected in warehouses in the West India Docks for six weeks. Banks often had to remind captains and dock officials that plants destined for the Royal Gardens should receive prompt attention and despatch. They came up the Thames to Kew on barges or by road if ships had docked at other ports. Kew's gardeners rushed the sickly ones into an 'infirmary' or a 'hospital hothouse', of which there were at least two in operation in 1800.

Kew's collectors played an essential role in Banks's strategy for discovering and developing the natural resources of Britain's expanding empire. Menzies was urged to seek coal deposits and minerals with commercial applications and to extend his investigations to "beasts, birds and fishes likely to prove useful as food or in commerce".[21] Plants as dyeing agents and trees suitable for cabinet-making or shipbuilding were high on Cunningham's list of priorities. Hove's mission to collect Indian plants for His Majesty's garden at Kew was a deliberate pretext to enable him to obtain secretly seeds of the finer varieties of cotton for experimental cultivation in the West Indies. The East India Company was not informed of this act of commercial espionage since both the British Government and Banks, quite rightly, assumed that its Court of Directors would object to any competition from West Indian planters. Banks contended that officially sponsored voyages should combine their objective of geographical discovery with the search for products for exploitation by British commerce. Such was his influence that government officials frequently consulted him on overseas affairs and gave him a free hand in the planning of

HMS *Investigator*'s survey of the Australian coast, a voyage that co-ordinated the interests of exploration, science and colonial trade.

Banks made no secret of his belief that exploration preceded colonisation. He was, as his contemporary Robert Hay, Permanent Under-Secretary at the Colonial Office,. admiringly declared, "the staunchest imperialist of the day". From the mid-1780s he attended meetings of the Privy Council Committee for Trade and Plantations as adviser on colonial and scientific matters, and in order to retain his valuable services, he was appointed a Privy Councillor in 1797. He was a founder member of the Association for Promoting the Discovery of the Interior Parts of Africa in 1788, an organisation which frequently met at his London residence. In 1791 he recommended Adam Afzelius as a competent botanist to the Sierra Leone Company, which administered this new colony of former African slaves. At the Company's request, Kew selected and supplied plants considered suitable for Sierra Leone's climate. A Kew gardener, David Lockhart, joined Captain Tuckey's exploration of the Congo in 1816.

With the failure of the *Nautilus* mission in 1785–6 to find a site for a penal settlement in West Africa, Banks's preference for Australia resurfaced. Following the independence of the former North American colonies, Canada and the West Indies refused to be dumping grounds for Britain's criminals, so the First Fleet, assembled in some haste, sailed for Botany Bay in May 1787 with the first convicts to be transported. When Australia yielded disappointingly little of commercial application during the next decade, Banks decided to give George Caley a chance, hopeful that

> he will find an ample field for his research, & where it is probable
> there are many objects both in the vegetable & mineral kingdoms
> hitherto undiscovered, that will, when brought forward, become objects
> of national importance, & lay the foundation of a trade beneficial to
> the mother country with that hitherto unproductive colony.[22]

All the great European powers – Portugal, Holland, Britain and France – competed for supremacy in India and ultimately for control of its immense wealth. Johann Koenig, a Danish surgeon in the employ of the British East India Company, was Banks's exemplar of a botanist always mindful of the commercial uses of vegetation. His replacement, Patrick Russell, enlisted Banks's support in the publication of an illustrated work depicting Indian plants of "established utility". William Roxburgh, who succeeded Russell, persevered with the project which ultimately appeared as *Plants of the Coast of Coromandel* (1795–1820) PLATE 17. As a commercial body, responsible to its shareholders, the East India Company encouraged its officials in India to identify plants with economic potential and to report on their uses by Indians. Banks recommended the intensive cultivation of a whole range of crops in India – cocoa, coffee, cotton, sugar and indigo – but although his advice was frequently solicited it was seldom acted upon. When the Court of Directors invited his comments on a viable tea industry in India, he responded enthusiastically, assessing it to be of "the greatest national importance". He saw nothing morally wrong in making the most of the raw materials of overseas possessions since the colonies themselves also benefited from this creation of wealth. He believed that universal prosperity would follow from the bold and imaginative utilisation of the world's natural resources, and that the movement of crops from one part of the globe to another and between colonies would enrich the world's larder of food and reduce famine.

The introduction of the breadfruit tree to the West Indies is perhaps the best-known instance of plant transfer, largely because of the *Bounty* mutiny. The ease with which Polynesians harvested the breadfruit had impressed Banks on the *Endeavour* voyage. Shortly after his return to England, Valentine Morris, a West Indian planter, questioned him about the feasibility of bringing the tree to the West Indies as an additional source of cheap food for plantation workers. This proposal, supported by the Standing Committee of West Indian Planters and Merchants, was backed by financial inducements from the Society of Arts, Manufacturers and Commerce. When there was no response to these initiatives, Banks was asked to help directly. Assisted no doubt by the example of the French who were introducing commercial crops to their West Indian possessions, he persuaded the British Government to despatch HMS *Bounty* to the Pacific in 1787. Its captain, William Bligh, in his memoirs of the voyage wrote:

> The object of all former voyages to the South Seas undertaken by command of his present Majesty, has been the advancement of science, and the increase of knowledge. This voyage may be reckoned the first, the intention of which has been to derive benefits from these distant discoveries.[23]

Here Bligh expressed the credo of his patron, Sir Joseph Banks, that exploration and commerce were complementary, not incompatible.

Much of the wealth of Europe's colonies came from their trade in spices, coffee and tea, tobacco and drugs. Entrepôts to assist this migration of plants were established; the Dutch had a garden at Cape Town as early as 1694, and about 1768 the French established a nursery on Mauritius to acclimatise plants in transit from the East. British participation began in 1765 when the Governor of the Windward Islands agreed to the proposal of his medical officer on St Vincent to establish a garden for "the cultivation and improvement of many plants now growing wild and the importation of others from similar climates".[24] To this garden came some of the breadfruit trees transported by Bligh in his second attempt, on HMS *Providence*. Banks rated these gardens essential components in the global transfer of plants, as he explained to Alexander Anderson, the Superintendent of the Botanic Garden at St Vincent.

> [They] might in the future be of great utility to the public both by the improvement of many valuable plants, etc, growing wild & neglected in the British colonies, & by the introduction of many articles of value in a commercial or medicinal view, only produced in foreign

settlements, & not to be procured by the British, but at very high prices, especially in time of war.[25]

Banks actively supported existing botanic gardens in the colonies and campaigned for new ones. The War Office consulted him about the management of the garden on St Vincent, which was then a garrison island. He monitored the progress of the botanic garden at Calcutta from its formation in 1786, and advised the Secretary of State for War and the Colonies in 1810 on establishing one in Ceylon "with a view to the increase of the resources of that Colony & an improvement of the science of botany in Europe". He always encouraged these colonial establishments to contribute to the communal prosperity of the Empire rather than their own parochial interests. Thus when Sir George Yonge became Governor at Cape Colony at the end of the eighteenth century, he set about revitalising its moribund garden with the intention of making it useful to Kew, now recognised as the senior partner in this network of gardens. Kew was the focus for the colonial outposts which also enjoyed direct links with each other. There was, for example, a mutually beneficial exchange between the Calcutta Botanic Garden and the West Indies.

It was either Banks's nominees or Kew's staff who managed these overseas gardens, usually a reward for previous services. James Wiles and Christopher Smith, the competent gardeners on *Providence*, received attractive postings at the end of the voyage: Wiles to the superintendency of the Liguanea Botanic Garden in Jamaica, and Smith as nurseryman at the Calcutta Botanic Garden. George Caley took charge of the Botanic Garden on St Vincent, and David Lockhart, a survivor from Captain Tuckey's fiasco, became the first superintendent of the Trinidad garden. When William Kerr died only two years after setting up the Botanic Garden in Ceylon, he was succeeded by another competent Kew gardener, Alexander Moon. They all usually repaid Banks's patronage by donations of plants to Kew. Such was Kew's reputation that the Dutch had no hesitation in appointing James Hooper, the gardener on Lord Amherst's mission to China in 1816, as their head gardener at the Buitenzorg Botanical Garden in Java.

Without doubt Banks was an imperialist, yet he was reluctant to let national interests take precedence over what he perceived were the impartial aspirations of the international scientific community. In this spirit he secured the restitution of several natural history collections seized from the French by British warships on the high seas. A notable example of his intervention concerned the confiscation of the collections of the French naturalist, La Billardière, when his ship, *La Recherche*, was intercepted by a British frigate. His plants, reptiles, fish and bird skins were offered to Queen Charlotte, who limited her choice to plants for her herbarium. When, however, the French botanist Antoine-Laurent de Jussieu appealed for their return, Banks persuaded the British Government to waive its rights, and the Queen to relinquish the plants "for the honour of the British nation and for the advancement of science".[26] His influence failed to achieve a similar concession for the French ship *L'Union*, captured in 1803 with tubs of plants and seeds collected by Joseph Martin during six years' sojourn in French Guiana, and destined for Madame Bonaparte's garden at Malmaison and the Muséum National d'Histoire Naturelle in Paris. Two British privateers which had captured the French vessel demanded prize money. Three barge-loads of surviving plants finally reached Kew where the old Cape House had been made ready for them. R.A. Salisbury praised Aiton's efforts in saving so many "after the neglect and rough treatment they had experienced from our British tars".[27]

As President of the Royal Society from 1778 until his death 42 years later, Banks exercised power and patronage and numerous scientific ventures benefited from his entrepreneurial skills. He advised the Board of Longitude, the Royal Mint, the Society of Arts and the Society of Antiquaries. The Royal Institution, dedicated to the diffusion of scientific and technical knowledge, was founded at his home in 1799.

Banks was an ideal committee man: impatient of bureaucracy, quick to make decisions, and a good organiser. His influential friends came from all walks of life: explorers, diplomats, administrators, military personnel, missionaries, businessmen and gardeners. He had a talent for communicating enthusiasm and enlisting support. He was assuredly a man of action rather than a scholar. Hardly the polymath that some have claimed him to be, he had a superficial knowledge of many subjects and, aware of his limitations, never hesitated to seek expert opinion when in doubt.

In Banks imperial leanings and an abiding interest in botany coalesced in a resolve to transform the Royal Gardens at Kew "into a great botanical exchange house for the empire".[28] He never ceased telling his correspondents that Kew was his "favourite establishment", and envied the expenditure that royal patronage had lavished on the Schönbrunn Garden in Vienna while Kew was managed with "well considered economy". Yet by encouraging donations of plants and by personal subsidy, discreetly given, he improved Kew's resources. W.T. Aiton gratefully acknowledged this support:

> But His Majesty's collection is most especially indebted to Sir Joseph Banks who has procured loads of seeds and plants from all parts of the world at his own expense.[29]

He demanded loyalty to Kew and a competitive spirit in all those who collected for it. Dr Clarke Abel was tactfully reminded that "it is desirable that as many of the new plants as possible should make their first appearance at the Royal Gardens".[30] This genuine concern for Kew comes out in his will. It stipulated that his assistant, Robert Brown, should continue advising the establishment on the identification of plants, and that Francis Bauer would go on receiving an annuity for as long as he remained the Gardens' botanical artist. Bank made Kew the premier botanical garden of his day, and although its fortunes declined after his death, the memory of its former prestige rallied the support of the scientific and horticultural community when the Treasury tried to dismember it during the 1830s.

CHAPTER NINE

Years of neglect

*W.T. Aiton — part of Kew Green enclosed —
John Smith — Kew's parsimony — William IV's
interest in Kew — Sir Jeffry Wyatville —
gradual decline of Kew — campaign for a
national botanic garden — official inspection of
royal gardens — Kew transferred to Woods and
Forests*

THE DEATH OF SIR JOSEPH BANKS diminished the dynamism, the sense
of purpose and direction that his presence had bestowed on Kew. His curator,
Robert Brown, now custodian of his former employer's collections which had
been bequeathed to the British Museum, continued to identify plants for Aiton.
A paternalistic concern for the Royal Gardens motivated Banks's physician and
friend, Sir Everard Home, a frequent visitor to Francis Bauer's gatherings
of scientists at his home on Kew Green. Sadly absent, however, was royal
patronage. No son could have been more unlike his father than George IV:
impulsive, unconventional, gregarious and extravagant, whereas George III
was circumspect, conservative, reclusive and frugal. The new King's genuine
interest in the arts manifested itself in a passion for building (something he
did share with his father) and in furnishing his new homes with furniture
and pictures selected with a connoisseur's flair. Having spent prodigiously on
improving Carlton House in The Mall where his grandparents, Prince Frederick and Princess Augusta, had lived, he ordered its demolition when its
small rooms cramped his elaborate receptions. His favourite architect, John
Nash, transformed Buckingham Palace, and in Windsor Great Park he
designed a cottage orné dwelling, thatched and mullioned, with a verandah
smothered in ivy and honeysuckle. At Brighton, Nash pandered to the King's
oriental tastes by designing the Brighton Pavilion. At Windsor, Queen's
Lodge, the modest retreat of George III and his family, disappeared in Jeffry
Wyatville's comprehensive recreation of a Gothic castle.

William Townsend Aiton's gardening expertise was conscripted in the
remodelling of the royal estates. Nash consulted him about planting schemes
at St James's Park and the Brighton Pavilion. Aiton landscaped the 40 acres
at Buckingham Palace, merging two small ponds into an impressive lake.
Windsor demanded his frequent attention, where he added a semi-circular

The House of Confucius was also moved – to the eastern end of the lake in 1758. Its chinoiserie decorations frequently required renovation. Lithograph by G.E. Papendiek c. 1820.

flower garden on the East Terrace, and assisted Nash in forming a garden in character with the rural charms of the Royal Lodge. Aiton's expansion of his duties earned him in 1827 a new title, the wording of which he himself had proposed: Director General of His Majesty's Gardens.[1] Thus with less time for Kew, the direct responsibility for its management fell on the four foremen in charge respectively of the Botanic Garden, the Pleasure Grounds, the Kitchen Garden and the Fruit and Forcing Department.

When he was Prince Regent, George IV had planned alterations at Kew. In 1819 he began negotiations for the purchase of Hunter House (now the Herbarium), its grounds of about six acres and also Meyer House on the corner of Ferry Lane. These properties were to be adapted as, or replaced by, a summer residence – the King's Lodge – and Kew Palace pulled down. The

George E. Papendiek's *Kew Gardens* c. 1820 records in 24 lithographs many of the buildings then at Kew. Some like the Aviary, Menagerie, Mosque and Gothic Cathedral had already disappeared. This is his sketch of the Doric Temple of Pan, built by Chambers in 1758 and eventually demolished in 1844.

The Temple of the Sun stood near the Orangery – a small *Ginkgo biloba* now marks the site. Built by Chambers in 1761 it had eight Corinthian fluted columns and a room with the sun depicted on a coved ceiling, with a frieze of bas-reliefs of signs of the Zodiac.

Chambers's Temple of Bellona, originally overlooking the Menagerie, was moved to its present position in 1803/4. In the 19th century it was also called the Temple of Minden.

western tip of Kew Green between the King's Lodge and the Royal Gardens
on the southern side (about five acres) was to be incorporated. Thomas Hardwick, a former pupil of Sir William Chambers and now Clerk of Works for
Hampton Court, Richmond and Kew, was instructed in 1821 to design
entrance lodges to the enclosed portion of Kew Green.[2] By Act of Parliament
in 1823 authorising this enclosure, George IV undertook to make new roads
on the south and north sides of the Green and along the riverside to the
Brentford Ferry. The path bisecting the Green was restricted to a carriage
road leading to the Royal Gardens. In 1825 a low wall with railings crossed
Kew Green from Ferry Lane on the north side to near Methold House (now
the Director's residence) on the south side. In the centre elaborate iron gates
were flanked by two lodges bearing the royal supporters, the lion and the
unicorn. Shrubbery was planted on the Gardens' side of the railings and lime
trees on the public side, some of which still survive. Methold House and the
contiguous houses west of it were to be pulled down and their sites added to
the Gardens, a proposal fortunately never implemented.

Before this enclosure and realignment of roads, the route to the Brentford
Ferry followed the path along the centre of the Green, past the walls of the
Botanic Garden and arboretum on the left and the opposite boundary walls of
Hunter House, lined respectively with avenues of chestnut and lime trees; a
turning to the right near the Orangery led to the river and the ferry. A small
wooden door in the arboretum wall which admitted the public to the Botanic
Garden[3] was replaced in 1825 by a more imposing double entrance – one for
the public and the other for workmen – immediately east of Methold House.[4]

Another acquisition of land was successfully negotiated in 1822. The lease of some 20 acres on which the Pagoda stood had expired in 1819; its owner, William Selwyn, now surrendered it to the Crown in exchange for land on the other side of the Kew Road.

In 1827 the last Royal edict affecting Kew during George IV's reign authorised the demolition of the Castellated Palace.

Kew suffered not only from the King's indifference but also from the Treasury's habitual economies. When in 1823 the funds allocated to maintaining Kew's two collectors in the field were severely reduced, James Bowie was recalled from the Cape, a decision deplored by Professor W.J. Hooker at Glasgow University:

> this indefatigable naturalist, after sending the greatest treasures, both
> of living and dried plants to the Royal Gardens, and in the midst of
> his usefulness, has, by a needless stretch of parsimony, been recalled.[5]

On his return Bowie was urged to become a collector for the Horticultural Society of London by its Secretary, Joseph Sabine, who gloomily predicted that "Kew Gardens will soon pass away from its present management."[6] Aiton, evidently not expecting this cut, had in the same year created an additional post of foreman to deal with the flow of plants from Bowie and from Cunningham in Australia. He promoted one of his most promising gardeners, John Smith, who took charge of the hothouses and propagation department.

Smith, the son of a Scottish gardener, was a journeyman gardener for several years before employment at the Royal Botanic Garden in Edinburgh under its Curator, William McNab, once a foreman at Kew. With a letter of introduction from McNab, he presented himself in 1820 to Aiton, who placed him in the Royal Forcing Garden at Kensington. Aiton usually tested new entrants in other royal gardens under his control before admitting them to Kew. Smith, having survived his probation, transferred to Kew's propagating department in 1822; a year later he took charge of it. The Botanic Garden, its nine acres enclosed by high brick walls, was a sort of *hortus conclusus*. Its arboretum, planted in Princess Augusta's time, had reached maturity. The herbaceous collection still displayed plants in Linnean order and most of the hothouses were lean-tos, heated by small flues. With its Director often at the Royal Lodge at Windsor, the pace of life was leisurely and it is not surprising that after only four years there Smith sought another post, but Aiton bribed him to stay with an increase in salary and a house in the Gardens. At that time Kew's garden labels in the herbaceous collection identified only genera and tagged species with numbers which related to names in a ledger. On his own initiative, Smith undertook to relabel all the plants on public display with their specific names, an exercise far from completion when an official investigation of the Gardens took place in 1838. He also named and rearranged the grass collection planted in circular beds.

While Allan Cunningham, now Kew's only collector, searched the flora of New South Wales, his brother Richard, a clerk at Kew, annotated an inter-

John Smith (1798–1888), gardener, foreman and ultimately Curator, during his 42 years' service.

leaved copy of the *Epitome* (1814) of the *Hortus Kewensis* for a new edition that Aiton hoped to publish. As evidence of the plants then in cultivation at Kew, Aiton in 1822 gave one of his young gardeners with artistic talent, Thomas Duncanson, the responsibility of drawing new acquisitions. This Duncanson continued to do full-time until illness forced his resignation four years later. Another Kew gardener, George Bond, maintained the pictorial record for a further nine years. Between them, these two amateur artists produced nearly 2,000 competent drawings, the majority by Bond. A number of their flower portraits appeared anonymously at intervals in *Curtis's Botanical Magazine* after W.J. Hooker became editor in 1827.

Signs of Kew's imminent decline emerged during the 1820s. Its former supremacy in the horticultural world was being challenged by other establishments such as the Horticultural Society of London which had leased 33 acres at Chiswick from the Duke of Devonshire in 1821, and new provincial botanical gardens like the one at Liverpool. John Claudius Loudon, the Royal Gardens' most persistent critic, conducted his campaign in issues of the *Gardener's Magazine*, a periodical he had founded in 1826. He disliked the long narrow passageway that led from the new public entrance on Kew Green into the Botanic Garden, preferring the broad vista that had immediately greeted a visitor entering by the former gate; he judged that the palms were desperately overcrowded in their glasshouse, and that many trees in the Pleasure Grounds needed remedial attention. He printed a letter from a correspondent who complained that at Kew "you have a man at your elbow all the time, to prevent you from *stealing*; and, besides, are hurried over in half an hour".[7]

Inadequate public transport during the 1820s as well as this unfriendly reception limited the number of people who came to Kew. Occasionally a floral novelty such as Napoleon's willow would draw the crowds. In 1825 a Kew gardener who had been growing vegetables on the island of St Helena for ships of the East India Company, returned with a cutting of the willow (*Salix babylonica*) near Napoleon's grave. Smith placed the twig under a bell jar and a few years later planted it in the open near the entrance on Kew Green. He remembered that "French visitors paid reverence to it by taking off their hats, some even kneeling below it."[8]

Foreign visitors still frequently included Kew in their itinerary. They were not always complimentary. When Professor Schultes of Landshut in Germany called in 1824, he failed to see Aiton who, as usual, had business at Windsor. He found it inferior to the botanical garden at Göttingen, the Grand Duke of Weimar's garden at Eisenstadt, the Imperial Garden at Schönbrunn and Malmaison in France.[9] He also discovered inaccuracies in plant labels. Kew's general air of neglect surprised Prince Pückler-Muskau on his visit in December 1826. The Frankfurt nurseryman, Jacob Rinz, touring London gardens and nurseries in 1829, rated the collections at Messrs Loddiges in Hackney and Lee in Hammersmith infinitely superior to Kew's, where only a few of the hothouse exotics met with his approval.[10] In his opinion the Liverpool Botanic Garden was better maintained. But another continental visitor in 1829, Fr. Philippar, viewed Kew more charitably.[11] The size of the

TEMPLE OF VICTORY

HERMITAGE

RUINED ARCH

CHINESE PAGODA, KEW GARDENS

CHINESE TEMPLE

Robinia pseudoacacia and other well-established trees impressed him; like Rinz, he admired the Monkey Puzzle tree, approving of the alpines and shade-loving plants which encircled it. The display of grasses in a series of concentric circles with the tallest in the centre caught his eye. He wrote at length on the glasshouses where as a general rule plants were grouped according to species.

Since Kew's reputed reluctance to dispose of plants and seeds provoked widespread condemnation from botanists, gardeners and journalists, perhaps we should pause at this point to consider the validity of the charge. The Revd William Herbert, Dean of Manchester, spoke for many frustrated gardeners when he denounced

> The illiberal system established at Kew Gardens by Sir Joseph Banks, whereby the rare plants collected there were hoarded with the most niggard jealousy, and kept as much as possible out of the sight of any enquirer . . . It was the narrow-minded doctrine of Sir Joseph Banks that he could only render the King's collection superior to others by monopolising its contents.[12]

He claimed that he had never received any assistance from "that odious and useless establishment" during his researches on bulbous plants. Frederick Scheer, a resident of Kew Green for many years, who knew the Gardens well, sprang to the defence of Kew's reputation.

> It was, we believe, the practice of Sir Joseph Banks to retain rare plants at Kew for one year after they had flowered, and then they were liberally distributed to learned societies and eminent men.[13]

Guidebooks sometimes described buildings at Kew that had been demolished or, as in this instance, misidentified them. The so-called 'Hermitage' is, in fact, the old Stone House. C.F. Partington: *National History and Views of London and its Environs*, vol. 1, 1834, p. 53.

The truth of the matter is hard to establish. When Professor Sibthorp requested plants for the Oxford Botanic Garden, Banks could not have been more cooperative, even suggesting that Sibthorp might find it helpful to base his needs on the *Hortus Kewensis*.[14] Banks readily acceded to J.F. Jacquin's overture in 1792 for an exchange of plants and seeds. William Curtis's *Botanical Magazine* praised "Sir Joseph Banks's well-known liberality".[15] But Baron Dumont de Courset's application was tactfully refused because Aiton "is too much engaged with the vast variety of exotics he is daily receiving to pay proper attention to your requests",[16] and Teylers Museum in Holland received a similar evasive reply.

Even persistent and influential applicants like the Marquis of Blandford met their match in Sir Joseph. The Marquis, an obsessive collector, began cultivating a garden at Whiteknights in Berkshire during the 1790s with a profligacy that eventually bankrupted him. George III had impulsively offered him surplus plants from Kew, a gesture that the Marquis interpreted as a right to demand any rarity he coveted. After Aiton refused to part with some fuchsias and specimens of Masson's stapelias, he complained to Banks, citing another affront.

> I fixed my eye upon the *Houstonia coccinea* [*Bonvardia ternata*], [of] which I counted 11 or 12 plants; & in the hearing of both L[y] B., my sister, & gentleman who was with me (after much refusal) he [i.e. Aiton] promised to send it with the rest of the plants in a day or two. The plants came but no *Houstonia*; I wrote repeatedly for it, conceiving it a mistake, but no answer came: about a year after I got it from the trade . . .[17]

Banks's reply was diplomatic but firm. He informed the indignant nobleman that the King had promised such plants as "could properly be spared" and that their difference of opinion rested on the interpretation of the word "properly". Whenever Kew had a surplus stock of any one species, the Marquis was assured that

> it is proper that it should be given to your Lordship, provided that the superiority which his Majesty's Garden has for some years held over the other Gardens of this country is not put in hazard by parting with too many species which Kew alone possesses. The amount of the danger of the superiority of Kew Gardens being lost is a matter which probably no-one but myself is able to judge.[18]

The acrimonious correspondence rumbled on for some years, but the Marquis had put his finger on a problem that constantly worried Sir Joseph Banks – plants that were exclusively Kew's were appearing for sale in nurseries. Disturbed by this evidence of pilfering from the Royal Gardens, he reviewed the matter in a document written about 1815 or 1816.[19] Whole plants as well as cuttings were being smuggled out of Kew, probably with the assistance of garden staff bribed by persons engaged in this profitable business. The *Gardener's Gazette* believed that Aiton preferred numbers rather than explicit names on plant labels because they made identification more difficult for thieves.[20]

When some choice Australian plants were stolen from the Botany Bay glasshouse in January 1824, Aiton was determined to make an example of the man accused of receiving them, Robert Sweet, Manager at Colvill's nursery. The case was heard at the Old Bailey in February but Sweet's unblemished reputation, vouched for by other nurserymen, secured his acquittal on technical grounds. The discovery at the same time that the keys to the Royal Gardens had been counterfeited necessitated changing all the locks.[21] Such precautions and even the likelihood of imprisonment were no deterrent, argued Allan Cunningham, who advised Aiton to exchange surplus plants with nurserymen.[22] During the investigation of all the royal gardens in 1838, Aiton maintained that it had always been his policy to give away plants on a reciprocal basis but, he reminded his interviewers, Kew was a private garden and not obliged to participate in such arrangements. The conservatory at Carlton House and the Royal Lodge at Windsor were constantly replenished by Kew, and the crowned heads of Europe were regular beneficiaries. Glasgow Botanic Garden received several consignments after W.J. Hooker became its energetic custodian, but few individual gardens benefited from this dispersal of stock, and consequently the litany of complaints intensified, assiduously orchestrated by J.C. Loudon. He fired his first broadside in his *Encyclopaedia of Gardening* (1822), mildly regretting Kew's parsimony. The first number of his *Gardener's Magazine* (1826) lamented "the peculiar system of exclusive possession which it is thought necessary to maintain in that national institution".[23] An entire article in 1827 was devoted to Kew's now legendary meanness.[24] The author contended that had Kew fallen in line with the practice of botanical and private gardens in a liberal distribution of its stock, then many of its rare plants which had died might still be available in other collections. J.C. Loudon's editorial urged a reform of all the royal gardens together with adequate funding. With a gradual but perceptible shift in Kew's policy during the 1830s, John Lindley was able to announce that "a liberal management has been introduced, and the collection is as accessible as that of other nations".[25] And even Dean Herbert, who had been so vitriolic, conceded that "I have incautiously admitted a stronger expression concerning the unpopularity of the principle on which that garden has been conducted than I should wish to have used."[26]

The closing years of George IV, now detested by many of his subjects, were spent in seclusion at Windsor, and on his death in 1830, it was hoped that his brother, the Duke of Clarence, would restore respect for the monarchy. William IV, nearly 65 when he ascended the throne, looked exactly what he was — a retired naval officer; patriotic, essentially good-natured and, although tetchy on occasions, possessing a genuine fondness for Kew. But it would appear that this affection for the place he had known as a boy did not extend to W.T. Aiton. During the early years of his superintendency, Aiton had more successfully established a rapport with the other young princes, especially the Duke of Kent, and subsequently enjoyed the patronage of the Prince Regent.

When a Captain Gardiner took charge of the Kitchen Garden at Kew without Aiton's knowledge, royal intervention soon removed him from the scene.[27] Aiton's authority, which under George IV had embraced other royal gardens, was now confined to Kew and Buckingham Palace. It was hinted at the time that William was settling an old score for some offence Aiton had committed years earlier. The *Gardener's Magazine* announced that Gardiner's son was tipped to take charge of the Pleasure Grounds at Kew,[28] an appointment that never materialised. Changes to the administration at Kew prevented the publication of the revised *Epitome* of the *Hortus Kewensis*, now completed, then deferred, and ultimately abandoned.[29]

In 1831 William IV revoked his brother's enclosure of the west end of Kew Green, had the two lodges removed, the road through the centre turfed, and access to Kew Palace directed along the road skirting the Thames. Hunter House, which George IV had intended to be his Royal Lodge, now became a residence of William's sister-in-law, the Duchess of Cumberland, for her lifetime. An apron of a garden and a porter's lodge which jutted on to Kew Green gave her greater privacy.[30] Several hundred acres of the former Richmond Gardens and the Old Deer Park were taken over by the Duke of Cumberland as a game preserve and for pasturage.[31]

Sunday opening of the Gardens was suspended, otherwise the public were admitted daily. Entry to the Pleasure Grounds, now at the disposal of the Duke of Cumberland, operated under some peculiar regulations; access was restricted, the wearing of a tall black hat and black or white 'choker' was obligatory – no coloured neckerchiefs were counternanced – and "dogs and gentlemen's servants" were barred.

William IV began the process of opening up the northern area of the Gardens by removing part of the boundary wall of the original Botanic Garden east of the Orangery. The walls behind the chestnut and lime walks leading from Kew Green were also pulled down, together with the perimeter wall at the east end of Kew Palace garden.[32] In 1834 the public was denied access to the lawn adjoining the Palace grounds by a combination of a ha-ha and railings stretching from the river near the site of the former Castellated Palace to the west wall of the Botanic Garden above the ice-house.[33] Ornamental shrubberies were to be planted in the grounds of the Palace which, according to John Smith, the King wished to reopen as a royal residence.

William IV is supposed to have regretted being too old to commission any building at Kew;[34] nevertheless, his architect, Sir Jeffry Wyatville, planned about 1834 a large extension to the south-west corner of Kew Palace. This unexecuted wing with 'Queen Anne' stylistic affiliations is one of 88 drawings the architect declared he had made for Kew. When the Duchess of Kent asked the King in 1836 for a country residence for herself and Princess Victoria, he offered her Kew Palace. She rejected it as "an old house quite unfit for the princess and me to occupy, being very inadequate in acccommodation and almost destitute of furniture".[35]

Wyatville, who had already built conservatories at Longleat, Belton House and Bretton Hall, was well-qualified to tackle a new Palm House for Kew.

Sir Jeffry Wyatville's proposed extension on the south front of Kew Palace. The existing entrance to the Palace was to be converted into a window. *Royal Institute of British Architects.*

Such a building had been mooted during the Regency and the reign of George IV; as a temporary expedient the roof of the existing one had been raised by four feet in 1828. In October 1834 the King chose a site for Wyatville's wooden structure. It was to be 200 feet long, 50 feet wide, with side aisles and a lantern 40 feet high, but it was never built. The inexpensive modifications to the Orangery were also put off. The King had been persuaded that the Orangery's effectiveness as a repository for plants could be improved by the insertion of large windows, similar to those on the garden façade, at both ends. Their position was chalked out on the stucco but it was not until Queen Victoria's reign that they were installed.

Tubs of orange trees were still housed in the Orangery in the 1820s, but by 1838 araucarias and Australian plants had been added.

One of John Nash's conservatories, here seen at Buckingham Palace, was transferred to Kew in 1836. The lake seen in the gardens had been designed by W.T. Aiton.

Palms were among the first occupants of the Architectural Conservatory and when they were transferred to the new Palm House in 1848, their place was taken by Australian plants.

Denied the chance to build a Palm House at Kew, Wyatville had to be content with adapting a conservatory transferred from Buckingham Palace in 1836, one of four that John Nash had designed in 1825. The two on the northern elevation of the Palace were unsuitably located for plants, so one was removed to Kew on the instructions of William IV, who carefully selected a spot where no trees would have to be chopped down. It has been claimed that its 12 Ionic columns came from Carlton House. Wyatville had intended there

The Architectural Conservatory, as it was called, came to Kew during the course of Edward Blore's work at Buckingham Palace. It is 80 feet long, 40 feet wide and 26 feet high. Sir Jeffry Wyatville's drawing shows A.M. Perkins's hot-water system. *Royal Institute of British Architects.*

to be 12 Portland stone columns on both the north and south faces but with stringent economy imposed on him, he substituted Bath stone pilasters and omitted any decorative carving on the plain surfaces.[36] Sundridge Park in Kent had pioneered the heating of glasshouses with hot-water pipes in 1816, and by the 1830s several systems were available. Kew's new conservatory had a system patented by A.M. Perkins four years earlier and tried out by both Sir John Soane in the Bank of England and J.C. Loudon in his conservatory in Bayswater. Two of Perkins's boilers at Kew circulated steam through hermetically-sealed small diameter coil-pipes, a system in use for some 30 years until superseded by hot-water pipes of a much larger diameter. Until, however, the reliability of Perkins's design was proved, Kew's architectural conservatory was filled with *Araucaria* and *Eucalyptus* rather than with tender exotics.

Wyatville supervised this conservatory conversion at the same time as he was erecting a 'Temple of Military Fame' at Kew. Within it a series of 18 cast-iron plates and two stone tablets recorded British victories from 1760 to 1815, thus complementing Chambers's Temple of Victory, which William IV had restored. King William's Temple, as it is now known, was almost finished at the time of his death in 1837. Its austere interior was relieved by a parade of busts by Chantry of George III, George IV, William IV, Frederick, Duke of York, and the Duke of Wellington; a tablet which British troops had attached to Cleopatra's Needle in Egypt in 1802 faced the entrance.

Sir Jeffry Wyatville designed King William's Temple, also known as the Temple of Military Fame and the Pantheon, in 1837. The two Francavilla statues once stood in its portico. The railing was erected to deter vandalism.

The decline in Kew's financial support which had begun in the late 1820s continued during William IV's reign: the annual average expenditure of £1,900 between 1824 and 1827 slipped to £1,460 during 1828 to 1831 and to £1,277 for the period 1832 to 1836.[37] The cost of Kew's maintenance was borne by the Board of Green Cloth (so-called from the green-covered table at which its business was originally transacted) under the Lord Steward who, presumably as a saving, ordered the recall in 1830 of Allan Cunningham, the Royal Gardens' only official collector. Cunningham returned from Australia in July 1831 to spend the next five years sorting and identifying his collections. He declined in 1832 the offer of the post of Colonial Botanist and Superintendent of the Botanic Garden at Sydney in favour of his brother Richard, for over 20 years Aiton's personal clerk. When his brother was killed by aborigines in 1835, Allan Cunningham took his place but resigned after only a few months, objecting to the menial nature of some of his duties. He died in Sydney in June 1839. The genera *Cunninghamia* and *Alania* and many species commemorate his distinguished achievements as a botanist and collector.

With Kew in recession, its rival, the Horticultural Society of London, formed in 1804, was in the ascendant. In 1821 it sent one of its gardeners, John Potts, on a round trip on an East Indiaman to China to collect plants, a voyage that was repeated by another member of its staff in 1823. It engaged George Don of the Chelsea Physic Garden to join a survey ship in 1821;

during the voyage he collected in Sierra Leone, Brazil and the West Indies. John Forbes collected for the Society in South and East Africa, before dying in Mozambique in 1823. David Douglas, one of their greatest collectors, operated in North America between 1823 and 1832. Beginning in 1836 Theodore Hartweg worked under contract exploring the flora of Mexico and part of South America.

The Horticultural Society's commendable initiative inevitably drew attention to Kew's neglect. Its many friends and wellwishers helped to replenish its stock: continental botanical gardens and institutions in Paris, Vienna, Göttingen and Moscow; David Lockhart at the Botanic Garden in Trinidad never forgot his former employer; Alexander Moon in Ceylon was equally loyal and generous. When the Kewite, George Aldridge, returned from Trinidad through ill-health, he brought a collection of orchids and ferns. Nathaniel Wallich at Calcutta continued to repay his debt to Banks with regular donations of South and South East Asian plants. Robert Brown at the British Museum still exercised some influence on Kew's behalf. Through his intervention, George Barclay, a young Kew gardener, joined HMS *Sulphur* on its survey of the west coast of America in 1835 to collect living and dried specimens. His equipment included an extending pole tipped with a hook to pull down plants beyond normal reach.

> You must not mind cutting down trees to get at the seeds or specimens . . . if the survey is continued as far as California you will meet with firs of a great hight [sic], some having large cones. Let no plant escape your notice but take care of yourself.[38]

There being no room on the small ship for plant boxes, Barclay was unable to collect many plants and few of his seeds germinated. His dried specimens went to Brown, who doubtless anticipated this when he recommended him for the voyage.

In 1837 John Armstrong, a former Kew employee, was kitted out by the Royal Gardens to collect plants at Port Essington in North West Australia, where he was to be the gardener in the new settlement. Quickly disillusioned with life as a colonist, he ended up in Timor cultivating rice. A case of dried specimens was all that Kew ever received from him.

The end of William IV's reign in 1837 coincided with a growing unease about the prevailing condition – and even the status – of the Royal Gardens. George Glenny, outspoken, belligerent and provocative, used his *Gardener's Gazette* to attack what he perceived to be Kew's mismanagement. "The state of the place is slovenly and discreditable, and that of the plants disgracefully dirty." His advice to Aiton was "reform – or quit".[39] His invective, which continued over several issues, found support from a correspondent to *The Times*, who diagnosed "niggardly expense and exclusiveness"[40] as the root cause of Kew's decline. But Aiton had his defenders, one of whom – "A Lover of Botany and Truth" – refuted the charges of neglect and applauded his policy of free admission to the Gardens.[41]

From this public debate we learn that the overcrowded hothouses desperately needed repair or replacing; that the spacious lake had shrunk to an insignificant pond, choked with mud and filth; and that Chambers's garden buildings were shabby. Even Kew's historic core, the original physic garden enclosed within high walls, did not escape criticism from the *Horticultural Journal*:

> The whole at present bears too close a resemblance to the botanic gardens
> of a university or a medical college, to cause any great interest in mere
> amateurs, or to excite the admiration of the public.[42]

Clearly Kew had lost its way; no longer guided by any positive policy or aspirations, it was only a matter of time before its demise unless progressive stagnation could be halted.

Uncertainty at Kew contrasted with an optimistic proliferation of botanical gardens elsewhere. University cities such as Oxford, Cambridge, Edinburgh, Glasgow and Dublin, industrial centres like Liverpool, Birmingham and Sheffield, and the historic towns of York, Colchester and Bury St Edmunds supported them, at least in name. There was no commonly accepted concept of such gardens. In 1725 Richard Bradley had envisaged the one at Cambridge propagating crops for Britain's colonies, as well as functioning as a teaching resource for medicine and agriculture. John Hill had advocated an educational role in 1758, having in mind an adaptation of the grounds of Kensington Palace where the public would be introduced to botany through lectures and appropriate displays. Soon after John Hope took charge of the Botanic Garden in Edinburgh, he founded in 1763/4 a Society for the Importation of Foreign Seeds and Plants as a way of stocking his garden with exotic plants of commercial value, seeing their provision as a *raison d'être*.

John Claudius Loudon reminded fellows of the Linnean Society at a meeting in December 1811 that while the capitals of Scotland and Ireland had botanical gardens, London lacked one "suited to its dignity".[43] He claimed that the collections at Cambridge and the arrangement at Liverpool put to shame those at Kew and the Chelsea Physic Garden. His idea of a "National Garden" deemed cultivated varieties of fruits, vegetables and flowers as important as botanical species. He advocated imaginative presentations by climatic affinities, by seasons and in a range of landscaped settings "to blend splendour, beauty and variety, with taste, science and utility". Ancillary services such as a museum, herbarium and library and public lectures would be mandatory. Such a garden, preferably extending over about 100 acres, could be supported by subscribers, by a society, or by a government department.

Loudon's enthusiasm reflected the mood of the time. In 1812 a new association issued a prospectus proposing "a National Botanic Garden, Library & Reading Rooms in the Regent's Park". It charitably offered to take Kew's plants, a gesture that predictably infuriated Banks.

20 Gothic Cathedral; 21 Gallery of Antiques; 22 Temple of Arethusa?; 23 Temple of Solitude? It is interesting to note some differences between this plan and that of 1763: the ha-has which encircled the lawns and the Palladian Bridge have disappeared. Below Love Lane, which separated the two Royal estates, can be seen some of 'Capability' Brown's re-landscaping of Richmond Gardens. *British Library*

PLATE 13 Castellated Palace. Aquatint by J. Gendall, 1819. This palace, designed by James Wyatt, stood more or less on the site of the present riverside carpark. Work began on its construction in 1801 but was halted by the permanent breakdown of George III in 1811. George IV ordered its demolition in 1827.

PLATE 10 Syon House seen from Richmond Gardens. Oil painting by Richard Wilson (1714–82) who executed a number of views in the vicinity of Richmond during the 1760s. 'Capability' Brown had destroyed Queen Caroline's riverside terrace in order to bring the Thames closer to the Gardens. *Neue Pinakothek, München*

PLATE 11 Lake and part of the Island in Princess Augusta's garden looking south to the Pagoda. Oil painting by Richard Wilson, 1762. Note the Palladian Bridge to the Island. *Yale Center for British Art, Paul Mellon Foundation*

A Plan of KEW GARDENS
the environs

THAMES

PLATE 12 Plan of Princess Augusta's garden and part of adjacent Richmond Gardens, c.1785. 1 Dutch House;
2 White House; 3 Orangery; 4 Temple of the Sun; 5 Great Stove; 6 Physic and Exotic Gardens; 7 Flower Garden
(Note that the adjacent Menagerie shown on the 1763 plan has disappeared); 8 Temple of Bellona; 9 Temple of Pan;
10 Temple of Eolus; 11 Smeaton's Water Pump; 12 House of Confucius; 13 Lake and Island; 14 Theatre of Augusta;
15 Temple of Victory; 16 Ruined Arch; 17 Alhambra; 18 Pagoda; 19 Approximate site of the Mosque;

It does not appear to me that any advantage to science would be the consequence of removing the Botanic Garden at Kew to the Regent's Park . . . The science of botany is best improved at a situation remote enough from the crowded populations of the metropolis to prevent the commerce of persons induced by idle curiosity alone to visit it.[44]

Nothing came of this proposal but agitation for a scientific establishment in the field of natural history persisted. In 1825 the Duchess of Somerset complained to Sir Stamford Raffles that whereas the Jardin des Plantes in Paris arranged public lectures, there was nothing in London promoting such activities and certainly not at Kew with its staff of twelve and little to show for their presence. A correspondent in Loudon's *Gardener's Magazine* lamented "the non-existence of a public botanic garden in the neighbourhood of London, in which the most perfect collection that the scientific connections and resources of the empire could furnish should be preserved and cultivated".[45]

In December 1837 Loudon received for publication an article on the formation of a public botanical garden.[46] The author, his identity concealed by the initials 'C.C.', deplored the absence of a botanical garden in London while every European capital had one. In urging that Kew Gardens be "ceded to the public" for such a purpose, he wanted it transformed into a centre for scientific research, with professional staff and an active programme of lectures. Its range of plants would be extended and a comprehensive arboretum formed, possibly by requisitioning 100 or 200 acres in Richmond Park. He believed that a government-maintained institution would be more likely to ensure permanence and public access than any society dependent on the support and whims of its members. He saw the accession of a new monarch as an opportune moment for Parliament to approve the creation of an enhanced Kew. One wonders whether he had been moved to write this article by rumours that the future of all the royal gardens was being currently considered.

Upon the death of William IV and the cessation of income from Hanover, the Treasury instituted a review of the funding of the Royal Household. The Board of Green Cloth managed the Royal Household under the Lord Steward. When the Treasury's preliminary investigation of the Board's finances disclosed rising expenditure, the royal gardens were identified as one area where costs might be reduced. On 30 January 1838 the Treasury appointed the Members of Parliament, Robert Gordon and Edward Ellice, to enquire into the management and expenditure of the Royal Gardens at Windsor, Hampton Court, Buckingham and Kensington Palaces, and Kew. Within a week, John Wilson, gardener to the Earl of Surrey (who was Treasurer to Her Majesty's Household), and John Paxton, gardener to the Duke of Devonshire at Chatsworth, formed a working party under the chairmanship of John Lindley, Professor of Botany at University College London and Assistant Secretary of the Horticultural Society of London, to report on the royal gardens.

Without the vision of Lindley it is doubtful whether Kew would have survived the Treasury's cost-cutting exercise. Like his friend William Jackson Hooker, he was a man of exceptional energy and stamina with an enviable capacity for sustained work. Both men were Norfolk-born and educated at Norwich Grammar School. Through Hooker, 13 years his senior, Lindley had obtained a temporary assistant post in Banks's herbarium and library. When Hooker declined the new post of Professor of Botany at University College London, Lindley was the obvious alternative. He also lectured at the Chelsea Physic Garden, wrote and edited prodigiously, and still found time to accept an invitation to advise the investigating committee.

The Commissioners of Woods and Forests informed him of his terms of reference on 8 February.[47] The kitchen gardens which supplied the Royal Household with fruit, vegetables and flowers were to be his first priority: the quantity and quality of their produce, the suitability of soil and location and the adequacy of their forcing houses; whether they should be maintained as they were, enlarged, or perhaps amalgamated, and the remainder relegated to "some other public purpose".

Of all the royal gardens only Kew was singled out for intensive examination. Lindley's brief instructed him to make recommendations on its future either as a garden serving both the Royal Household and the public or as a place "solely for the interests of science". Regarding the latter, Lindley had to bear in mind that it "should be efficient for the general extension of botanical knowledge, and the production of new and rare plants". Its facilities were to be accessible to British colonies and relations established with foreign botanical gardens. Everybody hoped he would find a solution to the contentious matter of the disposal of Kew's surplus plants.

Following a preliminary meeting with his two collaborators on 12 February, Lindley wrote to all the superintendents of the royal gardens. Exceptionally, his letter to Aiton required him to submit figures of income and expenditure, details of produce supplied to the Royal Household and elsewhere, and information regarding the receipt and despatch of plants and seeds at home and abroad. In strictest confidence Lindley also asked John Nussey of the Society of Apothecaries whether the Society would consider relinquishing its Physic Garden at Chelsea, given the assurance that Kew or some other public garden would provide a similar service for medical students. In return the Society would transfer its plants and guarantee an annual allowance to such a garden.

Lindley's working party inspected the Kitchen and Forcing Garden at Kew on Friday 16 February and returned the following Monday to look at the Botanic Garden. The walled Kitchen Garden of about 10 acres consisted of two units: Methold's Ground and Home Ground, separated by a paddock belonging to the King of Hanover. Fruit trees covered the walls and glasshouses much of the ground. In Methold's Ground a large vinery grew black Hamburgh grapes but the Home Ground had most of the glass: seven peach houses, two vineries, two cherry houses, three pine stoves and a range of pine pits, a mushroom house and many frames for vegetables. Lindley noted its reputation for good quality asparagus and congratulated John Aldridge on maintaining a creditable garden. He subsequently recommended that it be enlarged by four acres and entrusted with supplying the London palaces.

Aldridge, who supervised a foreman, 9 men working in the glasshouses and 15 allocated to other duties, assured Lindley that he had the right to hire or discharge men without prior permission. William Townsend Aiton, described as the "retired Director General of the Royal Gardens", now had his authority confined to the Botanic Garden and Pleasure Grounds at Kew and Buckingham Palace garden. John Smith was his principal foreman in the Botanic Garden and James Templeton in the Arboretum; the latter also acted as accountant and clerk. Two more foremen, six gardeners, four labourers and two watermen completed his establishment.

Lindley's minutes of proceedings for 19 February succinctly noted that he, Paxton and Wilson "minutely examined the houses, plants, etc. [at Kew] as far as the weather permitted". John Smith tells us that it was one of the coldest winters on record with deep snow in the Royal Gardens, hardly a propitious time to inspect any garden. However, Lindley and his team persevered, visiting every glasshouse, scrupulously inspecting the records and closely interrogating Aiton who, understandably, was on the defensive. Aiton stressed deficiencies in his funding, and he was acutely sensitive to questions about the distribution of plants and seeds. Not having ever received any directives, he had always given to those who could benefit the Garden in return, but, as he reminded Lindley, Kew was a private garden. Nevertheless, its traditional obligations to the colonies had always been honoured and their requirements fulfilled although such requests had been infrequent. Kew naturally had a duty to supply flowers to the Royal palaces and for special occasions. During George III's reign it had reserved flower-beds and hothouses just for nosegays and floral decorations for the Royal Household, a practice that ceased when the garden at Frogmore became the source of supply.

In little more than a fortnight Lindley completed his investigations and with characteristic efficiency submitted his report on 28 February. He had surveyed all the royal gardens but only his recommendations regarding Kew were eventually published.[48] The Botanic Garden, lying between the Kitchen Garden and Kew Palace, he noted, extended over some 15 acres (subsequently corrected by Sir William Hooker to 11 acres). Its small arboretum contained some good specimens of exotic trees. The plants in the haphazard arrangement of glasshouses, the largest still being Chambers's Great Stove, were in excellent health despite cramped conditions. The majority were unlabelled, a criticism also applicable to the plants in the open. While acknowledging the commendable efforts of the foreman, John Smith, to name them, Lindley judged it a task that demanded botanical expertise and access to a herbarium and library. He praised the horticultural skills of the gardeners but condemned the general management of the Gardens: no meaningful arrangement of plants; failure to maintain contact with the colonies; reluctance to dispose of surplus stock; and no discernible policy other than that of diligently raising the plants presented to it. Aiton had obtained permission to admit the public. "It is, however, not easy to discover what advantage, except that of a pleasant walk, has been derived from the privilege." Now that Kew Palace no longer served as a royal residence, the Botanic Garden had become a burden upon the Lord Steward's department.

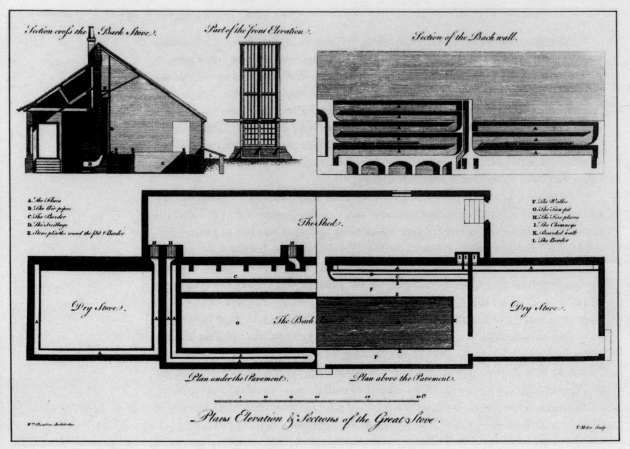

Section cross the Bark Stove. — Part of the front Elevation. — Section of the Back wall.

The Shed.

A. The Flues
B. The Air pipes
C. The Border
D. The Trellage
E. Stone plinths round the Pit & Border

F. The Walks
G. The Tan pit
H. The Fire places
I. The Chimneys
K. Boarded walk
L. The Border

Dry Stove.

Dry Stove.

The Bark Stove

Plan under the Pavement. — Plan above the Pavement.

Plans Elevation & Sections of the Great Stove.

Wᵐ Chambers Architect.

T. Miller Sculp.

In the reigns of George IV and William IV, Chambers's Great Stove, erected in 1761, was still the largest conservatory at Kew, 114 feet long. Its three sections contained succulents, palms, miscellaneous plants and "a few forced flowers for nosegays". W. Chambers: *Plans . . . of the Gardens . . . at Kew*, 1763.

In the event of its relinquishing Kew, the State should consider taking it over, making it "worthy of the country", transforming it into "a powerful means of promoting national science". If that were not possible, it should be abandoned.

An echo of Banks's imperialism sounds through the rhetoric of Lindley's justification for a National Botanic Garden, the destined nexus of all colonial gardens which would come within its orbit.

> . . . they should be all under the control of the chief of the garden, acting in concert with him, and through him with each other, reporting constantly their proceedings, explaining their wants, receiving their supplies, and aiding the mother country in every thing that is useful in the vegetable kingdom. Medicine, commerce, agriculture, horticulture, would derive considerable advantages from the establishment of such a system.

The service that he anticipated such a garden rendering to the Empire – that of giving information "upon points connected with the establishment of new colonies" and of providing "the plants required on those occasions" – was a goal that would be diligently pursued by three future directors of Kew: the two Hookers and Thiselton-Dyer.

The foundations for a National Botanic Garden were in place at Kew. To make Kew effective it needed more space (Lindley recommended appropriating at least 30 acres from the Pleasure Grounds), more glasshouses, systematic naming of all the plants, nurseries propagating plants for export, exhibitions and public lectures – "in short the Garden should be perfectly adapted to the three branches of instruction, exhibition and supply".

With a promptitude that matched the speed with which Lindley had submitted his report, Lord Surrey, Ellice and Gordon communicated their approval to the Treasury on 12 March 1838.[49] They recommended that the number of royal kitchen gardens be reduced to two – Windsor and Kew – with the one at Kew enlarged by incorporating the King of Hanover's paddock. Since there was no justification for leaving the Botanic Garden in its present state, they proposed its transfer to the Commissioners of Woods and Forests. Their desire to see it function as "an efficient institution for the promotion of botanical science throughout the Empire" indicated a complete acceptance of Lindley's thesis. Part of the consequential costs of setting it up might be met either by a more economical management of the Pleasure Grounds or by building on some of it. (For the next couple of years the integrity of the Kew estate was going to be under threat from several quarters.) They suggested that the scientific work of this reconstituted Kew might be directed by a board of trustees on which London University, the Colleges of Physicians and Surgeons, the Society of Apothecaries, the Linnean Society and the Horticultural Society of London would be represented. It was announced in the House of Commons on 6 July 1838 that the Linnean and Horticultural Societies had petitioned Parliament for the preservation of Kew.

The Treasury, which pondered the Committee's letter for some six weeks, referred it to the Commissioners of Woods and Forests for comment. The First Commissioner, Lord Duncannon, accepted most of the recommendations for the kitchen gardens but evasively declared that Kew's future required more consideration. It was not until 24 April 1839 that the Treasury received the Commissioners' offer to take over the Garden provided it was adequately funded.

Meanwhile a threat to Kew's destiny as the country's National Botanic Garden came from the Royal Botanic Society. In October 1838 the Society purchased the Inner Circle of Regent's Park, a plot of 18 acres, for landscaped gardens, a conservatory, hothouses, a museum and a library. This newly-formed organisation, managed by two ambitious joint-secretaries, quickly recruited an impressive roster of influential supporters and obtained a Royal Charter in a bid for respectability and status within the scientific and horticultural community. While the debate about a National Botanic Garden continued, it saw an opportunity to usurp that role for itself. In a memorial to the Treasury in February 1840 it eagerly accepted a Government proposal to take over Kew, a move emphatically rebutted by the Duke of Bedford in a letter to the Chancellor of the Exchequer. "The situation for a botanical establishment worthy of the nation is unquestionably Kew Garden."[50]

The Treasury never replied to the Woods and Forests letter of 24 April 1839. George Glenny in the *Horticultural Journal* for July 1839 dismissed the embryonic botanical garden in Regent's Park as "a wild, impracticable

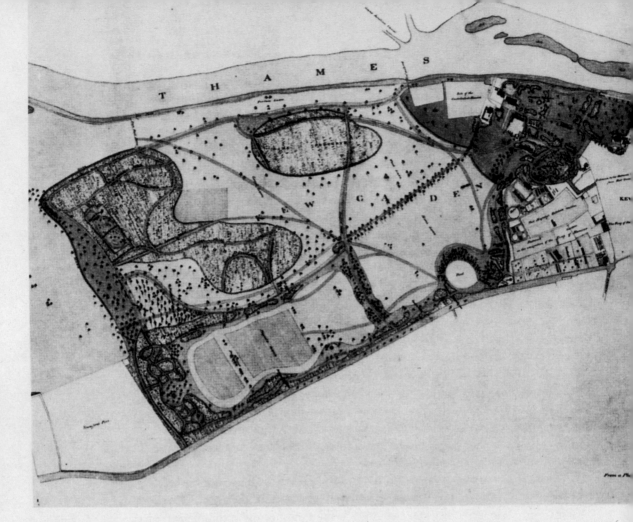

scheme". He detected a few improvements at Kew, for which he immodestly
took the credit. In December 1839 he floated the idea of converting
Whiteknights, where much of the original planting by the extravagant Lord
Blandford still survived, into a national garden. His constant denigration of
Kew, especially in the *Gardener's Gazette*, provoked John Smith into threatening instant dismissal of any of his gardeners discovered reading that proscribed
periodical.

The first half of 1840 was a period of rumour and innuendo, of Government duplicity, of public bewilderment and apprehension. At a meeting with
Robert Gordon, now Secretary to the Treasury, on 11 February,[51] John Lindley learned that it was the Government's intention to abolish the Botanic
Garden, to demolish the glasshouses, and to dispose of the plants to "some
public body" provided the public had access to them on two days a week. The
Horticultural Society of London, which the Treasury favoured, would be
allowed to occupy Kew until glasshouses had been built in their Chiswick
garden. Lindley responded by informing Sir Robert Peel that he intended
having the matter raised in Parliament. John Smith alleged that John Aldridge
had told him on 18 February of Lord Surrey's intentions to convert the Botany
Bay and Cape Houses into vineries.[52] Smith, however, was told by A.B.
Lambert ten days later that the offer of the plants to the Horticultural Society

of London (which they had unhesitatingly declined) had been a misguided act and that Kew was safe.

When Lord Aberdeen, strongly opposed to the dismemberment of Kew, sought clarification in the House of Lords on 3 March, Lord Duncannon assured him that the Government had no intention of breaking up the Gardens. Two letters in *The Times* from J.C. Loudon demanded that the correspondence on the purported offer to the Horticultural Society be published. According to A.B. Lambert, Queen Victoria now appeared to be taking a personal interest in the future of a garden that had been developed by her ancestors.[53] Probably recognising that the climate of opinion was against it, the Commissioners of Woods and Forests at last indicated to the Treasury their unconditional willingness to accept Kew. The Lord Steward's department committed itself to an annual subvention of £800 towards Kew's maintenance, rising to £1,000 when the personal allowance to Aiton ceased. After a request by Joseph Hume in the House of Commons on 4 May that the relevant part of the 1838 report be presented to Parliament, Lindley's recommendations on Kew were published eight days later. On 25 June the Treasury sanctioned the transfer of Kew – excluding the Royal Kitchen Garden, the grounds of Kew Palace and the Queen's Cottage – from the Lord Steward's department to Woods and Forests. Four days later the Prime Minister and Lord Duncannon approved the appointment of Sir William Jackson Hooker, Professor of Botany at Glasgow University, as the new Director of the Botanic Garden. However, it was not until November that W.T. Aiton indicated to the Commissioners that he wished to relinquish control of the Botanic Garden, but he agreed to stay in post until Lady Day, 25 March 1841.

CHAPTER TEN

Sir William Hooker to the rescue

*Hooker's early career – his efforts to obtain
a post at Kew – appointed Director –
enlargement of the small botanic garden –
Palm House – Decimus Burton – Richard
Turner – admission of the public*

DURING HIS YOUTH IN NORFOLK, William Jackson Hooker was an all-round naturalist with as much enthusiasm for birds and insects as for plants. His discovery at the age of 20 of a moss near Norwich was identified by Dawson Turner, banker, antiquarian and botanist at Yarmouth, as a new species, *Buxbaumia aphylla*. Soon after his twenty-first birthday, he was elected a Fellow of the Linnean Society, where he met Sir Joseph Banks, his assistant Robert Brown, and other prominent naturalists. In 1809, through Banks, Hooker obtained a free passage to Iceland, all expenses paid, but unfortunately lost his collections and drawings in a fire on the ship returning to England. In order to write his *Journal of a Tour in Iceland* (1811), Banks magnanimously insisted that Hooker use the notes of his own trip to Iceland in 1772. An unfortunate partnership with Dawson Turner in a brewery at Halesworth in Suffolk probably encouraged him to seek consolation in a study of British Jungermanniae. Banks backed his desire for foreign travel by arranging for him to join a new governor about to leave for Ceylon. In the meanwhile Hooker familiarised himself with the flora of South Asia by making simplified pen and ink sketches of flower paintings by Indian artists kept in the London headquarters of the East India Company. When unrest among the Ceylonese threatened to make travel on the island dangerous, Banks suggested Java, recently taken from the Dutch by British forces, as an alternative destination. Having Hooker as a collector for Kew was too good an opportunity to miss, and Banks undertook to persuade the Board of Trade to finance the expedition. But Hooker, sharing his family's apprehensions about the journey and especially the inhospitable climate, prudently withdrew, much to Banks's ill-concealed annoyance. He could not understand why Hooker allowed himself to be influenced by

the importunities of those who think they can guide you to a more
serene, quiet, calm & sober mode of slumbering away life than that

you proposed for yourself . . . I was about 23 when I began my peregrinations. You are somewhat older but you may be assured that if I had listened to a multitude of voices that were raised up to dissuade me from my enterprise I should have been now a quiet country gentleman ignorant of a multitude of matters I am now acquainted with & probably never attained to no [sic] higher rank in life than that of country Justice of the Peace.[1]

Clearly cosy domesticity and research appealed more to Hooker than the hazards of expeditions. He married Dawson Turner's eldest daughter in 1815 and spent the next five years immersed in cryptogamic botany. A worrying distraction, however, was the inadequate return from his brewery investments to support a growing family. With ruin seemingly imminent, he asked Sir Joseph Banks to inform him of anything offering the possibility of using his botanical expertise to supplement his income. It so happened that Glasgow University needed a professor of botany and, with Banks's recommendation, in 1820 Hooker was appointed. Lecturing and managing the small botanical garden still left him with time to found a couple of periodicals, and the books that he wrote during his exile in Glasgow (for that was how he perceived the time he spent there) included *Exotic Flora* (1823–7) *Icones Filicum* (1827–32), *Flora Boreali-Americana* (1829–40), four editions of the *British Flora* (1830–8), and *Genera Filicum* (1838–42).

Only two years after his arrival in Glasgow, Hooker began enquiring about posts in the south of England, the centre of Britain's scientific life. He desperately wanted to be part of it. Robert Brown and the Secretary of the Linnean Society advised him that he would find the salary of Assistant Keeper of Natural History at the British Museum unacceptable, and other friends assured him that only graduates of Cambridge University would be considered for the vacant post of Professor of Botany there. When he sought enlightenment from Brown in 1824 about a rumour that Aiton at Kew intended retiring, his hopes were dashed when Brown claimed the appointment for himself.[2] Yet another potential retirement offered a chance of returning to England when John Sims contemplated giving up the editorship of *Curtis's Botanical Magazine*. Hooker mistakenly thought the position was worth £300 a year, and eventually did succeed Sims, but only for £100 a year plus £96 for drawings – not enough to induce him to resign from Glasgow. In July 1827 he accepted the post of Professor of Botany at University College London with no requirement to take it up for about two years, then withdrew the following year as the salary was inadequate and there was no botanic garden attached. The vacant Chair of Botany at Oxford in 1834 was unavailable to him, as the Cambridge professorship had been – he was not one of their graduates.

Kew, however, still remained his goal and, if he were to achieve it, Hooker realised that he needed an influential backer. He turned to the sixth Duke of Bedford, whom he had known for some years. The Duke, well-informed on both agriculture and horticulture, had published elegant catalogues of his Woburn collections of grasses, ericas and willows, with botanical descriptions by his able gardeners, George Sinclair and James Forbes. Hooker had already dedicated a volume of *Curtis's Botanical Magazine* to his lordship when he wrote in 1833 soliciting his support. Not only was that unreservedly given but the Duke warmly recommended his protégé to the Duke of Argyll, then Lord Steward of the Royal Household, in whose patronage the directorship of Kew lay. His prospects as a candidate improved when he received a Hanoverian knighthood in 1836 from William IV for his services to botany. There is an unconfirmed anecdote that the King once provocatively asked him whom he considered the better botanist, himself or Dr Lindley. Hooker's discreet reply accorded that distinction to his friend, soon to be his rival for the coveted post at Kew.

Kew's foreman, John Smith, intermittently in touch with Hooker since meeting him in Glasgow in 1832, was the first to alert him to an imminent investigation of Kew. Learning that John Lindley would be a likely contestant should the post become vacant alarmed him. He felt bound to disclose his candidature when Lindley asked him for his support. Since Lindley had a formidable ally in the Duke of Devonshire, Hooker decided to persuade the Duke of Bedford, his son, Lord John Russell, and other acquaintances to present his case to the Prime Minister (Lord Melbourne) and the Chancellor of the Exchequer (Thomas Spring-Rice). His friend George Bentham, Secretary of the Linnean Society, assured him that if the Government "do make Kew a public garden it will be not worth his while [i.e. Lindley's] to take it though it may be worth yours".[3] Only the probability that Lindley would not consider a salary under £1,000 whereas he himself would accept £600 consoled Hooker. Another rival, but of modest stature, entered the arena when J.C. Loudon in the March issue of the *Gardener's Magazine* presented John Smith as "the fittest man in England for the Kew Botanic Garden". As Lindley had not definitely declared his intentions, Hooker introduced a personal appeal into the rivalry.

> I am however quite sure that a more moderate income at that place would satisfy my wishes rather than yours . . . I have reason to believe that my income is much less than yours . . . & I am willing to make some sacrifice to be enabled to return & spend the rest of my days among my friends & connections in England.[4]

Hooker's most active advocate continued to be the Duke of Bedford, determined to see Kew elevated into a national garden superior to the Jardin des Plantes in Paris. A letter from the Duke of Devonshire announcing that he was petitioning the Prime Minister and the Chancellor of the Exchequer on Lindley's behalf dismayed Hooker, and Robert Brown remained ambiguously impartial, reluctant to offend Lindley by favouring his rival.

Hooker, now completely despondent about his chances, seriously considered an offer of a professorial post from the newly constituted Royal Botanic Society, a move cautioned against by the Duke of Bedford who viewed its establishment in Regent's Park as a disincentive to Kew's conversion into a national botanical garden. The Bedford family campaigned vigorously on

Hooker's behalf; in October Lord John Russell saw the Prime Minister at Windsor Castle, and the Duke wrote to the Chancellor of the Exchequer extolling Hooker as being "with the exception of De Candolle, the first botanist in Europe". The Chancellor's reply suggested that the Duke had made a convert but the Prime Minister still remained unconvinced.

In mid-January 1839 Hooker had an interview with the Chancellor of the Exchequer, during which he learned that Lindley's proposals for Kew "which he wished to superintend will not be listened to & the appointment which he had carved out for himself will come to nothing".[5] "I think the salary will not be adequate to Lindley's wishes," he hopefully predicted to his father-in-law, "& I frankly told Mr S. Rice what my present income was & what I would be satisfied with here." His resolve to leave Glasgow was sustained by the University Senate's withdrawal of an annual bounty of £100 and its medical staff's questioning the relevance of botany to their students. He learnt from Spring-Rice before his elevation to the peerage as Lord Monteagle that no financial assistance would be forthcoming for Kew. A rumour that Aiton wished to remain in post deepened his gloom. An encouraging letter from the Duke of Bedford on 16 October 1839 looking forward to Kew being "placed on a respectable footing with yourself as the fittest & most efficient person at the head of it" was followed four days later by the sudden death of the Duke. Sir William had lost a friend and a powerful advocate. Before October was out a letter from Bentham confirmed that Lindley was still in the race and would not decline any reasonable offer.

The disclosure in February 1840 that Kew's plants had been offered to the Horticultural Society of London and the Government's vehement denial in the House of Lords on 3 March offered Hooker no comfort. Reluctant to remain passive, he published a eulogy of the 6th Duke of Bedford, discreetly drawing attention to his lordship's efforts to convert Kew into a national garden.[6] It was, in fact, an exercise in self-promotion and, although privately circulated, Hooker entreated the 7th Duke of Bedford to present a copy to Queen Victoria so "that my wishes may be known to H. Majesty".[7] But he hesitated to send one to Lord Melbourne, still uncertain of his reaction. Now vacillating between hope and despair, he believed that even if Aiton retired his post would go to "a common gardener" – obviously a peevish reference to John Smith. In May Lindley's recommendations were at last published and in June the Treasury ratified the Royal Garden's transfer to the Board of Woods and Forests and Land Revenues. It was hinted that Hooker would be offered the post provided his salary came from a "contingency fund". Lindley, who generously congratulated him, speculated that his disapproval of the dismemberment of Kew in his report had adversely affected his own chances:

> I forever was aware that the stand I made & the opposition I created
> would destroy all possibility of my receiving any appointment.[8]

On 19 November Aiton indicated his desire to relinquish the management of the Botanic Garden by the end of the year, given an assurance that he could remain in his official residence near the corner of Kew Green. He transferred the small Botanic Garden to Hooker's charge on 26 March 1841 but still retained control of the Pleasure Grounds, and insisted on taking with him all the administrative records, a decision since deplored by all historians of Kew. Aiton enjoyed a generous pension of £1,000 a year; Hooker reluctantly accepted £300 a year plus a housing allowance of £200. Such a derisory salary could never have tempted John Lindley, who was Hooker's equal in scholarship, competence and experience. Nevertheless, Hooker was perhaps the better choice, having to a greater degree the charm, the tact and the ability to compromise – vital attributes for dealing with suspicious civil servants. Lindley's role had been to save Kew for the nation and, coincidentally, for Hooker.

The outsider in the contest, John Smith, "looked very gloomy & uncomfortable", so Hooker told his father-in-law, his expectations having been boosted by the *Gardener's Magazine*'s support. No one, however, disputed his competence as a gardener and his promotion to Curator in December 1841 with an annual salary of £130 and a house was well deserved.

Hooker had already chosen his local residence, a plain three-storey house standing in 7½ acres of ground, only ten minutes' walk from the Gardens. It took three months to ship his possessions from Glasgow to London docks where they were transferred to lighters and brought up the river to Kew. Five rooms of West Park, formerly known as Brick Farm, were allocated to his herbarium and another three crammed with books.

With his domestic arrangements well in hand, Hooker took up his new

West Park, Mortlake where Sir William Hooker resided when he came to Kew in 1841.

appointment on 3 April at an age when most men would be anticipating the pleasures of retirement. The Botanic Garden of 11 acres[9] was bounded on its northern side by the gardens of the houses on Kew Green; to the east its limits were defined by the long wall of the Royal Kitchen Garden; on the west the lawns of Kew Palace skirted it; and at its southern extremity lay the Pleasure Grounds, still in Aiton's charge. The gate opened in 1825 after the enclosure of part of Kew Green by George IV still functioned as the main entrance for the public who proceeded along a shrub-lined path past Napoleon's willow, now a mature tree, to the arboretum which still retained many of Princess Augusta's trees within its five acres. The Monkey Puzzle tree (*Araucaria araucana*), known by some as 'Sir Joseph Banks's pine', was still being cosseted during the winter months by a wooden frame covered with mats. Flower-beds, some displaying ferns and mosses on a scattering of rocks, a grass collection laid out in a large bed of four concentric circles, a tank of aquatic plants and a random assortment of stoves and greenhouses occupied the rest of the Botanic Garden. The grass collection was adequately labelled, but the rest could only be identified by checking their numbered tags in a manuscript catalogue.

The ten major glasshouses, heated by fires producing deleterious soot, were in a deplorable state, especially the Palm House where the trees had to be severely pruned to prevent their bursting through the roof. One large house grew the Australian flora which overflowed into three other houses, casually mingling with the flowers of the Cape. The three compartments of Chambers's Great Stove housed specimens from several continents. Only the architectural conservatory, transferred from Buckingham Palace, had modern heating apparatus installed. There was no library, and the only herbarium, consisting of unsorted boxes of dried plants from Bowie and Cunningham, together with Francis Bauer's artistic record of Kew plants, had been moved to the British Museum by Robert Brown; even the Gardens' administrative papers had been taken away by Aiton. Blessed with a resilient nature, Hooker always found some crumb of comfort; "there is really a fine collection of plants," he told his father-in-law.

As he had no board of trustees to advise him nor any mandate from the Commissioners of Woods and Forests, he chose to adopt the main proposals of Lindley's report as his basic objectives. These he interpreted as adequate funding, enlargement of the grounds, an imaginative building programme, and a flexible policy to satisfy the needs of science and the demands of recreation. Beautifying the grounds and presenting popular exhibitions would certainly win public support. Fortunately his First Commissioner, Lord Duncannon, never enthusiastic about Kew's transfer to his department, was soon replaced by Lord Lincoln, and the excellent relations the Director established with him, and with his successor Lord Morpeth in 1846, smoothed the progress of his reforms. Hooker who had impressed both Commissioners with his ability boasted to his father-in-law that he knew how to manipulate them:

> I endeavour to ascertain what the feelings of the Commrs are before I
> carry out any great alteration & carry out to the utmost.[10]

He publicly expressed his appreciation of their cooperation by dedicating a volume of *Curtis's Botanical Magazine* to each of them.

Members of the Royal Family received his assiduous attention. A week before he took up his post at Kew he besought Lord John Russell, son of the late Duke of Bedford, to draw the Prince Consort's attention to Kew.

> Were his Royal Highness to foster an interest in that noble establishment it would soon stand as high in public esteem as it did in the time of George III and Sir Joseph Banks.[11]

Hooker ostentatiously commemorated Kew's royal connections by placing the Queen's coat of arms in the Orangery's pediments in 1842, and added the date of the building, giving it wrongly as 1751 – it was corrected in 1868 to 1761. It is not known when the heraldic shield of Frederick and Augusta was added but it was in place when E.W. Brayley published his *Topographical History of Surrey* in 1850. The shield may have been put there by William IV who according to F. Scheer (1840) added the letter 'A' for Augusta over the two doors. After the Queen had readily sanctioned his application for land in the royal Pleasure Grounds in 1841 and 1842, he confidently forecast that "as far as Her Majesty is concerned I believe I can have every reasonable thing my own way".[12] When Sir William acknowledged a gift of a Japanese plant from the Prince Consort, he took the opportunity of sending his Royal Highness extracts from the letters of Joseph Hooker written on the Antarctic voyage of the *Erebus* and *Terror*. This marked the beginning of his persistent campaigning and scheming for a suitable appointment for his son, preferably at Kew.

John Lindley's vision of a reconstituted Kew Gardens depended, in the first instance, on more space and more glasshouses. These became the new Director's priorities, and before 1841 was over he had obtained Lord Lincoln's approval for the transfer of four acres between the Orangery and the architectural conservatory. By May 1842 he had extracted a promise from the Office of Works that "a noble Palm House" would be erected just beyond the limits of his latest encroachment into the Pleasure Grounds. By mid-1843 his expansion had taken him as far south as the Pond, with ample space for the new Palm House. A high rabbit-proof wire fence now curved in a wide arc from the Unicorn Gate to the private grounds of Kew Palace, a physical division between the Botanic Garden and the Pleasure Grounds. Internal boundary walls were knocked down to open up views, rampant shrubbery thinned, lawns extended and new walks formed. The hardy plants were the first to be considered for an arrangement in systematic order – another of Lindley's recommendations – and a display of the British flora replaced a garden of American plants.

In 1845 the aged W.T. Aiton decided to retire, and thus in early July the remaining portion of the Pleasure Grounds of about 178 acres of grass and woodland, and also the Deer Park (*c.* 350 acres), passed to Hooker's control. The Deer Park, which remained in his charge for only five years, was let out to a local farmer for pasture. He removed a three-quarter-mile-long wooden fence separating the Pleasure Grounds from the Deer Park, substituting a ha-ha and iron railings, thereby opening up a view towards Isleworth. This enlargement of the modest Botanic Garden did not pass without protest from some notable local residents. The Duchess of Cambridge expressed concern about her "fishing-ground", a stagnant pool of water, all that remained of the eighteenth-century lake. Hooker knew how to mollify her but the irascible King of Hanover, formerly the Duke of Cumberland, proved a more formidable opponent. He objected strongly to all these sweeping changes in the Gardens.

> I am sadly pressed for time and sadly annoyed by the King of Hanover, who swore with horrid imprecations that I should *not* carry the contemplated alterations of the Garden into effect. He sent for me but I was perhaps fortunately at Ryde. As soon as I returned I waited upon him but he was gone into London to complain to Lord Lincoln.[13]

William IV had given his brother the Pleasure Grounds and the Deer Park for pheasants and forage in 1831. Although his accession as King of Hanover in 1837 took him out of the country he only gradually surrendered his Kew and Richmond estates: the Deer Park and 44 acres of the Pleasure Grounds in 1844 and much of the rest in 1845, retaining less than eight acres as a pheasant ground; that, too, he relinquished in 1848. Hooker could now proceed without hindrance or censure.

When a kitchen garden was laid out at Frogmore, the Royal Family had no further need for the one at Kew, whose 10 to 14 acres were thereupon transferred to Hooker in June 1846.[14] Demolishing one of its walls, he filled it with a collection of hardy herbaceous plants arranged according to Jussieu's natural classification in irregular-shaped beds.[15] A catalogue of these plants compiled by the foreman, James Niven, was published in 1853 to facilitate exchanges with other botanical gardens.[16] Only the private grounds of the Queen's Cottage and Kew Palace now remained outside Hooker's orbit.

Kew's new management ushered in an era of substantial and significant donations – in 1841 A.B. Lambert's *Opuntia* collection and John Parkinson's *Echinocacti*, formed while he was British consul in Mexico. Although gratefully received, such gifts exacerbated the acute congestion in the glasshouses, making their enlargement and the building of new ones imperative. The doubling in size of a small dry stove and uniting it with an adjacent span house in 1841 initiated an intensive building programme, but financial constraints sometimes compelled the Director to adapt houses he would have preferred demolishing. At the same time efficient hot-water pipes gradually replaced antiquated flues. He gained additional space by despatching what was left of the citrus fruit in the Orangery to Kensington Palace and improved its effectiveness by installing the large windows at either end that had been first contemplated in William IV's reign. He even thought of covering the Orangery with a double-span glass roof to introduce more light. Chambers's elegant building now functioned as a repository for plants too large for other houses.

This orchid house, 80 feet long, displayed the collections of the Duke of Bedford and J. Clowes. It had two boilers with hot-water pipes running beneath the slate shelving. C. M'Intosh: *The Book of the Garden*, vol. 1, 1853, p. 409.

Hooker's father-in-law, the privileged confidant of Kew's progress, learned in early July 1843 that the entire orchid collection of the late Duke of Bedford was destined for Kew. To meet the urgent need of approximately 400 square feet for the orchids, a lean-to house underwent a rapid conversion into a span-roofed house, fitted with stone and slate shelves. Princess Augusta's Great Stove had housed a couple of tropical orchids but the rest of the two dozen described in Hill's *Hortus Kewensis* (1768) were hardy British and European species. Of the 15 foreign orchids listed in W. Aiton's *Hortus Kewensis* (1789), *Encyclia fragrans*, *E. cochleata* and *Phaius tankervilleae* PLATE 9 were among the first to be successfully coaxed into flower. Kew's orchid population had expanded to 115 species, representing 48 genera collected mainly in the West Indies, South East Asia, the Cape and Australia, when the second edition of the *Hortus Kewensis* appeared in 1810–13. Allan Cunningham's contributions from Australia and David Lockhart's from Trinidad during the 1820s were potted and placed on a back shelf of a propagating house, and from 1836 tropical orchids were allocated a separate house. For a long while aerial or epiphytic orchids defied gardeners' expertise. Grown in pots and plunged into a tan bed they might be persuaded to yield a flower before dying. A breakthrough came in 1813 when a gardener at Claremont in Surrey flowered *Aerides odorata* by suspending it in a basket of bark and moss. Sir Joseph Banks succeeded with cylindrical wire baskets, and a lone specimen at Kew fortuitously created its own congenial habitat on the rear wall of a stove. Another variation in cultivation practices was suggested when David Lockhart's epiphytes were received still clinging to fragments of branches.

In 1846 the superb orchid collection of the late Reverend John Clowes of Manchester, who had his own collector in Colombia and Venezuela, joined the Bedford bequest. Despite this spectacular increase in Kew's orchid population their numbers gradually dwindled until by the time of Hooker's death in 1865 there were fewer than there had been 15 years earlier. The essential requirements of temperature, humidity and ventilation had not yet been fully understood by Victorian gardeners and losses were consequently high.

All scientific and charitable institutions in receipt of public funds were required from 1844 to submit annual statements of their activities and expenditure. Hooker's first report, compiled in response to this request, covered the three and a half years since his appointment in 1841. He quoted judiciously from John Lindley's seminal proposals and then paraded his own progress with evident satisfaction: the gradual penetration of the adjacent Pleasure Grounds by his Botanic Garden, the repair and enlargement of ramshackle glasshouses, the employment of a plant collector, and the increasing popularity of Kew Gardens with visitors. But his greatest triumph was the announcement of a new palm house, a "noble structure, which if carried into complete execution, will be second to none in Europe".

Sir John Hill had listed six species of palms in Princess Augusta's garden in 1768; the number had increased to ten by 1789 and to 24 by 1813. A lean-to house, 60 feet long but a mere 16 feet wide, which served as their shelter, had its height extended by another four feet in 1828 to delay the inevitable contact with the glass roof by the more vigorous trees. Some palms were subjected to the moist heat of the tan pit in the Great Stove. Sir Jeffry Wyatville designed a more spacious palm house at the behest of William IV, who chose a site for it half way between the Orangery and the Pond. Wyatville's plan never left the drawing board. When the conservatory from Buckingham Palace was re-erected at Kew in 1836, some of the palms were lodged there. The Kew collection had been overtaken by that of the Loddiges nursery at Hackney. The *Gardener's Magazine* for 1840 urged the Lord Steward's department to emulate the new Great Conservatory at Chatsworth by building a comparable house for the "fine old palms of Kew".

Hooker resurrected Wyatville's plan of a wooden structure and Parliament voted £2,000 in its 1842–3 session and another £3,000 during the following session for a new palm house. When Wyatville's design had been rejected, Mr Robinson, Clerk of Works for the Kew district, submitted his concept of a palm house, 200 feet long, 100 feet wide and 55 feet high with a carriageway through the centre like the one at Chatsworth that Hooker had seen. A model of it was admired by the Queen and the Prince Consort when they visited Kew Gardens in October 1843. They expressed a hope that this splendid building would be visible from Kew Palace, a comment seized upon by Hooker who had been instructed to conceal it among the trees. At the same time the Curator, determined that his own ideas for a palm house should not be ignored, contributed his interpretation of Kew's needs.[17]

The concept and design of plant houses had been slowly evolving over the past two centuries. Early orangeries were simple buildings fitted with wooden shutters during the winter and heated by an open fire, but by the eighteenth century they had been transformed into a desirable architectural acquisition for every aspiring landowner, an elegant repository to show off rare and exotic blooms. Heating flues were inserted into the solid north wall and light came only through large south-facing windows. Glasshouses for kitchen and forcing gardens were purely functional without any architectural pretensions. When the industrial revolution ushered in new building materials, wooden glazing bars were replaced by slimmer wrought iron of greater strength. Such technological changes coupled with investigations into optimum growing conditions fundamentally changed the shape of glasshouses early in the nineteenth century. Now their metal framework could be glazed on all sides. Curved roofs, now a possibility with the use of light and malleable iron, were favoured by Sir George Mackenzie in his lecture to the Horticultural Society of London in 1815. The leading innovator, J.C. Loudon, invented in 1816 a wrought iron sash bar which could be shaped to the contours of these curvilinear glasshouses which were capable of spanning greater distances than had been possible with wood. At his home in Bayswater Loudon experimented with a novel form of roof construction which he called 'ridge and furrow'. He anticipated the massive structures at Chatsworth and Kew in his design for a palm house at the Birmingham Botanical Garden, unfortunately never erected. Loddiges nursery had the largest curvilinear glasshouse in the country, but by the 1830s a proliferation of curved roofs and glass domes confirmed the availability of new engineering skills. Joseph Paxton, the Duke of Devonshire's head gardener at Chatsworth, brought distinction both to his employer and himself with the Great Conservatory, built between 1836 and 1840. He incorporated Loudon's ridge and furrow roofing to span nearly an acre of ground, landscaped with ponds and rockwork. Improved methods in glass technology which enabled the production of curved panes, the repeal of the glass tax in 1845, and a growing competence in building techniques prepared the way for the Palm House at Kew.

Richard Turner, Ireland's leading glasshouse builder with premises in Dublin, was well experienced in curvilinear design when he came to London in January 1844 seeking commissions. While trying to convince the Royal Botanic Society of the merits of his design for a winter garden in Regent's Park, he learnt of Kew's plans for a new Palm House. He promptly petitioned the First Commissioner of Woods and Forests, who referred him to Hooker. Exuding great confidence, three mornings in consultation with the Director did the trick for Turner. "He knows more about hothouses & greenhouses & the best principles of heating them than any man I ever met with," an appreciative Hooker told his superior.[18] With impressive speed Turner submitted plans and an estimate at a meeting at the Office of Woods and Forests to which the architect, Decimus Burton, had been invited as a consultant.

Burton had been the supervisory architect for the Great Conservatory at Chatsworth and was currently architect to the Royal Botanic Society. Versatile and accommodating, he was equally familiar with neo-classical, Italianate Renaissance and the latest developments in metal and glass technology. He had consulted the writings of Sir George Mackenzie, J.C. Loudon and T.A. Knight before designing the curvilinear conservatory at Grove House in Regent's Park; he competently applied ridge and furrow glazing to a glasshouse at Gleavering Hall, Suffolk; he added a domed conservatory at Venn House in Somerset; and had profited from his collaboration with Paxton at Chatsworth. Lord Lincoln had chosen wisely in commissioning him to advise on building and landscaping at Kew.

Only a fortnight after the meeting with the Commissioner, Turner delivered detailed specifications of his design. Burton, while appreciating Turner's technical competence, disapproved of his "Ecclesiastical or Gothic style" which indulged in exuberant ornamentation, needlessly inflating costs. He countered with a plan dictated by the purity of simplified form and, predictably, Turner found fault with it: he argued that Burton's semicircular roof would deny plants precious sunlight during the winter (Turner preferred pointed arches); the design had overlooked ventilation sashes; and trussed arch supports would be needed to compensate for the absence of upright pillars for the aisle roofs. Hooker, prompted by his Curator, desired a central area unimpeded by pillars. Architect and engineer returned to their drawing boards and it was Turner who made the breakthrough by devising a means of spanning 50 feet, using the tensile strength of wrought iron, thus dispensing with supporting columns. More changes and modifications emerged from their consultations before all their differences were resolved and the plan of the building we know today had evolved. Undoubtedly Turner's substitution of lighter wrought iron for cast iron was his most significant contribution; he is also credited with the swivelling sash of the clerestory and gallery levels and the sliding sashes of the curvilinear roof. In addition, he also supervised the changes in the heating system, which eventually took the form of hot-water pipes below an iron grating floor, thus ensuring a more efficient circulation of heat.

The positioning of the 12 boilers below the Palm House, connected with an underground tunnel for transporting fuel and also carrying smoke ducts leading to an adjacent chimney, repeated a similar arrangement at Chatsworth where Burton had been consultant architect. The chimney, 107 feet high, was disguised as an Italian Romanesque campanile, a design Burton had to modify because of costs. Standing 550 feet east of the Palm House, an uneasy presence

Daguerreotype of the Palm House nearing completion. *Kew Library.*

Watercolour of the Campanile, designed by D. Burton, and erected in 1847 by Thomas Grisell who also built the Palm House foundations. *Kew Library*.

in the Gardens, it functioned not only as a smoke stack but also as a source of water. A water tank, located near its summit and filled by a steam-driven pump, provided water under pressure to spray the tallest plants in the Palm House through perforated pipes. Since the smoke from the furnaces never completely dispersed, a chimney was eventually discreetly installed in both wings of the Palm House.

The choice of glass, the protective skin of the Palm House, exercised many minds. Crown glass was rejected in favour of sheet glass, not only stronger but also capable of being manufactured in larger pieces; its unevenness, however, tended to concentrate the sun's rays, likely to cause leaf burn. Blinds or shades did not seem a feasible solution for such a large structure. When John Smith recalled that W.T. Aiton had used 'Stourbridge Green' glass for glazing at Kew, Robert Hunt of the Museum of Economic Geology, an authority on light transmission through different media, was called in. After testing many samples of green glass, he eventually recommended one coloured with copper oxide producing a pea-green tint. Thus, the inherent elegance of the Palm House was disfigured by a green canopy. Unfortunately this was an innovation copied in other glasshouses. Thirty years later, when air pollution effectively screened the sun, the practice lost favour and for a while, during the gradual process of reglazing, the glasshouses at Kew presented a patchwork of conflicting coloured and clear glass.

One cannot say that the Palm House project progressed smoothly and without controversy, certainly not in its early stages. Even its location provoked disagreement. The Commissioners wanted it hidden. The Curator, who had envisaged "a grand walk" from the Temple of the Sun to the Pagoda, favoured flanking this path with the Palm House on one side and a future Temperate House on the other, both sited just north of the Pond.[19] Turner, instructed by Hooker, staked out a site near the other glasshouses. Burton chose a position on the north side of the Pond with the adjacent mound flaunting his free-standing chimney. His disparaging reference to the mound's Temple of Eolus as a "dilapidated summerhouse"[20] suggests that he had its removal in mind. Hooker's later preference for the west side of the Pond, which was scheduled to form an attractive feature in the new landscape, ultimately prevailed.[21]

The Curator pointed out that this site had once been covered by a large lake of which the Pond was a sad remnant, and warned of the likelihood of flooding in the boiler room. The Director and Burton ignored the warning, and one suspects that Smith enjoyed their discomfiture when flooding actually occurred in November 1848. Pumps were in constant use to lower the water level, which by May 1849 had reached the tunnel. A well which was sunk failed to divert the water, and in 1853 the floor of the boiler room was raised. This simply served to reduce the amount of draught to the flues.

In October 1845 the first rib of the Palm House was in place; thereafter it grew slowly, delayed for a while by a strike of the Irish work force, until in October 1848 the last coat of paint was applied. The House outdid the Great Conservatory at Chatsworth in size: 362 feet 6 inches long, 100 feet wide in the central portion, and 63 feet from metal floor to the lantern. Turner had desperately wanted the contract for, as he told one of the Commissioners,

> I *wish* to build my *fame* upon this structure at *Kew*, which *will be* unequalled as *yet*, by very *far* and not likely to be surpassed.[22]

Disregarding the possible financial consequences of such ambition, his losses were serious, at one time bringing him perilously close to ruin.[23] He was even denied the consolation of public recognition of his substantial contributions to the building. The *Illustrated London News* in 1845, the Kew Gardens official guide in 1847 and Charles M'Intosh in *The Book of the Garden* (1853) gave Burton unqualified credit for its construction. As late as 1880 the architect Maurice B. Adams, in a letter to *Building News* for 20 March, defended Burton's reputation as the putative designer. Turner was relegated to the subordinate role of a builder who had merely followed his architect's plans.

However, an examination of all relevant archives reveals how much Burton was indebted to Turner's engineering skills and ingenuity. Burton exercised a classical restraint on Turner's tendency to decorative excess but, thankfully, did not entirely inhibit him. His scrolls and plant forms and the ubiquitous

sunflower motif endow the ironwork with vivacity, even frivolity. The puritanical proclivities of Burton were counterpoised by Turner's instinctive ebullience.

It is perhaps Kew's most beautiful building; seen at sunset on a summer's evening from the far side of the Pond, it becomes an insubstantial bubble enveloping the silhouettes of tall trees, a memorable symbol of Victorian competence and confidence. A few years later Turner received one of the only two 'special mention' awards in the competition for the 1851 Great Exhibition building, and in 1854 he gave Lime Street Station in Liverpool a glass roof with a breathtaking span of 153 feet. But he never again quite matched his performance at Kew.

Even before the workmen had finally left, tropical and subtropical plants were being transferred from the old Palm House, which was then demolished, and from the architectural conservatory. The latter was subsequently filled with

OPPOSITE: Wood engraving of interior of the Palm House where many of the plants were then grown in large pots and tubs.

BELOW: The Palm House, terrace and Pond with the Water Lily House on the right. "One of the boldest pieces of 19th century functionalism – much bolder indeed, and hence aesthetically much more satisfying than the Crystal Palace ever was". B. Cherry and N. Pevsner: *London 2: South*, 1983, p. 510.

specimens of the Australian flora. The largest palms were slowly moved on rollers across the lawn and hauled up the steps of the Palm House with a windlass. The spreading leaves of the cycads helped to fill the central area where all the plants were initially congregated. By mid-August, Queen Victoria had honoured Kew Gardens with three visits, "enchanted" with all that she saw.

Notwithstanding royal approbation and popular acclaim, the Curator remained unconvinced about the suitability of a building which reminded him of a gaunt railway station or "some dockyard smithy". The perforated iron floor especially offended him; a patch of soil at the base of each pillar for climbers and four brick mould pits for the tallest palms were the only concessions he got from the Director. Everything else was placed in a teak tub or a pot until 1854, when Hooker relented, allowing John Smith to remove part of the grating to make three beds, eight feet wide, for more palms. Further panels of the iron floor disappeared during the winter of 1859 for trees that had outgrown their containers. Removed from their constricting tubs, trees flourished. Blossoms opened on palms that had never flowered in Europe before and in 1863 two trees pressing hard against the roof had to be felled. Smith's climbers – *Bignonia*, *Passiflora*, *Piper* and *Clerodendrum* – smothered pillars and gallery rails, and formed floral arches over paths. Shortly after the appointment of Joseph Hooker as Assistant Director in 1855, a programme of vigorous thinning was instituted. The two Hookers, father and son, ignoring the protests of their Curator, ordered climbers to be cut back and other plants which Smith said had "become the pride of the house" suffered a similar fate.

Palm House and the Water Lily House with the Campanile rising above the trees.

The mode of proceeding was Sir W. Hooker would fix his eye upon a plant, ask its history, then say 'Away with it' and in a moment the foreman's big knife made the bark hang in ribbons. This was the signal for the men to break it up and convey it and the box in which it was grown to the rubbish yard. Plant after plant followed in the same way with apparently as much indifference as if they had been common laurels. Sir W. Hooker was very excited on these occasions and did not seem to see the mischief he was doing. I occasionally said 'What a pity' but this had no effect in stopping the knife.[24]

During this time of intensive activity and change, the Gardens were critically appraised by the horticultural press. The overcrowded condition of many of the houses was still deplored although the number of glasshouses, mainly lean-tos and the conversion of peach, grape and pineapple houses in the former Royal Kitchen Garden, had doubled in just under a decade. The new Palm House naturally received unstinted praise and so did the museum but, above all, improvements in public access to the Gardens got the most enthusiastic welcome.

Even when it was still a private royal garden, the King's subjects had clamoured for admittance. Limited access had been permitted during the reigns of George III and IV and William IV, a facility enjoyed largely by local residents and those who could afford the cost of getting to Kew by the Brentford coach to Kew Bridge or, during the summer months, by hired boats. A correspondent to the *Gardener's Magazine* in 1828 despaired of finding any botanical garden in the neighbourhood of London where he could study plants at leisure. Unable to visit the Chelsea Physic Garden, he was put off by the excessive vigilance of Kew's gardeners. From 1825 a forbidding mahogany door on Kew Green had dared hesitant visitors to knock. "You entered unwelcome, you rambled about suspected, and you were let out with manifest gladness shown at your departure."[25] John Lindley, who made that comment, nevertheless courteously acknowledged in his 1838 report W.T. Aiton's determination to keep his Botanic Garden open daily except Sundays. Lindley probably knew that Aiton had resisted a proposal from the Lord Steward to restrict admission to two days a week and then only at the discretion of the Board of Green Cloth.

When Sir William Hooker became Director, visitors were no longer escorted by gardeners, a gesture prompted as much by the need for a more efficient deployment of labour as by a desire to placate public agitation. He maintained Aiton's hours of opening the Botanic Garden from 1 p.m. to 6 p.m. in summer, and until dark in winter but "respectable individuals", unaware of these hours and having travelled some distance, were allowed in before official opening time. By the close of his first year in office, over 9,000 people had passed through the gate on Kew Green. From 1846 Decimus Burton's grand entrance, flaunting the royal coat of arms within a flourish of ornamental foliage, enticed the public along a promenade into an open landscape of lawns, shrubs and flower-beds with the dominant profile of the Palm House taking shape.

People flocked in greater numbers to Kew attracted not only by the reorganisation taking place there but by the improvements in public transport. Steam boats which formerly went straight on to Richmond now had a reason to stop at Kew. In 1846 the London and South Western Railway line reached Richmond, and in 1853 a new line from Willesden to Brentford with a station at 'Kew Junction' brought Kew Gardens within easy reach of North London suburbs. People came to Kew in organised parties, competing with their banners, drums and trumpets. In July 1861, a contingent of the South Middlesex Cadet Volunteers, 200 strong, preceded by their band, marched with their muskets at the slope right up to the Main Gates. On the gate attendant's insistence that regulations forbade music in the Gardens, the cadets left their musical instruments in a heap at the gate, marched into the Gardens and performed their manoeuvres in an unmilitary-like silence.

When the Gardens were transferred to another government department in 1850 the annual attendance figure had reached nearly 180,000. A few metropolitan police officers had patrolled the grounds since 1845 but after the Crimean War army pensioners swelled their ranks to control the flood of Londoners to one of their favourite resorts.

CHAPTER ELEVEN

Kew transformed

Arboretum planned — W.A. Nesfield —
parterres around Palm House — medical garden
formed — increase in flower-beds — Lake
created — Water Lily House — Temperate
House

EVEN AFTER THE RETIREMENT of W.T. Aiton in 1845, public access to the Pleasure Grounds, fenced off from the Botanic Garden, continued to be confined to Thursdays and Sundays from Midsummer to Michaelmas; except for minor adjustments to these times of opening it was not until 1864 that visitors could wander through the new arboretum on any day of the week. An arboretum is the term applied to a collection of trees grown to display species and varieties. Henry Compton, Bishop of London, created what was probably the first British arboretum in the grounds of Fulham Palace. There, during the last quarter of the seventeenth century and the first decade of the next, he assembled an unrivalled collection of trees and shrubs, many imported from North America. Such arboreal collections became fashionable in Georgian England, a particularly notable one being that of the Duke of Argyll whose obsession earned him the tribute of 'tree-monger' from Horace Walpole.

The site of the two royal gardens at Kew and Richmond, a tediously flat piece of terrain, demanded some theatrical gestures in landscaping to relieve its pervasive blandness. So mounds were raised, hollows and slopes were shaped, a stretch of water was added, and curtains of greenery — contrived silhouettes and perspective views — were fashioned with trees. Garden designers had their favourite trees. Charles Bridgeman preferred elms which he regimented in straight avenues, planted thickly on mounds, and insinuated into the woodlands of Richmond Gardens. 'Capability' Brown's repertoire exploited oaks, beeches and elms with occasional birches and Scots pines and the fastidious disposition of a few dominant cedars. As trees at Kew have died, a routine calculation of their annual rings reveal the oldest to be oaks, elms and beeches planted during the reigns of George II and his grandson. Brown sensibly retained some of Bridgeman's original planting when he relandscaped Richmond Gardens. Some sweet chestnuts (*Castanea sativa*) just south of the Lake, a tulip tree (*Liriodendron tulipifera*) in the Azalea Garden, and the Cedar of Lebanon (*Cedrus libani*) terminating a Palm House vista are probably survivors of this eighteenth-century planting.

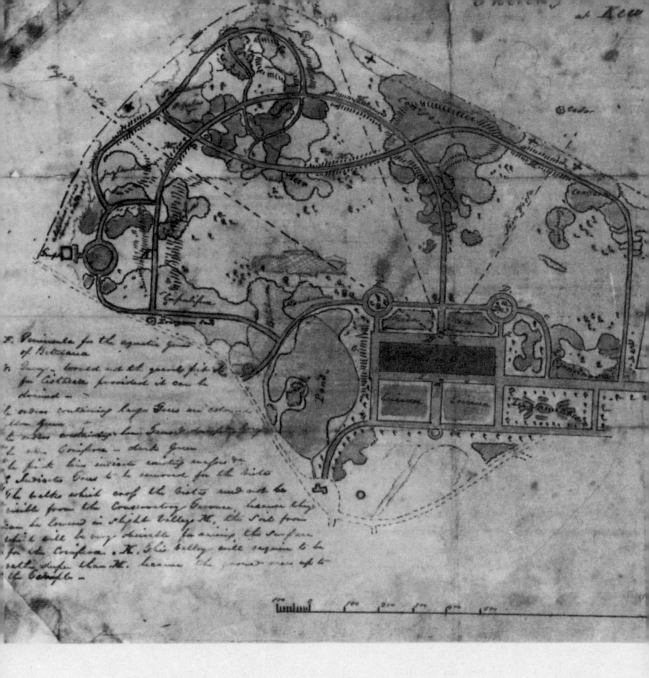

Brown selected trees according to practical and aesthetic principles. The five-acre arboretum in Princess Augusta's Botanic Garden is the Royal Gardens' first attempt to display trees for botanical reasons. Lord Bute's discriminating selection of trees from the arboretum of his late uncle, the 3rd Duke of Argyll, gave distinction to a growing collection which numbered 488 trees and shrubs in 1768.[1] A manuscript record of 1773 lists 791 species. Rarities like the Monkey Puzzle tree (*Araucaria araucana*) from Chile and curiosities like Napoleon's willow added popular appeal. Kew had a fondness for the Cedar of Lebanon. They were generously planted about the Pagoda, and one in the old arboretum unfortunately destroyed the exquisite Temple of the Sun when it was blown down on 28 March 1916. A fine red oak (*Quercus rubra*), one of

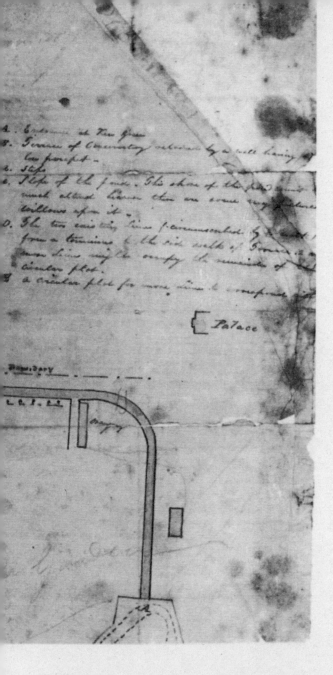

the largest in the country, also a casualty in that violent storm, was discovered to have been 170 years old.

This inexorable disappearance of ancient trees over the years included three notable Scots pines in the southern part of the Gardens in 1942, two old walnut trees near the Broad Walk in 1943, and a magnificent horse chestnut in front of the Orangery in 1951. The *Sophora japonica*, supported by metal crutches, and the disintegrating *Robinia pseudoacacia* are now in the final stages of decrepitude, but the *Ginkgo biloba*, possibly from the Duke of Argyll's garden, still appears likely to enjoy many more years of vigorous growth.

When John Smith, as a newly appointed foreman, took charge of the small arboretum in 1823, it resembled "a large forest crowded with underwood and

W.A. Nesfield's sketch of the proposed Arboretum showing the Palm House north *of the Pond. Dotted lines indicate the vistas. He advised that the Pond "must be much altered". February/March 1844. Kew Library.*

it was the resort of hundreds of birds".[2] *Ailanthus platanus*, *Gleditchia* and
Populus alba, wreathed in mistletoe, soared to over 60 feet and even other trees
reached a creditable 40 feet. Tender trees were trained against the arboretum's
boundary walls until William IV had them demolished on the north and west
sides. Although his intention was to introduce more air and light to this dense
planting, Smith complained that as a result many splendid specimens, deprived
of their brick windbreaks, were blown down. The King never failed to show
his visitors the Monkey Puzzle tree in the centre of the experimental ground.
In 1831 W.T. Aiton tried but failed to obtain the transfer of some acres from
the Pleasure Grounds to enlarge the arboretum.[3] John Lindley in his 1838

report rightly condemned the arboretum as being too small and desperately overcrowded. J.C. Loudon regretted that this historic arboretum now grew only a quarter of what could be seen in the stock of the Horticultural Society of London and Loddiges' nursery.[4] The *Florist's Journal* dismissed the Pleasure Grounds as "simply a park, consisting of open glades, interspersed with clumps and masses of trees, many of them stately".[5] Apart from thinning and some cosmetic treatment, little could be done to Kew's arboretum without additional space. Sir William Hooker finally acquired it from the Pleasure Grounds in the summer of 1843. He ordered trees and shrubs from British and continental nurseries and, so the Curator tells us, prepared a scheme in which specific sites

Nesfield's sketch plan of the Arboretum, 1845. A taxonomic grouping of trees and shrubs was required. The pinetum is placed within the wire fence boundary of the small Botanic Garden. Nesfield had wanted a large oval parterre to terminate Pagoda Vista. *Kew Library*.

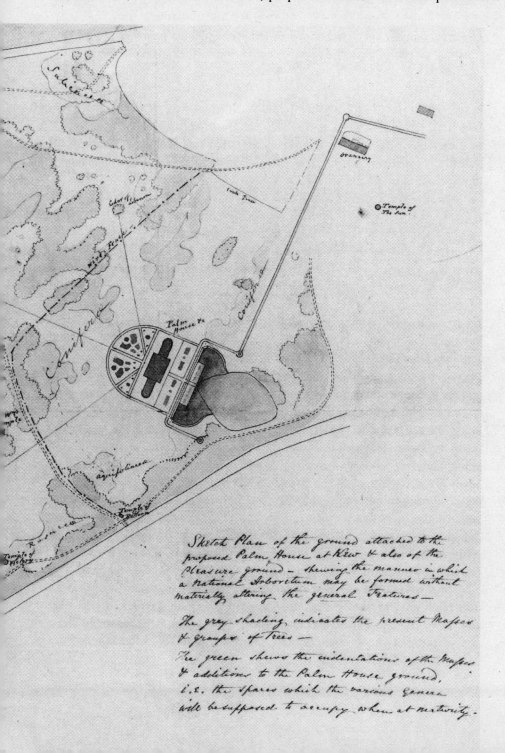

were allocated to genera.[6] The following January the First Commissioner of Woods and Forests told Hooker that Nesfield had been chosen to draft a design for the new Arboretum.

William Andrews Nesfield had deserted watercolour painting for landscape gardening in the 1830s when he designed period gardens for his brother-in-law, Anthony Salvin. He had a predilection for French parterres, then enjoying a revival in Britain, and a fondness for topiary, but no experience in creating arboreta. Although Hooker was impressed by his pertinent observations when they inspected Kew together he still had some reservations about his suitability.

> He perhaps favors too much the formal or what he calls the 'geometrical' arrangement, which to a certain extent, with so noble a piece of ground may be desirable. But I trust he has too much good sense to carry it too far.[7]

For his part Nesfield had second thoughts about a brief which in specifying a taxonomic display of trees might inhibit his landscaping skills. He pleaded for some flexibility,[8] submitting a preliminary plan to Hooker, who predictably objected to Nesfield's "highly ornamental parterre or geometric garden" and to the excessive use of gravel for the walks.[9] Three vistas were to radiate from the west side of the Palm House situated on the *north* side of the Pond. Nesfield proposed to sink those walks which crossed the vistas so as not to mar the view from the Palm House Terrace. He massed trees in families while preserving the "park-like character" Hooker had requested.[10] This plan had to be revised when it was decided that the Palm House should not terminate the principal walk from the entrance on Kew Green.[11]

Nesfield had a competitor when a Chiswick nurseryman, Robert Glendinning, formerly head gardener at Bicton in Devon where he had been largely responsible for laying out its arboretum, submitted a proposal in December 1844. The Board of Woods and Forests offered Hooker, still undecided about Nesfield's fitness for the project, the opportunity of employing Glendinning either as a replacement for Nesfield or as his assistant.[12] Nesfield, clearly annoyed by the Director's lukewarm support, intimated his desire to withdraw.[13] Burton, however, reluctant to lose him, drafted a letter to the Board making it possible to recall Nesfield, but should Hooker still prefer Glendinning, Burton advised him first to consult John Lindley.[14] Unwilling to commit himself, but with the Broad Walk designed (Victoria Walk was the name originally chosen for it), Hooker solicited Burton's assistance in amending Nesfield's plan now that the new site for the Palm House had been approved.[15] Burton agreed but requested the retention of Nesfield "upon whose judgement in these matters I place much reliance".[16] The Board of Woods and Forests concurred, authorising Burton to engage Nesfield.

In May 1845 Hooker heard rumours of Aiton's imminent retirement and of the likelihood of the rest of the Pleasure Grounds passing to his control. To improve the chances of that happening, he got Nesfield "to make a report to the effect that the present Botanic Garden cannot be what it ought to be

PLATE 16 Watercolour of dahlias, attributed to Margaret Meen. Dahlia seeds were sent from Mexico to the Botanic Garden in Madrid in 1798. This drawing may have been done at Holland House in Kensington where dahlias were grown in 1804. Margaret Meen also drew plants at Kew and taught flower painting to the Royal Family. *Royal Botanic Gardens, Kew*

No: 2219 Corypha elata. R.

PLATE 17 *Corypha elata*. Watercolour by an anonymous Indian artist. Encouraged by Sir Joseph Banks, William Roxburgh, a surgeon and botanist in India, commissioned native artists to draw the local flora, especially plants of economic importance. The library at Kew houses over 7,000 flower drawings by Indian artists, of which this is an example.

Tab. 11.

Tusfilago palmata vol. 3 pag. 188.

PLATE 14 *Petasites palmatus.* Watercolour on vellum in 1767 by Georg Dionysius Ehret (1708–70) who based it on a dried specimen brought back from Newfoundland by Joseph Banks. The plant was subsequently introduced by J. Fothergill in 1777 and grown at Kew. Ehret's painting is reproduced in vol. 3 of the *Hortus Kewensis* (1789). *Natural History Museum, London*

PLATE 15 *Paeonia tenuifolia*. Flower collage by Mrs Mary Delany (1700–88). She took up "flower mosaics", as she called them, in her seventies, skilfully reproducing every part of a flower in matching coloured paper, and mounting her compositions on black paper. Her specimens came from well-stocked gardens, and George III instructed Banks to send her floral novelties from Kew which supplied her with plants for 84 of her pictures. *Department of Prints and Drawings, British Museum*

unless the Pleasure Grounds are considered one with it".[17] Nesfield also told
Burton that any "pictorial arrangement" in the Arboretum would require a
further extension of the Botanic Garden.[18] In confidence he informed Burton
that without that enlargement he wanted nothing further to do with Kew.[19]
Hooker immediately wrote a conciliatory letter placating Nesfield, and Burton
was relieved that "so good an understanding happily exists between us".[20]

"Poor Aiton, with tears in his eyes, gave me possession of the pleasure-
grounds on Saturday," Hooker wrote to his father-in-law on 9 July. "In
future not a tree is to be cut down for profit, only when necessary for improving
the beauty of the place."[21] He received Nesfield's 'Report on the formation of
a National Arboretum at Kew' in just under a fortnight.

Nesfield took into account aesthetic as well as botanical considerations. He
decided that the placing of trees in "broad masses and endless detached groups",
the attention paid to "irregularity of outline", and the creation of vistas and
glades need not conflict with the Director's botanical and educational objectives.
He proposed two vistas from the *west* side of the Palm House, one south to
the Pagoda and the other west towards the Thames near Syon House, the latter
to be terminated by an obelisk or some other suitable architectural feature.
When a shorter vista to a Cedar of Lebanon was added later, a goose foot or
patte d'oie was formed. He calculated that this plan, with accommodation for
a thousand trees and space for their growth, would absorb 150 of the 178
acres newly acquired.[22] Some of the sprawling woodland with its ground cover
of brambles, ferns and colonies of wild flowers remained as the King of Hanover's game preserve.

In the autumn, Nesfield presented a large-scale plan consolidating all his
proposals. The staking of walks and vistas began the following February, and
trees scheduled for felling or, as in the case of a Lucombe oak, for removal
to another site, were indicated on a master plan. With this preparatory work
under way, Burton submitted in June 1846 details of the final plan to the
Commissioners. The Director had approved Nesfield's suggestion that all
flower-beds and the pinetum should be placed in the Botanic Garden within
the wire fence boundary. First stages would tackle enlarging the Pond, laying
out the Palm House parterres, opening up the three vistas, planting the
pinetum, and removing undergrowth and dead and worthless trees.

Thinning this dense mass of trees, the majority beech, elm and oak, began
at the Lion Gate, progressing gradually towards the river. This reduction of
woodland revealed that only four forlorn specimens of the once magnificent
cedar grove near the Pagoda had survived. Pagoda Vista, the first of the trio
to be completed, was eventually flanked by clumps of genera in the Rosaceae
and Leguminoseae families. This taxonomic grouping which still dictates the
pattern of planting in the arboretum facilitates the study of related species.
With more than 2,000 species and over 1,000 varieties and hybrids planted,
Hooker felt justified in 1849 in calling it a National Arboretum.[23] Certainly
the horticultural community welcomed it as a national asset and showed its
approval with generous donations: in 1850 the Bicton arboretum gave a comprehensive collection of willows, and two Scottish nursery firms, Lawson of Edin-

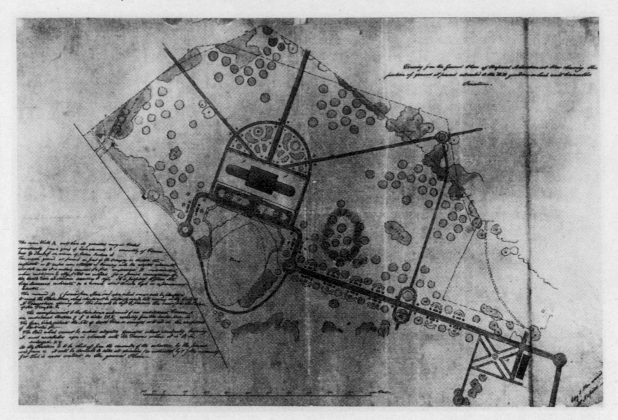

The Broad Walk leading to the Palm House, now on the *west* side of the Pond. Nesfield proposed a plot with diagonal paths in front of the Orangery; flower-beds and Deodars flank the walk; Crab Mound lies just north of the Palm House; the Pond has been reshaped. May 1845. *Kew Library.*

burgh and Turnbull of Perth, added to Kew's holdings of Scottish trees and shrubs.

The radiating arms of Nesfield's three vistas conferred cohesion, space and visual stimulus on the Arboretum. Pagoda Vista, 2,800 feet long, was demarcated by scarlet thorns, mixed with cypresses, juniper and thuja, and backed by deodars, intended eventually to form the permanent avenue. Syon Vista, started in 1851 and completed the following year PLATE 25, extended nearly three-quarters of a mile, a broad gravelled walk, bordered by deodars alternating with limes. Surplus earth from this walk, which was made to a depth of 18 inches with coarse shingle and sand, supplied the core of material for Mount Pleasant at the western extremity.[24]

With the completion of the Palm House in 1848, Nesfield concentrated on the details of his design for the Botanic Garden. During 1845 and part of 1846, a gravelled walk 25 feet wide had been laid down. Decimus Burton's grand processional way, his Broad Walk, starting at his Main Gates on Kew Green, took an obligatory right-hand turn at the Royal stables and the Palace lawn and headed straight for the Pond. As it invited some gesture of floral display, Nesfield placed small clusters of flower-beds on both sides, each cluster backed by a crescent bed (the latter still survive), and connected by a line of *Cedrus deodara*. This uniformity was interrupted on one side by an old Turkey oak which all agreed should not be sacrificed. Nesfield broke the undeviating

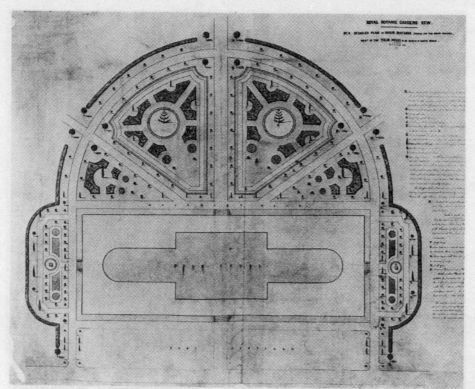

Nesfield's plan of the shrub parterre west of the Palm House, 1848. Two specimens of *Araucaria araucana* dominate the circular beds, and a clipped boundary hedge reaches the north and south parterres. He planned six parterres for the terrace facing the Pond. *Kew Library*.

straight line of the Broad Walk with transverse looped paths half way along its length. Soil excavated during the construction of the Broad Walk formed Crab Mound just north of the Palm House. On it he planted conifers so that the assertive shape of the Palm House remained obscured until visitors reached the large circular flower-bed at the southern end of the Broad Walk; however, a transverse opening was to be made in the trees so that the Temple of Bellona would be visible from the Palace. And it can be no accident that the positioning of the campanile intended it as an eye-catcher at the end of the Broad Walk. Nesfield levelled the uneven ground in front of the Orangery where he thought a plot divided by paths with an *Araucaria* at its centre could serve as a display area for tubs of exotics during the summer.

The mound formed in the early 1750s from soil dredged from the lake had been originally clothed with laurel. Now it was stripped of its tangle of shrubs and small trees, grassed, and the ruinous Temple of Eolus on its summit replaced by a stone replica, designed by Burton, but without its revolving seat. As the destination of the Broad Walk and the hub of the Arboretum, the Palm House deserved an appropriate setting. Nesfield responded to the challenge with intricate parterres, remodelling the Pond to link it with the east side of the Palm House, and providing unimpeded perspective views of his three vistas on the west side. He prepared several designs and consolidated a consensus of opinion in two final plans in January 1848.[25] The two main parterres bounded the east and west sides of the Palm House with linking ones at the north and south ends, an encirclement of symmetrical beds. Box borders and stone kerbs shaped the geometrical patterns; white gravel paths defined areas of grass; each flower-bed was allocated "one kind of plant for the sake of colour"; a parade of evergreen shrubs delimited three large plots in front of the Palm House. This floral structure of the Pond parterre was to be strengthened with 22 large urns or vases which Richard Turner undertook to cast in iron after Nesfield's design. The parterre on the west side was to be encompassed by a semi-circular hedge of holly, box or yew (yew was chosen), formally clipped and never to exceed four feet in height. Openings in this hedge, flanked by standard variegated holly with perfectly straight stems and globular heads, were aligned on the three vistas. Three gravel paths within the hedge perimeter terminated the lines of the main vistas. Flower-beds of dwarf American plants (hence the name of American Garden sometimes given to this parterre) were clustered around two *Araucarias*, one in each half of the semi-circle. A vase on a mound dominated both the north and south parterres which were protected by alternating Portuguese laurel and Irish yew on their grass verges.

One can imagine that Nesfield had some difficulty in converting a doubtful Director to an acceptance of his designs, but in November 1848 he had obtained his approval. Encouraged by this, he then commended the attractions of topiary or "any natural forms which are most quaint & formal . . . the more you ring the changes on the spiral, the round, or the pyramid like Versailles, the better".[26] He tried to convince an unenthusiastic Director that the "Royal Botanic Gardens should be the Head Qrs for every department of gardening art".[27]

The 2½-acre Pond was dredged and deepened during 1846 and 1847, preparatory to widening it at its western end to form a sheet of water the length of the Palm House. In 1847 and 1848 about 200,000 bricks from demolished glasshouses were used to construct the Pond walls and flights of steps.[28] Other surplus materials must have been salvaged as dressed stone has since been discovered in its walls.

Nesfield's vision of flower-beds laid out in complex patterns, of shrubs geometrically clipped, and of huge vases dominating his design was never fully realised. Such incongruity in a traditional botanical garden compelled him to heed Hooker's antipathy. The gardener and writer Charles M'Intosh looked in vain for a bravura display of fountains, parapets and balustrades which he had expected from the leading exponent of formality. Nesfield's skills as a painter influenced his perception of garden design. He attempted to modify the taxonomic order in the arboretum by arranging trees and shrubs according to size, and by manipulating their shapes into harmonious contours. The relationship between Burton and Nesfield was not unlike that between Burton and Turner. In both projects, Burton directed operations, backed by the Commissioners and the Director, while his partners contributed technical and practical experience: with Turner it was engineering skills, with Nesfield horticultural expertise. Burton admitted to the Board of Woods and Forests that

as I do not pretend to a knowledge of arboriculture, my object would
be confined to the suggesting of such general arrangement of the paths,
sites for groups of trees, vistas, etc. as might appear to me to be the
best with regard to convenience and the appearance of the locality . . .[29]

Nesfield worked at Kew under many constraints. The opportunity for greater
freedom of expression came with his next major commission, the designing of
the Royal Horticultural Society's gardens at South Kensington in 1860–1: a
tour de force of intricate parterres of box and gravel, terraces, canals, water
basins and even a maze.

While Nesfield laid out the Arboretum, Hooker smartened up other parts
of Kew Gardens. Rubble from the Castellated Palace, demolished in George
IV's reign, still littered the seven-acre site, hidden behind ugly hoardings. By
1847 the ground had been cleared, levelled, grassed, planted with shrubs and
the river brought back into view. The remains of a venerable elm, reputed to
have been planted by Queen Mary I and a favourite with her sister, Princess
Elizabeth, was left and suggested a name for this new feature – Queen Elizabeth's lawn. Hooker never missed any opportunity to link Kew with its royal
past.

In 1844 Hooker had thought of turning the Castellated Palace site into a
garden for medicinal plants.[30] A year earlier he had been encouraged by
rumours that the Society of Apothecaries might be willing to transfer plants
at the Chelsea Physic Garden to Kew, provided the Royal Botanic Gardens
agreed to cultivate plants of interest to its members. He urged the Board of
Woods and Forests to sanction a medical garden at Kew serving London
University, medical schools and hospitals as well as the Society of Apothecaries.[31] But he also had an ulterior motive which he confided, as always, to
his father-in-law. "The more important the Garden becomes in the eyes of
the Government, the better chance I think there may be of an opening for
Joseph at some future day."[32] Hooker, now in his late fifties, urgently desired
to secure a suitable appointment for his son, if possible as his eventual successor
at Kew. Lord Lincoln, knowing that the Commissioners for Metropolitan
Improvements hoped to construct an embankment between Vauxhall and Battersea bridges, a development that might affect the Chelsea Physic Garden, proposed to the Society of Apothecaries that Kew might accept the cost of
maintaining and expanding their collections if their garden were transferred
to the Commissioners for Metropolitan Improvements.[33] The Society turned
down this offer on the grounds that Kew was too far removed from London.[34]
And with this impasse the matter rested, until the Commissioners for Metropolitan Improvements convened a meeting on 30 April 1845 to consider the role
of the Chelsea Physic Garden "and its present fitness for the objects for which
it was originally established". Kew's Director, well known for his "interest in
the promotion of medical botany", was invited to attend.[35] He left the discussion
confident of Kew's claim to be best qualified as London's medical garden: the
imminent rail connection with Brentford would nullify criticism about Kew's
inaccessibility, and through his extensive international correspondence there

should be little difficulty in importing rare medicinal plants.[36] Lord Lincoln, too, shared his optimism, believing that the contents of the Chelsea Physic Garden would be transferred or the Government would sanction "forming here [i.e. Kew] a medical garden on a national scale".[37] In the event neither possibility materialised. In 1851, when the Queen sanctioned the transfer of the kitchen garden and paddock belonging to the late King of Hanover to the Gardens, Hooker resolved to convert part of this new ground into a medical garden. It came into being two years later, but was not as ambitious as he had hoped.

In his report to Parliament for 1849, Sir William Hooker announced that now his primary objectives for the Gardens had been achieved, in future he would focus on improved maintenance of the grounds and on increasing the collections. These aims, however, were deflected by the consequences of governmental administrative changes. In 1851 the Board of Woods and Forests split into 'Woods, Forests and Land Revenues' and 'Works and Public Buildings'. With the transfer of Kew Gardens to the latter department, the Deer Park remained with Woods, Forests and Land Revenues, passing from Hooker's control. Thirteen acres were transferred from the Deer Park to the Queen's Cottage grounds in 1851 to produce more satisfactory boundaries. Both the Queen's Cottage and Kew Palace remained royal property.

Hooker now reported to Lord Seymour, the First Commissioner of Works and Public Buildings. Although a man with scientific interests and therefore sympathetic to Kew's programme, Seymour had to be scrupulous in approving expenditure since his department did not receive revenues directly. For instance, he refused Hooker permission to purchase items from the combined sale of the two Aiton brothers in the summer of 1851. At the same time he imposed economies, using the fallacious argument that Kew's labour costs were higher than those of Kensington Palace, which had a larger acreage. He also ordered sites being prepared for flower-beds to be restored to grass.

> It is Lord Seymour's intention to make early arrangements for depasturing the Royal Pleasure Grounds with sheep, and it is his lordship's hope, by these arrangements, to effect a material reduction in the amount of your present charge for labour.[38]

Lord John Manners, who followed Lord Seymour as First Commissioner in 1852, introduced a new interpretation of Kew's role, one which his successors, Sir William Molesworth and Sir Benjamin Hall, imposed on a despondent Director. This was Kew as a place of public recreation. "Molesworth is a queer beast & takes so little interest in the gardens that he has never once paid us a visit since he has been in office,"[39] Hooker complained to his botanist friend George Bentham. When Molesworth became Secretary of State for the Colonies in July 1855, he was replaced by Hall, whose Parliamentary Bill in the same year to improve the local management of London had established the Metropolitan Board of Works. This Board possessed the authority to establish parks and open spaces for London's inhabitants. Hall was a prominent figure in the movement to ameliorate the appalling conditions of the capital's deprived classes. Public parks which had their origin in the Report of the Select Committee on Public Walks in 1833 were created as lungs in congested streets. The new First Commissioner saw Kew Gardens as an integral part of this movement, a resort for healthy relaxation. And Hooker could hardly object, at least not in public, since in his annual report for 1847 he had commented approvingly that "Health, pleasure and instruction seem to be the main guiding motives of the visitors to the Botanic Gardens."

Following an unexpected visit to Kew Gardens in July 1856, Hall reprimanded the Director for drab flower-beds, untidy lawns, and neglected walks.[40] Three months later another inspection provoked even stronger criticism. He insisted on more flower-beds, the planting of ornamental shrubs like laburnum and lilac, and added ominously, "it is very probable that I shall require geraniums and other half hardy plants to put out in some of the clumps in St James's Park and Kensington Gardens next year; I must therefore request that you will have ample provision."[41] Hooker protested that the number of flower-beds had been increased (28 according to his annual report), and more were being prepared for the next summer when a larger stock of bedding plants would be available.

> I have a double interest in making the garden gay and ornamental; I admire the effect, and I believe that our thus gratifying the public is the surest way of making the estimates palatable.[42]

Hooker, suitably deferential, nevertheless politely reminded his superior that "I am sure too the importance of this establishment as a botanical garden will not be lost sight of."

This exercise in prettifying Kew was achieved at the expense of eliminating beds devoted to species and by destroying "many old and unique specimens",[43] their places usurped by calceolarias, verbenas, geraniums, fuchsias and other flamboyant blooms. The return of Lord John Manners in 1858 to the post he had held six years earlier offered no promise of a reprieve for Kew which in 1859 had laid out 400 flower-beds and propagated about 40,000 plants to fill them. The appointment of an additional foreman to supervise this seasonal floral display provided the only consolation.

This transformation did not pass without comment during the debate on Civil Service Estimates in the House of Commons on 18 July 1859. Sir Joseph Paxton deplored Kew's metamorphosis "into a gaudy flower show", trying to compete with the commercial venture at Sydenham's Crystal Palace. Lord John Manners, however, "was certain that where one person was interested in the botanical specimens, 100 were attracted by the flowers". The next Commissioner of Works, W.F. Cowper engaged Nesfield to advise on adding colour to London's parks. "Even Kennington Common, which now aspires to the name of Park, has its bordering of flowers, as bright as the smoke and vapour from an adjoining factory will let them be."[44]

At the behest of Sir Benjamin Hall in 1855, Kew also supplied trees to the metropolitan parks.[45] Elm, plane and lime trees were grown for this purpose in a four-acre nursery in the private grounds of Kew Palace. In 1856 4,000 trees and ornamental shrubs went to Battersea Park and Hyde Park, and Victoria Park also benefited from Kew's propagation. In 1860 Kew was required to extend the range of trees it grew. Donald Beaton praised this successful tree nursery in the *Cottage Gardener*: "I never saw such fine healthy-looking and such well-selected stock in all my experience."[46] By 1865 more than 25,000 trees and shrubs had been despatched from this and Kew's domestic tree nursery. But either Kew had been too successful in cultivating trees or demand had drastically fallen off, because early in 1866 it had 10,000 surplus elms. Since commercial nurseries were not interested in purchase or exchange, a public sale of elm saplings took place on 9 March. Thereafter Kew concentrated on evergreens and ornamental shrubs which the parks preferred.

Kew had a so-called American Garden behind the Palm House but another was laid out in the northern part of the Arboretum during the early 1850s. American gardens, fashionable from the late eighteenth century, grew peat-loving ericaceous plants, kalmias, andromedas and the Virginian swamp honeysuckle from North America within a formal framework of grassed walks. It was here, presumably, that Kew's hardy rhododendrons and azaleas were displayed before 'Capability' Brown's Laurel Vale (now Rhododendron Dell) provided a more dramatic setting.

Another hollow, also man-made but not as part of the landscaping, presented Hooker with a chance to have a lake, a feature he had always desired. In 1848 he had told the Board of Woods and Forests that he wanted "an open flow of water through a portion of the pleasure grounds".[47] With gravel still being excavated, much of it for terracing the new Temperate House, the Director formed a stretch of water of 4½ acres, appropriately very near the site of the former duck pond in Queen Caroline's garden. Underground culverts connected with the tidal Thames filled it for the first time in 1861. It sits snugly in the landscape, not visible until one is close to it. The skilfully controlled planting around its margins and the four wooded islands made from surplus gravel, reveal only glimpses of the water from the perimeter path.

The horticultural press generally responded appreciatively to every new building erected, to changes in the landscape and to innovative planting schemes. The *Gardeners' Chronicle* reported such activities briefly, seldom raising objections, but readers of the *Cottage Gardener* received penetrating assessments from Donald Beaton, a regular contributor. As head gardener at Shrubland Park in Suffolk he had made himself an authority on the bedding out of plants in co-ordinated colour schemes. His prompt action in disclosing to the press the threatened disposal of Kew's plants during the late 1830s had endeared him to Sir William Hooker who, he now claimed, had consulted him about floral arrangements.

> They made a very bold start at first, when they began to flower-garden here [i.e. Kew], a few years back, and I gave them all the help I could in kinds of plants.[48]

KEW TRANSFORMED · 183

But his apparently excellent relations with both the Director and Curator in no way inhibited his frank expression of opinion. Insensitive colour clashes, a disregard of graded heights in flower-beds, or a badly sited vase would incur his censure. He particularly disliked the arrangement in the Herbaceous Ground, "huddling together the most ugly plants of all the principal orders, after the manner of a flower garden".[49] Rather than display such a density of different species, he recommended just a few typical of each genus. Joseph Hooker, now deputy Director, had partly heeded his advice when he informed General Munro that

> I have destroyed many plants lately and especially in the hardy her-
> baceous ground where we are resolved to have much larger masses of
> each species and to keep only such species as will grow satisfactorily.[50]

When Beaton came to Kew he inspected everything, the Arboretum as well as the Botanic Garden, exotics as well as hardy plants, but the flower-beds on the Broad Walk and around the Palm House got most of his attention. His observations on the floral displays along the Broad Walk were qualified by knowing that they had never been intended to adorn this wide thoroughfare; it was to have been escorted by a taxonomic grouping of trees in parallel lines on either side – "a splendid and quite a new idea in landscape gardening".[51] Because of Sir Benjamin Hall's insistence on more floral displays, the Broad Walk received the promenade style of bedding – that is the shape of beds and the choice of flowers exactly the same on both sides. Beaton denounced the insipid circle of grass at the Pond end of the Broad Walk.

> What is wanting at the top end of the great walk here is some object
> to hide the lake till you are at the edge of it . . . a large vase with
> flowers in it, or a flagstaff, would form a central object to the eye
> coming up to the walk, but neither of them would hide the lake.[52]

He was therefore happy to report in 1862 that a raised bed, planted in panels of geraniums and verbenas, and prominently capped by a large flower vase, had displaced the offensive grass patch.

The bare grass plots in the centre of two compartments of the Palm House terrace also irritated him. Nesfield had intended them as settings for monumental vases but Beaton preferred the flower-beds which eventually filled them. He recommended the removal of most of the Irish yews and the "lumpy-headed" standard hollies on the terrace. He liked the semi-circular garden behind the Palm House divided into two segments, each identical with an *Araucaria* encircled by beds of calceolarias, verbenas, petunias, dwarf dahlias and geraniums – Tom Thumb and Punch dwarf geraniums were grown to perfection at Kew. "The flower gardens at Kew were the great social question round London all this summer," he wrote during 1859.[53] Not only were the plants impeccably displayed but they were all correctly named.

Princess Mary Adelaide, daughter of the Duchess of Cambridge, discussing the gardens with Craig, the foreman, Sir William Hooker meekly listening. Drawing by Ella Taylor; 'Reminiscences of happy days at Cambridge Cottage, 1856–72', plate 14. *Reproduced by gracious permission of Her Majesty The Queen.*

The flower-beds for the million will astound some people; but everything is done at Kew to attract the million.[54]

His approval embraced the Arboretum where the Queen's Cottage garden had been replanted, a lake was taking shape and the Temperate House was rising. Apart from misgivings about the Palm House – an "ugly building", he thought it – he had nothing but praise for the glasshouses where the plants responded so well to expert care. The conservatory for ornaments (House no. 10) was "the best greenhouse in all England", and the Victoria House was "worth going the length of the Kingdom to see".

Most visitors to Kew made straight for the Victoria House to marvel at the size of the South American water lily. The first European to see it was a botanist employed by the Spanish government in Peru in 1801; a French botanist in Argentina sent dried specimens to Paris in 1828; in 1832 a botanist from Germany found it growing in the Amazon region. Britain provided another location when Robert Schomburgk, travelling in British Guiana on behalf of the Royal Geographical Society, saw it on the River Berbice in January 1837 – "a vegetable wonder", he called it. From his specimens and drawings John Lindley identified it as a new genus and named it *Victoria regia* in honour of the Queen.[55] The *Gardener's Magazine* hoped that the living plant would soon be seen in "an aquarium worthy of Her Majesty" at Kew, but transporting its seeds in a viable state proved difficult. A plant collector in Bolivia, Thomas Bridges, brought seeds packed in wet clay in June 1846. Kew bought 25, of which only two vegetated, formed embryonic leaves and then promptly died. Hooker devoted the entire issue of the first number of *Curtis's Botanical Magazine* for 1847 to this remarkable water lily which, now royally christened, became an even more desirable acquisition for Kew.

If it could be said, in reference to the royal ancestor of Queen Victoria, the Consort of His Majesty George III, that the Strelitzia was peculiarly appropriated to Her, because of the patronage which she gave to Botany, by improving and embellishing the Royal Gardens of Kew, much more does the name of Victoria claim to be handed down to posterity on similar grounds; seeing that Her Majesty has been graciously pleased to make these Gardens available to the public enjoyment, and even to endow them with a liberal provision for that especial purpose.[56]

Three further despatches from South America in 1848 trying different methods of packaging all failed: roots bedded in soil in a glazed Wardian case, dry capsules containing the seeds, and seeds in a bottle of river water. The first of four successful consignments – seeds in small phials of clean water – arrived in February 1849; by the end of March Kew had half a dozen vigorous plants, and by midsummer had raised 50 in a large tank in a former tropical propagation house.

The Duke of Devonshire sent his gardener, Joseph Paxton, to bring one back to Chatsworth where a heated tank in the Great Conservatory had been prepared for it. This had to be enlarged when the leaves continued growing; finally in November 1849 it bloomed and its first flower was presented to Queen Victoria. The Duke of Northumberland at Syon House, another recipient of Kew's seedlings, also flowered it while Kew's stock died without a single blossom. Poor lighting and the impurities of Thames water were blamed for Kew's

Victoria regia (now V. amazonica). Lithograph by W.H. Fitch. Curtis's Botanical Magazine, 1846, plate 4275.

Water Lily House built in 1852 to house the *Victoria amazonica*. Richard Turner provided the ironwork but no evidence has been found that confirms he was its designer.

failure. However, planted in the same tank in 1850, it flowered in June and continued its floral display right up to Christmas. The Director's plan for a house, 100 feet long, with two tanks – one for the *Victoria regia* and the other for aquatics – was vetoed by the Commissioner of Works. Instead, the Commissioner's district manager supervised the building of a span-roofed structure 44 feet square for a circular concrete tank, 36 feet in diameter, with triangular tanks in each corner for other aquatics. At first an extension pipe from the flues of the neighbouring Palm House supplied the heating but a separate boiler was later installed. The structure was finished by December 1852, Richard Turner providing the ironwork and Messrs Bird the foundation and masonry walls.

Although the nympheas and nelumbiums grew well in this new house, it did not suit the *Victoria*. The *Gardeners' Chronicle* in December 1856 blamed the Office of Works for an ill-conceived and poorly ventilated building. In 1858 the giant water lily had an additional home in a square slate tank in one of the smaller houses while much of its former abode was transformed into a tropical habitat of white, blue and red water lilies, ferns, papyrus and hanging gourds.

Sir William was one of several authors who took advantage of public interest, publishing in 1851 a slim folio of W.H. Fitch's bold and assertive lithographs of the *Victoria regia*. Designers exploited its distinctive shape, incorporating it into mantelpieces, chandeliers, gas brackets and household fittings. One enterprising craftsman even marketed a papier-mâché *Victoria regia* cradle.

Kew also coveted another spectacular plant, the Burmese tree *Amherstia nobilis*, judged by Nathaniel Wallich, Superintendent of the Calcutta Botanic Garden, to be unsurpassed "in magnificence and elegance in any part of the world". When the Duke of Devonshire saw it featured in the first part of Wallich's *Plantae Asiaticae Rariores* in 1829, he sent his gardener, John Gibson, to India in 1835 to collect it. Gibson returned with a specimen but this time Paxton's horticultural skills failed to coax it into flower. The three Amherstias received by the East India Company in 1846 and shared among Kew, Syon House and the Horticultural Society of London also failed. Mrs Louisa Lawrence, who received a healthy plant from the Governor-General of India in 1847, was the first to flower it and graciously presented one pendulous inflorescence of vermilion flowers to Queen Victoria and another to Sir William Hooker for illustration in *Curtis's Botanical Magazine*. Her tree came to Kew in 1854 where it continued to flower until its transfer to the Palm House three years later killed it.

The sculptural qualities of succulent plants have always had their admirers – plump cacti, architectural agaves, pebble-like *Lithops* and many others with unusual or bizarre shapes. Shrewdly aware of the publicity value of such plants, Hooker gladly accepted some *Echinocacti* from Mexico during the mid 1840s. One in particular, *Echinocactus platyacanthus*, 4½ feet high, nearly 3 feet in diameter and weighing almost a ton, featured in the *Illustrated London News* in 1846. In 1855 a larger glasshouse enabled Kew to display more effectively its growing collection of succulents.

Hooker now had his sights on a glasshouse for semi-hardy plants – the vegetation of South Africa, Australasia, northern India, parts of South America and the temperate regions of southern Europe. His campaign began with intimations of this aim in his annual report for 1853, repeated regularly ever more persuasively in subsequent years. In 1855 he counselled prompt action if trees dispersed among the Orangery and small houses in the Gardens were to be saved from decapitation, deformity and ultimate destruction. In 1856 he increased the pressure by announcing that one of the two remaining smoke-flued greenhouses had been demolished in anticipation of a new conservatory being built. Persistence and subtle variation on emphasis always distinguished his tactics. In 1857 he declared that only when he had a temperate house would "the national establishment be perfect". A sympathetic editorial in the *Gardeners' Chronicle* for 16 January 1858 may have encouraged him to present specific requirements for a 'Winter Garden' to the First Commissioner of Works.[57] He demonstrated a prudent regard for economy in building it: wood instead of iron for the rafters, sashes of well-seasoned pine, brick instead of ornamental stone, no staircases and gallery, a straight roof rather than a more expensive curvilinear one, an earthen floor rather than the Palm House's perforated iron plates and minimal heating. "We should be quite content with the most simple and unadorned structure, provided it is at once of a form agreeable to the eye and suited to the most perfect cultivation of the plants in question." He calculated that a building designed as a parallelogram, 400 feet long and 100 feet wide, would prevent any congestion of vegetation and a height of 50 feet would contain most trees. At last won over, Lord John Manners promised that if he were still in office in 1859 he would provide for such a house at Kew in the Civil Service Estimates.[58] In the event, the financial vote was presented to Parliament by his successor, H. Fitzroy, but before he had left the Board of Works Manners had commissioned Decimus Burton as architect for the new conservatory. Burton's brief required utility as his primary consideration, the central range to be a parallelogram flanked by two wings capable of being added at any time, and a maximum height of 60 feet.[59] The discussion on the Estimates in Parliament on 18 July 1859 met with some opposition from an unexpected quarter: Sir Joseph Paxton thought the Temperate House could be built for £10,000 rather than the £25,000 being requested.

With Government approval secured, Hooker chose a site in his new Arboretum where the hardy trees and shrubs would be of the "same character", so he claimed, as those in the Temperate House – a somewhat idiosyncratic decision. Joseph Hooker later wrote that although it stood too close to Pagoda Vista, its precise position had been determined by a desire to avoid felling some of the finest beeches, chestnuts and oaks in the Arboretum.[60] Gravel and sand from the new lake under construction formed a podium about six feet high. Burton's design incorporated Hooker's desiderata: a straight sloping roof with windows large enough to admit rain and the free circulation of air during the summer, wooden sashes as better draught excluders than iron and easily repaired by the garden staff, and a decorative stone cornice to conceal functional eaves.

The contractor William Cubitt and Company, which had submitted the lowest bid, employed the structural engineer Jabez James of Lambeth to prepare detailed specifications. Cubitt, with economy in mind, substituted elliptical wrought iron arches placed on cast iron pilasters for Burton's semi-circular arched ribs springing from the floor. He also devised iron slides on the roof for the upper sash windows to lie on the glass lower down. Hooker had been prepared to dispense with a gallery but one was nevertheless installed 30 feet up from the ground. In accordance with Kew's practice, the building was glazed with green-tinted glass.

In 1861, the two octagons below which the boilers stood were erected. In the following year the central range completed all that the Treasury would permit, although foundations for the two wings had been partially laid and some of the ironwork made.

As soon as the octagons were ready the Director lost no time in transferring the contents of the 'New Zealand House' to them. This had been Chambers's 'Great Stove', built in 1761, the repository of the Cape introductions of Francis Masson and James Bowie and, latterly, of the flora of New Zealand. Now a

The Temperate House as
Sir William Hooker
would have seen it —
without the two wings
which were added much
later. *Gardeners'
Chronicle*, 5 August 1876,
supplement 9.

century old, its ruinous state and its position blocking views of the Temple of the Sun from the south made it an obvious candidate for demolition. Its site near the Orangery is now marked by an iron frame supporting the wisteria that once draped its eastern end. Australian plants which had been grown in the architectural conservatory near the Main Gates since 1848 went to the Temperate House as soon as the central range was ready to receive them. This architectural conservatory acquired the name of Aroid House when specimens of Araceae from the rainforests of South East Asia and South America and other tropical climbers filled it. The Orangery's tubs of "unhappy trees", (as the *Gardeners' Chronicle* described them), and some of the palms in the now crowded Palm House joined this migration of stock to the Temperate House. Except for paths, the whole of the spacious interior of the central range was laid out in 20 oblong beds: lines of araucarias, palms, tree-ferns and other tall plants in the centre, side beds of rhododendrons, acacias, camellias, magnolias, etc., and smaller plants in pots and boxes on benches. It was opened to the public in May 1863, two-thirds finished.

Sir William Hooker protested at not having been consulted about leaving the building in its incomplete state. Not only would it be "an eye-sore", but his plan for a large display of colonial plants would not be possible. Playing on his age and poor health, he ended his minute to the Board of Works on a plaintive note: "in the natural course of events, I shall not again be entrusted with so important an undertaking."[61]

CHAPTER TWELVE

Scientific consolidation

*Death of Aiton – Museum – Hooker's
personal herbarium – Director's official
residence – Herbarium housed in Hunter
House – rivalry between Kew and British
Museum*

ON 9 OCTOBER 1849, in his 83rd year, William Townsend Aiton died
in a house on the corner of Kew Green. The plaque in St Anne's church where
he is buried describes him as "late Director-General of all the Royal Gardens".
He had served his apprenticeship under his father, whom he had succeeded
in 1793, and his competence as a garden designer had attracted private and
royal commissions. His candidature for fellowship of the Linnean Society of
London had been sponsored by such distinguished botanists as J. Dryander,
A. Menzies, S. Goodenough, R.A. Salisbury and J. Sowerby. A few short
articles in the *Transactions of the Horticultural Society of London* and his collaboration in the revision of the *Hortus Kewensis* represent his entire literary output.
Whether he performed his managerial functions at Kew efficiently is open to
question. Alexander Anderson in St Vincent complained to Sir Joseph Banks
that "Mr Aiton never writes to me nor tells me his wants as his father did."[1]
Wallich told Sir William Hooker that "poor old Aiton is such a bad correspondent",[2] and his procrastination sometimes irritated Banks. His jealous protection of Kew's resources, a stance he had inherited from Banks, annoyed the
gardening fraternity, but Sir William Hooker apparently had little difficulty
in obtaining plants from him for the Glasgow Botanic Garden, and frequently
acknowledged his donations of material or loan of illustrations for *Curtis's
Botanical Magazine*. He failed, however, to dissuade Aiton from removing all
the official papers on his retirement, nor after his death did Hooker persuade
the Treasury to let him have Aiton's house as a residence for an annual rental
of £100.

John Smith, our sole authority for the fate of Aiton's correspondence and
other papers, including the draft of the revised *Epitome* of the *Hortus Kewensis*,
affirmed that they had been burnt by John Aiton.[3] Only the Bond and
Duncanson flower drawings and 13 volumes of plant records escaped the confla-
gration, and these were presented to Kew by W. Atwell Smith, John Aiton's
heir, who also gave a portrait of William Aiton, then attributed to Zoffany
but now recognised as being by Engleheart. Lord Seymour refused to sanction

Kew's purchase of botanical items in the library of W.J. and J. Aiton, auctioned by Messrs Foster and Son on 4 September 1851.[4]

Sir William Hooker, now in his late sixties, desperately wanted to live nearer the Gardens. In June 1847 he had rejected the offer of Lady Napier's house on Kew Green, which he considered too small. In December of the same year he moved a large part of his herbarium from West Park to a cottage in the Gardens. When Sir George Quentin, former riding-master to George III and his family, died on 15 December 1851, his Georgian house (now 49 The Green) became the Director's official residence. Its stables were demolished and the adjacent former entrance to the Botanic Gardens, now superseded by Decimus Burton's impressive gates, was closed.

Hooker was now more accessible to advise on projects in progress at Kew. One particular aspiration, the formation of a museum, had already been realised. Museums were just one manifestation of a public appetite for knowledge, a trend which promoted the growth of local societies, mechanics' institutes and free libraries. Kew's contribution was a museum of economic botany, the first of its kind in the country. Hooker had prepared the way by presenting Kew with over a hundred drawings he had used in lectures at Glasgow University.

> My object in presenting the best portion to the Garden is to induce Ld
> Lincoln to form a museum in wh such things may be deposited & then
> I think a herbarium will be required, & if once begun it will not be
> on a trifling scale.[5]

A brick building in the Royal Kitchen Garden which served as a fruit store, gardeners' mess room, packing room and residence for a foreman was suggested as temporary accommodation for a museum until a purpose-built structure could be designed. Laying out his own collection of textiles, drugs, gums, dyes and timbers, accumulated over a quarter of a century, on some trestle tables in this building, Hooker invited the Commissioners to inspect them. Having convinced them of their potential value to botanists, manufacturers and merchants, he was authorised in the summer of 1846 to instruct Decimus Burton to adapt part of the building as a museum. The architect modified the central room, installing a skylight and a gallery, and furnished it with glazed wall and table cases. The Secretary of State for Foreign Affairs instructed British consuls abroad to send suitable items to the new Museum, and the Admiralty invited the Director to contribute instructions on collecting specimens to its *Manual of Scientific Enquiry* (1849). His own collection and material donated by the Curator formed the nucleus of the Museum which soon attracted other presentations. The Curator's son, Alexander Smith, who had volunteered to arrange these collections, was officially employed in June 1847 and in 1856 became its first custodian.

Hooker saw the Museum, which opened to the public in 1848, as complementing the living collections in the Gardens by exhibiting examples of products derived from them. He boasted to George Bentham that Queen Victoria

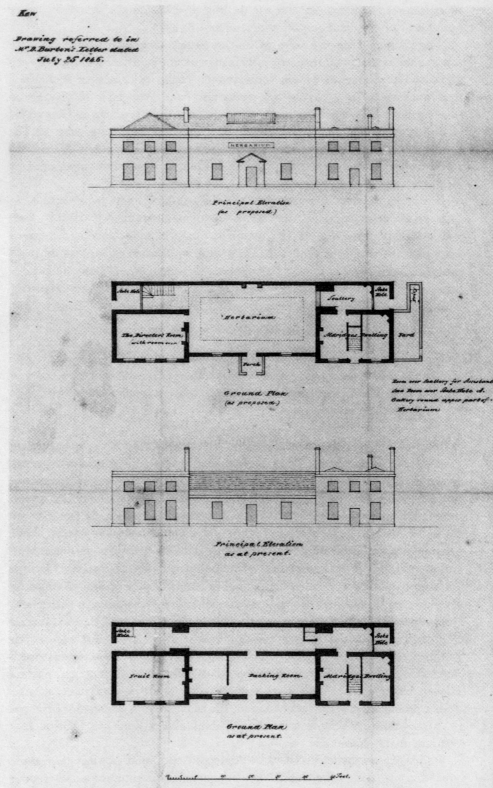

Decimus Burton converted a fruit store into a herbarium which was subsequently made into a museum. This sketch is the only record of the appearance of the original fruit store (third drawing from the top). *Kew Library*.

had shown particular interest in the Museum during her visit in July 1848. John Forbes Royle, late Superintendent of the Saharampur Botanic Garden, and now the East India Company's Correspondent relating to the Vegetable Productions of India, planned a similar institution although restricted to India – a Museum of the Raw and Manufactured Products of India. Others which emulated Kew included the Royal Botanic Garden in Edinburgh, the Royal Institution in Liverpool, and the Jardin des Plantes in Paris. In his 1851 annual report Hooker referred to his creation as a Museum of Vegetable Products; his next report called it a Museum of Economic Botany. It was a principal beneficiary during the disposal of material from the Great Exhibition in 1851 but Joseph Hooker feared that its "great accumulation of scientific objects . . . [were] gradually being consigned to oblivion in favour of showy articles."[6]

It now filled the entire building, exhibits even crowding the corridor. As soon as the Director obtained the tenancy of Sir George Quentin's home, he planned another museum close behind it.[7] Lord Seymour had casually suggested "the centre of the Crystal Palace coming to Kew",[8] an idea perhaps fortunately thwarted by the removal of the entire structure from Hyde Park to Sydenham. In April 1855 the First Commissioner of Works, Sir William Molesworth, chose a site between Kew Palace and Hanover House for the new Museum, a decision rescinded by his successor, Sir Benjamin Hall, who insisted on placing it at the eastern end of the Pond,[9] a location Hooker disliked but meekly accepted. It was designed by Burton in what Nikolaus Pevsner disparagingly described as "utilitarian minimum-classical" style. The *Gardeners' Chronicle* for 1856 was more forthright: "Seen from the road . . . it is frightful, seen from the garden it is mean as well as misplaced." Sir Joseph Paxton thought that "it resembled a third-rate lodging-house".[10] Instead of being an elegant riposte to the dominating presence of the Palm House, it stands as a salutary reminder that even distinguished architects suffer from lapses of good taste.

This additional Museum, officially opened in May 1857, provided almost twice as much space as the old one. The original arrangement of exhibits by commodities – oils, gums, resins, fibres, etc. – was replaced by a taxonomic grouping: Dicotyledons and Gymnosperms in the new building and Monocotyledons and Cryptogams in the old one. Objects were displayed in glazed mahogany cabinets on the three floors of what was now called Museum No. 1, flower paintings, engravings and portraits hung on the walls, while outside on the Pond there floated a canoe made of paper birch by Canadian Indians. International exhibitions continued to be generous donors – for instance, those held in Paris in 1855 and 1867, and especially the one at Kensington in 1862 which added an outstanding collection of colonial timbers and cabinet and furniture woods for which there was no room in either of the two existing museums. Fortunately the Orangery, now empty after its contents had been transferred to the new Temperate House in 1863, provided sufficient space for a Timber Museum.

The Director had originally intended the Museum to house his personal

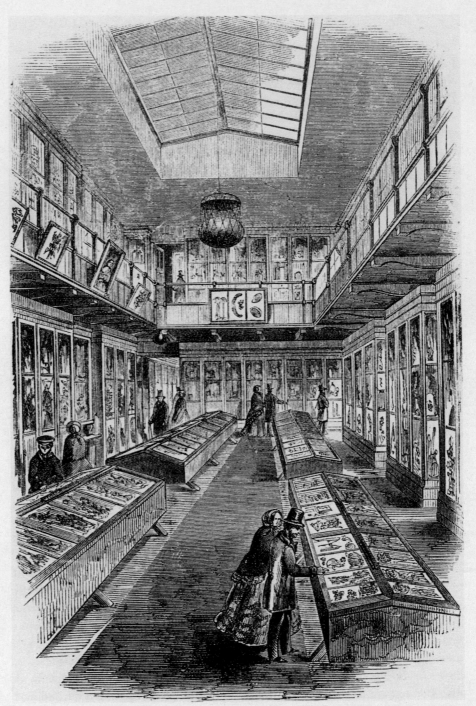

Wood engraving of the Museum's interior. *Museum of Economic Botany; or, a Popular Guide*, 1855, frontispiece. See also plate 22.

herbarium as well. As John Lindley observed in his 1838 report, the identifying and naming of plants depended on "an extensive herbarium and considerable library". When Sir Joseph Banks was alive Kew could rely on access to his herbarium and library in Soho Square, and Robert Brown, who took charge of them after his death in 1820, dealt with W.T. Aiton's taxonomic problems. Lindley's official investigation revealed that during Banks's lifetime all dried plants automatically went to his herbarium, Kew receiving living plants and seeds.[11] Even after 1820, herbarium material from Kew's collectors still ended up at Soho Square, but duplicate sets from James Bowie and Allan Cunningham were received at Kew where they were mounted on paper and shelved in a shed adjacent to Aiton's office. Following the announcement of Sir William Hooker's appointment in 1840, Robert Brown, presumably with Aiton's consent, transferred the lot to the British Museum.[12] The only herbarium Brown did not remove belonged to Aiton who kept it at home in a cabinet in a room which he called his 'museum'. It consisted of plants collected by Robert Teesdale, a gardener at Castle Howard in Yorkshire, bought by Aiton in 1805, and a selection of *Hortus Kewensis* plants. These were sold at a public auction after his death.

With no official herbarium at Kew, Hooker generously made his own available to staff and visitors. Perhaps a pause should be made at this point to

A selection of some of the objects acquired by the Museum: a Tibetan teapot, Sikkimese cups and Indian lacquerware collected by Joseph Hooker; a clay cooking pot and anklet made of seeds from Richard Spruce in the Amazon; an African basket donated by the American botanist, Asa Gray. *Photograph by A. McRobb.*

define a herbarium and its function. Since any botanical garden can only grow a fraction of the world's vegetation, it relies upon a comprehensive repository of 'museum' specimens for study. Gardens of Kew's stature not only cultivate plants but identify and classify them as well. The discipline of taxonomy, or systematic botany, which establishes relationships in the plant kingdom needs a museum of plants – or herbarium – with dried specimens mounted on standard-sized sheets of paper. Such dried plants, which if properly stored and handled can last almost indefinitely, preserve structural features of the shape and disposition of leaves, the inflorescence, and the fruit; any large plants are represented by carefully selected fragments. Specimens collected from different locations demonstrate the geographical range of a species and its variability under diverse conditions. Fleshy or fragile flowers which cannot be pressed and mounted on paper are kept in bottles of liquid preservative, and large fruits

The Museum facing the Pond, built by Burton in 1857.

and seeds are stored in boxes. Specimens upon which the original descriptions of plants are based are known as 'types', of which Kew now possesses several hundred thousands. This precise identification of plants is an essential preliminary to their use in commerce, medicine and agriculture. Kew's pioneering work in publishing colonial floras, initiated by Hooker, could not have progressed without the resources of its herbarium.

The annual allowance of £200 Hooker received from the Treasury had enabled him to rent a house large enough to accommodate his herbarium and library, perhaps the biggest and best in private ownership at that time. Realising that together they represented the most valuable asset he could leave to his son Joseph, Hooker was understandably anxious that they should stay intact, preferably at Kew where he hoped his son would eventually succeed him. The transfer of his herbarium and library to the Gardens became a priority, not only for personal convenience but also to convince his Commissioners that a herbarium was an integral part of the establishment. Decimus Burton's report in June 1846 considered the adaptation of the old fruit store as "a temporary museum for a dried collection of plants, fruits, etc. which should certainly be established in a botanic garden of this eminence" (one detects Hooker's influence here).[13] Burton's plan dated August 1846 provided a central room, designated the herbarium, with the Director's office and a gardener's dwelling on either side. Although the plan was approved by the Office of Woods and Forests in September,[14] a later revision substituted a museum for the herbarium.

During the late summer of 1846 Hooker tried to transfer the expense of maintaining his herbarium to the Gardens' vote but some mounting paper and presses were all he managed to get. "I told them I would be willing to give the herbarium if they would make my son my assistant & successor," he told his father-in-law in confidence. "My offer will be put on record but I dare not anticipate a very favourable result."[15] Lord Morpeth, the First Commissioner, while sympathetic, could not guarantee that Joseph would be appointed Director in due course nor with the same salary. Remarkably resilient, seldom deterred by any rebuff, the Director continued scheming. When his father-in-law wanted to give him his herbarium, Hooker urged him to offer it to Kew to impress the Commissioner.

> Its being offered by you at once to the Garden would tell better for the esteem in which you hold the Garden & Museum as a great public scientific establishment, & is setting a powerful example to others. You may indeed say if you choose it that you have been somewhat influenced by my being here in the donation, as well as by your sense of the value of such an establishment as this.[16]

Naturally Joseph Hooker viewed his father's tactics with acute concern. He had declined the post of Superintendent at the Botanic Garden at Peradeniya in Ceylon in 1849, preferring to gamble on a suitable appointment at Kew. Nevertheless, he had misgivings about his father using his inheritance — the herbarium — as a bribe or bait.

> I do not understand what my mother says about the condition that the
> herbarium is wanted to be thrown open to the public; that sounds most
> unfair; of course I am not prepared to advise, but I firmly believe it
> would be losing my only hold on H.M. Commissioners, it would be
> virtually making the herbarium over to the Crown, they would have
> all the good of it, in perpetuity. I could never touch it, except with
> their sanction, or still less make it any claim to my own establishment,
> or sell it, if driven thereto by any unexpected event, or permanent
> ill-health.[17]

He urged his father to persuade the Commissioners to give him W.T. Aiton's
house in the Gardens, still unoccupied, as a base for sorting his Indian collections, a move that might improve his chances of getting financial assistance
from the Treasury to publish the results of his Indian expedition.[18] Hooker's
request for the house failed, and Joseph and his fellow Indian traveller, Thomas
Thomson, were forced to sort their collections in the discomfort of a long shed
at the back of the Orangery.

When the Director acquired the tenancy of Sir George Quentin's house on
Kew Green, the First Commissioner agreed to build a museum incorporating
an herbarium close to it. In the meanwhile the gift of William A. Bromfield's
herbarium and library in the summer of 1852 highlighted the urgent need for
official herbarium accommodation. It so happened that in 1851, the zoologist
Richard Owen had been offered one of the late King of Hanover's houses on
Kew Green as a 'grace and favour' residence, an offer he declined, preferring
Sheen Lodge in Richmond Park. Probably this prompted the Director to
petition the Office of Works in September 1852 for several rooms in the King
of Hanover's house for Bromfield's collection and part of his own herbarium
and library until the new museum/herbarium was ready.[19]

The King of Hanover's residence on the north side of Kew Green had been
one of two houses purchased by George IV. One of them had belonged to
Peter Theobald, a wealthy local benefactor; after the death of his widow in
1796, its ownership passed to John Maud and then to Robert Hunter. Tradition
has it that Hunter and William Aiton had left Lanarkshire together as young
men to seek employment in England. Hunter became a successful city merchant, maintained his friendship with Aiton, and died at Kew in 1812. George
IV purchased this house from Hunter's eldest son and the adjacent property
in 1820, authorised repairs and improvements, and then sold them to the Office
of Woods and Forests in July 1823,[20] a transaction subsequently forgotten until
many years later.

This sale was certainly not brought to the attention of William IV who,
still believing them to be royal property, presented the houses to his brother,
Ernest, Duke of Cumberland and his wife in 1830 or early 1831. Sometimes
reviled as Queen Victoria's 'wicked uncle', the Duke differed markedly from
his siblings: taciturn where they were garrulous, tenacious where they
were irresolute, and far more intelligent. A cavalry officer in his youth, he
matured into a defiant Tory whose only liberal act was to support women in a

Hunter House, part of which became available in 1852 for the Herbarium. The building on the right is Hanover House.

Parliamentary debate on the Adultery Bill which aimed to prevent the marriage of the guilty partners to a divorce. In defence of the opposite sex he declared that "So few men are inclined to marry the woman they have seduced, that it would be cruel to deprive the females of this last hope." Disfigured by a sabre cut across his face suffered during a battle at Tournai, and with an ill-concealed contempt for the conventions of political behaviour, he became one of the most feared and hated men in the country. He attracted malicious gossip: it was, for instance, widely believed that he had killed his valet; that his wife had murdered her two previous husbands; and that he had enjoyed incestuous relations with his sister, Princess Sophia. Innocent of much of the calumny he had to endure, in 1837, at the age of 66, he ascended the throne in Hanover where his subjects gave him the respect denied him in England. Thereafter his visits to Kew were infrequent but for some years he insisted on retaining his game privileges in the Pleasure Grounds.

On 30 November 1852, the Director learned that he could occupy for a limited period the ground floor of the late King of Hanover's house,[21] temporary accommodation that eventually became permanent. The Director having agreed to public access to the collections, Allan A. Black was appointed curator of the Herbarium in 1853.

The need for this post was fully justified in 1854 with the presentation of George Bentham's herbarium and library, ranking next in size to Hooker's. The two men had met in Glasgow in 1823, became friends, exchanged duplicates as they added to their herbaria, and competed in the purchase of books for their respective libraries. As Secretary of the Horticultural Society of London, Bentham had convened an emergency meeting of its Council in 1839 when the Treasury threatened to disperse Kew's plants. He applauded the transformation of the Gardens under Hooker and in 1845 told him that he had bequeathed his herbarium to Kew in his will, making Joseph an executor. Hooker perceived that such an important bequest might improve prospects of a future appointment at Kew for his son. With covert guidance from Hooker, Bentham disclosed his bequest to the First Commissioner in 1850, making it conditional on adequate accommodation being available at Kew. However, his wife's illness and the expense of maintaining his collections forced him to present them during his lifetime, given an assurance that they would always be freely accessible to him. With official acceptance of his gift in March 1854, he moved from Herefordshire to Westminster, visiting Kew daily, a ritual he observed for some 30 years.

Bentham's and Bromfield's donations formed the nucleus of Kew's official herbarium, but although housed in the same building as Hooker's, they were kept separate. Apart from their scientific worth, Hooker calculated their monetary value at around £10,000. Wishing to be relieved of the financial burden of keeping his own herbarium, he indicated to the Office of Works in November 1854 his willingness to sell it to the State at a fair valuation. This offer was declined, but in 1855 he at last realised his most cherished ambition, the appointment of his son as his assistant. In October Joseph moved into the Georgian house on the south side of Kew Green (no. 55), now a contented man but for the unresolved matter of his father's herbarium. "Daddy and I get on famously in Garden matters," he told W.H. Harvey.

> The destiny of the Herb^m is a mute [sic] point between us. I would, between ourselves, that he left it to the state, or would allow me to give it to Kew after he dies: he likes neither proposition, and looks on it as *good coin* to be left to me; which it can never be. This *entre nous*. The subject has been dropped for years, but is a very serious one to me, for legally the state I fear could claim a good bit, and virtually my destiny is cast with it.[22]

From 1857 the Director introduced the practice of listing in his annual reports the names of distinguished botanists who consulted the Kew herbaria – it was good publicity. For that year they included Dr Grisebach from Germany engaged on the *Flora of the British West Indian Islands*, Dr Nylander from Paris studying lichens, Professor Vriese of Leiden prior to his departure to Java, and J. Kirk before joining David Livingstone's African expedition. J. Lindley, Revd M. Berkeley, T. Moore and W. Mitton were British botanists regularly visiting Kew.

This accessibility of Kew's herbaria to all genuine researchers made the institution a deserving candidate for other collections. In February 1858 Joseph Hooker obtained permission from the East India Company to examine its vast collections of dried plants received over many years from its officials in India and territories further east. In the cellars of its headquarters in the City he found specimens still in unopened chests, others in disorder and nearly all affected by damp and insects, the result of 30 years of neglect. The Company readily accepted his offer to identify and list them, and in July eleven waggon-loads reached Kew, among them the plants collected by William Griffith during the Afghan campaign and in Bhutan, Burma and Bengal, by Hugh Falconer in the Himalayas and Tibet, and by other travellers in northern India. After six years' intermittent work on this daunting task, Joseph Hooker reported to the India Office, the successor of the East India Company, that nearly 400,000 duplicate specimens had been distributed between the principal British and continental herbaria.[23] Among other notable accessions received during Sir William Hooker's directorship must be mentioned Thomas Thomson's Indian plants, William Borrer's British herbarium, the Australian plants of Allan Cunningham, the Mauritius plants of P.B. Ayres, F. Boott's collection of specimens, manuscripts and drawings relating to carices, and W.J. Burchell's immense gatherings made in St Helena, South Africa and South America.

Kew, now established as a major centre in the British Isles for taxonomic research, had only one rival, the British Museum in Bloomsbury. Mutual suspicion, degenerating at times into open warfare, marred relations between the two institutions until the early years of this century. Even Sir William Hooker's amiability and diplomacy made no impression on Robert Brown. The vigilant custodian of the British Museum's botanical collections was possessive, secretive and reticent. "I think I have never seen in the world so grasping a man as Brown," Hooker complained to Dawson Turner after a confrontation with this dour Scotsman. "He grasps at everything & gives literally nothing."[24] At the end of his life, Sir Joseph Hooker remembered, still with resentment, that "During the long struggle for the maintenance of Kew Brown gave no aid, & when my father told him of his candidature, all he said was 'I hope we shall not quarrel.'"[25] Kew's openness, contrasted with Brown's policy of restricted access, won it public support. The *Gardeners' Chronicle* for 22 May 1847 advocated the transfer of the botanical department of the British Museum to Kew, a recommendation prompted by the recent formation of a Royal Commission to investigate the "constitution and government of the British Museum", and, in particular, the relationship between the Department of Printed Books and its other departments. Its interrogation of witnesses considered the consequences of a dispersal of part of the natural history collections. "Do you think the collection of recent plants might be removed to Kew, or elsewhere, without detriment to the other collections?" Sir Philip Egerton asked Professor Richard Owen, the Superintendent of the Natural History collections.[26] Owen's reply was equivocal. He conceded that Kew

> appears to be the most appropriate place for a national museum of that department of natural history; but no great amount of space, I apprehend, would be gained by the removal of the Banksian Herbarium, and a great evil were the present easy opportunity of consulting the eminent botanist in charge of it to be done away with.

The "eminent botanist" (Robert Brown) never evaded answering a similar question. His reply was unambiguous, his opposition perfectly plain. In answer to Monckton Milnes, he dismissed Kew as being "much too distant from the metropolis" for the convenience of botanists.[27] Some scientists signed a memorial rejecting outright the notion of divorcing the natural history collections from other parts of the British Museum. In its issue for 3 November 1849, the *Gardeners' Chronicle*, still loyal to Kew, asked "And why have a botanical museum at all in dirty, smoky London? Why even a herbarium?" Nothing could shake its conviction that the British Museum's botanical collections should be deposited at Kew. Joseph Hooker in Darjeeling heard all about this inconclusive debate with incredulity.

> The idea you mention . . . of possibly removing the herbarium from thence to Kew, has startled me very much; because, of course, Mr Brown *must* object; and how that obstacle could be got over quite passes my comprehension.[28]

The obstacle was not surmounted and the *status quo* prevailed. Brown could not be transferred elsewhere since Sir Joseph Banks's will made him the sole curator of the Banksian collections at the British Museum. Joseph Hooker confided to W.H. Harvey in 1852 that "Lord Seymour talks largely of a great public herbarium & library at Kew & also the Brit. Mus. coming when Brown dies."[29] Sir William Hooker never missed a chance to remind the Office of Works of Kew's superiority.

> If I am asked why it is that these valuable collections are offered here rather than to the British Museum; those I mean of the late Dr Bromfield & of Mr Bentham, the answer is obvious because it is notorious with what liberality everything here is made subservient to the public use.[30]

And the Director's annual report for 1856, in Joseph Hooker's words, "pitched it very strong about the uses of the Herbarium as a scientific adjunct to the Gardens".[31] Loyal *Gardeners' Chronicle*, edited by John Lindley, continued the agitation.

> The public lives in the hope of one day seeing the whole of the botanical collections now at the British Museum transferred to Kew.[32]

Joseph Hooker told T.H. Huxley that the Royal Commission of 1848/9 had secretly resolved to abolish the botanical department as soon as Brown died.[33] Whether or not this was true, plans apparently prepared during Brown's lifetime were put into effect with unseemly haste when he died on 10 June 1858. Just four days later Anthony Panizzi, the Principal Librarian of the British Museum, informed the Treasury that his Trustees had been

> induced to institute an examination into the question whether it might be expedient or otherwise to remove the botanical collections from the Museum as it presents a case in some degree peculiar.[34]

They urgently wanted the space occupied by the botanical collections for other exhibits.

A sub-committee, appointed by the Trustees, held their first meeting on 16 June when both Hookers, Lindley and Owen were questioned.[35] Sir William Hooker, the first witness, recommended the removal of the botanical collections to the cleaner air of Kew where, moreover, the herbarium was larger and better organised. His son and Lindley concurred. Neither did Sir Richard Owen, who acknowledged that Kew fulfilled the functions of a national botanical institute, raise any objections; speaking as a palaeontologist, however, he proposed that both places should have a collection of fossil plants. The only dissident was J.J. Bennett, Brown's assistant, who got a most perfunctory examination. While the omens looked good for Kew, Joseph Hooker revealed his qualms to Huxley.

> I do freely say that we at Kew do not want the Brit. Mus. herbarium here at any price. It is no use to us — & if it be the means of breaking up the Brit. Mus. Nat. Hist. collections or withdrawing support from them, I shall deeply regret its coming here: but as an honest man I must say (with every working botanist), that it is for the interests of botanical science it should come here — it would take 20 years & as many thousand pounds to make the B.M. herbarium anything like ours here & *there are no men to do it*.[36]

But when the sub-committee reconvened, other witnesses — George Bentham, Arthur Henfrey and Hugh Falconer — had doubts about the transfer. Bentham wanted a herbarium of sorts in London while recognising the advantage of housing Banks's herbarium at Kew. Henfrey agreed. After an earnest exposition on the role of botanical gardens, Falconer concluded that both institutions needed a herbarium. Letters from Charles Darwin and Sir Charles Lyell opposed the move to Kew. Taking into consideration that Sir William Hooker's herbarium was still his personal property and that Kew had no permanent building for an herbarium, the sub-committee advised against the transfer.

But T.H. Huxley, not willing to let the matter rest, sought support and signatures for a memorial to the Government. He convinced Darwin, who had been wavering, of the advantages of dispersing the natural history collections among metropolitan institutions. W.H. Harvey, G. Bentham, J. Lindley, A. Henfrey and, of course, T.H. Huxley were among the nine signatories in a memorial to the Chancellor of the Exchequer on 18 November 1858.[37]

The publicity resulted in a Parliamentary enquiry in 1860. Both Panizzi and Owen still supported the transfer but Bennett, now promoted to Brown's post at the British Museum, restated his objections, producing his trump card, an earlier letter from Sir William Hooker conceding that the absence of adequate funding, staff and a suitable building at Kew militated against the move. In 1862 the Trustees of the British Museum finally resolved that the natural history collections must leave Bloomsbury, a decision sanctioned by an

Act of 1878 and eventually implemented in the construction of a splendid Romanesque building in South Kensington, opened to the public in 1881.

In the course of his cross-examination by the 1858 sub-committee, Bentham remarked pithily that a "botanical library is useful without a herbarium, but not a herbarium without the library", an epigram wholeheartedly endorsed by the Hookers. An attempt to establish a library at the Royal Gardens had been made several decades earlier but the precise details are unclear. Joseph Hooker perpetuated a story that has never been authenticated: that when Robert Hunter's house on Kew Green was purchased, Sir Joseph Banks intended part of it to accommodate a botanical library and herbarium; and that it was only his death in 1820 that aborted a project that had already got to the stage of shelving one of the rooms on the ground floor.[38]

What are the facts? The Prince Regent began negotiations to acquire Hunter's house in July 1819.[39] As George IV, following the death of his father on 23 January 1820, he completed the purchase of the house in April or May 1820.[40] Sir Joseph Banks died in June, too soon to obtain permission to convert a room and furnish it as a library. A more plausible explanation is provided by an obituary of Francis Bauer[41] which states that "at the suggestion of Sir Everard Home, George the Fourth resolved to establish a Botanical Museum at Kew". Home had been Banks's physician, and as Francis Bauer's friend, naturally concerned about his future now that his benefactor was dead, he may even have suggested him as librarian of this 'Botanical Museum'. Sir James Everard Home, Home's son, presumably wrote from personal knowledge when he told Brown that

> The shelves for the books had been prepared. All the botanical books of the King's library were to be sent to Kew, and Mr Bauer was to be the librarian with a salary (£200 per an.) All this was unfortunately put to a stop by a dispute respecting the land, the Woods and Forests claiming it as crown property and the idea was given up. Mr Bauer first learnt by seeing the artillery waggons drive off with the books that had been already put there.[42]

Since we now know that George IV had sold Hunter House to the Office of Woods and Forests in 1823, this account rings true. Joseph Hooker's annual report for 1875 mentions that the shelving was still *in situ* at Kew.

As W.T. Aiton, who had his own library, never attempted to form an official one, Sir William Hooker, when he became Director, generously made his own books available. Bromfield's small but well-chosen collection of about 600 volumes and Bentham's library of 1,200 standard texts provided the beginnings of an official library. During the 1850s the Treasury allowed Kew to spend £100 a year on book purchases.

Since accurate flower drawings are an indispensable aid to plant identification, the Director welcomed the presentation in 1854 of about a thousand drawings done by Indian artists for John F. Cathcart during his service in the Bengal Civil Service. In the same year W.A. Smith donated over 2,000 drawings of Kew plants executed by George Bond and Thomas Duncanson. In 1858 Joseph Hooker persuaded the East India Company to lend more than 2,000 drawings by Indian artists under the direction of William Roxburgh of the Calcutta Botanic Garden. All these drawings eventually found their way to the library, which also became the repository for Walter Fitch's many drawings published in *Curtis's Botanical Magazine*.

In his eightieth year, Sir William Hooker drafted a memorandum[43] for the attention of the First Commissioner of Works after his death. It outlined the genesis and growth of his herbarium, amassed over a period of 60 years, and stressed his acute anxiety about its future. He pointed out that his son who would inherit it lacked the financial resources to maintain it, and would be reluctant to sell it piecemeal, an act which "would be reprobated by the whole scientific world". Sir William hoped that the Government would purchase it at a reasonable valuation decided by six eminent naturalists, representing both parties. Naturally he expected his herbarium and other collections to remain at Kew.

After his father's death in August 1865, Joseph duly notified the First Commissioner of his father's wishes, and in March 1866 submitted an independent assessment of the worth of the entire collection: £5,000 for the herbarium, £1,000 for the library and £1,000 for correspondence, manuscripts, portraits and miscellaneous items – a modest valuation, indeed. A memorial was signed by eight distinguished botanists – even Richard Owen advised the Commissioner to accept the offer – and in October, with the Treasury's approval, Sir William's herbarium and library at last belonged to the nation. Many years later Sir Joseph still regretted his father's decision to persuade Bentham to present his herbarium to Kew, thus jeopardising the future purchase of his own herbarium by the State. "The prime object," he told W. Thiselton-Dyer, "was to keep Bentham at work at Kew & this was the obvious means."[44]

CHAPTER THIRTEEN

Overseas activities

Kew's plant collectors — the global transfer of
plants — cinchona introduced to India —
publication of colonial floras — W.H. Fitch
— library facilities and lectures for student
gardeners — death of Hooker

THERE WERE NO KEW COLLECTORS working abroad when Sir William Hooker became Director in 1841. George Barclay had returned to England in February 1841 after five years with HMS *Sulphur* on a survey of the Pacific seaboard of America and the Sandwich Islands. His substantial collection of herbarium specimens went to Robert Brown at the British Museum while all Kew received were seeds, and most of these failed to germinate. With such disappointing results, Hooker had no wish to re-engage him for another mission.

It was not until early in 1843 that he decided to find another collector, prompted, perhaps, by the active collecting policy of the Horticultural Society of London which, after the Treaty of Nanking in 1842, had lost no time in recruiting Robert Fortune PLATE 19 and had him on a boat bound for China in February 1843. Hooker justified his going into partnership with Lord Derby and sharing the costs by arguing that the spoils of one collector "would much exceed what was required for one establishment".[1] So Joseph Burke, who had already collected live animals for Lord Derby in South Africa, went to California via the Rockies in 1843. When a few seeds were all that he could show for three years' absence, the contentious matter of his wages required a legal decision following his employers' reluctance to pay him.

Hooker was better rewarded by William Purdie, whose services were jointly hired by Kew and the Duke of Northumberland. Trained as a gardener under William McNab at the Edinburgh Botanic Garden, his proven competence at Kew made him a promising candidate to explore Jamaica for floral novelties, especially epiphytic plants and cacti. But before the end of the year Hooker expressed disappointment with the paucity of his consignments. Purdie's performance was defended by McNab's son, Gilbert, a doctor in Jamaica. "Sir Wm Hooker has very little idea of the difficulties a person has to encounter collecting in this country."[2] Even greater hardships were endured by Purdie when he went to Colombia in May 1844 to find the ivory-nut palm (*Phytelephas macrocarpa*), used by wood turners to fashion heads for canes and

PLATE 18 Drawing room on the first floor of the Queen's Cottage, at one time also known as the Swiss Cottage. Shaped like a tent, the walls of this room are painted with bamboos, nasturtium and convolvulus. These floral designs were attributed by Edward Croft Murray, Keeper of Prints and Drawings at the British Museum, to Princess Elizabeth on the evidence of her flower decoration at Frogmore. The work was carried out in 1805. *Historic Royal Palaces, Hampton Court*

PLATE 19 "Fortune's Double Yellow Rose". Watercolour by Kew's botanical artist, W.H. Fitch (1817–92). Fitch lithographed his drawing in *Curtis's Botanical Magazine*, plate 4679 (1852). The plant collector, Robert Fortune, claimed he found the rose in 1845 in a Chinese mandarin's garden, its "masses of glowing yellowish and salmon-coloured flowers" covering an old wall. *Royal Botanic Gardens, Kew*

PLATE 20 *(opposite)* Wax model of branches with hazel, cob and filbert nuts, made by Mintorn and purchased by Kew in 1899. The Mintorn family were well known for their wax replicas of flowers and fruit. Kew's museum has other examples of their work and that of other wax modellers.

PLATE 21 Watercolour by W. H. Fitch of cattle grazing in the Old Deer Park. Fitch is better known as a botanical artist but he was also a competent landscape painter in watercolour and gouache. Two of his landscapes which were reproduced in black and white in the *Gardeners' Chronicle* are included in this book.

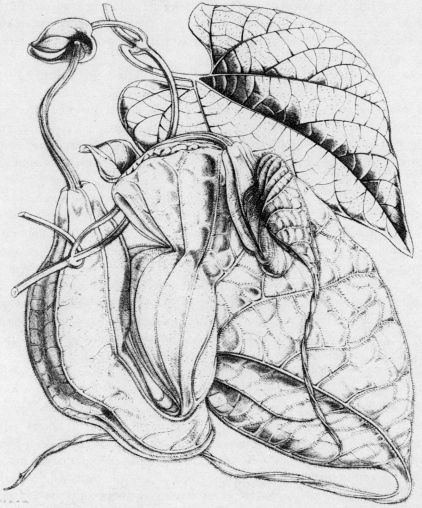

Aristolochia grandiflora
var. *sturtevantii*
lithographed by W.H.
Fitch. *Curtis's Botanical
Magazine*, 1848, plate
4369. This plant was
raised at Kew from seed
sent by William Purdie in
Jamaica.

umbrellas, and as a substitute for ivory chessmen, billiard balls and buttons.
He sent Kew specimens of this and another unusual palm, *Ceroxylon alpinum*,
whose wax coating on its trunk, mixed with tallow, produced excellent candles.
On one occasion during torrential rain in the mountainous interior, one of his
mules, loaded with orchids, miraculously survived a fall over a precipice. In
the meantime the Office of Woods and Forests queried Purdie's expenses,
demanding his recall at the end of 1844. While the Director disputed this
decision, Purdie tirelessly continued collecting, but the question of his future
employment was fortuitously resolved by his appointment as Government Botanist and Superintendent of the Botanic Garden at Port of Spain in Trinidad in
1846. There he remained until his death in 1857, from time to time showering
Kew with floral tributes.

In the West Indies Purdie met another Kewite, Berthold Seemann, a German who had been preparing himself at the Royal Botanic Gardens for his
chosen career as a plant collector. Having been recommended by Hooker for
the post of naturalist on HMS *Herald*, he was en route to Panama where he

joined the ship in January 1847. He sent several cases of Panamanian plants to Kew before the vessel undertook its survey of the American west coast and the Arctic regions. His dried plants were delivered to Hooker when the *Herald* returned to England in 1851.

In 1852, the same ship departed for the South Seas with William Milne, a gardener at the Edinburgh Botanic Garden, as an assistant naturalist. Hooker had contracted him at a salary of £70 a year to collect for Kew, but was disappointed with his efforts: just a few consignments of living plants and some packets of seeds. However Milne had found – on Norfolk Island – the umbrella palm (*Hedyscepe canterburyana*), which became a popular ornamental in this country.

The Indian flora had never been particularly well represented at Kew. Sir Joseph Banks had sent Anton Hove there but collecting for the Royal Gardens had not been his first priority. So in September 1847[3] Hooker sought his First Commissioner's permission to despatch a collector to South Asia for two years, nominating his own son. With the approval of Lord Seymour, who was an amateur scientist, Joseph Hooker, still on the payroll of the Royal Navy, was authorised by Lord Auckland, First Lord of the Admiralty and former Governor General of India, to go to India on reduced naval pay for two years provided he spent a third year in Labuan, North Borneo, assessing its agricultural potential. When the Borneo trip was cancelled, Joseph gained an extra year in India. He reached Calcutta in January 1848 and the first packets of his seeds reached Kew in the autumn. He was an ideal choice: as naturalist on the *Erebus* and *Terror* voyage of 1839 to 1843, he had acquired practical experience in collecting; he was a botanist and competent artist; and his social status guaranteed him privileges and assistance denied to less exalted collectors. Having first acclimatised himself in Bengal and East Nepal, he explored the mountains of Sikkim, surveying and mapping as well as collecting. Kew's annual report for 1850 noted the receipt of 21 baskets of Indian orchids and new species of rhododendrons – "twenty-two of these kinds have already been reared at the Royal Gardens". Joseph named some of his discoveries after friends – *Rhododendron auklandii*, *R. campbellii*, *R. hodgsonii* and *R. thomsonii*, for example. The first species of this spectacular genus to flower at Kew was *R. ciliatum* in 1852. Those species considered hardy were planted in sheltered parts of the Gardens – Queen's Cottage grounds and 'Capability' Brown's Hollow Walk, soon to be known as Rhododendron Dell and still a popular feature with visitors every May. Donald Beaton, who called it the 'Sikkim of Kew', rated the Gardens' display of rhododendrons as being among the best in the country. Tender species cultivated under glass were eventually moved to the new Temperate House. W.H. Fitch, Sir William Hooker's artist on *Curtis's Botanical Magazine*, drew their flamboyant flowers, and 30 of his drawings were reproduced in *Rhododendrons of Sikkim-Himalaya* (1849–51), dedicated by Hooker to his neighbour, Princess Mary Adelaide, "whose taste for the pleasure of a garden . . . has made her feel peculiar interest in the great national establishment at Kew". Seeds of Joseph's more tender rhododendrons, distributed throughout the south west and in western Scotland where they

Rhododendron dalhousiae. Lithograph by W.H. Fitch, based on a drawing by Joseph Hooker. *Rhododendrons of Sikkim Himalaya* (1849–51), frontispiece. Discovered by Hooker in Sikkim in 1848, it first flowered in this country in 1853.

could be safely grown outdoors, contributed to the development of the British woodland garden. But Hooker had an ulterior motive in sending his son to India. He hoped it would enhance his reputation, thereby improving his chances of a post at Kew. A letter to George Bentham makes this abundantly clear.

> . . . Lord Seymour is quite ready to receive my report on what Joseph has done, & I must then show the necessity of his being employed with a salary in the publication of these great collections & it will be my object if possible to get him in the staff of the Garden . . .[4]

The Admiralty, which had made the Indian trip possible, again came to Kew's assistance by providing a berth on one of their ships for Charles Wilford, an assistant in the Herbarium. Having learned that the British Government was to present a steam yacht to the Emperor of Japan, Hooker hoped that this gesture of goodwill might persuade Japanese officials to look favourably on the presence of a plant collector in their country. Wilford was chosen because his competence and diligence had impressed the Director although the Curator had reservations about his experience. In 1857 a troop ship took him to Hong Kong to join the yacht, and after three years in Japan he was to transfer to HMS *Actaeon*'s survey of the Chinese coast. Although his first consignment – herbarium specimens collected in Hong Kong – was well received, Joseph Hooker reminded him that plants, bulbs and seeds were to be given priority. With no response to their repeated instructions and advice, the Hookers expressed growing incredulity and exasperation.

> My surprise & astonishment I need not say daily increases that we hear nothing from you & received nothing from you. Your last letter was your unfinished journal about Formosa . . . Since that, not a parcel, not a seed, not a word . . . Yet you continue to draw bills, without giving us any notice or any account whatever how this money is spent.[5]

Another letter demanding to know why he had sent neither plants nor objects for the Museum threatened his recall. Still no word from Wilford who had in the meantime, so the Hookers learnt from another source, visited China and sailed up the Yangtse river. Although the Admiralty knew that Kew were about to dismiss him, it surprisingly appointed him botanist on the *Actaeon* in 1860. Within a year it, too, was disillusioned with his performance and dispensed with his services.

Wilford was replaced by Richard Oldham, a young gardener working in Kew's Succulent House with some knowledge of botany. In addition to the thorough briefing he received from the Director, the Admiralty reminded him that his duties included searching the Japanese islands for timber trees suitable for shipbuilding. The Director cautioned him to avoid the company of Wilford, still out in the Far East: "his idleness & misconduct have brought him into great difficulties".[6] Like Wilford, he had a three-year contract to explore Japan and the China Seas. By August 1861 he was on board HMS *Actaeon* in

Tokyo Bay. One wonders if Hooker's unfortunate experience with Wilford contributed to his unreasonable, even vindictive, treatment of Oldham. The collector was grieved that "the tone of Sir William's letter is so harsh & unfavourable towards me, & that he makes no allowance for the trouble & disappointment which I have all along had to encounter".[7]

The Director in fact had no grounds to complain of his collections nor of the care he took to pack plants in glazed cases, seeds in soil, and herbarium specimens in stout boxes. In January 1863 he transferred to the *Swallow*, which had taken over *Actaeon*'s surveying duties, having impatiently waited for the ship in Nagasaki since the autumn of 1862. He had wanted to pay a brief visit to Yokohama but Hooker had insisted he stayed at Nagasaki. Invited by the British consul in Formosa to explore the island, whose flora was virtually unknown, he incurred Hooker's censure for accepting without his permission.

> It is unfortunate that from my first arrival in China I have never been allowed to exercise my own discretion as to my movements, & have (entirely on account of this) lost much time & gone to places where I could really do nothing.[8]

Hooker, like Sir Joseph Banks before him, objected to collectors departing from a specific itinerary or a pre-arranged programme even if there were justification for doing so.

When Hooker ordered him back to England at the end of 1863, the Treasury grant nearly exhausted, Oldham expressed a wish for a post of superintendent of a colonial botanical garden or a permanent appointment at Kew. "I regret exceedingly that you should be silent on this subject,"[9] he wrote as he tendered his resignation, still hopeful, however, that the Director would introduce him to nurserymen and gardeners who might commission him to collect for them. Only 26 years old, he died at Amoy in November 1864. Kew's annual report for that year noted that some 13,000 Japanese and Korean dried specimens from Oldham had already been received. *Elaeagnus multiflora*, *Rhodotypus kerrioides* and *Styrax japonica* were among the notable living plants he introduced, but some of his discoveries were overlooked, and later collectors got the credit for their introduction.

Oldham succumbed to dysentery, a fate shared by another gardener, Charles Barter, who had left Kew in 1851 for a foreman's post at the Royal Botanic Society in Regent's Park. With Hooker's influence, he got the post of collector on the Niger expedition under W.B. Baikie in 1857, sending Kew numerous herbarium specimens before his untimely death in 1859. Hooker found a replacement, the German, Gustav Mann, recently taken on as a gardener at Kew. Later the same year he left for West Africa but, unable to join up with Baikie in the interior, collected in the mountain ranges of the Cameroons, and on the island of Fernando Po. He also visited Gabon. Frequent consignments of both dried and living plants reached Kew until his return in 1862; the following year he joined the Indian Forest Service.

As Professor of Botany at Glasgow, Sir William Hooker had incessantly solicited plants for the university's Botanic Garden, even going to the expense of printing a leaflet describing methods of collecting plants which he distributed freely to cooperative travellers. He issued a similar leaflet after he became Director but Kew's collectors still received additional instructions before their departure. A journal was required of them, and all dried specimens had to be accompanied by informative labels that noted the colour and smell of flowers and their habitats – "wet or dry ground, wood, or meadow or cultivated ground". Ornamental plants always featured high on their desiderata lists but those of economic importance were to be sought as well. A small collection of relevant books formed part of a collector's equipment but they did not always answer his needs. Oldham desperately wanted a good Japanese flora: "I am unable to find out the names of plants which I have never seen before."

When Hooker considered the recruitment of collectors in 1843, he calculated the total cost of hiring such a person to be around £600 a year, a sum that included £100 for his salary with possibly another £100 for additional items and contingencies. He appealed to Wilford and Oldham to exercise "prudence, economy and judgement" in their expenditure, but Oldham complained that his allowances were wholly inadequate in Japan.

> Here permit me to state that the *only* way of botanising well in the south of Japan, is to take rooms or a house, get 2 or 3 assistants, instruct them how to help you, & when the most intelligent of them understands the work, get a passport for him from the Customs House, provide him with money, papers, etc. & send him into the interior, trusting to him to select the best localities, pay his expenses as he goes, & let him bring home his collections on pack horses or in the best way he can.[10]

French and Russian collectors who operated in this way were presumably receiving more generous allowances.

Oldham was Kew's last full-time collector. More plants were now being acquired through Hooker's international network of correspondents, and from other botanical gardens. Nurserymen reciprocated the Director's advice on exploration and his naming of their collections with items from their stock. Officials in the British colonies forwarded choice plants or curiosities for the Museum, a trend that acknowledged Kew's growing prominence in imperial affairs. For example, Berthold Seemann's analysis of "the resources and vegetable products of Fiji", a report presented to Parliament in 1862, attracted settlers to those islands. For many years the Indian Army depended on Joseph Hooker's accurate maps of the eastern Himalayas.

An imperial role for a rejuvenated Kew Gardens, co-ordinating the activities of botanical gardens in the colonies "aiding the mother country in everything that is useful in the vegetable kingdom", had been advocated by John Lindley. "From a garden of this kind, Government would always be able to obtain authentic and official information upon points connected with the establishment of new colonies."[11] With the Gardens now enlarged and better equipped and organised, the Office of Woods and Forests sympathetic, the public enthusiastic, and the monarchy supportive, the Director was emboldened to consider the realisation of Lindley's prophecy. He prepared the way by drawing attention to the contributions Kew had made to colonial prosperity.

> . . . there scarcely exists a garden or a country however remote, which has not already felt the benefit of this establishment. All our public gardens abroad – those of Ceylon, Mauritius, Sydney & Trinidad; & cultivators of the soil, Governors of our own colonies, & consuls are supplied with various products of such other climes as may be deemed suitable to them . . .[12]

In his annual report of 1849 he reminded his superiors of his constant endeavours to introduce plants useful "in commerce, in medicine, in agriculture, in domestic economy." Reiterating in his 1850 report that "It has, indeed, been our especial object to cultivate what may be useful and valuable for our colonies," he proceeded to give examples: tussack-grass from the Falkland Islands, "the finest of all grasses for agricultural purposes"; 'Chinese grass' used in calico manufacture; 'Para grass'; deciduous and evergreen beeches of Tierra del Fuego; the 'lace-bark tree' of Jamaica; and African oak or teak, "yet til now unknown to science". In 1851 Kew sent Wardian cases filled with useful plants to New Zealand, Tasmania, India, Trinidad, Jamaica, Sierra Leone and British Honduras. Subsequent annual reports regularly recorded this growing traffic in the despatch and receipt of Wardian cases, boxes of hardy plants, and parcels of seeds. Kew's Museum of Economic Botany, bulging with manufactured goods of vegetable origin from the 1851 and 1862 International Exhibitions, provided a free information service to manufacturers and businessmen.

With Kew's rehabilitation after years of neglect, Hooker was able to persuade shipping companies like the P. and O., commercial bodies like the East India Company, and even the Royal Navy to transport the Gardens' plant cargoes free of charge. Through his readiness to advise on botanical and horticultural matters, Hooker established profitable relations with many colonial governors. Sir Henry Barkly, successively Governor of British Guiana, Jamaica, Victoria, Mauritius and the Cape of Good Hope, became a firm friend. Wherever he served he remembered Kew with donations of plants and seeds and items for the Museum. Hooker, as a token of appreciation, dedicated a volume of *Curtis's Botanical Magazine* to Sir Henry "who, amidst the many arduous duties attendant upon high office, has patronized and encouraged horticulture and botany in our colonies".

The Director was also consulted about senior appointments in colonial botanical gardens. Walter Hill at Brisbane, William Purdie in Trinidad and George Gardner in Ceylon owed their selection as superintendents to his patronage. When Hooker told Dawson Turner that he believed he would have the disposal of the vacant post at the Sydney Botanic Garden, he wondered momentarily whether he should nominate his own son.[13] And he could be peeved when his opinion was not automatically sought.

Earl Grey has appointed his own gardener (as far as I know without
consulting anybody) to the Bot. Garden at Mauritius.[14]

In response to a request from the Governor of South Australia, Kew shipped
out Cork oaks (*Quercus suber*) in 1864; Queensland and Victoria also received
consignments. Varieties of pineapples from Kew stock were cultivated in Brisbane, Ceylon and West Africa.

When Joseph Hooker visited lonely Ascension Island, then a naval station,
in 1843 only one tree had survived an acute lack of water. However, in 1865
the island's commandant informed the Admiralty that it now "possessed thickets
of upwards of forty kinds of trees besides numerous shrubs . . . through the
spreading of vegetation the water supply is excellent, and the Garrison and the
ships visiting the island, are supplied with abundance of vegetables of various
kinds." This transformation owed much to Kew's choice of trees and shrubs
likely to withstand strong sea winds and to its provision of Bermuda grass,
which became the island's staple fodder. Kew also improved the vegetation of
neighbouring St Helena.

The best known and best documented example of the transoceanic migration
of plants in which Kew participated during Sir William Hooker's tenure of
office was the transfer of cinchona from South America to India. Beginning
in the seventeenth century cinchona had been used for treating malaria, and
after two French chemists had discovered how to extract the quinine-yielding
alkaloid in 1820, the plant was coveted by those European countries with
tropical possessions where malaria was prevalent. Sir Joseph Banks had recommended its cultivation in India; in 1816 the surgeon, George Govan, had
designated it a desirable crop for the proposed botanical garden at Saharanpur;
John Forbes Royle made a similar recommendation in his *Illustrations of the
Botany of Himalayan Mountains* (1835). During the 1840s the French received
some cinchona seed from their collector in Bolivia (H.A. Weddell), and from
one of the French seedlings the Dutch tried unsuccessfully to propagate it in
Java in 1851. Late the following year the Dutch had their own man in South
America, and when he succeeded in getting some specimens to Java in 1854,
he was rewarded with the directorship of the world's first cinchona plantation.

The East India Company remained indifferent to the Governor General
of India's plea to send a collector to America, and also to Royle's assurance
"that after the Chinese teas, no more important plants could be introduced into
India". In December 1852 Royle entrusted six precious cinchonas raised from
Weddell's seed to Robert Fortune who was about to depart on a plant hunting
expedition in China. Five reached Calcutta safely; two died there and the three
survivors succumbed in an experimental garden in Darjeeling. In 1858 some
British businessmen in Ecuador urged the Board of Trade to take the lead in
introducing cinchona as a commercial crop in India and the colonies. The
India Office, a new department of state which succeeded the East India Company, now discredited after the sepoy rebellion in India, was more receptive
to the proposition. It was responsible for the health of a large standing army
in India where the annual cost of quinine in the Bengal Presidency alone

exceeded £40,000. Costs might be substantially reduced by growing cinchona in India, and so when one of its junior clerks, Clements R. Markham, offered his services as a plant collector, the India Office unhesitatingly accepted since he had travelled in Peru, could speak Spanish and some native tongues, and claimed to be able to identify several species of cinchona.

Markham was told to consult Hooker before he finalised his plans. In October 1859 the India Office formally invited Kew to participate in the project,[15] and approved Hooker's nomination of Richard Spruce, a British botanist already in South America. Spruce and his companion, Robert Cross, a Kew gardener, were to collect *Cinchona succirubra* in Ecuador; Markham and John Weir concentrated on *C. calisaya* in Peru and Bolivia, and G.J. Pritchett, not a botanist but well-informed on Peru, would search for *C. nitida* and *C. micrantha* in the central regions of the country. Sir William's annual report for 1860 proudly announced his association with "one of the most important horticultural operations in which as Director of this establishment, it has been my privilege to cooperate". Kew's pivotal role was to receive the cinchona seed, and the India Office funded the building of a double-forcing house in the Gardens to germinate them. A few of Markham's first consignment of *C. calisaya* were diverted to Kew but the bulk proceeded to Ootacamund in the Nilgiri Hills in South India where a Kewite, William McIvor, planted and lost them. He had better luck with Spruce's *C. succirubra*, which he nursed into a population of some 2,000 healthy plants by June 1861. The Botanic Garden at Peradeniya in Ceylon established a mountain nursery at Hakgala solely for cinchona, and Kew sent a gardener, Mr M'Nicoll, to take charge of it. Successful plantations were formed at Darjeeling in the Himalayas and in the West Indies but all these species had a disappointingly low quinine content.

Another cinchona species with a much higher quinine yield (although this was not known at the time) had been collected in Bolivia by an Indian servant of Charles Ledger, an alpaca wool trader. Ledger sent a box of its seed to his brother, George, who offered it first to Kew where Sir William Hooker had just died and Joseph was ill. The Curator declined to purchase it but he could hardly be criticised for this action since the Government, having judged the cinchona transfer successfully concluded, had just decided to discontinue "the auxillary establishment" at Kew.

However, the Dutch Consul-General in London bought a pound of the seed which was despatched to Java where the seedlings were in due course grafted on to *succirubra* stock, and *C. ledgeriana* – the name given to the new species – having the highest quinine content of all commercial cinchonas, soon established Dutch supremacy in the world market. For Britain the cinchona venture was not an unqualified success, but it had enhanced Kew's reputation by confirming its dominance in the botanical activities of the Empire, and also its ascendency over colonial botanical gardens which increasingly looked to Kew's director and staff for guidance and support.

During the Victorian era the Kew Herbarium and other major British herbaria gave priority to cataloguing the Empire's vegetation. Its effective commercial utilisation depended upon precise identification, and Kew with its extensive collections of the colonial flora led the way in their global botanical survey. When colonial officials and settlers looked in vain for published lists of plants intelligible to laymen, Sir William Hooker conceived the idea of a series of colonial floras – books enumerating the plants of a region or country. He himself had compiled such a flora under the authority of the Colonial Office – *Flora Boreali-Americana* (1829–40) – based upon collections made in the "northern parts of British America". Joseph Hooker was writing the floras of the Falkland Islands, New Zealand and Tasmania from collections made on his Antarctic voyage. But these were expensive publications, exhaustive in their detail, essentially reference manuals for professional botanists. What Sir William had in mind when he approached the Secretary of State for the Colonies in 1857 was a number of cheap, octavo volumes, "scientific yet intelligible to any man of ordinary education".[16]

For the first volume he selected the British West Indies, whose flora was well represented in the Kew Herbarium; moreover, the botanical gardens in Jamaica and Trinidad could be relied upon to give assistance, and he felt confident that colonists there would welcome such a guide. That his proposal received a sympathetic hearing owed much to his friend, the amateur botanist John Ball, then Under-Secretary of State in the Colonial Office. With the Colonial Office's permission, Dr A. Grisebach of Göttingen undertook the *Flora of the British West Indian Islands* (1859–64). In 1859 Sir William Denison, Governor General of Australia, proposed an ambitious survey of the natural resources of the British Empire.[17] We do not know Sir William Hooker's response when he was asked to nominate a botanist to serve on a scientific commission to be set up to supervise the scheme, eventually abandoned through cost. But sensing the time was propitious, he mentioned his more modest and less expensive objective of a series of colonial floras in his 1859 annual report. He announced that floras of Hong Kong and Australia were being considered and that his project would be of "immense benefit to the Colonies and the Mother Country and raises this establishment [Kew] to a position pre-eminent for practical utility in all that concerns Scientific and Economic Botany, as well as Horticulture". Denison, whose survey would have encompassed botany, zoology, geology and meteorology, was perhaps consoled to be informed that

> extensive collections on the botany of several of the Colonies are in the hands of the accomplished and energetic Director of the Royal Gardens at Kew, and that by order of the Government, publications of some of these Colonial Floras are in the course of being produced, under his superintendence.[18]

George Bentham, a relentless workaholic, readily accepted the challenge of producing a flora of Hong Kong. Sir William had persuaded the Cape Government to contribute to the cost of publishing at least the first volume of *Flora Capensis* (1859), and when the Treasury refused to finance a flora of Australia, his advocacy gained the financial backing of the states of New South Wales, Queensland, Victoria and South Australia. With the *Flora Hongkongensis* completed in 1861, Bentham retired to a room in the Kew Herbarium where, with the indispensable cooperation of F. von Mueller at the Melbourne Botanic Garden, he tackled *Flora Australiensis*, a monumental work in seven volumes, "the first flora of any large continental area", according to Joseph Hooker.

Joseph still cherished an ambition to continue his flora of India, the first volume of which had appeared in 1855,[19] only to be abandoned when the East India Company declined to continue its sponsorship. With the support of Sir William Denison, now Governor of Madras since 1861, Sir William Hooker proposed to the India Office a ten-volume flora of British India, based largely on Kew's Herbarium, with his son and Thomas Thomson as joint authors. In 1863 the India Office not only agreed to subsidise its publication but also guaranteed the purchase of 100 copies of each volume. In the same year the Director gave the Colonial Office a résumé of what had already been achieved in producing these floras, and recommended other colonies for a similar treatment.[20] Thus Sir William pioneered a tradition in the publication of overseas floras, still a primary activity at Kew today.

The Director scrupulously reported Kew's involvement in colonial enterprises in his annual submissions to Parliament. He also gave it prominence in *Curtis's Botanical Magazine*. The nurseryman, William Curtis, who had launched it as a monthly periodical in February 1787, aimed it at "such ladies,

Walter Hood Fitch (1817–1892). Botanical artist.

Scaevola sericea. Curtis's Botanical Magazine, 1827, plate 2732. Athough the plate is inscribed 'W.J.H. del' (*i.e.* Sir William Hooker), the drawing is, in fact, by the Kew gardener, George Bond. Bond drew it at Kew in 1826 from a specimen grown from seed sent by Allan Cunningham in Australia in 1824.

gentlemen and gardeners, as wish to become scientifically acquainted with the plants they cultivate". Each issue featured three hand-coloured engravings of attractive hardy and exotic plants, many of them recent introductions. When Hooker became editor, and for a while its artist, he extended its range of plants to include a few of economic importance. In 1827, his first year as editor, 13 of the plates were derived from drawings done at Kew, probably by the gardeners George Bond and Thomas Duncanson. While he held the chair of botany at Glasgow, he had recognised the potential talent of Walter Hood Fitch, an apprentice pattern-drawer in a local calico mill. He trained Fitch as a botanical artist PLATE 19, employing him as an illustrator for *Curtis's Botanical Magazine* from 1834, and when he migrated to Kew in 1841 Fitch came with him. Fitch's situation was not unlike that enjoyed by Francis Bauer who had been employed by Sir Joseph Banks; he was, in effect, the Gardens' resident artist but his salary was paid by Hooker and not by the State.

Extracts from Hooker's international correspondence often enlivened the text in *Curtis's Botanical Magazine*: experiences or botanical observations of a plant collector, a superintendent of a botanical garden, a government official or a member of the armed forces. Shortly after taking up his post at Kew, Hooker's relations with the proprietor of the magazine deteriorated so much that he seriously considered resigning as editor and launching a rival periodical with a title such as *Kew Gardens* or *Plants of Kew*. Fortunately a change of ownership coincided with a new series of the magazine in 1845 which linked its future with Hooker's management of Kew. This was confirmed in a new subtitle: "comprising the plants of the Royal Gardens of Kew, and of other botanical establishments in Great Britain." A vignette of the new Palm House, then under construction, became its logo. Hooker resurrected the slim supplement he had initiated in 1835 – *Companion to the Botanical Magazine* – and during the four years he continued it, he wrote about the history of Kew, made some additions to the *Hortus Kewensis*, reported William Purdie's botanical mission to the West Indies and America, and listed the latest floral discoveries in Australia, Ceylon and Borneo. The Curator, who contributed a list of ferns cultivated at Kew, also from 1849 added cultivation notes to the taxonomic descriptions of the plants illustrated in the magazine. Subsequent editors of *Curtis's Botanical Magazine*, usually the director or a senior member of staff at Kew, followed Hooker's precedent of selecting many of their choice plants from Kew's stock.

Hooker was a compulsive editor, the progenitor of several new periodicals: *Icones Plantarum* in 1837, *Botanical Miscellany* in 1830 which progressed through a sequence of titles – *Journal of Botany* (1834), *London Journal of Botany* (1842) and *Hooker's Journal of Botany and Kew Garden Miscellany* (1849). When he had contemplated a change of title in 1848, he turned to his father-in-law for advice.

> I wish you could help me to a title of a sort of Journal of Botany in intimate connection with Kew Gardens. I have thought much of what you have lately said respecting a more popular Journal, & how far I could introduce matter relating to our more curious & instructive living plants of the Garden, & almost endless objects now likely to be brought together in the *Museum* of vegetable origin – histories of Caoutchouc, Gutta Percha – & palm and willow platting, the innumerable kinds of paper of vegetable substance, etc., etc. to replace the drier monographs of genera, etc. of the former Journal [*London Journal of Botany*]. That Journal will be given up unless I consent to pay for the plates. I think it *possible* that the Woods & Forests may do this if I make it a *Kew Garden book*. I have indeed made the application on the ground that it is there I want to publish an account of what Joseph is doing & of the objects he will (we hope) introduce to Kew [from India].[21]

He wondered whether to call it *Annals of Kew Gardens* or just *Kew Gardens* but finally settled for *Hooker's Journal of Botany and Kew Garden miscellany*. Before it ceased in 1857 he had published data on the 'vegetable ivory palm', 'rice paper plant', 'African oak', 'Eboe nut', 'Tallow-tree and insect wax of China', 'Camphor-tree of Borneo', 'Soap plant of California', ferns of Hong Kong, recent acquisitions to the Museum, and several obituaries. It may have inspired William Thiselton-Dyer when he started a *Bulletin of Miscellaneous Information* in 1887.

Sir William evidently appreciated the value of publicity, turning it to his

Sir William Jackson
Hooker (1785–1865).
Marble bust by Thomas
Woolner, 1859.

own and Kew's advantage. Instructed by the Office of Woods and Forests to
produce a guide book, he promptly complied. *Kew Gardens, or a Popular
Guide to the Royal Botanic Gardens of Kew* (1847), attractively decorated with
wood-engravings by Fitch, had exhausted 21 revised reprints by 1863. The
first guide to exhibits in the Museum appeared in 1855, again illustrated by
Fitch.

Obviously the Director's aims and achievements relied upon a competent
and supportive staff, among whom the gardeners were a key component. Kew
had never lacked applicants who regarded a horticultural apprenticeship in a
royal garden as a desirable qualification. Recruits like John Smith, with more
than a basic knowledge of botany, were exceptional; most came from the county
establishment of farmers, village tradesmen, gardeners and gamekeepers

interested only in the practical side of horticulture. A few were Germans drawn to Kew through the Royal Family's connections with Hanover. When Smith came to Kew in 1822, no library facilities existed for gardeners who wished to study, although W.T. Aiton would occasionally lend his own books. J.C. Loudon, who passionately believed that every young gardener should have access to a library during his training, regretted its absence at Kew;[22] and the 1838 report on the royal gardens recommended that each one should be equipped with a small collection of books. After his arrival in 1841, Sir William Hooker ensured at least a few basic texts were available and five years later the Office of Woods and Forests voted £10 for books, later increased to an annual sum of £20. The conversion of John Smith's former cottage in 1848 provided a reading room for the gardening staff in the evenings. Gardeners were also allowed half an hour after breakfast solely to familiarise themselves with Kew's enormous variety of plants. In 1853 the First Commissioner proposed a course of summer lectures, primarily for gardeners, but open to the public as well. Nothing appears to have been done about this until 1859 when Daniel Oliver, a young botanist recently taken on by the Director, volunteered to lecture gardeners on elementary botany, chemistry and meteorology. After 1861 these were held in a new reading room adjoining the Director's Office. Most of these student gardeners left after 18 months for posts in public and private gardens, at home and abroad – Kew's annual report for 1850 mentions destinations as far away as the West Indies, South Africa, India and Australia.

In 1864, suffering from failing eyesight, John Smith retired after 42 years' service, more than half of them as Curator. The acute disappointment Smith must have felt when, on W.T. Aiton's death he failed to be appointed Director, certainly affected his relations with Hooker. That his judgement and experience in matters that were his direct concern were frequently ignored distressed him. He interpreted Kew's refusal to purchase his personal herbarium of the Gardens' plants and his fern herbarium of some 2,000 species as yet another instance of management's spite. 'Old Jock', as he was called by his staff, "was not of a sympathetic nature" since "he required no relaxation himself . . . and gauged other people's wants by his own."[23] Yet he could be genuinely concerned about the career prospects of his young gardeners. In 1850 in an attempt to encourage them to study, he offered a copy of Hooker's *British Flora* as a prize for the best collection of dried plants gathered in the neighbourhood of Kew. Joseph Hooker paid him a well-merited tribute.

> That the Royal Botanic Gardens maintained any position in a scientific
> establishment in the interval between the death of Sir Joseph Banks in
> 1820 and the appointment of the new Director in 1841, was wholly
> due to the unaided exertions of Mr Smith.[24]

When in his 81st year Sir William Hooker died on 12 August 1865, he left behind him a botanical garden that was admired and loved for a variety of reasons. Many visitors appreciated it only as a superior public park, a role imposed upon it by several Commissioners of Works who had insisted on an extravagant display of colourful flower-beds. Hooker, while reluctantly conceding that this might be a legitimate function, constantly told his superiors and the public that Kew had scientific obligations as well. His annual reports for some years after 1855 reviewed the Gardens' activities under two headings: "a place of beautiful recreation" and the "Educational, Instructive and Scientific Department". He reminded his audience that Kew grew plants distinctive for "their beauty and ornamental qualities, or for the services they render in commerce, in medicine, in agriculture, in domestic economy".[25] Under Hooker Kew had become an entrepôt for plants that were new, rare or useful, received from many sources, and dispersed to nurserymen and gardeners in this country and to appropriate institutions and individuals overseas.

He made botany accessible to the masses through a Museum and informative guides to its contents and to the Gardens; he met the demands of serious researchers with an unsurpassed Herbarium and Library; he enhanced Kew's international reputation by assiduously courting influential people here at home and throughout the world; his improvements to the Gardens, coinciding with better travel facilities, transformed Kew into a horticultural mecca for the metropolis. In the year of his death annual attendance figures exceeded half a million.

Like his great exemplar, Sir Joseph Banks, he exploited a talent for friendship; through charm and tact he won the support of powerful politicians like the Foreign Secretary (Lord John Russell), the Colonial Secretary (the Duke of Newcastle), and the First Lord of the Admiralty (the Duke of Somerset). He never ceased cultivating royal patronage but sometimes, it must be said, as much for the advancement of his son as for the benefit of Kew. His life was regulated by a routine of work which began daily at eight and finished with private research about midnight. All his obituaries solemnly observed the passing of a great man but his son regretted that they failed

> to make clear the immense debt botany owes him. They do not state boldly that Banks required too much deference, that Sir J[ames Edward] Smith was still more narrow-minded, that Brown actually & actively stifled the science, & that but for my father & Lindley botany would actually have been nowhere in Britain.[26]

Two other notable figures who had played a part in the reconstitution of Kew also died in 1865: Sir Joseph Paxton in June and John Lindley in November. Both had served on the eventful 1838 working party, and both had continued promoting, through the periodicals they edited and, in Paxton's case, in the House of Commons as well, the interests of a royal garden they had helped to save.

CHAPTER FOURTEEN

Botanic garden or public park?

*J.D. Hooker as Director – early career – his
personal interest in garden design – dislike of
excessive floral display – T-Range – opposes
extension of public opening hours – Temperate
House Lodge – additional public entrances –
public agitation for removal of boundary wall
– makes concessions in times of opening*

ON 1 NOVEMBER 1865 Joseph Hooker became Director. From a small boy he had been groomed by his father for a career in botany; even as a medical student at Glasgow University any spare moments were spent in his father's herbarium. Through his father's connections he had been appointed Assistant Surgeon and Botanist in 1839 on the *Erebus* and *Terror* expedition to the South Polar Seas, and six substantial volumes on the *Botany of the Antarctic Voyage* (1844–60) had established his reputation as a taxonomist and plant geographer. After his return to England in 1843, still undecided about his future, the vacant curatorship of the Sydney Botanic Garden in 1844 might have been his had he wanted it. Instead he gave a series of lectures at Edinburgh in 1845, hoping to improve his chances of succeeding to the chair of botany there. When this manoeuvre failed he accepted the post of palaeobotanist with the Geological Survey in 1846. Sir William Hooker, who was behind the appointment, perhaps thought that since the Survey came under the Commissioners of Woods and Forests it might make it easier for Joseph to join the Kew staff at some opportune moment. But his son, preferring to travel, went to northern India on reduced naval pay collecting plants for Kew Gardens and objects for its museum. In reclusive, mountainous Sikkim he gathered rhododendrons and alpines, sketched its hostile landscape, mapped the terrain, notified Darwin of interesting observations on animal behaviour, and began a love affair with the Indian flora that sustained him in his old age. While botanising in the Himalayas he seriously considered, but declined, the superintendency of the Botanic Garden at Peradeniya in Ceylon.

> I must say that it is not the position I should seek, though I could prefer it ten thousand times to a professorship in the University of Glasgow, or to lecturing. In a pecuniary point of view, it is better than I might get in England for many years.[1]

Magnolia campbellii by Joseph Hooker: one of more than 1000 botanical and landscape sketches he made in India. "I always carried my notebook and pencil tied to my jacket pocket, and generally walked with them in my hand. It is impossible to begin observing too soon, or to observe too much . . ." J.D. Hooker: *Himalayan Journals*, vol. I, 1854, p. 247.

He aspired to a position at Kew, encouraged by his father's promise to canvass the Office of Woods and Forests on his behalf. However, the idea of a lectureship based on Kew, which had been floated, held absolutely no appeal for him. When Lord Seymour, the First Commissioner of Woods and Forests, expressed a desire to see Joseph Hooker installed in some capacity at Kew, Hooker's supporters extolled their friend's achievements, and the "importance of receiving to Her Majesty's Botanical Establishment at Kew his future services in the following up the investigations so ably commenced".[2] In the event, the Admiralty agreed to retain his services on half pay for three years while he wrote up the results of both the Antarctic voyage and the Himalayan expedition. With the expiry of this agreement in sight in 1854, Joseph Hooker informed Asa Gray at Harvard that "my own plans are quite unsettled and I sometimes think seriously of giving up Kew and living in London and writing for the press".[3] Having failed to obtain one of the official residences around Kew Green in December, he moved to 3 Montagu Villas on the top of Richmond Hill, only to move again six months later to 55 Kew Green after getting what both he and his father had long sought, the post of Assistant Director. As an old man in 1900 Joseph still gratefully remembered his father's determination to secure his future.

> Latterly his untiring efforts to get me on with the Admiralty, India Office, Treasury, Woods and Forests, and a host of Ministers, is quite bewildering. He thought far more of me than of himself.[4]

Yet despite the satisfaction of having attained this goal, he gradually became disenchanted with the administrative routines of Assistant Director. Not surprisingly, therefore, a Government of India's commission in 1863 to write a flora of the sub-continent attracted him.

> Pay would tempt me, but only because it would hold out a prospect of early retirement from the struggle of scientific work for one's livelihood, and shaking the dust off my feet at the Govt. and Kew Gardens – but for God's sake let this go no further. I regard succession to my father with horror. Not that a better scientific place exists in the world, except my own.[5]

This rejection of Kew may have emanated from an awareness of his unsuitability for management, of his impatience with Civil Service procedures.

His problems began as soon as he accepted the post of Director, conditional upon his agreeing to certain changes in the establishment. The post of Assistant Director disappeared – the Office of Works claimed that there was no one suitable to fill it; that of Curator of the Pleasure Grounds was also abolished and his duties transferred to the Curator of the Botanic Garden. Daniel Oliver, Keeper of the Herbarium and Library, had the Museum added to his responsibilities. As a conciliatory gesture, Hooker got a few additional subordinate posts: a scientific assistant, a foreman, and a storekeeper, and the wages of gardeners and labourers were marginally increased. He told Charles Darwin that it was not a promising start.

> I am up in heaps with work, and find I shall have a desperate fight to get scientific assistance. I will not give in however. I am prepared to improve the Gardens enormously and will do so, but if the scientific character of the establishment is to go down one iota, I shall intimate that I only hold the post with a view to retirement when able.[6]

Hooker may thus have pursued an active interest in horticulture as an agreeable escape from the frustration and tedium of official negotiations and endless minutes. He usually rose early to inspect the Gardens before tackling his administrative chores. "I am very busy out of doors six hours a day and delighting in my occupation,"[7] he tells Darwin. It was no secret that he thoroughly enjoyed landscaping.

> I am getting very proud of the Gardens, in which I really have worked
> tremendously hard now for two years.[8]

He relied for professional advice on his new Curator whose name was confusingly also John Smith, "a capital man to work with", he informed Asa Gray. Initially they concentrated on replanting barren patches in the grounds and creating new walks, both gravelled and grassed – one of the first was a path between the Lion and Unicorn Gates. They cut both straight and curving paths through the Arboretum to form designated routes. Inevitably mature trees were felled to achieve this objective but the Arboretum probably suffered much more from the severe winter of January 1867 than from Hooker's impetuous enthusiasm. He laid out seven avenues radiating from the Pagoda composed of Irish yews, Japanese cypresses and junipers flanked by belts of laurel which also encircled the Pagoda. These avenues and vistas usually comprised trees of one genus or belonging to the same family: Thorn Avenue (1868), Cedar Vista (1871), Acacia Avenue (1872), Holly Walk (1874), Sweet Chestnut Avenue (1880). Maples were paraded along the walk to the Brentford Ferry Gate. Cedar Vista thrusting confidently west of the Pagoda, opening up a prospect of the Thames and the meadows of Syon House, and Isleworth Vista (1874), focusing on the distant tower of Isleworth church, aimed to unite Kew Gardens with the landscape beyond its boundaries. Though Hooker seemed to favour precision and formality in garden design, he and the general public disliked Nesfield's gravelling of Syon Vista which was replaced by grass in 1882. This maze of paths and walks, largely carved out in Joseph Hooker's time, made the Arboretum more accessible to visitors.

Trees were felled not only to make these walks but also to replace them with generic clusters. When Hooker began implementing his father's concept of a classified arrangement of trees and shrubs in an area between the Temperate House and the Pagoda, he announced in his annual report for 1868 his intention to proceed in a similar fashion throughout the Arboretum "whether for picturesque effect or for purposes of reference and instruction". By 1873 the transformation was well advanced: birches and elms congregated near the Brentford Ferry Gate; poplars marched towards Rhododendron Dell; limes flanked the path to the flagstaff; and magnolias had a neighbourly relationship with Berberis Dell.

Conifers were kept apart in a pinetum. The *Hortus Kewensis* tells us that Kew Gardens had 36 species of conifers in 1789 and 56 in 1813. In those days they were scattered throughout the grounds; Cedars of Lebanon, for instance, encircled the Pagoda, crowned 'Mossy' Hill, and invaded Hollow Walk. Although Sir William Hooker continued to disperse conifers, he had formed Kew's first pinetum in the late 1840s on lawns north and west of the

Palm House. It was superseded in 1871 by his son's more ambitious pinetum on the south side of the Lake. Conifers had always interested Joseph Hooker; when he went to Syria in 1860 he studied the Cedar of Lebanon in its native habitat, and the conifers of North West America were one of the reasons for his going there in 1877. By 1872 he had assembled a community of 1,200 conifers, some transferred from his father's pinetum, with Old and New World species of the same genera facing each other.

But the magnificent Himalayan cedar (*Cedrus deodara*) rejected Kew's hospitality. On Nesfield's advice, deodars lined the Broad Walk and also Pagoda Vista where their roots were protected by beds of shrubbery. They did not thrive. Douglas firs replaced the forlorn deodars on the north side of Syon Vista in 1871, and in 1882 a short avenue of them between the North Gallery and the Temperate House yielded to Atlantic cedars. They grew slowly and, according to Joseph Hooker, suffered from cold winters and springs. A few, however, survived and a particularly good specimen can still be admired between the Visitor Centre and the Pond.

Hooker told Sir Henry Barkly that he was "trying to make our very ugly lake an ornamental piece of water with a gang of 50–60 navvies".[9] His father's conception, the Lake divided the Arboretum as a sweep of space and a glint of water among a mass of trees. The son extended it by about half an acre, softening its edges with vegetation.

The Lake and Rhododendron Dell are Kew's biggest man-made excavations; another large one, Berberis Dell, lies just north of the flagstaff and in fact was originally known as Flagstaff Dell. One of the Gardens' many undulations that owed their origin to gravel workings, the Director and Curator in 1869 had recognised its potential as a sunken garden. After enlargement by the removal of more gravel, its sides were shaped and planted with shrubs, predominantly Berberis, which suggested the name for this new feature.

Other changes to the Gardens included an extension of the rock garden near the ice-house with a hardy fernery and the substitution of rectangular plots for the curving beds in the Herbaceous Ground. In 1873 a small tank at the northern end of the Herbaceous Ground housed hardy aquatics and a larger one, 80 feet long, replaced it in 1879.

It is inevitably the fate of most gardens to undergo constant renovation and change. Kew is a palimpsest of successive layouts by professional designers, innovative gardeners and impulsive directors. For instance, the mound near the Pond, covered with laurel in Princess Augusta's day, was transmuted into a grassy knoll in the 1840s, then became a playground for children. When the scars of countless sliding feet made it an eyesore, it underwent another metamorphosis in 1882 – a wild garden, planted with spring bulbs, put out of bounds to the public and gradually evolving into a woodland garden.

In 1882 it was the turn of the American Garden to be altered and replanted. Formed by Sir William Hooker during the early 1850s, it was reshaped into two concentric circles, though still called the American Garden on the key plans to the Gardens until the 1890s when the public came to know it as the Azalea Garden.

Little by little fragments of the landscape sanctioned by Sir William Hooker were disappearing. The Curator reconsidered Nesfield's labour-intensive parterres around the Palm House. The 1880 annual report announced plans for their remodelling, ostensibly because they were frequently waterlogged; the elaborate flower-beds facing the Pond surrendered their narrow gravel paths and box edging to six grass plots, four of which displayed flowers. The *Gardeners' Chronicle* applauded the disappearance of these "miserable little beds". An Irish ivy hedge, now one of the oldest in the country, demarcated the front of the terrace. The weed-infested gravel paths in the great semicircular garden behind the Palm House were turfed.

As noted earlier, confusion and uncertainty about Kew's legitimate function arose during the the 1850s when successive Commissioners of Works ruled that the Gardens had to cater for "pleasure seekers", as Joseph Hooker dismissively called them. He claimed that at one time nearly three-quarters of Kew's annual vote was spent in building public gates and public conveniences, on salaries for additional gatekeepers and police patrols, on garden ornaments, and on seasonal changes to the flower-beds. He had inherited the annual ritual of bedding out forced upon his reluctant father. From 1866 Kew not only

Carpet bedding with the Broad Walk lined with Deodars in the background. *Gardeners' Chronicle* (8 October 1870, p. 1344), while approving this circular flower-bed as a focal point, thought its floral patterns too elaborate.

supplied bedding plants for London's parks, but also for "the poorer inhabitants of London", a social obligation honoured for some 20 years. In vain did Joseph Hooker protest to the Treasury that "Kew never was regarded as one of the Parks and never should be: its primary objects are scientific and utilitarian, not recreational."[10] He detested "the present passion for a blaze of gaudy colours along our garden paths",[11] a fashion denounced by William Morris as "an aberration of the human mind". But his superiors compelled him to compete with the floral extravaganza of the metropolitan parks, and in 1870 Kew even succumbed to carpet bedding.

While the *Garden* sympathised with the Director's doubts about such floral frivolity, it nevertheless pontificated that "it is a most pernicious and foolish notion which supposes that a botanic garden should not be ornamental in the highest sense".[12] This particular periodical and its proprietor, William Robinson, kept Kew's bedding schemes under constant review. On one occasion it recommended the addition of yuccas, lilies, irises and grasses to the conventional range of plants; it disapproved of the tendency to overcrowd beds and the absence of any flair in display. But after several years of monitoring Kew's efforts, Robinson conceded that large-scale bedding out ought to be left to pleasure gardens and parks.

Shortly after his appointment Joseph Hooker initiated a programme of reorganising the Palm House: paths were made in the wings, flowering plants were placed on shelves, coconuts and mangoes grew in a specially prepared hotbed, palms were cut back to give more space and light to their neighbours, and bare beds were covered with ferns and aroids. One of the glories of the Palm House, *Livistona australis*, sent to Kew by Allan Cunningham, had to be felled in 1876 when it reached the glass roof. By 1882 Kew's population of palms had grown from the ten listed in the *Hortus Kewensis* (1789) to 420. The Director claimed that only the botanical gardens at Herrenhausen in Hanover and at Buitenzong in Java had more palms than Kew. He also boasted that only the Imperial Gardens at Schönbrunn in Vienna had a better collection of aroids.

During his first year in office, Joseph Hooker converted the glasshouse erected in 1852 for the *Victoria regia* (now *V. amazonica*) into an Economic Plant House for a selection of medicinal and culinary plants. When a new range of glasshouses went up in 1868–9, two compartments were allocated to tropical and sub-tropical economic plants. A cluster of eight fruit houses, each with its own furnace, in the former Methold Garden had for some years served as temporary and wholly inadequate accommodation for tender herbaceous plants. By 1869 they were all replaced by this new house with a single stokehole. The T-Range, so called because of its resemblance to the letter 'T', had the *Victoria amazonica* (which had never fared well in its original house), a cool and a warm compartment for orchids, and begonias, temperate ferns, gesnerias and economic plants in other sections.

Heating deficiencies in the glasshouses were also rectified. Replacing the 12 boilers in the Palm House with just six was a notable improvement. The Office of Works engineer who designed them also supervised a partial distribution of nearly four miles of hot-water pipes, taking them up to the gallery of the Palm House.

The T-Range, completed in 1869, was the biggest glasshouse development during Sir Joseph Hooker's directorship.

All this activity in the Gardens received the attention and approval of the press. In 1873 the *Edinburgh Review* published a long, appreciative account and the *Gardeners' Chronicle* in August 1876 devoted a 14-page supplement to Kew. But Hooker's regime did not proceed without some adverse comment, especially from William Robinson. He thought that the habit of planting trees everywhere, sacrificing spacious lawns, had made Kew "devoid of any picturesque beauty".[13] Long, assertive vistas through a density of trees could never replace "a few acres of cool grass".[14] Unnecessary straight lines and repetitive patterns were anathema to the man who favoured informality and pioneered the wild garden. He disliked the rash of kidney-shaped beds along the Broad Walk, and Kew's partiality for shrubs in long, straight beds intensely irritated him. While he accused Hooker of lapses in horticultural taste, others deplored the Director's inflexible attitude to any appeal for an extension of the Gardens' opening hours. That particular crusade and the agitation for the removal of the Gardens' high boundary wall were to become a *cause célèbre*, the subject of many public meetings, frequent questions in Parliament, and hostile or bemused newspaper comment.

The public had always enjoyed access to the Botanic Garden, though perhaps in earlier days with little show of enthusiasm from the staff; to stroll through the Pleasure Grounds had been a rarer treat, limited to Thursdays and Sundays. When Sir William Hooker took over the Botanic Garden in 1841 he continued to keep it open daily, Sundays excluded, from 1 p.m. to 6 p.m. or to dusk in winter. After inheriting the Pleasure Grounds in 1845 he continued to restrict its admission to two days a week. Even this, so he told his father-in-law, he did reluctantly.

> I could wish these gardens were not thrown open on Sunday. But they always have been so it will be impossible to take this privilege away from the public: after all it is better for them than going to the public house.[15]

A letter in the *Gardeners' Chronicle* in 1847 provocatively recommended that those members of the public "in search of nothing but rough amusement" should be restricted to the Pleasure Grounds, whereas those "quiet and desirous of gaining information" could enjoy the superior facilities of the Botanic Garden. Here we have the first intimations of a public debate on the perception of the class and calibre of Kew's clientele. "The vast majority of those who visit both [the Botanic Garden and Pleasure Grounds] are there for recreation and not for study," declared a correspondent in the *Gardeners' Chronicle* in 1848. Another complained of crowds filling the glasshouses and swamping the Gardens. A contributor to the *Florist* in 1849 urged an earlier opening time.

> 'Tis to tempt them away from the gin-palace, the public house, and the beer-shop, that we would have these delightful Gardens opened at nine o'clock; ay, and we would have in the beautifully kept ladies' cloak-room a building where they would have an opportunity of partaking of any refreshments they might bring.

Temperate House Lodge, 193 Kew Road. Designed by W. Eden Nesfield in 1866, it is a very early example of the so-called Queen Anne style.

The general public, who were denied access until 1 p.m., naturally questioned why students of botany and horticulture and flower painters enjoyed the privilege of being admitted in the mornings. Starting in 1851 the Pleasure Grounds opened daily from 1 p.m. to 6 p.m. from May to September. A further concession followed in 1853 when the Office of Works ordained that both the Botanic Garden and the Pleasure Grounds should open on Sundays from 2 p.m. to 6 p.m. during the summer months, a regulation that Sir William Hooker almost certainly disliked. Now only Christmas Day remained sacrosanct. But still the public grumbled, demanding to be admitted in the mornings. One letter to *The Times* for 5 October 1859 accused the authorities of favouring privileged local residents by allowing them in early. And opening the Arboretum (the former Pleasure Grounds) every day of the year, except Christmas Day, from March 1864 did little to diffuse the vociferous resentment.

Joseph Hooker's stubborn refusal to concede an earlier opening time was prompted by a fear that it would mark yet another step towards Kew's transformation into a public park, and a dilution of its scientific integrity. When he succeeded his father the public entered by the Main Gates on Kew Green, the Brentford Ferry Gate, and the Lion and Unicorn Gates in the Kew Road, the last two being entrances only to the Arboretum, where gates in the internal wire fence gave access to the Botanic Garden. The proposed extension of the London and South Western Railway to Kew with the certainty of yet more visitors to the Gardens made the provision of another gate absolutely essential. The site chosen in the boundary wall opposite the Temperate House was the nearest to the anticipated location of the new railway station. It was to be a grand carriage entrance, suitable for receiving royalty, and the Lion Gate, a short distance along the Kew Road, would be closed.

A lodge adjacent to these imposing double gates was designed by William

Eden Nesfield, the son of the landscape gardener. He had trained as an architect under his uncle, Anthony Salvin, entered into a partnership with Norman Shaw in 1863, and had experimented with several architectural styles before the Kew commission in 1865. The Temperate House Lodge, as it is now called, declared a vague kinship with the period of Queen Anne, in a style later to be made fashionable by Norman Shaw. Others have detected an indebtedness to English brick buildings from the 1630s to 1660s, of which Kew Palace is a notable example. Its steeply pitched roof surmounted by a bold chimney stack reminded Maxwell T. Masters of "a large handbell". Nesfield embellished it with rubbed brick pilasters, embossed roses on the lead lining the dormer windows, and added the royal coat of arms on two sides of the chimney. This idiosyncratic building spawned a similar but more exuberant lodge when Nesfield worked at Kinmel Park. The foreman of the Arboretum moved into the lodge but the Queen's Gate, finished in 1868, never opened; the station, which opened in 1869, was unexpectedly built half a mile nearer Kew. In 1871 the broad gravel path specially laid from the gate to the Temperate House was grassed over and planted as an avenue of alternating deodars and Douglas firs.

George Engleheart, a local property developer, desirous of enhancing the value of an estate he was building near the new station, offered to pay for another gate into the Gardens provided it faced his new road (Kew Gardens Road). Until the Victoria Gate opened, Cumberland Gate, erected at Engleheart's expense in 1868, was the nearest to Kew Gardens station.

In response to a petition from residents across the river in Isleworth, Hounslow and the neighbourhood, the Isleworth Ferry Gate opened in 1872. The number of people using the Brentford Ferry Gate fell when the toll on Kew Bridge was abolished, visitors preferring instead to enter by the Main Gates on Kew Green.

The extension of the railway network throughout the capital made Kew Gardens accessible to a more varied and disparate clientele. No longer was it largely the resort of local people, the prosperous middle class, and earnest botanists and gardeners. It now rated as one of London's most popular attractions for the poor of the East End. The Director tried to classify visitors in his 1871 annual report: those seeking information, the "industrial class" (by which he loosely meant school teachers, mechanics and artisans), plant collectors, colonials on home leave, botanists and horticulturists, and "mere pleasure or recreation seekers . . . whose motives are rude romping and games". Saturdays, it would seem, were preferred by the "professional and upper classes" who arrived in their carriages; the "trading classes" travelled to Kew Green in omnibuses on Sundays, and "artizans" chose Mondays, often in parties organised by trade unions and charities. Crowds swelled to an invasion on Bank Holidays, an innovation decreed by Act of Parliament in 1871. The Lord's Day Observance Society tried unsuccessfully in 1874 to close museums, parks and Kew Gardens on Sundays. In 1874 the Richmond Vestry petitioned the Office of Works for an earlier opening of Kew Gardens.

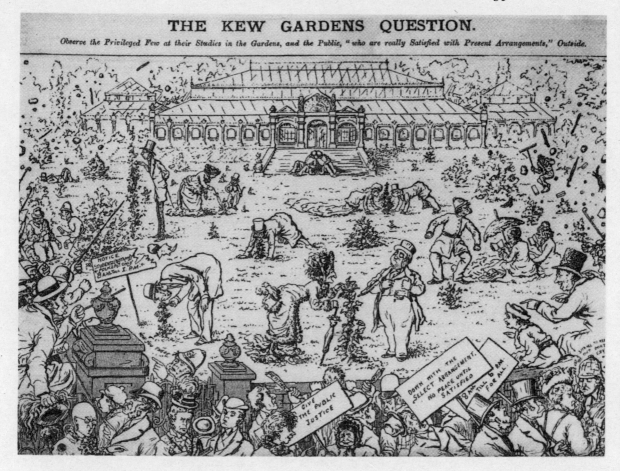

THE KEW GARDENS QUESTION.

Observe the Privileged Few at their Studies in the Gardens, and the Public, "who are really Satisfied with Present Arrangements," Outside.

Large numbers of excursionists came from London to the Gardens, and great was their disappointment at finding that the gardens were not open till the afternoon. Many of these excursionists were poor people, who only obtained a holiday once or twice a year.[16]

> The general public were denied access to the Gardens in the mornings, ostensibly to allow botanists and students to work outdoors undisturbed by crowds.

This concern for the welfare of others was cynically dismissed by the Director as a camouflage for their own interests. He responded that the Gardens would be over-run by "swarms of nursery maids and children who inhabit the innumerable villas that have sprung up around Kew and Richmond".[17]

Local frustration voiced at a meeting convened in the Kew Institute in August 1877 led to the Kew Gardens Public Rights Defence Association being launched with two objectives: the opening of the Gardens at 10 a.m. and the removal of the boundary wall in the Kew Road. Sir Joseph Hooker (he had been made a Knight Commander of the Star of India in July 1877) remained unmoved by the protestors' argument that the botanical gardens at Edinburgh, Dublin and Liverpool coped satisfactorily with pre-noon opening. He insisted that the presence of the public in the mornings would interfere with the maintenance of the grounds, would make out-of-doors botanical study difficult, and would most certainly incur extra expense.

Even drunkenness and disreputable behaviour were blamed on Hooker's obduracy, according to the *Richmond and Twickenham Times*:

> . . . a large number of people came down in the morning [of the last Bank Holiday] expecting admission to the Gardens, but finding the gates shut betook themselves in true British fashion, to drinking and dancing, and then some 2 hours later sought to refresh exhausted nature by falling asleep in the grass.[18]

The Gardens' high boundary wall in the Kew Road similarly inflamed passions. When the Richmond Vestry sought to have it lowered as early as 1844, they castigated it as "a great evil and unsightly object". The Vicar of Richmond compared walking along it to "a tour around the prison at Clerkenwell". So imagine the intensity of local resentment when in the summer of 1877 about three feet were added to it between the Cumberland and Unicorn Gates. Kew justified this action on the pretext that the wall was so low that the Gardens' labourers habitually scaled it at all times of the day to drink in a nearby public house. Furthermore the heightened wall would conceal a view of the public conveniences in the Gardens from the upper windows of the houses now under construction on the opposite side of the road. Sir Joseph does not make it clear whether his concern was for the modesty of residents or for users of the conveniences; probably the latter since he had little sympathy for local residents who, he felt, thought only of the value and surroundings of their own properties. He stressed the practical benefits of this notorious wall: it protected the Gardens from east winds, excluded dust and litter from the roads, and screened passing traffic. To add weight to his argument, he commended it as an element of good garden design.

> It is one of the first principles of landscape gardening to conceal boundaries and produce an effect at once of privacy and interminableness [sic]. There is no better means of accomplishing this than by a wall backed by well arranged shrubberies.[19]

The public, absolutely unimpressed and unmoved by this reasoning, continued to demand its demolition and replacement by iron railings. "Is the raising of the wall a sine qua non?" an exasperated Secretary of the Office of Works asked the Director. "The neighbourhood is pestering us to death about it."

In October 1877 Hooker relented a little but not enough to make it appear that the opposition had gained a victory. As a tactical gesture he opened the Gardens at 10 a.m. on Bank Holidays, making it clear, at the same time, that other work on those days would suffer as a result. His reluctance to make concessions emanated partly from obstinacy but also from a genuine fear that Kew's future as a botanical garden would be in jeopardy.

> If opened the whole day the Gardens will be regarded as a Park. Park-licence will insinuate itself & demands for luncheons, pic-nics & bands of music will follow.[20]

After his distressing encounter with A.S. Ayrton in 1872 (see Chapter 15), he had good reason to be on the defensive, seeing in every complaint or criticism unfortunate consequences for Kew.

When the First Commissioner, well briefed by Hooker, was asked in the House of Commons in January 1878 whether Kew Gardens could open at an earlier hour, he replied that "the change demanded would involve the consideration whether Kew could continue what it originally intended to be — a scientific utilitarian institution — or merely a resort for pleasure seekers." The Director himself could not have put it better. In the meantime the Linnean Society got up a memorial to the Office of Works signed by "professors, teachers and students, cultivators of natural science, and examiners for the public services", unanimously opposing any amendment to the current hours of opening as they would no longer be able to pursue their privileged studies in the mornings "uninterrupted by ordinary pleasure-seeking visitors". Another memorial expressing similar sentiments came from the Fellows of the Royal Society, the Royal Geographical Society and the Royal Horticultural Society. In any serious conflict Hooker always rallied his supporters. But tempers were becoming frayed, language more immoderate, and the Director openly accused of despotic behaviour. Even the national press turned against him.

> The truth is that Sir Joseph Hooker, like most other officials, is pos-
> sessed with an exaggerated sense of his own importance and of that of
> the institution over which he presides.[21]

The Member of Parliament and great plantsman, Sir Trevor Lawrence of Burford Lodge, Dorking, who sadly found Hooker "full of what I cannot but call inveterate prejudice on this question", chaired a public protest meeting in Richmond on 19 February 1879, and on 6 March moved in the House of Commons that "it is desirable that the Royal Gardens at Kew should be opened

In July 1877 the boundary wall between the Unicorn and Cumberland Gates was raised in order "to prevent workmen climbing over the wall". But the public suspected that it was a defiant reaction by Kew to their clamour for its removal.

This 'Indignation meeting', held on 28 October 1882, was one of several public protests against restricted opening hours and the notorious boundary wall.

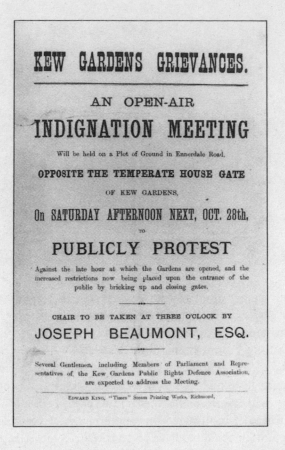

KEW GARDENS GRIEVANCES.

AN OPEN-AIR
INDIGNATION MEETING

Will be held on a Plot of Ground in Ennerdale Road,

OPPOSITE THE TEMPERATE HOUSE GATE

OF KEW GARDENS,

On SATURDAY AFTERNOON NEXT, OCT. 28th,

TO

PUBLICLY PROTEST

Against the late hour at which the Gardens are opened, and the increased restrictions now being placed upon the entrance of the public by bricking up and closing gates.

CHAIR TO BE TAKEN AT THREE O'CLOCK BY

JOSEPH BEAUMONT, ESQ.

Several Gentlemen, including Members of Parliament and Representatives of the Kew Gardens Public Rights Defence Association, are expected to address the Meeting.

EDWARD KING, "Times" Steam Printing Works, Richmond.

to the public at 10 a.m. on week-days, with such reservations as may be found expedient." He deplored Hooker's reference to "a swarm of filthy children and women of the lowest class [who] invaded the Gardens" and his "serious charge against the people that they resorted to the woods for immoral purposes in great numbers".[22] The First Commissioner of Works repeated the official line that earlier opening would add to Kew's costs and alter its character. Though the motion was lost, the campaign never faltered. In 1882, following rumours that the Queen's Gate was to be bricked up, the Lion Gate closed, and a new gate opened nearer the Cumberland Gate, a deputation informed the First Commissioner that the trustees of the Selwyn estate had purposely made the road opposite the Queen's Gate a wide avenue and had chosen red bricks for its houses to harmonise with the Temperate House Lodge. The Commissioner, unable to meet their objections to removing the Queen's Gate, promised to consider replacing it with a low wall and railings. This was a prudent compromise, as was the advancing of opening time from 1 p.m. to midday beginning on 1 April 1883. When Sir Joseph Hooker retired two years later, William Thiselton-Dyer, who was himself not a particularly tactful negotiator, took on the responsibility of pacifying local discontent.

PLATE 22 The Georgian fruit store and foreman's residence, which stood in the former Royal Kitchen Garden, was converted by Decimus Burton into Kew's first museum in 1848. It was truly a "cabinet of curiosities"; plants and their products, artefacts and drawings, displayed in a somewhat chaotic fashion. It survived almost intact until the early 1960s when this photograph was taken, a museum which time had endowed with a certain period charm.

PLATE 23 The interior of the Marianne North Gallery. When it was opened in June 1882 it displayed 627 paintings, all carefully grouped under countries by the indefatigible Miss North. An extension at the rear of the gallery in 1883 enabled her to add another 221 paintings; 16 large paintings on canvas were later removed. Originally the paintings were not glazed.

PLATE 24 *Brugmansia arborea*. Oil painting by Marianne North of "Flowers of Angel's Trumpet" which she drew during her visit to Brazil. It is interesting to compare her work with that of another adventurous flower painter, Margaret Mee (see plate 32). Both occasionally added a bird or small animal to their composition; here we have a humming bird. There is greater delicacy and precision in Margaret Mee's work but both painted with evident enjoyment. *Royal Botanic Gardens, Kew*

PLATE 25 View from above the Palm House looking down the length of Nesfield's Syon Vista, with the Rose Garden in the foreground and the roof of the Temperate House on the left. *Photograph by A. McRobb*

PLATE 26 The Pagoda and Heath Garden in winter. *Photograph by T. Harwood*

CHAPTER FIFTEEN

The Ayrton controversy

*Ayrton appointed First Commissioner – in
conflict with Hooker – scientists appeal to the
Prime Minister – Richard Owen sides with
Ayrton – controversy debated in Parliament –
extension to the Herbarium –* Index Kewensis
– Jodrell Laboratory

AN ACRIMONIOUS DISPUTE with one of Gladstone's ministers and with the
naturalist, Richard Owen, threatened Kew's survival as a scientific institution.
Gladstone, who had made the reduction of public expenditure the bedrock of
his election campaign in the autumn of 1868, oddly appointed A.H. Layard
as his First Commissioner of Works. A cultivated man who collected Italian
paintings and had excavated Nimrud, Layard seemed temperamentally more
suited to commissioning stately public buildings than to implementing a policy
of retrenchment. His recommendation that Joseph Hooker should attend the
Botanical and Horticultural Congress at St Petersburg was rejected by the
Treasury, the instrument of the Prime Minister's "thrifty administration".
Frustrated by official parsimony and thwarted in his plan to locate the
law courts, the Natural History Museum and other government buildings
along the new Thames Embankment, Layard gladly left the Government
after only a year in office for the more congenial post of British Minister in
Madrid.

His replacement was Acton Smee Ayrton, Financial Secretary at the Treasury, a transfer welcomed by the Chancellor of the Exchequer, who considered
him a disruptive element in his department. Ayrton was born in Kew, a fact
which may help to explain his obsessive interest in the Royal Botanic Gardens.
He had briefly practised law in Bombay before being elected Liberal Member
of Parliament for Tower Hamlets in 1857. His impetuosity, indiscretion and
brusqueness secured him many enemies, including the Queen, who never
forgot nor forgave his tactless remarks in public on her secluded life after the
death of the Prince Consort. He referred disparagingly to "architects, sculptors
and gardeners" during his election campaign, and subsequently confirmed his
antipathy by enforcing economies in the designs of the new law courts, the
National Gallery and the Natural History Museum, and by trying to demote
Kew to the category of a public park. Gladstone had rewarded his loyalty with
a ministerial post in the Treasury before putting him in charge of the Office

of Works, where, remarked the Chancellor wryly, he would "be quite in his element wrangling about coals, candles and furniture". His creation of a new post of Director of Works and other establishment changes in his department were sensible reforms, and his firm control of expenditure was not unreasonable, but his indifference to public opinion and insensitivity in staff management provoked resentment.

One of his first acts after becoming First Commissioner was to reprimand the Director at Kew, not an auspicious beginning to his relations with Joseph Hooker. It was, in fact, a prelude to reducing his status and challenging his authority. Sir William Hooker had endured the imposition of disagreeable directives from Sir William Molesworth and Sir Benjamin Hall, but his

One of a number of satirical comments on the confrontation between Ayrton and Hooker.

FUN.—August 10, 1872.

A PARTING WORD.

Dr. Bull:—"YOU'RE MORE TROUBLE THAN ALL THE REST OF THE BOYS PUT TOGETHER—WITH YOUR BULLYING AND STUPIDITY;—I'VE A GREAT MIND TO EXPEL YOU! MIND YOU BEHAVE BETTER AFTER THE HOLIDAYS, OR I WILL!"

authority as Director had never been questioned. He could, for instance, consult contractors of his own choosing for buildings and equipment, a practice which his son naturally continued. In October 1867 Joseph Hooker had been advised by the Office of Works that responsibility for heating and ventilation of the glasshouses and the museums devolved upon him, the actual work being supervised by a surveyor. With Ayrton's reorganisation of the Office of Works in 1870, a Director of Public Works now assumed responsibility for all projects, buildings, maintenance and repairs, introducing, whenever necessary, competitive tendering. When, however, Hooker learned of this procedural change through one of his subordinate staff and protested about not being officially informed, he received a curt reply from an assistant secretary, requesting him "to have the goodness to govern yourself accordingly".[1] Hooker, stung by this offensive remark, angrily retorted

> this arbitrary act [is] a further indication of that disregard of the Director's office, or want of confidence in himself, of which, I have had such conspicuous proofs since, and only since, the accession of the present First Commissioner.[2]

At the same time he sent a formal complaint and a copy of this letter to the Prime Minister, an action he had contemplated six months earlier.[3]

> After upwards of 32 years spent in the public service, at home & abroad, without a suspicion of mistrust on the part of my many official superiors, I have had since Mr Ayrton's accession, to submit to various arbitrary measures which, though compromising my position & authority, have been concealed from myself, & become known to my subordinates, through them alone I have first been made cognisant of them.[4]

In confidence, he told his friend, T.H. Huxley:

> If it leads to my resigning Kew Directory [sic], I shall be sorry & glad. I have no stomach for this sort of worry, though I shall now fight to my stumps, if I get no more sleep for a month.[5]

When the Office of Works demanded that he produce evidence of their alleged disregard of his authority, Hooker cited several instances in a letter that revealed his long-suppressed anger and frustration.[6]

He recalled that the First Commissioner had visited Kew Gardens on Sunday, 18 December 1870, without making any attempt to see him, in order to discuss with the Curator his duties, offering him promotion to a new post he had in mind, that of 'Surveyor or Secretary of the Parks'. The Curator, cautioned not to reveal this conversation to the Director, was summoned next day to Hyde Park to supervise the laying out of the grounds around the Albert Memorial. It was not until the following day that Hooker was officially informed of this proposed secondment, necessitated by the Hyde Park Superintendent's illness. Hooker also resented not being consulted about structural modifications to the Museum and the installation of a new heating system. Certain that Ayrton would deliberately misrepresent the facts, he thought it prudent to send a copy of this letter to Gladstone, assuring him that he was "unconscious of any feelings of personal animosity against Mr Ayrton". In a memorandum rebutting Hooker's accusations, Ayrton concluded that

> It would be incompatible with the position and duties of the First Commissioner to allow himself to be a passive agent in Dr Hooker's hands; not to see or hear or do anything except through Dr Hooker, and in the manner he might think fit to prescribe . . .[7]

Although Gladstone seemingly respected Hooker, he had an uneasy relationship with scientists, and scant regard for their work. "If . . . the State does nothing for science, it cannot be helped, nor need it be lamented," he told the Institution of Civil Engineers at a function in 1872. The historian W.E.H. Lecky reckoned "That the whole great field of modern scientific discovery was out of his range." So instinctively the Prime Minister sided with Ayrton, attempting to placate Hooker with a bland assurance that his minister had no desire "to disregard your feelings or withhold the consideration due both to your person & your office".[8]

But Hooker now had no intention of letting the matter rest. A former Under-Secretary of State for the Colonies, his friend John Ball, counselled him "to make it quite clear that you are not merely or mainly complaining of personal slight but that you are fighting for Kew".[9] In two letters to Gladstone, Hooker challenged Ayrton's interpretation of the facts and at the same time requested to be put in touch with one of the Prime Minister's secretaries or some other appropriate official as a channel of direct communication.[10] Hooker thus obtained authorisation to liaise with Gladstone's secretary, Algernon West, who, having served in the Admiralty and the India Office, was known to him. In thanking West for arranging a meeting, Hooker confessed that

> I am at a loss what to say as to my future position under a Minister whom I accuse of evasion, misrepresentation and misstatements, in his communications to the First Minister of the Crown; whose conduct to myself I regard as ungracious and offensive & whose acts I consider to be injurious to the Public Service, & tending to the subversion of discipline.[11]

In need of advice and allies, Hooker consulted the Chancellor of the Exchequer, the Duke of Argyll at the India Office, and George Bentham in his role as President of the Linnean Society. The Astronomer Royal dissuaded him from seeking independence for Kew through a Board of Visitors. "Neither I nor anyone I know will regret to see you drag the body of the prostrate Ayrton behind your chariot wheels," wrote John Ball, "but the first thing to do is to lay him prostrate – and I doubt if you have taken the best way."[12] Gladstone delayed making any final decision by offering first to receive a deputation from Hooker.

In confidence, West raised Hooker's hopes by the disclosure that consideration was being given to a plan which would change his relations with the Office of Works, but Hooker confided to his friend Asa Gray at Harvard that he was "contemplating resignation . . . If I can keep this house I would retire tomorrow for I am sick of the Government & want quiet to work at Gen[era] Plant[arum]."[13] When he learned that this plan to remove Kew from the Office of Works had been abandoned, he proposed a 'federation' of scientific bodies, including Kew, answerable to a minister or government department. Gladstone referred Kew's dispute to a Committee of the Cabinet consisting of the Marquis of Ripon, Lord Halifax and Edward Cardwell. On 13 March this committee considered a memorandum from the Director, who requested a restoration of the authority he had enjoyed under the previous First Commissioner.[14] However, Hooker's meeting with this committee merely produced an ambiguous statement from the Treasury that did absolutely nothing to resolve the impasse.

> Mr Ayrton has been told that Dr Hooker is in all respects to be treated
> as the head of the local establishment at Kew; of course, in subordination
> to the First Commissioner of Works.[15]

Hooker, who had never doubted that Ayrton intended destroying Kew as a scientific institution, now suspected the motives of all politicians, an opinion shared by his scientific colleagues who resolved to present a memorial to the Prime Minister urging him "to give the Director a substantive function".[16] "They shall know", wrote John Tyndall, a leading figure in this revolt, "that men of science can use a sledge hammer."[17]

This memorial, presented to the Prime Minister by John Lubbock, MP on 20 June, had the backing of the Presidents of the Linnean Society, the Royal Institution, the Royal College of Surgeons, the Royal Geographical Society, and the Treasurer of the Royal Society, as well as eminent scientists such as Charles Darwin and Sir Charles Lyell.[18] At great length it extolled the achievements of the Hookers at Kew, deplored the indignities suffered by the present Director, and concluded that "The resignation of Dr Hooker . . . would we declare be a calamity to English science, and a scandal to the English Government." An edited version was released to the press which with few exceptions took Hooker's side. Ayrton, already an unpopular minister, was pilloried for his "brazen cynicism" (*Globe*), his "manner so uncourteous as to amount to a direct insult" (*Daily Telegraph*), his "desire for a sort of official omnipresence and omnipotence" (*Economist*), and for his attempt "to turn the beautiful Gardens at Kew into a common park" (*Greenock Telegraph*). Expressions of support came from the Royal Horticultural Society, the Royal Society of Edinburgh, the Royal Botanic Society, and from James Bateman on behalf of a specially convened meeting of botanists and horticulturists.

One has to admit to a reluctant regard for Ayrton's indifference to public clamour, his determination not to yield to pressure from whatever quarter. He presented a plausible defence to the Prime Minister[19], and pointed out that the imminence of a new natural history museum at South Kensington justified constraints on expenditure at Kew. Provocatively, he wondered "whether the sum now spent on the collections, library, and establishment for botany at Kew, ought to be expended in completing and improving the establishment at Kensington, or be saved", a course of action doubtless endorsed, if not suggested, by Richard Owen, Keeper of the natural history departments at the British Museum. When a printed selection of the minutes and letters concerning what some newspapers called the "Kew scandal" was presented to the House of Commons on 25 July, it included a paper on the botanical departments at the British Museum and Kew, written by Owen at Ayrton's request. Its hostile appraisal of Kew presented Hooker with another and probably more dangerous adversary.

Richard Owen, recognised by many of his peers as Britain's foremost naturalist, enjoyed the patronage of the Royal Family. An eminent comparative anatomist and palaeontologist, he rejected Darwinian theories and, as an apostle of traditional science, gained the support of Gladstone. With the Prime Minister's backing he became a Companion of the Bath in 1874 and, on his retirement a decade later, a Knight Commander of the Bath. In 1856, at the age of 52, he was appointed the first Superintendent of the natural history collections in the British Museum. When the Trustees of the British Museum seriously contemplated the transfer of Banks's herbarium to Kew in June 1858, Owen concurred and his presidential address to the British Association for the Advancement of Science in September commended Sir William Hooker's competent administration. About this time he began campaigning for the physical separation of the natural history collections from the British Museum, a prospect opposed by T.H. Huxley, Charles Darwin and other radical scientists. Joseph Hooker identified himself with these men and that may explain Owen's first misgivings about Kew. His suspicions grew in 1868 when Joseph Hooker, now Director and openly contemptuous of what he judged as poor management of the botanical collections at the British Museum, advocated the transfer of part of them to Kew.[20] Such a move, had it occurred, might have threatened Owen's vision of a new Natural History Museum in South Kensington.

Owen was reassured in 1870 when Parliament voted £6,000 to prepare a site in South Kensington for his museum, but nevertheless he felt apprehensive about the outcome of a Royal Commission on Scientific Instruction and the Advancement of Science which began its deliberations in the same year. In April 1871 William Carruthers, who had a few months earlier succeeded J.J. Bennett as Keeper of the Department of Botany, gave evidence to the Commission. In the issue of *Nature* for 23 March an anonymous article on 'Botanical Museums' (attributed to George Bentham) presented a *modus vivendi* for the two botanical collections, each retaining its own specialities and methods of organisation. Hooker, in a conciliatory letter to Carruthers on 20 July, endorsed that point of view. Although Carruthers actually favoured two separate herbaria, he probably reasoned that the best mode of defence was to suggest to the Commission the transfer of Kew's collections to the British Museum.

A similar line had been adopted by Owen when Ayrton consulted him privately in May 1872 for his views on Kew. Owen's 'Statement', delivered eight days later, provided just the sort of ammunition the beleaguered First Commissioner wanted. Kew's "chief application", wrote Owen, "is in the instructive pleasure of the public, its scientific one is mainly in economical relations". By this he meant that Kew had an essentially practical role: the introduction and naturalisation of new and useful species, experiments in hybridisation, and participation in agricultural development in the colonies. Its purely scientific needs, he said, could be met by the British Museum which already offered a similar advisory service to the Zoological Society in Regent's Park. During Banks's life his personal herbarium had dealt with Kew's taxonomic problems, and after his death Robert Brown advised Kew until the appointment of Sir William Hooker. Owen thus refuted Kew's case for its own herbarium, and his parting shot – that the new museum at South Kensington would have sufficient space to absorb Kew's collections – was a malicious gesture. Hooker remonstrated against the discourtesy of presenting this document to Parliament without giving him an opportunity to comment upon it.

Lord Derby, a former Tory Prime Minister, mildly protested in the House of Lords on 29 July about this disregard of the Director at Kew, his moderation dictated perhaps by the placatory Treasury minute of 24 July defining the duties of the Director. Hooker agreed to withdraw some derogatory remarks about Ayrton provided his response to Owen's misleading 'Statement' was laid before Parliament. This took place on 8 August, the day scheduled for discussing the Kew affair in the Commons.[21] Hooker's 'Reply' naturally repudiated Owen's interpretation of Kew's functions, dismissed his allegations of the Royal Gardens' former dependency on Banks's herbarium and staff, deplored the disparaging remarks against his father, and vigorously defended Kew's current policies.

Ayrton tried to belittle the seriousness of the conflict by reminding the House that Kew's entire annual expenditure amounted to a mere £31,200 out of a vote of £1,200,000 at the disposal of his department. He believed he had shown great forbearance in order "to satisfy a person occupying so subordinate a position". Hooker's duties as Director required him to obey every instruction issued by the First Commissioner, and he recalled that Sir William Hooker had been subjected to "the most peremptory orders by the First Commissioner, even in matters which might be supposed peculiarly under his control, as to the cultivation of the gardens and laying out of the grounds". To the accusation that he had colluded with Owen, he retorted that the Director had given evidence to the Royal Commission without his permission. In winding up the debate Gladstone drily observed that few of Hooker's supporters were present. Several speakers had mentioned Hooker's hypersensitivity and the Prime Minister added his contribution.

> I must say that I think scientific men, as they are called by the exclusive appropriation of a title which I must protest against, have a great susceptibility.

He refuted the charge of evasion levelled against his First Commissioner and exhorted both men to forget their differences: "my right hon. Friend exercising his rule with mildness and Dr Hooker doing his duty in subordination to my right hon. Friend".

It was hardly a well-argued debate and certainly inconclusive. Sir John Lubbock, who presented Hooker's case, had little enthusiasm for it. Ayrton, however, held the floor with a polished performance that portrayed himself as the "helpless victim of a scientific tyrant" (*Saturday Review*). Gladstone had tactfully presumed that illness had prevented Hooker submitting a statement that would end the controversy, a reference to an apology that Ayrton was determined to get. A letter from the Director on 10 August "withdrawing any imputations that may be or may have been regarded as of a personal character . . . in my letter to Mr West of 30 October" did not satisfy him. Ayrton demanded the retraction should include as well the insinuations on his official acts as First Commissioner. George Bentham entreated Hooker not to yield, "a course which would stultify all that you and your friends have done and render nugatory any efforts to restrain his attempts to degrade science".[22] On 2 September Hooker conceded an ambiguous apology "for all expressions that are considered to be incompatible with official subordination". Thereafter a tense truce prevailed between the two protagonists.

Richard Owen came back to the attack with his 'Remarks' on Hooker's 'Reply'. Now he accused Kew of acquiring dried specimens collected on government expeditions at the expense of the 'national herbarium' at the British Museum.[23] Ayrton covertly helped his campaign by sending him 'original papers' relating to Kew.[24] William Carruthers at the British Museum argued in *Nature* (3 October 1872) for the continuance of two separate herbaria, his own for the "scientific botanist" and that at Kew "for the use of the gardens", but were a single national herbarium to be formed he had no doubt that it should be located at the new Natural History Museum. Hooker complained to Asa Gray that he found Carruthers "thoroughly nasty . . . ; he makes a boast of the Kew collection going to the British Museum & gives out that the Government have settled that".[25] Hooker's method of counter-attack was to propose to the mycologist, the Revd M.J. Berkeley, that a memorial be drawn up. He would himself be excluded from the signatories: "it is very important that I should be able to say that the memorial was drawn up entirely independent of me, and that I had no hand in it".[26] Fifty-four eminent botanists and horticulturists supported a recommendation that the British Museum and Kew should maintain separate herbaria.[27] The Government protested that it had no intention of moving Kew's collections to South Kensington. The acrimony and mutual suspicion persisted between the two institutions but August 1873 signalled a turning point for Hooker: he had breakfast with Gladstone – a gesture of reconciliation, he hoped – and Ayrton was transferred to the post of Judge Advocate General. Ayrton resigned with the rest of the Cabinet in March 1874 and lost his seat in the ensuing general election.

The Royal Commission's final report in 1874 approved the retention of the two herbaria, that at Kew preserving its systematic arrangement while

South Kensington adopted a geographical sequence. Plants collected on official expeditions were, in the first instance, to go to Kew for sorting and distribution, with a set being automatically earmarked for South Kensington.

For a while Owen pursued his vendetta in the columns of the *Garden* whose editor, William Robinson, sided with him in the dispute. Nevertheless, Owen supported Hooker's nomination for President of the Royal Society in 1873.

> True it is, I have differed from the Director of Kew Gardens on an administrative question, the calm discussion of which was unfortunately troubled by a personal complication of the Director with the then Minister of Works, which justly roused the warmest sympathy of Dr Hooker's scientific friends.[28]

In his obsession with public economy, Ayrton had behaved tactlessly and Hooker had reacted with wounded pride; the abrasive character of both men had made conflict inevitable. Gladstone's secretary, Algernon West, concurred that "Ayrton had an evil tongue, but I confess I thought him the more reasonable man of the two."[29] Hooker's intransigence precipitated a crisis which raised some fundamental issues: the power of politicians, the responsibilities of civil servants, and the status of scientists. At times, it has to be said, some of the participants conducted themselves in a demeaning and petty fashion.

Notwithstanding the uncertain future of Kew's Herbarium, the Director's report for 1873 proposed a new building for its expanding collections. Following a favourable outcome of the Royal Commission, Hooker pursued the matter more vigorously with Lord Henry Gordon-Lennox, his new First Commissioner. His requirements were modest: a rectangular two-storey building, about 120 by 40 feet. The Duchess of Cambridge, a resident of Kew Green and a zealous guardian of royal property, opposed any building in the grounds of Hunter or Hanover House or any extension to the house itself. Since she had the Queen's support, the Director was instructed to locate the Herbarium in a public part of the Gardens. He rejected a site in the Melon Yard, and doubted whether Hunter/Hanover House could ever again be a royal residence now that it was subjected to obnoxious smells from the gas works and soap factory in neighbouring Brentford. When this tactic failed, Hooker was resigned to abandoning the project, at which point it was discovered that Hunter/Hanover House was no longer royal property.[30] Now there were no grounds for royalty refusing the extension of the existing herbarium in Hunter House.[31] The demolition of the drawing-room, south room and kitchen at the rear of the building in 1876 preceded the construction of the first wing to the Herbarium, functional and somewhat austere in its design, with two galleries projecting from its perimeter walls.

The specimens in the Herbarium were eventually rearranged according to the families and genera described in the *Genera Plantarum* (1862–83), a massive undertaking, jointly edited by Joseph Hooker and Kew's principal benefactor, George Bentham. Linnaeus's artificial classification had been superseded by the more natural systems of Jussieu, de Candolle, Endlicher and Meissner, but these in turn now required revision. Based largely on the resources of the Kew herbarium, Bentham and Hooker compiled concise descriptions of seed-bearing plants, but the pressure of official duties allowed Hooker to contribute only 76 of the 200 plant families. The arrangement of the plants in the Herbaceous Ground and the grouping of trees in the Arboretum followed their classification, which also defined the range of genera in the *Index Kewensis*, a record of the botanical names of seed-bearing species of plants, unquestionably the most important publication ever compiled at Kew.

Both Hookers had relied on its predecessor, the second edition of Steudel's *Nomenclator Botanicus* (1840), now in urgent need of revision. When Hooker appealed in his 1880 annual report for help in updating it, Charles Darwin, who had experienced the confusion of synonymy in plant names, responded with the offer of financial assistance. He had first met Hooker just before the Antarctic voyage of HMS *Erebus* and *Terror*, and their acquaintance matured into a lifelong friendship. Hooker was privy to Darwin's speculations on evolution, and he and Lyell had arranged the joint presentation of papers by Darwin and Wallace at the Linnean Society in 1858. Darwin valued his advice on botanical matters, his critical judgement and his constant encouragement, and his offer to subsidise a revision of Steudel was a practical expression of his gratitude for the services the Director and Kew had rendered him during

The first wing added to Hunter House in 1877, now a period piece of Victorian design.

his many years of research. Benjamin Daydon Jackson, a competent botanical bibliographer as well as Botanical Secretary of the Linnean Society, who had been asked in December 1881 to undertake this revision, suggested expanding its scope by citing the publications in which the plant names first appeared. It aimed to record botanical names of seed-bearing plants published between 1735 and 1885, that is, from Linnaeus's *Systema Naturae* (1735) for genera and his *Species Plantarum* (1753) for species. Work on surveying relevant literature started in February 1882 and when Darwin died a few months later his family generously took on the expense of publishing it. Hooker had considered calling it *Index Darwinianus* or *Nomenclator Darwinaeus*,[32] but eventually settled for *Index Kewensis*.[33] After ten years' intensive labour by Jackson and his clerks and Hooker's proof-reading of about 375,000 specific names, the first volume appeared in 1892 and the fourth in 1895. An indispensable tool for plant taxonomists, it has been kept up to date with regular supplements, and in the year of its centenary the contents of more than 20 volumes were transferred to a compact disc, thereby expanding its research capabilities.

The Library was accommodated in a couple of rooms in the front of Hunter House. The elegant table on the left had been used by George Bentham during his many years of research at Kew.

When the Royal Commission on Scientific Instruction interrogated the Director in 1872 about Kew's experimental research programme, he admitted few facilities existed. The Commission in due course recommended that

> it is highly desirable that opportunities for the pursuit of investigations in Physiological Botany should be afforded at Kew to those persons who may be inclined to follow that branch of science.[34]

Most British botanists at that time devoted their efforts to identifying and classifying plants of the Empire and to the compilation of regional floras, whereas their European colleagues, inspired by the discoveries of de Bary, Hofmeister and Sachs, became plant anatomists and physiologists. Thomas

The Jodrell Laboratory with part of the Herbaceous Ground or Order Beds in front of it.

Jodrell Phillips-Jodrell, an active promoter of scientific research in this country, had already founded a chair of animal physiology at University College London when he offered to pay for a laboratory and its equipment in Kew Gardens provided its Herbarium and Library had acquired a permanent building.[35] This public-spirited offer, completely unexpected, so the Director told the Office of Works, would enable Kew to investigate "the effects of blights, insect ravages & diseases of plants".[36] With the First Commissioner's consent and £1,500 from Jodrell the builders began erecting the laboratory in 1875, but because of potential fire hazards, not as an annexe to the Herbarium as originally intended. A site was chosen in the Melon Yard, conveniently near to the resources of the Herbaceous Ground and the propagating houses. Even before its completion in 1877, Professor John Tyndall moved in, eager to continue his studies on organisms of putrefaction in the relatively pure air of Kew. The Jodrell Laboratory, named after its benefactor, was a drab single-storey brick building which, if it were not for its location, could easily have passed for a municipal branch library or a workingmen's institute. Two rooms served for chemical and microscopical research and a smaller one for gas analysis.

The Director, a traditional taxonomist, gladly delegated responsibility for the laboratory to his new Assistant Director, William Thiselton-Dyer. The latter was better qualified, having assisted T.H. Huxley as demonstrator at South Kensington. For the next 16 years he encouraged botanists to use its facilities in research that ranged from the purely theoretical to investigations with practical applications. B. Sanderson and S.H. Vines examined insectivorous plants; F.O. Bower, a frequent visitor for eight years, published 15 papers as a result of the time he spent at Kew; Sir John Lubbock's pioneering work on the morphology of seedlings began there; and F.W. Oliver took advantage of being the son of the Keeper of the Herbarium to gain access. The Jodrell Laboratory became, in the words of C.R. Metcalfe, one of its Keepers, "an important channel through which an interest in the structure and physiology of plants was restored" in Britain.

CHAPTER SIXTEEN

Forging colonial links

Thiselton-Dyer as Assistant Director –
greater involvement in colonial botany and
agriculture – introduction of rubber to India
– improved methods in transporting plants –
Marianne North Gallery – Fitch resigns –
Hooker retires

WHILE THE AYRTON AFFAIR dragged on, the scientists' memorial to the
Prime Minister reminded him of the Empire's indebtedness to Kew Gardens:
"the records of the Colonial and India Offices will show of what immense
importance the establishment at Kew has been to the welfare of the entire
British Empire".[1] Sir William Hooker's imperial quest had been taken up by
his son not only as a legitimate function of Kew but also as a means of enhancing
its status. Joseph Hooker made sure that his annual reports kept politicians
and senior civil servants well informed of the demands the colonies made on
his institution. In 1869 he reported that the cost of freightage of plants
despatched to Jamaica alone exceeded £100, and that requests for cinchona and
Ipecacuanha from the East and West Indies and Mauritius outstripped supply.
Since the post of Assistant Director had been abolished in 1865, the burden
of this international service had fallen upon the Director who, in October
1874, requested the appointment of a secretary.[2] Not only was the First Commissioner, Lord Henry Gordon-Lennox, sympathetic but the Treasury decided
that the additional post should be graded "Assistant to the Director" with the
same salary that Hooker himself had received as Assistant Director.[3] Furthermore, no objections were raised to his nominee, William T. Thiselton-Dyer.
Astonished as well as delighted, Hooker reported the news to George Maw,
a botanical friend:

> . . . this Treasury goes into matters of administration in a very differ-
> ent spirit and way from the last and the reforms in my department are
> most important. Kew is made a separate *department* of the O[ffice] of
> W[orks] under me, altogether apart from the Parks, etc.[4]

On 12 June 1875 Thiselton-Dyer learnt that he had been appointed. He
was already well known to the Director who for about three years had been
employing him as his private secretary for a few hours a day, four days a

week. Impressed by his abilities, he even invited him to contribute to the *Flora of British India*. His protégé, still in his early thirties, already had a distinguished academic record: Professor of Natural History at the Agricultural College, Cirencester (1868), Professor of Botany at the Royal College of Science, Dublin (1870), and, with Hooker's support, Professor of Botany at the Royal Horticultural Society (1872); he had also collaborated with H. Trimen in writing a *Flora of Middlesex* (1869). He strengthened his rapport with Hooker when he married his eldest daughter, Harriet Ann, in 1877.

His presence in Kew's administration soon became apparent. He immediately took charge of Kew's responsibilities in the transfer of rubber plants and seeds from South America to India. The annual reports which he compiled from 1876 gave greater prominence to news on crop trials throughout the Empire. J.D. Hooker had set the pattern in his rather attenuated annual reports. In 1867 he had reviewed Kew's distribution of economic plants to the colonies: cork oaks to the Punjab, Valonia oaks to the Cape, *Ipecacuanha* to India, Ceylon, South Africa and Trinidad. In 1869 he noted the despatch of large quantities of seeds of trees and shrubs to New Zealand, Australia, South Africa and the USA. The colonies were eager to cultivate the best Manila and Havana tobaccos; with seed obtained from Kew, Natal had developed a prosperous tobacco industry, supplying good quality cigars to the home market. Recipients of Hooker's annual reports learnt that the West African oil palm (*Elaeis guineensis*) flourished in Queensland, New Zealand, India, Ceylon and Jamaica; that the Queensland nut (*Macadamia ternifolia*) was growing in the West Indies, South Africa, India, and Singapore; that many pineapple varieties reached tropical plantations via Kew – one popular variety was known as the 'Kew pine'. Kew could not satisfy demands for *Eucalyptus globulus*, especially from West Africa. The seeds of 'sheet bush' (*Pentzia virgato*), shipped to the Botanic Garden at Adelaide in 1864, proved to be an additional fodder plant for deserts and – an unexpected bonus – imparted an attractive flavour to mutton. In 1883 Adelaide sent its seed to the Saharanpur Botanic Garden in northern India – just one example of the independent traffic in plants and seeds between British colonial gardens. Kew, in its search for new industries for the colonies, recommended sericulture as a "light occupation for women and children"; in connection with this project it acquired seeds of mulberries from northern India in 1877 for experimental cultivation in Australia's botanical gardens.

When a quarter of a million acres of coffee plantations in Ceylon had been devastated by a coffee leaf rust, Kew introduced Liberian coffee to the island in 1873, hoping that it would be immune to the disease. The apparent superiority of this newly-discovered variety precipitated a demand from coffee producers that Kew could not meet from its limited supply. Since, however, several English nurseries were importing the seed from West Africa, Kew left it to them to fulfil such requests. Once a crop had become established, it was normally Kew's policy to leave the development and replenishment of stock to the trade, keeping only a small supply for exchange with other botanical gardens.

Ceylon's coffee 'leaf disease', caused by a minute fungus, *Hemeleia vasta-*

trix, naturally received attention in Kew's annual reports. The Director, with reports at hand from other coffee-producing countries, suggested lines of investigation to the Peradeniya Botanic Garden in Ceylon. Daniel Morris, a former student of Thiselton-Dyer, and destined to be Assistant Director at Kew, spent several years in Ceylon studying the fungus. Kew's remit now included plant pathology: opium poppy disease in India, a borer which attacked sugar cane in Demerara, and 'rust' which afflicted it in Queensland. When insect pests destroyed the vegetable gardens on Ascension Island, the principal source of food for the small garrison there, the Admiralty automatically turned to Kew for assistance.

Ceylon had faced economic disaster when disease destroyed its coffee plantations; ruin again threatened it when world prices for cinchona, another staple crop, slumped. Though Ceylon was saved by its tea industry, Kew cautioned against dependence on a monoculture, and sent cocoa varieties from Trinidad and mahogany from South America, and recommended a range of crops to ensure a balanced commercial economy.

The introduction of new crops incurred an element of risk and uncertainty – the effect of climate, soil and other factors could not always be predicted. Kew failed in India with *Ipecacuanha*, the standard drug for dysentery. A specimen of this South American plant, raised at Kew in 1864, went to the Calcutta Botanic Garden in 1866. From this one plant 300 were propagated in Sikkim, but the slow growth of the rootstock from which the drug is extracted and its low yield made it an uneconomic enterprise. But trials were completely successful in Singapore where four plants were sent in 1885; after only two years the first commercial consignment of Malayan *Ipecacuanha* appeared on the London market.

Rubber, however, completely redeemed Kew's reputation. News of this extraordinary substance had filtered back to Europe after Columbus's second voyage to America in 1493, but it remained a botanical curiosity until 1839 when Nelson Goodyear invented vulcanisation, turning it into a tough and durable material. A display of solid rubber tyres attracted attention at the Great Exhibition in London in 1851 and manufacturers quickly found other uses for it. Bolivia, Brazil, Peru and Venezuela, the source of three commercial rubber trees – *Castilla elastica*, *Hevea brasiliensis* and *Manihot glaziovii* – enjoyed a rubber boom. Richard Spruce, a solitary British botanist and plant collector in Amazonia, who had witnessed the reckless destruction of thousands of trees through indiscriminate slashing and over-tapping of the latex, presented Kew with its first herbarium specimens of rubber trees in 1854, and described in *Hooker's Journal of Botany* in 1855 how they were tapped. Another article on rubber in the same journal in 1868, this time by James Collins, the Curator of the Pharmaceutical Society's museum, came to the attention of Clements Markham.

With the cinchona transfer a success, Markham resolved in 1870 "that it was necessary to do for the India-rubber or caoutchouc-yielding trees what had already been done with such happy results for the cinchona trees".[5] The India Office, still Markham's employer, commissioned Collins in 1871 to survey the present state of the world's rubber industry. A copy of his report,

Hevea brasiliensis.
Watercolour by Barbara
Everard, 1947. *Kew
Library.*

published in 1872, which recommended the introduction of *Castilla* and *Hevea* species into India, reached Joseph Hooker on 7 May 1873 for comment.[6] The India Office wanted the Director's opinion on the commercial exploitation of India's indigenous rubber tree, *Ficus elastica*, and on whether *Hevea* seeds obtained from Brazil should first be germinated at Kew and the seedlings despatched to India. Hooker favoured the introduction of *Hevea*, far superior to *Ficus elastica*, and proposed writing to a correspondent in Brazil about collecting seeds. His letter does not reveal his correspondent's identity but it could have been H.A. Wickham, a coffee planter who a year earlier had offered his services to Kew. Meanwhile, on 2 June, James Collins had met Charles Farris, recently returned from Brazil with some 2,000 *Hevea* seeds. The India Office, which had undertaken to bear the entire cost of the rubber operation, purchased these seeds. Kew managed to germinate only a dozen, half of which were taken in a Wardian case to India by George King, returning to his post of Superintendent at the Calcutta Botanic Garden. When none of them survived in Bengal or Sikkim, it was agreed that all future consignments should go to Ceylon. There, in the opinion of the Inspector-General of the Indian Forest Department, the climate was more favourable.

Wickham, after prolonged negotiations with the India Office about whether he should despatch plants or seeds and also about his fee, subsequently missed the 1875 collecting season. Robert Cross, who had collected cinchona for the India Office, had no quibbles about the merits of plants or seeds and left for Panama in May 1875 to find *Castilla elastica*. Despite a shipwreck off the coast of Jamaica, Cross salvaged several thousand seeds and three boxes of cuttings for Kew. Unfortunately *Castilla* seeds, like those of *Hevea*, have a brief viability and none germinated. The cuttings, however, survived and plants were distributed to Ceylon, West Africa and Java. Six months later

Sir Henry Alexander Wickham (1846–1928).

Silver salver presented by the Rubber Growers' Association to Kew in 1911 to commemorate the Gardens' role in the establishment of the Malayan rubber industry. *Kew Museum.*

Cross returned to South America, this time to Brazil for *Hevea*. His haul on this trip amounted to just over a thousand *Hevea* seedlings bedded in boxes of decayed leaves mixed with wood ash – of which only three per cent were saved – and some *Manihot glaziovii* cuttings. His contribution to the introduction of *Hevea brasiliensis* was, through no fault of his, insignificant.

Cross's departure once more to the Amazon in June 1876 coincided with Wickham's arrival in England with a consignment of *Hevea* seeds. H.A. Wickham, a restless wanderer who had traded in bird plumage in Central America and had tried coffee-growing in Brazil, had first introduced himself to Kew with some plants "of possible interest to the British Empire". The sketch he had made of the foliage and fruit of *Hevea* in his book, *Rough Notes of a Journey through the Wilderness from Trinidad to Para* (1872) lent some credence to his claim to be knowledgeable about rubber. Now remembered as the man who "smuggled" rubber seeds out of Brazil, it is his recollection of the event that admits the deception perpetrated on the Customs officials at Para.[7] Aware of the difficulties Markham had experienced in getting cinchona out of Peru and anticipating similar bureaucratic delays at Para, he declared his crates to be carrying "exceedingly delicate botanical specimens specially designated for delivery to Her Britannic Majesty's own Royal Garden at Kew". The subterfuge succeeded and once the ship had cleared the port it was all speed to England before the seeds perished. But W. Dean contends that not only did Wickham romanticise the incident but, more pertinently, there were no regulations prohibiting the export of rubber seeds in force at that time.[8] We have no record of Cross experiencing difficulty in getting his *Hevea* out of Brazil. In his old age, Wickham embellished the story, adding the presence of a Brazilian gunboat in the harbour to suggest an element of personal danger. Ridley, who knew this rather flamboyant character, believed he deliberately fostered the legend. Leaving aside the legality of the affair, one may ask whether it was an honourable thing to take the natural resources of a country and in so doing deprive it of some commercial advantage. One defence is that all the colonial powers participated in this global transfer of plants, including Portuguese Brazil which surreptitiously obtained seeds of *Coffea arabica* from French Cayenne in the early eighteenth century. Later in the nineteenth century, Brazil even attempted to cultivate India's rubber tree, *Ficus elastica*.

The rapid deterioration of rubber seeds required prompt action when Wickham's cargo reached Liverpool. We have only his word for the statement that Hooker arranged for a special goods train to meet the ship. But we do know that on 14 June 1876 the precious seeds – more than 60,000 of them – reached Kew Gardens and were sown the next day in pans covering an area of over 300 square feet. Thiselton-Dyer, in charge of the operation, years later recalled that

> We knew it was touch and go, because it was likely the seeds would not germinate. I remember well on the third day going into the propagating houses where they were planted and seeing that by good luck the seed was germinating.[9]

Nearly four per cent germinated and in August 1,900 healthy plants were transferred to Wardian cases for shipment to Ceylon. About 90 per cent of the plants survived the voyage and were planted out during the 1877 monsoon in a new garden at Henaratgoda. Two cases also went to Singapore in order to give the seedlings "an additional chance of becoming established in the East".[10] Other recipients were Jamaica, Mauritius and Queensland.

With confirmation that Brazilian rubber had been successfully introduced in Ceylon and Singapore, Kew requested permission to discontinue propagating *Hevea* for India.[11] When Henry Trimen followed Thwaites in 1880 as Director at Peradeniya, the original seedlings had grown into 300 slender trees, some 30 feet high. In June 1882 he presented Kew with a sample of rubber from *Manihot glaziovii* obtained from Robert Cross's seed. As Ceylon's planters were still hesitant about investing in a new cash crop, Trimen had accumulated an embarrassing surplus of rubber seeds. He disposed of them to Singapore, Penang, Fiji, Borneo, Jamaica, Queensland, Java and the German East Africa Company. Another rubber boom followed John Dunlop's marketing of the pneumatic tyre in 1888 and the expansion of the car industry. For a while rubber superseded tea as Ceylon's main crop. Kew's annual report for 1882 announced that the "task initiated by the India Office has now been successfully accomplished". The three important rubber-producing trees of South America were now established in the East, and by the outbreak of the First World War rubber production in Malaya and the Dutch East Indies had outstripped that of Brazil.

Three Kew gardeners, the first being H.J. Murton in 1875, had gradually transformed a park in Singapore into a botanical garden. Kew's first delivery of rubber plants in 1876 had all died but another lot in 1877 succeeded, the source of three-quarters of the *Hevea* stock in the peninsula. H.N. Ridley, who took charge of the botanic garden in 1888, having heeded Hooker's exhortation to concentrate on rubber, could rightly claim to be the founder of the Malayan rubber industry. He rejected the Brazilian method of rubber tapping, which frequently killed the trees, devising instead a safer procedure which also increased the yield of latex. Rubber was also grown successfully in other British possessions. Species of an East African rubber vine (*Landolphia*) were distributed by Kew to colonial planters with moderate success. Kew never withheld its stock of seeds from competing nations: the Dutch East Indies, Germany's colonies in East Africa, and Portugal's Mozambique established or replenished their rubber plantations with Kew's assistance.

By the mid-nineteenth century success in importing and exporting living plants and seeds was still greatly dependent on developing reliable methods of transporting them on long sea voyages, and on effective cooperation between colonial botanical gardens and Kew. Casualties were particularly heavy with consignments of *Hevea* seeds, whose oily content often became rancid, thus inhibiting germination. Kew's annual report for 1875 regretted that barely one per cent of seeds despatched from South America were surviving the journey. The botanic gardens at Calcutta and Peradeniya[12] agreed that seeds should be absolutely ripe and fresh, but had different views on methods of packing: in bags of slightly moist moss, in soil, in dry charcoal, coated with clay, or simply loose in well-ventilated boxes. Both recommended storage in a cool and airy part of the ship. No such precautions were taken with the 11,500 *Hevea* seeds sent from Ceylon to Singapore in 1888. Packed into gunny sacks, most of them nevertheless survived. Clearly there was an element of luck in transporting seeds but by the turn of the century charcoal had largely superseded other ingredients for packaging.

At the close of the eighteenth century the mortality rate of plants in transit had become so serious that Sir Joseph Banks had personally investigated the problem. Within 20 years of his death a system had been devised which guaranteed a reasonable expectation of success. In 1835 a ship brought from Australia some healthy plants in a tightly sealed box with a glass roof which was never opened and the plants never watered during the entire voyage. This box, soon to be universally known as a Wardian case, had been invented by Dr Nathaniel Ward. He had discovered that provided there is sufficient light to allow photosynthesis to take place, the transpiration of plants saturates the air in the sealed container with water vapour and this, condensing on the glass, moistens the soil in a continuous cycle. Sir William Hooker congratulated Ward that his cases "have been the means in the last fifteen years [i.e. 1836–51] of introducing more new and valuable plants to our gardens than were imported during the preceding century".[13] They were used in the cinchona and rubber transfer. Thirty-eight of these Wardian cases and a gardener to look after them safely transported nearly 2,000 rubber seedlings from Kew to Ceylon in 1876. Losses were still incurred: 88 out of a consignment of 414 trees and shrubs despatched to New Zealand in 1876 died, and in 1877 Liberian coffee plants packed into two Wardian cases were all dead on arrival in Queensland. Modifications to packaging procedures and to the positioning of the cases on board ship gradually reduced the failure rate.

Towards the end of his time in office at Kew William Thiselton-Dyer reviewed the profitable rubber trade in Ceylon and Singapore and, as he told the Colonial Office, "the success is a striking illustration of the utility of properly organised botanic establishments without which the enterprise could not have been carried through the experimental stage".[14] Sir Joseph Banks and subsequent directors had recognised the need for botanical and experimental gardens in the outposts of Empire. Without the facilities of such gardens the traffic in plants and seeds would not have progressed so smoothly. Kew maintained a paternal interest in their development, protesting to appropriate authorities whenever their functions were in danger of being changed, usually by conversion into a public park. While Kew willingly offered them professional advice, it encouraged greater self-reliance, especially in the acquisition and dispersal of plants. Joseph Hooker had this independent status in mind when he wrote in his 1878 annual report that

> I am anxious, however, to see the botanic gardens establish to a greater extent than at present a chain of independent interchanges, which would increase their own usefulness and enormously facilitate the work which is done at Kew.

It would, he observed, obviously be easier to send plants to Fiji from Ceylon than from Kew.

Greater independence did not necessarily mean a lessening of Kew's influence, which in any case survived in the Kew-trained gardeners who ran many of the colonial gardens. The Colonial Office and colonial governors almost invariably consulted the Hookers before filling senior appointments in these botanical gardens; often candidates were chosen from Kew's changing population of newly-trained gardeners. When Kew had nobody suitable, there was usually someone known to the Director. For instance, it was Joseph Hooker who nominated J.F. Duthie, then Professor of Natural History at Cirencester Agricultural College, to fill the vacant post of Superintendent at the Saharanpur Botanic Gardens in northern India in 1876. In 1880 Hooker, presumably with the help of his Assistant Director, drafted *Suggestions for the information of colonial governments about to appoint superintendents of botanic gardens, and for the guidance of the superintendents themselves* for distribution by the Colonial Office to every colony. Hooker, who enjoyed cordial relations with most colonial administrators, told Sir Henry Barkly, then Governor at the Cape, that

> I have had more than ever to do for the Colonial Office of late, and especially with W. India colonies, St Helena and the Straits Settlements, and Lord C. threatens me with Gibraltar where he talks of establishing a good garden.[15]

When Sir Henry was Governor of Mauritius, Kew sent a competent propagator there to oversee the preservation of its forests. After the British occupied Cyprus in 1878, it was at Kew's instigation that an officer of the Indian Forest Service surveyed the forest resources of the island. Kew's orbit of influence embraced missionaries as well as government officials, businessmen and traders. Henry Venn, Secretary of the Church Missionary Society from 1841 to 1873, turned to Kew for guidance on the introduction of new crops into West Africa and on the horticultural training of Africans. Hooker could legitimately claim that his institution had become "the botanical head-quarters of the British Empire and its dependencies",[16] an assertion echoed by Thiselton-Dyer when he addressed a meeting at the Colonial Institute in May 1880. On that occasion he described Kew as "a sort of botanical clearing-house or exchange for the Empire".[17] In recognition of Hooker's contributions to colonial prosperity, the Treasury in 1882 agreed to raise his annual salary from £800 to £1,200 over the next five years.[18]

The floristic survey of the Empire, initiated by Sir William Hooker, maintained its momentum during his son's regime, aided by the expanding resources of Kew's herbarium. The seventh and final volume of George Bentham's *Flora Australiensis* appeared in 1878. The first volume of *Flora of Tropical Africa*, another of Sir William's projects, came out in 1868 and its editor, Daniel Oliver, produced a third volume in 1877. J.G. Baker, an assistant in the Kew Herbarium, compiled a *Flora of Mauritius and Seychelles*

(1877). The Director himself concentrated on his *magnum opus*, the *Flora of British India*, the first volume of which was published in 1875.

The museums also reflected Kew's commercial interests; medicinal plants and their products – *Ipecacuanha*, cinchona, opium, etc.; samples of leaf tobacco; beverages such as tea, coffee and cocoa; specimens of paper; hats made of vegetable fibres. The Timber Museum, housed in the Orangery, received in 1878 more than a thousand specimens of timber from the Indian Forest Department.

OPPOSITE:
Marianne North (1830–1890) painting at Grahamstown, South Africa.

In 1880 the collections of the former East India Company's Museum, now no longer wanted by the India Office, were shared out among the British Museum, the South Kensington Museum (now the Victoria and Albert Museum) and Kew Gardens. In 1858 Joseph Hooker had rescued the herbaria of a number of distinguished Indian botanists from the East India Company's notorious cellars; now Kew was being offered what was left of its botanical collections and, as an inducement to accept them, also a sum of £2,000 to house them. In addition, one of the curators of the India Museum, M.C. Cooke, was transferred to Kew for a limited period. Kew also accepted 36 tons of specimens of Indian woods, most of which were discarded when it was discovered there was no inventory of identification and provenance. Kew, however, welcomed the permanent loan of 3,359 flower paintings, mainly by Indian artists, which complemented the Roxburgh drawings Joseph Hooker had borrowed in 1858. The India Office financed the building of a wing at the back of the Museum near the Pond and the installation of a more elegant staircase in 1881.[19] At the same time the original Museum near the Rock Garden was extended at its west end by another 30 feet and a corrugated iron shed, also from the India Office, provided a room for packing parcels and temporary storage.[20]

Pictures, portraits, historical prints of Kew Gardens and flower paintings had always been displayed in the museums but the Marianne North Gallery was the first building devoted entirely to paintings. Miss North had acquired a taste for travel from her father, the Member of Parliament for Hastings, whom she accompanied on tours of Europe and the Middle East. She received lessons in flower painting from a Dutch woman artist and from Valentine Bartholomew, Flower-painter-in-Ordinary to Queen Victoria. The Horticultural Society's gardens at Chiswick provided her with apprentice pieces, and on one of her visits to Kew Sir William Hooker graciously presented her with some blossoms of the *Amherstia nobilis*. On the death of her father – "the one idol and friend of my life" – she began her travels at the age of 40, and furnished with letters of introduction, sought exotic examples of the world's flora to paint. Her first solo expedition in 1871 took her to Jamaica and North America, her second to Brazil; in 1875 Japan was her destination, whence she returned home via Sarawak, Java and Ceylon; India stimulated her to produce well over 200 paintings of buildings as well as of the vegetation. She acquired a speed of execution that enabled her to complete a picture in a day and, it has to be confessed, the results sometimes show it. She would rapidly sketch an outline of her subject in pen and ink on stout paper, her oil paints coming straight from the tube, seldom mixed with any medium. With a palette of

bold and assertive colours, she painted with evident enjoyment, but one wishes, at times, she had applied more restraint, a little more sensitivity. Nevertheless, her output is impressive, a tribute to determination and enthusiasm, a fulfilment of her ambition to record "plants in their homes". Her plants are usually depicted in an ecological setting, occasionally populated by a solitary insect or some inquisitive small creature, and sometimes against a theatrical backdrop of mountains.

She wrote to Sir Joseph Hooker in August 1879[21] offering Kew not only her accumulation of paintings but also a gallery in which to house them. She modestly conceived it as "a rest house for the tired visitors" with perhaps a cottage attached where a resident gardener's wife could serve "tea or coffee & biscuits (nothing else) in the gallery at a fair price". Her only stipulation was a room which she could use as a studio. Kew readily accepted her generous offer (resisting the proposal to serve refreshments, something Hooker had always opposed). She rejected a site behind the Temperate House since it would incur felling some "noble trees", preferring the Queen's Gate with the Temperate House lodge as its neighbour.

She hoped her architect friend, James Fergusson, would be able to design a gallery "rather Indian in its outline". She had in mind, too, an "arcade" or verandah around the outside of the building as a shelter from rain and a support for climbing plants. Fergusson, an authority on the native temples of India, where he had lived for some ten years managing an indigo business, was also a prolific author on architecture, his most enduring work being *A History of Architecture in all Countries from the Earliest Times to the Present Day* (1865–7). With Miss North's Indian predilections, he was an ideal choice to design her gallery. He gave her the verandah she craved, and his structure has faint echoes of the bungalow-type dwellings that Europeans built in India. This commission enabled Fergusson to demonstrate his theories on the lighting of Greek temples with large clerestory windows high above the paintings. Miss North took charge of hanging her pictures, arranging them in geographical order above a dado of 246 vertical strips of different timbers PLATE 23. She added floral sprays on the dull gold surrounds of the doorways, and a deep cove below a narrow gallery received a classical motif. With Africa the only continent not represented in her pictorial record of the world's flora, she dutifully took herself there a few weeks after her gallery had been formally opened to the public in July 1882. Her last journey – to Chile in 1884 – contributed the Monkey Puzzle tree. A small studio had been added to the gallery in 1883 for her own use and that of any *bona fide* artist. The gallery houses 832 paintings delineating over 900 species of plants. Although its tight mosaic of images usually overwhelms all newcomers, its arrangement should be preserved for its uniqueness and as a fitting memorial to the extraordinary woman who conceived it.

Marianne North painted without any of the constraints imposed upon a botanical artist like W.H. Fitch, who observed plants with a disciplined regard for scientific accuracy. Joseph Hooker, an admirer of his talent for "seizing the natural characters of plants", had retained him as artist for *Curtis's Botanical Magazine* after taking over the editorship. He dedicated the 1869 volume of

the magazine to Fitch: "The accomplished artist and lithographer of upwards of two thousand five hundred plates already published in the Botanical Magazine." Fitch had become a freelance artist in 1860, gambling on his ability to attract commissions. He illustrated botanical monographs and textbooks and contributed plates to gardening magazines. The clarity of his line-drawings for George Bentham's *Handbook of the British Flora* (1865) ensured the book's popularity for many years. Flamboyant and exotic flowers such as rhododendrons, orchids and the giant American water lily *Victoria amazonica* had an ideal interpreter in Fitch who always painted with confidence and verve. Despite demands for his services, the uncertainties of employment probably exacerbated his resentment towards Kew, which, he believed, had treated him unfairly. He claimed that he had allowed Sir William Hooker to retain his drawings for *Curtis's Botanical Magazine* after 1860 on the understanding that he received an official allowance or honorarium in addition to the fee he got from the publisher of the magazine. Joseph Hooker continued adding Fitch's drawings to Kew's collections, disputing the artist's right to any payment for them. And so in 1878 Fitch declined to undertake any more work for *Curtis's Botanical Magazine*, an unfortunate conclusion to an association which had lasted for more than 40 years. Maybe a twinge of guilt prompted Sir Joseph to obtain a Civil List pension of £100 a year for Fitch in 1880. Fitch acknowledged this gesture of reconciliation with a wish that "our misunderstanding

Marianne North Gallery. Designed by James Fergusson and opened in July 1882.

(which I sincerely regret) has left no doubt on your mind of my respect for yourself as I had for your father".

His sudden departure left Hooker without an artist for the magazine. His daughter Harriet, now Thiselton-Dyer's wife, who had received some tuition in flower painting from Fitch, temporarily filled the breach; his sister-in-law, Mrs Anne Barnard, also came to the rescue; and he trained a distant cousin, Matilda Smith, who eventually became the magazine's principal artist, contributing over 2,300 plates before her retirement in 1922.

The demands upon him as Director, Civil Service bureaucracy, public complaints about times of opening of the Gardens, and endless reports, minutes and letters made Sir Joseph long for retirement. "Kew is becoming more toilsome than ever," he moaned to Darwin, and the "ordinary correspondence, etc. gets more extraordinary every year", he complained to Maw. Thiselton-Dyer, his most meticulous deputy, despaired of his complete lack of order in office procedures – inadequate records and a chaotic filing system. "Official papers drifted everywhere – Curator's Office, Herbarium, Museum. He never had the remotest idea what had become of them."[22]

> When I was Assistant Director, I thought the best thing I could do was to set Sir Joseph free to complete the work he had on hand. For some ten years beyond an hour which he gave me in the morning, he was free for the rest of the day and I practically carried the whole Garden of Kew on my shoulders.[23]

Even allowing for Thiselton-Dyer's frequent indulgence in hyperbole, it is abundantly clear from an examination of the archives that he dealt with most of the administrative chores. The Director relied on his judgement and loyalty. Doubtless his Assistant Director assisted in drafting a 'Memorandum relative to the requirements of Kew Gardens', submitted to the Office of Works in December 1881.[24] In presenting a case for more clerical assistance and for the conversion of a temporary foreman post into a permanent one, the memorandum paraded Kew's achievements and especially its involvement in colonial affairs. The Director reminded his superiors that the India Office had made him their 'Reporter on Vegetable Products' and that other government departments frequently consulted Kew.

> Kew has become the botanical centre of the work & literally carries on all economic & scientific botanical work of the Empire, under the direction of the various departments of the State.

He assured the First Commissioner that these overseas commitments never interfered with the maintenance and improvement of the Gardens. The complex of glasshouses known as the T-Range had been built, and flower-beds had been drastically reduced and replaced by plants "more suited to the character & objects of the garden & grounds". This well-argued case got the Director an extra clerical post and an unsought increase in his own salary.

Hooker, content to leave administration and the supervision of the Jodrell Laboratory to his Assistant Director, concentrated on the Herbarium and the Arboretum which, he tells Harry Bolus in South Africa, "I am rendering as perfect as I can." Though this landscaping was evidently a source of great satisfaction to him, both Thiselton-Dyer and W.J. Bean, the foreman of the Arboretum, viewed his improvements with some apprehension. Hooker was "exceedingly short-sighted", Thiselton-Dyer reminisced to David Prain.

> This had an unfortunate effect on his garden work. He planted out every view and vista, Oliver père sadly lamented to me. This might also be put down to sheer 'cussedness' but it was not so, & he was a man of sound taste. It was simply he could not see what he was doing. Bean and I gradually undid it all.[25]

Now in his 69th year, suffering from poor health, committee meetings wearied the Director and, according to Thiselton-Dyer, "he got excited over trifles". Hooker was well aware that his deputy chafed under the division of duties — "it produces many little inconveniences and cross purposes".[26]

In October 1885 Thiselton-Dyer learned that Hooker was firmly resolved to retire.

> I am not at ease in respect of the performance of my own duties. You have relieved me of all administration of the Garden, have undertaken the Curator's duties of cultivation & have kept up & enormously increased the Indian, Foreign, Colonial & Home correspondence, extended the Museum, & conducted endless petty & often vexatious jobs; leaving me the Arboretum in the hopes that I should further find time to proceed with the Indian flora.[27]

He retired at the end of November 1885 and moved to Sunningdale in Berkshire, at last able to devote all his time to taxonomic work, giving priority to his *Flora of British India* of which four volumes had already appeared, and to editing *Curtis's Botanical Magazine*. There he adopted a routine of being at his desk by seven in the morning and, with only brief interludes for lunch and a light supper, working until eleven in the evening, except on those days when he visited the Kew Herbarium. It was a productive retirement. With the completion of the Indian flora in 1897, he finished Trimen's *Handbook to the Flora of Ceylon* and summarised the progress of Indian botany for the *Imperial Gazetteer of India*. He died in 1911 at the age of ninety-four, still studying Indian plants.

He had travelled extensively during his long life, enriching our gardens with rhododendrons and alpines from the Himalayas. He cherished the friendship of Charles Darwin, whose researches he helped and whose conclusions he confirmed through his own investigations. A distinguished taxonomist and phytogeographer, his achievements, so his son-in-law affirmed, might have been even greater had he enjoyed "a modest independent income like Bentham" and thus been spared the distractions of official duties.

Sir Joseph Dalton Hooker
(1817–1911).

It is always tempting with the benefit of hindsight to seek confirmation of character in portraits. The one of Hooker as a young man by George Richmond (who always flattered his sitters) projects a dreamy and sensitive personality. The photograph of the mature man in 1868 by Julia Cameron reveals tenacity and dominance with a hint of arrogance – not the face of a man who suffered fools gladly. Darwin admitted he was "testy", and his son-in-law that he was "hot-tempered". He inherited his phenomenal energy, stamina and resolution from his father, but whereas Sir William was always prepared to compromise, his son could be stubbornly inflexible. In their dealings with difficult First Commissioners Sir William behaved with diplomacy while Sir Joseph responded with defiance. But such differences had never threatened the secure bond of affection between father and son. When the Linnean Society presented Sir Joseph with a specially struck gold medal in 1898, he took the opportunity to acknowledge publicly his indebtedness to his father.

> I inherited from him my love of knowledge for its own sake, but this would have availed me little were it not for the guiding hand of one who had himself attained scientific eminence; who by example, precept, and encouragement kept me to the paths which I should follow, launched me in the fields of exploration and research, liberally aided me during his lifetime, and paved for me the way to the position he so long held at Kew with so great credit to himself, and benefit especially to our Indian and Colonial possessions.[28]

For 44 years the Hooker had directed Kew and produced policies that are still relevant today.

CHAPTER SEVENTEEN

An autocrat at Kew

*W.T. Thiselton-Dyer as Director — Rock
Garden — Arboretum improvements — Queen's
Cottage grounds — Kew Palace and demolition
of its stables — Temperate House completed —
Refreshment Pavilion — public demand for
earlier opening continues — Women gardeners
— student gardeners — Kew's relations with
Natural History Museum — Jodrell
Laboratory staff — Cambridge Cottage — Kew
transferred to Board of Agriculture*

WHEN WILLIAM THISELTON-DYER became Director in 1885 it was,
as he himself later remarked, more a change of status than of responsibility
since for much of the preceding decade he had been *de facto* administrator.
The Hookers had restored Kew, and although the momentum of reform had
slowed down, Thiselton-Dyer, now his own master, developed activities more
congenial to himself such as strengthening imperial links and indulging a
penchant for landscaping.

His first major landscape project had been the creation of the Rock Garden,
in 1882 while he was still Assistant Director. Rockwork appeared in British
gardens during the early eighteenth century, usually in association with grot-
toes. What seems to have been an elaborate piece of rockwork, studded with
pebbles and shells, was designed by Aaron Hill for his garden in Westminster.
Philip Miller's *Gardener's Dictionary* of 1768 commended ornamental rockwork. In 1773, as we have seen, the young Joseph Banks presented both the
Chelsea Physic Garden and Kew with some of the lava used as ballast on the
ship bringing him back from Iceland. Chelsea made a rock garden near the
statue of Sir Hans Sloane with their material, and Kew constructed a moss
garden which impressed a French visitor in 1784.

> As the lavas are full of cavities, fissures, and roughnesses, and are
> likewise spongy, and capable of imbibing and long retaining water, it
> was resolved to form thick borders of them, more or less elevated
> round the verges of a shady piece of ground, appropriated to this
> moss-garden, which is unique of its kind.[1]

This moss garden may have survived until 1841, when it was referred to as "rock-works beds, that is, beds having small masses of stone distributed over them in different parts to imitate rock, among which mosses, ferns, etc. may grow".[2] In August that same year the *Gardeners' Chronicle* reported a rockery being replaced by a flower-bed, but also that a rockery of hardy ferns still surrounded two sides of the British Garden. Apparently Kew's gardeners laid out "heaps of rubbish commonly designated rockwork" which the *Gardener's Magazine* abhorred.

> In a pleasure-ground, a rockwork is chiefly to be considered as a pictorial feature; in a botanic garden it ought never, in our opinion, to be resorted to, except for such rock plants as will not thrive on the general surface of the garden.[3]

Kew's rock gardens were settings for mats of ivy and British and North American ferns; there was such a display on the north-facing wall of the Botany Bay house, and inside one of the glasshouses rockwork supported a healthy growth of tender ferns. In 1867 alpines were planted in a rockery of Reigate sandstone on the north side of the ice-house. William Robinson, a pioneer of natural gardening and the author of *Alpine Flowers for English Gardens* (1870), who denounced "the hideous piles of rubbish that go by the name of 'rockwork' all over the country",[4] declared that "Few public gardens show worse examples

This type of rock garden at Kew was condemned by William Robinson. *Garden*, 16 March 1872, p. 379.

of the traditional rockwork than Kew." In September 1872 Joseph Hooker added another rock garden for plants he had collected in Morocco. Robinson continued to grumble, criticising this time Kew's practice of scattering old bricks in flower-beds; 'brick-rubbishesque' he called it, but he had nothing but praise for the replanting of the ice-house rock garden – "it is now looking better than any rockery I have seen this season".[5] It had been such a success with visitors that during 1873/4 it was enlarged, the population of alpines increased to 560, a wire fence erected to protect them, and as an extension to the rock garden, a new hardy fernery was planted the whole length of a 40-yard winding path.

In the late summer of 1881 enthusiastic growers of alpines addressed a

Rock Garden about 20 years after its creation in 1882 by Sir William Thiselton-Dyer.

memorial to the Office of Works urging a much larger rock garden at Kew, one that would incorporate bogs, pools and running water. While the First Commissioner considered their petition, on 21 October 1881 George Curling Joad, the owner of one of the largest collections of alpines in the country, died, leaving nearly 3,000 plants to Kew, a bequest conditional on their immediate transfer. The existing rock garden being manifestly too small for such an acquisition, the Treasury granted £500 for the construction of a larger one which the memorialists had wanted. A fringe of stones around the lake or an adaptation of the ha-ha skirting the Palace lawn were ideas rejected in favour of a site between the T-Range and the Herbaceous Ground, whose gardeners would be responsible for the new rockery. Thiselton-Dyer, who had accepted the challenge of designing it, was determined to avoid "something uncouth and obtrusive – at the worst, even cockneyfied",[6] preferring an effect "suave and ample" (whatever that meant). He decided to create a valley on this narrow, long and flat piece of ground; a winding path, some 540 feet long, would simulate a rocky watercourse, typical of those found in the valleys of the Pyrenees. Insufficient funds for its construction were supplemented by gifts of Cheddar limestone and well-weathered Bath oolite, and by a thorough excavation of the forlorn remains of the old Stone House, built by the young sons of George III. Large tree roots, much favoured by Thiselton-Dyer, buttressed curving banks along the path. An unpretentious waterfall fed a bog garden, conifers were planted on a couple of mounds, and a screen of *Rhododendron ponticum* and small box provided protective seclusion. The whole operation

STONE HOUSE, RICHMOND GARDENS.

ENTRANCE TO THE CAVE, RICHMOND GARDENS.

MERLIN'S CAVE IN KEW PLEASURE-GROUNDS.

The remains of former buildings at Kew were salvaged for the Rock Garden and other projects. G.E. Papendiek drew (a) the old Stone House and (b) the Cave, the latter being perhaps the remnants of the Hermitage. The same building is erroneously described in *Leisure Hour* for 1862 as 'Merlin's Cave' (c).

was completed in just three months and in June 1882 the public could enjoy Kew's latest feature. The superseded rock garden near the ice-house now displayed the hardy fern collection, substantially increased by W.C. Carbonell's legacy, and gifts from fern growers like E.J. Lowe, E.F. Fox and A.M. Jones.

Wood engraving of a drawing by W.H. Fitch of two of the islands on the Lake. Sir W. Thiselton-Dyer postulated that such "islands should be heavily wooded with well disposed clumps of trees. These give effects of light and shadow on the water which are often in striking contrast." *Garden*, 6 January 1906, p. 2.

We learn from W.J. Bean and W. Dallimore, both senior gardeners under Thiselton-Dyer, that he held very positive views on landscaping. He believed that both scientific and aesthetic criteria could be accommodated – the public had a right to see ornamental trees and flowers as well as those solely of botanical interest. Paths had only a utilitarian function, a means of public access, and as many as necessary should be provided so long as they were straight and

Water Lily Pond off
Cedar Vista, created in
1897.

direct. Water should always be an integral part of a composition, never a dominant feature. His sensitive treatment of the Lake, fashioned by Sir William Hooker and improved by his son, is a practical demonstration of his tenet that such a stretch of water should be revealed gradually to the visitor through clumps of trees; its surface fragmented by thickly wooded islands; its margins irregularly contoured by promontories and softened by low vegetation; its banks shaped and gently sloped to the water's edge.[7]

The Lake had been formed from an old gravel working; another disused gravel pit off Cedar Vista, made water-tight and heated by waste warm water from the nearby pumping station, grew tender water lilies and other aquatics.

Thiselton-Dyer appreciated what the Georgians called a prospect but panoramic views were not possible in an enclosed garden like Kew. In 1891 he sought permission to open a view into and beyond the Old Deer Park[8]: a glimpse of the Royal Observatory could, he claimed, be gained by the thinning of some plantations; the removal of merely one branch of a tree would disclose the tower of Isleworth church. When the northern part of the Old Deer Park was considered in 1900 as a possible site for the National Physical Laboratory, he feared it would mar the view across the Park from the Kew Road. He had an instinctive eye for cosmetic treatment. He made a cutting through a high bank at the southern end of the Herbaceous Ground in order to give greater prominence to the Mound and Temple of Eolus.

But it was the Arboretum – "strangely unkempt and ill-cultivated", according to Bean – that presented him with irresistible challenges and scope to impose coherence, to introduce precisely defined spaces, vistas and paths. Trees that had to be left while the new Arboretum was being planted by Sir William Hooker – oaks, beeches and elms – and now dwarfing the species collections, were ruthlessly felled in their hundreds, the Director completely impervious to public outcry. The *Saturday Review* spoke for many when it deplored the destruction of

OPPOSITE:
Decimus Burton's replica of Chambers's Temple of Eolus on the Mound which Sir W. Thiselton-Dyer later turned into a wild garden.

> many noble trees of a century's growth . . . and if we may judge from what appears, nothing but a vulgar little taste in landscape-gardening is the motive for these changes; and the end of them will be the transformation of Kew Gardens into something like a pretty tea-garden. It is reported that the Director loves not trees in masses.[9]

Thickets of laurel and *Rhododendron ponticum* were uprooted and masses of berberis, forsythia and other flowering shrubs replaced them. Rhododendron Dell was thinned and replanted. Pruning became a passion with him. All paths were grassed and existing lawns relaid to satisfy his demand for immaculate order. Nothing escaped his censorious eye. After one of his frequent inspections of the Gardens he sent Dallimore a reprimand:

> seat [on] Palace lawn should be moved; broken branch on tree near Brentford Gate; bottle under bush end of Syon Vista; patch of nettles in wood near Bamboo Garden; weeds in ha-ha near Isleworth Gate; several docks in wood behind Stables; pale off fence in Stable Yard; brush out of place in Stable; patch of *Impatiens parviflora* in shrubbery near Stable Yard; brown marks on messroom table; wheel marks on lawn near picture gallery; grass near Pagoda marked by horses.[10]

A disused gravel pit near Rhododendron Dell filled with several years' accumulation of rubbish disappeared during the Director's programme of improvements. Its shallow depression offered convenient shelter for a collection of bamboos then surviving in poor soil near the Temperate House PLATE 29. During 1891–2 Bean organised their transfer to this new location, planting out some 40 kinds, mainly from India and Japan.

Another gravel pit, used for burning rubbish and as a dump for refuse from the tea pavilion, was transformed in 1895–6 into a sunken garden for rambler and bush roses. During this conversion, fragments of painted plaster were uncovered, confirming it as the site of Chambers's Alhambra. Another display of roses covered a wooden pergola between the Rock Garden and the Herbaceous Ground.

The high wire fence stretching from the Unicorn Gate to Kew Palace had separated the Botanic Garden and the Pleasure Grounds since 1843. In 1895

it was dismantled, finally reuniting the two areas. In the same year another barrier was removed when Queen Victoria gave permission for the 4½-acre lawn in front of Kew Palace to be added to the Gardens. Its boundary ditch or ha-ha was filled in, a step welcomed by the Director who believed it had been used for "immoral purposes".

The rest of the Palace grounds to the west of the lawn still remained private until 1898 when the Queen ceded it and the Queen's Cottage grounds of 37 acres to Kew. Seldom ever visited by any member of the royal family, the grounds around the Cottage had become a wilderness with trees lying where they had fallen. It was a condition of the gift that this area should remain in its present natural state and, indeed, apart from some judicious pruning and tidying the Director also wanted to preserve its "great sylvan beauty . . . much admired as a scenic adjunct to the Royal Gardens".[11] The naturalist, W.H. Hudson, and the Linnean Society on behalf of all ornithologists urged as few alterations as possible to this suburban haven for birds. In view of this public concern, only about a third of the grounds, of no particular merit, was incorporated into the Gardens and a railed grass path from the Isleworth Ferry Gate enabled the public to see the Cottage from a distance. At the time of this transfer, some pictures and furniture from the Cottage were removed to Windsor Castle.

Thiselton-Dyer had designs on another piece of royal property at Kew, none other than Kew Palace itself. This royal residence had been shut up after Queen Charlotte's death, George IV had momentarily considered pulling it down, and William IV got Sir Jeffry Wyatville to plan an (unbuilt) extension to it. But it received no more than essential repairs: in 1845–6 some of its

W. Rutton said that this wooden porch was attached to Kew Palace in 1801–2, but he does not give his authority. It had been installed when G. Papendiek drew it c.1820. D. Burton assured Sir W. Hooker that he would "continue to wage war against the shocking bad wood porch of the Palace". It was eventually removed about 1880.

chimneys were rebuilt, and in the 1880s that part of the service wing nearest to the Palace, now in a "ruinous state", was demolished. In June 1890 the Director investigated the possibility of the empty Palace being converted into a museum, perhaps devoted to the history of the Gardens.[12] His father-in-law, seven years earlier, had suggested that "it could be put to good use as a Museum of Vegetable Pathology in connection with education in forestry".[13] Queen Victoria agreed in December 1896 to the acquisition of Kew Palace by Kew as a museum provided no alterations were made to the room where Queen Charlotte had died.[14] Since a forestry school had been established at the Royal Indian Engineering College at Cooper's Hill, and the Institute of Surveyors now intended to display British timbers in their new building, Thiselton-Dyer abandoned the idea of a Forestry Museum in Kew Palace, preferring his original notion of the history of Kew Gardens as a principal theme. The transfer had been agreed when an obstacle arose in the guise of Mrs Wheatstone.

Mrs Wheatstone, for many years in the service of the Duke of Cambridge, had in 1865 been given the old lodge in front of the Herbarium as a residence. The Director, absolutely convinced "that the building, while in Mrs Wheatstone's occupation, was used as a brothel",[15] resolved to get her out. After it had been declared unfit for human habitation, Mrs Wheatstone, as an old retainer, got Kew Palace lodge as a grace-and-favour residence. There she lived with her 'brother' who some years earlier had been imprisoned for assaulting a young girl. With the Palace about to be opened to the public, the Director refused all responsibility for it until Mrs Wheatstone was ejected from the lodge. "It is becoming a pitched battle between this woman and the Government," he declared, and when she opened a public tea garden behind the lodge in the summer that was the last straw. His intemperate language antagonised the Board of Works which refused to move Mrs Wheatstone in view of her advanced age. In a peremptory note the Board warned the Director that

> If you are not prepared to take charge of Kew Palace, it will be placed together with surrounding buildings in charge of the Board's Surveyor.[16]

This provoked a defiant response:

> But to have a woman of Mrs Wheatstone's known character, placed on my staff, I will not.

The upshot of the contretemps was that the Surveyor did indeed take charge, converting some of the Palace outbuildings into a Works Depot in 1899. It was recommended that when Miss Hodgson, custodian of the Queen's Cottage, terminated her tenancy of one of the outlying buildings of the former White House, it might be given to one of the Gardens' foremen, and the adjacent Georgian kitchen block converted into a messroom for the labourers.

Standing just east of the Palace in picturesque symmetry with these outbuildings were the stables, which obstructed a view of the Palace from the Main

Gates. The Director wished to demolish them, and an opportunity to do so arose in 1905 after a violent storm had blown away part of their roof. He immediately ordered the felling of an adjacent large elm to expose and emphasise the extent of the damage, then summoned an official of the Board of Works, persuading him that the stables had to go. In no time he had the site levelled and grassed. It was his last exercise in landscaping at Kew. When he summarised his achievements in his valedictory minute to the Secretary of the Board of Agriculture in December 1905, he reminded him of the removal of "obstructive fences" and that "grass drives facilitating the movements of large crowds have been driven through the woods, and sylvan scenery of great beauty has been made accessible". He could also have mentioned the greater informality he had introduced – replacing box-edged beds and gravel walks on the Palm House terrace by less complex displays, planting a wild garden on the Mound near the Pond and another on the Flagstaff Mound, introducing narcissi in clumps throughout the Gardens, everywhere encouraging a more naturalistic presentation of herbaceous plants, trees and shrubs, especially roses. His rejection of elaborate summer bedding won the gratitude of William Robinson, who dedicated the 1898 edition of his *English Flower Garden* to William Thiselton-Dyer "under whose direction the Royal Garden at Kew has

The Alpine House after it had been enlarged in 1891.

become a beautiful flower-garden while more and more useful as a botanical one". Camille Pissarro, one of many artists attracted to the Gardens, wrote to a friend of a visit in June 1892.

> I am at Kew, profiting by the exceptional summer to throw myself wholeheartedly into *plein air* studies in this wonderful garden of Kew. Oh! my dear friend, what trees! what lawns! what undulations of the ground.

The contents of glasshouses as well as hardy plants received the Director's assiduous attention. While still Assistant Director, he had taken an interest in exotics.

> The Palm House, which is Dyer's special pet, is magnificent, and he has gone in for Cycads, and by correspondence all over the world got together a wonderful collection of them.[17]

Thiselton-Dyer had many of the glasshouses reroofed, reglazed or their heating improved. In 1887 he had a house built for alpines, situated not far from his Rock Garden; it was enlarged in 1891. The Temperate Fern House was replaced by a light iron span-roofed construction. The acquisition of Cooper-Forster's notable collection of filmy ferns in 1888 warranted a new house for them in 1892. In 1897 the collection of insectivorous plants (*Nepenthes* sp.) found another home in an extension to the T-Range. Two orchid houses, also in the T-Range, were completely reconstructed in 1898 to conform with modern cultivation techniques.

While this building and renovation went on, the Temperate House stood, reproachfully, in the unfinished state in which it had been left in 1862: a central block and two octagons. Some of the ironwork for the wings had lain in the contractor's yard for many years before being broken up. The Director raised the possibility of its completion in October 1891, but it was not until June 1894 that Joseph Chamberlain, a knowledgeable gardener and friend of Kew, and at that time Secretary of State for the Colonies, used his influence with the Chancellor of the Exchequer to get the Treasury to sanction the erection of one of the wings. The south wing or Mexican House was opened in July 1897, and in May 1899 the public were also admitted to the north wing or Himalayan House. The Temperate House, now with an area of 1⅜ acres the largest glasshouse in the world – twice the area of the Palm House – displayed most of its plants in beds of soil with pots positioned on a narrow bench around the central block. The vegetation of Australasia dominated the central area. The unheated Himalayan House suited plants from northern India, China and Japan which required cooler conditions. Cornish gardens supplied mature specimens of Himalayan rhododendrons, some derived from Sir Joseph Hooker's seeds. Both wings made a feature of rockwork. The octagons which connected the three main compartments had citrus fruit in one and standard bays in the other.

The Succulent House, erected in 1855, was replaced in 1904 by a lighter construction.

The completion of the Temperate House and the award of a knighthood made 1899 Thiselton-Dyer's *annus mirabilis*. One has a sneaking suspicion that he derived just as much satisfaction from getting rid of green glass. Kew had ceased using green glass in some houses in 1886 but cautiously continued to glaze the Palm House and ferneries with it. But when the Director saw how ferns flourished under a canopy of clear glass in Sir Trevor Lawrence's garden at Burford, he determined that Kew should follow suit. And when the plants showed no ill-effects, he sought permission in 1894 to banish green glass altogether, to be rid of its "unsightliness".[18]

All these improvements, together with better transport facilities and cheaper fares, made Kew Gardens extremely popular with the public. They came, said

the Director, mainly from the "lower middle classes . . . The upper classes, with the exception of individuals with horticultural or scientific tastes, take, I think, but little interest in Kew." The *Gardener's Magazine* (29 October 1887) defined Kew's role as twofold: "to promote the scientific study of plants" and "the entertainment of the public". One facility visitors had always wanted was a restaurant or tea-room since regulations forbade bringing food into the Gardens. The horticultural writer Shirley Hibberd had proposed turning the grounds at the Queen's Cottage into a public picnic area. Eventually a refreshment pavilion situated between the Temperate House and the Marianne North Gallery was opened in 1888. It was an unobtrusive building that looked vaguely like a Swiss chalet, offering tea and 'cold collations' during the summer months. Most visitors, however, found it too remote now that there was no intention of opening the Queen's Gate. This gate, between the Marianne North Gallery and the Temperate House Lodge, had never opened since its erection in 1868. It was now moved to a position opposite Lichfield Road so as to be nearer the revised location of the railway station at Kew. Formally opened on the Queen's birthday, 27 May 1889, it was appropriately designated the Victoria Gate.

The Kew Gardens Public Rights Defence Association lost the battle to keep the gate on its original site. But its crusade for earlier opening hours had been

The Refreshment Pavilion, opened in 1888; and burned down by Suffragettes in 1913.

taken up by Richmond town council and some determined individuals. A deputation of councillors and local residents met the First Commissioner of Works and the Director of Kew Gardens in November 1892 to listen to the familiar litany of excuses for retaining the present hours: the need to preserve the exclusivity of the Gardens in the mornings for scientific research, and the necessity to avoid the extra costs that earlier opening would incur. Undeterred, the protesters kept trying. A debate in the House of Commons on 9 April 1897, despite a vigorous defence of Kew by its ally, Joseph Chamberlain, confirmed that not all Members of Parliament were convinced by the Government's wish to maintain the *status quo*; one MP described Kew's attitude as "pure cussedness". In the face of mounting pressure the Gardens were finally opened from 10 a.m. between June and September in 1898 with an assurance that should the experiment prove successful it would be made permanent.

Sir William Thiselton-Dyer's peremptory manner, his lack of tact, and his inability to make concessions gracefully, got him into trouble not only with MPs but also with royalty. He reprimanded the old Duke of Cambridge when he rode his horse on newly-gravelled paths in the Gardens, and he stopped the practice of sending boxes of cut flowers to the Duchess. When the vicar of St Anne's church on Kew Green, who was also a chaplain to Queen Victoria, obtained permission to hold a bazaar in the Queen's Cottage, Sir William attempted to prevent a carriage with a royal passenger driving through the Gardens instead of approaching the Cottage through the Old Deer Park. He said it contravened the Park Rules and Regulations. Princess Mary Adelaide, Duchess of Teck and daughter of the Duke of Cambridge, also complained that her carriage was not allowed to enter the Gardens.

An authoritarian by nature, Thiselton-Dyer demanded unquestioning loyalty and obedience from his staff. He saw the wearing of uniforms not only as a symbol of status but also as a means of imposing order on the establishment. The constabulary were already in uniform and he had an inspector's uniform made for himself, the carters worked in conventional breeches and brown leggings, and the stokers were distinguished by blue uniforms with red piping. The journeymen-gardeners who had worn fustian, broadcloth or cord, according to their liking, were from 1899 compelled to report for duty in blue serge suits and grey flannel shirts with turned-down collars. The volunteer crew of the Gardens' fire brigade, however, resisted all his attempts to get them into uniform. He had a delicate problem with three women gardeners, recruited in 1896 from the Horticultural College for Women at Swanley in Kent. He made it an essential condition of their employment that they wore clothing unlikely to provoke their male colleagues: the luckless trio consequently laboured in brown bloomers, woollen stockings, waistcoat, jacket and peaked cap. Sir Joseph Hooker in retirement absolutely disapproved of their appointment.

> At one time some women (not ladies in any sense of the word) gardeners
> were employed at Kew but there are none now. Sir Joseph says he could
> not possibly recommend any lady to go there. She would have to work

Two women gardeners in the Rock Garden, c. 1898.

with the labouring men, doing all they have to do, digging, manuring, and all the other disagreeable parts of gardening. Then there is the work in the hothouses; the men, I believe, work simply in their trousers, and how could a lady work with them.[19]

One would really like to know whether their presence did distract men from the work and especially their studies. The journeymen-gardeners or students were offered lectures on elementary botany, physics and chemistry after work on most evenings, held from 1898 in a large room adjoining Descanso House; they also had access to a collection of horticultural books. "I very much doubt, however," opined Thiselton-Dyer, "whether these advantages add very much to the capacity of our men as gardeners."[20] Lectures were poorly attended since they were held in the students' own time. They petitioned for them to be included within working hours but a growing discontent about their wages presented a more serious grievance.

Young gardeners, seeking a brief appointment at Kew to improve their expertise, had to be between 20 and 25 years of age, with not less than five years' employment in private gardens or nurseries, part of which had to be experience of plants cultivated under glass. Their working day lasted from 6 a.m. to 6 p.m., followed by lectures or study in the library (the latter, though optional, was noted on their course record). For this they received 18 shillings a week with an additional allowance for any Sunday duty. Those who satisfactorily completed the course became eligible for promotion to sub-foreman as vacancies arose, and for any appointments at home or in the colonies and India. Kew's gardeners earned less than employees in the royal parks although they were partly compensated by free tuition, paid leave and sick pay benefit. This disparity in wages was brought to the attention of the First Commissioner of the Office of Works on 13 March 1891 by a delegation representing the Richmond Borough Council and local residents. Three years later Kew's labourers received a shilling a week increase but the gardeners got nothing. The British Gardeners' Association, formed in June 1904 "to promote the interests of gardeners, regulate their wages and hours of working", elected William Watson, an Assistant Curator at Kew, as its Honorary Secretary. This being a national body, Kew's gardeners formed the Kew Employees' Union in February 1905, as a branch of the United Government Workers' Federation to deal with their particular grievances. William Purdom, whom they had elected as their dynamic secretary, petitioned the Board of Agriculture, as Kew's parent body, for a substantial improvement in wages. "The fact of the matter is that the majority of Kew employees do not receive a living wage."[21] Thiselton-Dyer was not at all pleased.

> The memorial from the 'Kew Employees Union' has been submitted to me through the Curator. I have seen Mr Purdom, the Secretary, and have told him that I decline to receive it officially or to lay it before the Board [of Agriculture] for the following reasons:
>
> I am not prepared to recognise any body which comes between me and my staff or professes to speak on their behalf . . .
>
> PS. I have also told Purdom that I regard his communications with the Board as a breach of discipline . . .[22]

Furthermore, he unequivocally announced his dislike of trade unions in a letter to the *Gardeners' Chronicle* (10 June 1905):

> I do not think that it would be possible to bring the scattered units of the gardening profession under the control of anything like a trades-union. If it were, I think anyone would be shy of engaging a gardener who had at his back an organisation proposing to interfere between employer and employed . . .

With determination on one side and obduracy on the other, the unrest produced an inevitable confrontation. When a public meeting at Kew in October 1905 was addressed by Labour MPs, William Purdom and the Chairman of the Kew Employees' Union were summarily dismissed by the Director. After the intervention of some MPs and the Government Workers' Federation, the President of the Board of Agriculture ordered their reinstatement without loss of wages.

It needed only one visit to Kew on 19 May 1896 by Miss Beatrix Potter to size up the Director as "a radical if no Trades Unionist".[23] She had been studying and drawing fungi for some years and through her uncle, the chemist Sir Henry Roscoe, had secured permission to visit the Herbarium to discuss her methods of growing spores. She briefly met the Keeper of the Herbarium, J.G. Baker, a "slim, timid-looking old gentleman" with "an appearance of having been dried in blotting paper under a press". With the irreverence of youth she remembered Sir William with amusement.

> Mr Thiselton-Dyer puffed his cigarette, vituperated the weather, the rate of wages, discoursed vaingloriously upon his Establishment and arrangements and his hyacinths, better than the Dutch.

She returned some three weeks later to resume her talks with the mycologist G.E. Massee, who had, she reckoned, "passed several stages of development into a fungus himself". With ill-concealed irritation, Thiselton-Dyer declined to see her drawings or to consider her theories on spore culture, referring her disdainfully to Cambridge. Her paper, 'On the germination of the spores of Agaricineae', was presented to a meeting of the Linnean Society on 1 April 1897 by Massee himself, evidently won over by this very persuasive young woman.

J.G. Baker supervised about six assistants trying to cope with the avalanche of dried plants from all parts of the world. They worked industriously in monastic seclusion, withdrawing to a solitary oil lamp in the library when it became too dark in the Herbarium, gas lighting still being viewed as a fire hazard. At closing time, the Keeper, bearing this lamp, solemnly led the way out of the building. T.A. Sprague, who had joined the staff in 1900, recalled that their pay at that time was only £80 a year, insufficient to attract any but the most dedicated botanists. There is an anecdote that W.B. Turrill, Keeper about a half a century later, liked to tell.[24] One feels sure it is apocryphal, but nevertheless it says something about the Director's reputation. The Herbarium staff, we are told, asked Thiselton-Dyer to support their claim for an increase in salary. Having heard they found clothing expensive, he is reported to have bought a second-hand suit in Richmond for seven shillings and sixpence, put it on and paraded around the Herbarium to demonstrate how cheaply one could live. However, in 1890 he willingly forwarded their application for a pay award to the Office of Works, strengthening their case by pointing out that the botanists at the British Museum with identical duties enjoyed higher scales of remuneration.

This was one of the many causes of friction between the British Museum and Kew. Co-existence had degenerated into rivalry. William Carruthers, Keeper of Botany at the British Museum since 1871, had fought pugnaciously during the Ayrton controversy in defence of his department, which he believed Kew coveted. He had always suspected Sir Joseph Hooker's motives and intensely disliked Thiselton-Dyer. In this ongoing confrontation he had an eager but imprudent ally in his assistant, James Britten, a former assistant in the Kew Herbarium. Britten's brief stay there, apparently an unhappy experience, had converted him into Kew's most vocal critic. He continually attacked and ridiculed Kew in two periodicals which he edited – *Journal of Botany* and *Nature Notes*. His letter in the *Daily Chronicle* for 19 January 1893 charged Thiselton-Dyer with plagiarising extensive unacknowledged quotations from the *Journal of Botany*, in Kew's *Official Guide to the North Gallery* (this had, in fact, been compiled by Botting Hemsley and not the Director). His favourite target was the *Kew Bulletin of Miscellaneous Information* whose articles, he maintained, were anonymous because the Director did not want to give credit to his staff as contributors. And always referring to the Director as plain 'Dyer' was deliberately provocative. A public apology for malicious reporting forced from this maverick editor did nothing to reconcile the two national herbaria. Kew accused the British Museum of not honouring a long-established practice of sharing duplicate specimens and Carruthers retaliated in a similar vein. In 1890 Thiselton-Dyer tried to resolve the impasse by recommending the formation of a committee, to be appointed by the Treasury, to lay down guidelines for both establishments in order to minimise overlap of work and overstaffing.

When Carruthers retired in 1895 his replacement was another Scot, G.R.M. Murray. But it was a decade before Thiselton-Dyer's proposal was taken up. The Treasury convened a Departmental Committee on Botanical Work to investigate the reorganisation of Kew and the botanical department at the British Museum, and to report any changes necessary "to avoid duplication of work and collections of the two institutions".[25] A number of the issues raised had already been considered by the Royal Commission in 1847–8, the sub-committee of the British Museum Trustees in 1858, the Select Committee of the House of Commons in 1860 and the Commission on Scientific Information and the Advancement of Science in 1871–5. It was a case of *déjà vu*, particularly the pros and cons of the fusion of the two herbaria. As one would expect, Thiselton-Dyer's presentation was the longest and most exhaustive. He extolled Kew as a place of research, dismissing the British Museum (Natural History) as merely a repository. He reminded the Committee of Kew's commitments to the Empire. He could not resist a swipe at James Britten: "an accusation of bad faith towards my staff was launched against me by a member of staff of the Natural History Museum". After 14 sittings, the examination of 18 witnesses, and the perusal of relevant documents, a majority report of the Committee concluded that the two herbaria should be united at Kew under the aegis of an advisory board with representatives from the Colonial Office, Foreign Office, and India Office, the Royal Society, and the British Museum. It also recommended that the Director at Kew should be the official adviser to the Government on all matters regarding botany. It was all very satisfying to Sir William Thiselton-Dyer.

The Committee had advised the Director to give urgent attention to fireproofing the Herbarium. Some steps had already been taken. In 1898 the Keeper's adjacent residence (the building on the corner of Ferry Lane, now known as Hanover House) had been partially demolished because its proximity to the Herbarium presented a potential fire hazard. It would, however, be impossible to fire-proof the Herbarium without first removing all its existing cabinets. Another wing, besides providing temporary accommodation for the contents of the existing one, would alleviate the desperate shortage of space.

> I cannot control the expansion of Kew Herbarium because I cannot control the expansion of the Empire. The scientific investigation of new territories follows their accretion.[26]

Kew's enhanced reputation following the publication of the Committee's report in 1901 eased the negotiation for a new wing which was erected in 1902 in conformity with current fire regulations. In 1903 the original wing was gutted, fire-proof floors laid and its galleries widened by 18 inches. The highly ornamental ironwork around the galleries most likely disappeared during this alteration. In 1904 electric lighting was installed throughout the building.

The burden of administration left the Director little time for the Jodrell Laboratory, which after its first intake of researchers was disappointingly under-used. He therefore welcomed the unpaid services of D.H. Scott, lecturer at the Royal College of Science, South Kensington, who was appointed Honorary Keeper in October 1892. A man of independent means, Scott paid the salaries of three assistants: G.T. Gwilliam, W.C. Worsdell and L.A. Boodle. His reputation as a plant anatomist and physiologist attracted a younger generation of botanists to Kew, and his own interest in Coal Measure plants briefly established the Jodrell Laboratory as a national centre for palaeobotanical research. As Thiselton-Dyer admitted in retirement, the research carried out at Kew under Scott's guidance could have been pursued elsewhere but the Jodrell Laboratory's "provision of facilities with a sympathetic atmosphere may have operated as a determining influence".[27]

Within a stone's throw of the Jodrell Laboratory stood the home of the 2nd Duke of Cambridge, whose death in 1904 brought Kew additional accommodation and several acres of ground. Lord Bute had lived there while he advised Princess Augusta on the improvement of her Kew estate. In 1772 the young princes, William and Edward, lodged there when George III purchased the property. His son Adolphus Frederick, whom he had created Duke of Cambridge in 1801, received Cambridge Cottage as a residence in 1806 while the Duke of Cumberland got King's Cottage next door. A favourite with his parents, the Duke of Cambridge (unlike his brothers) had no obvious vices: cultured, amiable and, in his dotage, a trifle eccentric. In 1813 he represented the Prince Regent in Hanover, becoming Governor-General there next year. On the death of William IV in 1837, when the Duke of Cumberland ascended the throne of Hanover, the Duke of Cambridge returned to England.

As Ranger of Richmond Park the Duke could have resided in the Old Lodge, but as that was in a ruinous state he chose Cambridge Cottage. In 1839 the sanctioning of a new east wing gave him a drawing-room and additional bedrooms and in 1840 a *porte cochère* dignified the modest brick exterior. The Duchess of Cambridge continued in residence after her husband's death in 1850, and her younger daughter, Princess Mary Adelaide (the future mother of Queen Mary) frequently offered Sir William Hooker unsolicited

and unwelcome advice on the arrangement of the flower-beds in the Botanic
Garden. On the death of the Duchess in 1889, her son George, the 2nd Duke
and Field Marshal, moved in. When the Director opposed the Duke's attempt
in 1891 to annexe a portion of Kew Gardens, he was placated with a plot in
the grounds of Kew Palace. A forestry museum at Kew had been envisaged
by Sir Joseph Hooker, and Thiselton-Dyer saw Cambridge Cottage as a means
of realising this aim. In December 1904, on the Duke's death, the King agreed
that this royal property could be used "as a Museum of Forestry, as quarters
for the staff of the Gardens, and for other cognate purposes." The Director
wanted the museum to include pomology and diseases of farm and garden
crops as well as forestry. It was also his intention to maintain the late Duke's
garden as a *hortus conclusus* within its brick boundary walls. But Thiselton-
Dyer's retirement in 1905 delayed the implementation of any of his plans for
this latest acquisition.

The development which held out promising though unpredictable prospects
for Kew's future came at the end of Thiselton-Dyer's administration. The
official review of the respective botanical roles of the Natural History Museum
and Kew Gardens in 1900 very likely prompted the confidential discussions
between the Chancellor of the Exchequer, the First Commissioner of Works
and the President of the Board of Agriculture on the desirability of transferring
the Royal Botanic Gardens to the Board of Agriculture. The Director had not
been invited to these deliberations but a step towards a transfer had already
been made in 1900 when the Treasury authorised Kew to undertake research
on behalf of the Board of Agriculture which now perceived mutual benefits
in a permanent partnership. The Board believed that their assumption of
responsibility for Kew could eliminate duplication in some fields of research

– certain botanical investigations, the examination of seeds, plant introduction, and the propagation of new varieties were cited.[28] Furthermore, Kew's expertise could be of immense value to farmers and horticulturists.

Thiselton-Dyer raised no objections apart from having doubts about the applicability of the Parks Regulation Act under the new regime. The Treasury approved the transfer provided the Foreign Office, the Colonial Office and the India Office continued to enjoy their special relationship with Kew. The Office of Works still retained responsibility for the maintenance of Kew's buildings when the transfer took place on 1 April 1903. The initiative had not come from Thiselton-Dyer, but in retirement he claimed some credit for the outcome.

> But I saw clearly that a break with the Philistine traditions of O.W. [Office of Works] must be to the good & that the new atmosphere could not conceivably be other than an improvement on the old.[29]

CHAPTER EIGHTEEN

Imperial Kew

Kew's links with colonies strengthened —
advice given on introduction of new crops —
network of botanic stations created

SIR WILLIAM THISELTON-DYER is remembered today not for his rationalisation of Kew's administration nor for the landscaping of its grounds, but rather for emphasising its commitment to colonial prosperity. The Hookers had identified plants of economic importance for government departments and merchants, and now under Thiselton-Dyer, Kew published the results of its activities and recommendations, making the institution a centre of economic intelligence. When Thiselton-Dyer addressed an audience at the Royal Colonial Institute in 1905, he paid tribute to the enlightened policies of Joseph Chamberlain as Secretary of State for the Colonies and to the cooperation of senior civil servants such as Sir Robert Herbert and Sir Robert Meade, "who really dragged me into the work by their intensive interest in the material development of our smaller colonies, and induced me to study the question, and to place more intimately at their disposal the resources of Kew".[1] His remark to the Governor of Madras that "we at Kew feel individually the weight of the Empire as a whole, more than they do in Downing Street"[2] was an absurd claim, but Kew certainly exercised a pervasive influence through the directors of the larger botanical gardens at Calcutta, Peradeniya, Jamaica, Singapore and Mauritius, through the superintendents of the smaller ones, and through the gardeners who manned the network of modest botanic stations. All of them turned automatically to Kew for the supply of plants and seeds, and for guidance on cultivation, diseases and marketing. "Colonial Gardens are therefore not isolated," affirmed the director of Public Gardens and Plantations in Jamaica, "but are branches of our Agricultural Department as wide as the British Empire itself."[3]

Many of the personnel of these colonial outposts had either been trained at Kew or owed their appointment to its director. These personal links fostered by expatriates were encouraged by Kew. "Our men when they leave Kew always count upon keeping up their connection with us by correspondence," Thiselton-Dyer proudly told Sir George Goldie.[4] They were normally engaged on a renewable three-year contract. Not all were satisfactory or accepted by local administrators and planters; some resigned before the termination of their contracts; others sought posts elsewhere overseas. India recruited many of them; in 1893 about 50 Kewites were working in government and private gardens or on tea, cinchona and cotton plantations. In 1910 Kew could claim about 160 former employees serving in Asia, Africa, Australia and America, the majority in senior posts.[5] During their home leave the Colonial Office expected them to pursue research or enquiries relevant to their duties abroad.

There can be no doubting the dependency of the Colonial Office, the Foreign Office and the India Office on Kew's expertise. The search for new or improved vegetable fibres at the close of the last century demonstrates this relationship. For example, following a request from the Colonial Office in November 1889, Kew despatched a consignment of Egyptian cotton seed to West Africa where indigenous cotton, used mainly for making native cloths, was rated inferior. In March 1890 a sample of cotton from Sierra Leone came to Kew for an opinion on its commercial value.

Any new fibre-yielding plant invariably excited commercial expectations. A sample of the fibre of a local plant known as 'Kanaff', forwarded by the British vice consul in Persia, arrived at Kew for identification. The taxonomic skill of Kew's botanists was again tested with a specimen of 'Bolobolo' fibre from Lagos. On behalf of flax growers in Northern Ireland Thiselton-Dyer asked the Foreign Office to obtain seed of Siberian flax, reputed to be taller than ordinary flax and "capable of yielding a succession of stalks from the same root for many years". He supported a proposition that banana stems, discarded after the fruit had been harvested, might be converted into a useful fibre. On learning that none of the Colonies had suitable machinery to process the fibre of Ramie or 'China Grass' (*Boehmeria nivea*), he urged the Foreign Office to obtain details of a Spanish machine, said to be capable of decorticating the fibre on a "profitable scale".

Kew was consulted about the commercial potential of Mexico's Sisal hemp. The Governor of the Bahamas believed it might revitalise the islands' stagnant economy. For this particular project Kew identified *Agave* species with superior fibre and examined various methods of cultivation. When planters experienced difficulties in cropping the plant Kew suggested remedies. For a few hopeful years it looked as though a new industry was burgeoning in the West Indies, but British East Africa eventually emerged as one of the world's primary sources of Sisal hemp grown from stock initially supplied by Kew.

The success with *Hevea brasiliensis* never deflected Kew's quest for other sources of rubber:

> Colonial rubber is being often received at Kew, and submitted for examination, showing that this branch of economic botany is far from being neglected.[6]

Hevea spruceana shipped to Kew from Demerara was shared between the botanical gardens at Calcutta, Peradeniya and Brisbane. Other species of plants from which a latex or elastic gum might be extracted included the tropical African woody climber *Landolphia*, a Chinese species of *Eucommia*, *Kickxia* from West Africa, *Fosteronia* from British Guiana and Jamaica, and *Alstonia* from Fiji. The value of rubber exports from the Gold Coast exceeded half a million pounds in 1898, making it the world's third largest rubber producer from its own resources.

Much of Africa remained Europe's *terra incognita*; even by the end of the eighteenth century little was known of its natural resources. This abysmal ignorance of its topography led to the formation of the Association for Promoting the Discovery of the Interior Parts of Africa (the African Association), of which Sir Joseph Banks was naturally a founding member. The Sierra Leone Company, granted a charter of incorporation in 1791, engaged a botanist to advise it. Groundnuts were a significant export of Gambia in the 1830s. Palm oil was exported from the Niger delta. The exploitation of the vegetation of West Africa depended upon the initiative of officials like Alfred Moloney, appointed the first Governor of Lagos in 1886 and the author of a *Sketch of the Forestry of West Africa* in 1887. Before the end of his first year in office, he suggested to the Colonial Office that Lagos should have a botanic centre or station similar to those recently established in the British West Indies. This station would grow crops, recommended by Kew, for distribution to villagers who would receive some training in basic agricultural techniques. In making his case, he acknowledged the help Kew had already provided. James MacNair, a gardener at the Hope Nurseries in Jamaica, was nominated by Kew to take charge of the botanic station and Kew supplied him with plants and seeds. A comment in the *Kew Bulletin* persuaded the Governor of the Gold Coast to follow Lagos's lead:

> If once the natives inhabiting magnificent lands in this Colony were taught to cultivate economic plants in a systematic manner for purposes of export, the material wealth of the Gold Coast might be enormously increased.[7]

Land was requisitioned in 1890 on the slopes of Aburi and Kew despatched William Crowther to plant coffee and cocoa, and together with the West Indies provided seed. By 1892 Crowther had cleared and planted much of the site and softened its functional severity with avenues of oranges and Royal Palms, and a small flower garden. "The charms of a scientifically laid out and cultivated garden such as can be seen nowhere else in the Colony" impressed the medical officer in charge of the local sanatorium.

"I know how much the Gold Coast Colonies have profited by your knowledge and experience and readiness to promote land culture." This compliment to Thiselton-Dyer came from Lord Aberdare, Chairman of the Royal Niger Company. "We shall only be too glad to share in these benefits, & to cooperate with you in extending the usefulness of the great institution over which you preside."[8] The Company had already formed a garden at Asaba in 1888 and four trial plots in 1889–90; two Kew gardeners – George Woodruff and Harold Bartlett – supervised them briefly before their untimely deaths.

These botanic stations had proved their usefulness by 1893, the year in which the Gold Coast began exporting cocoa. Their European gardeners were given permanent appointments after a meeting of the four West African Governors at the Colonial Office in September 1893. It was agreed at this meeting, which the Director and Assistant Director attended, that branch botanic stations or local depots were to be staffed by native gardeners trained at Kew or in the West Indies. Crowther visited the West Indies during the winter of 1893 to observe the management of botanic stations there and to bring back plant and seeds for trials. Botanic stations were set up in Gambia in 1894, at Zomba in the British Central African Protectorate and at Freetown, Sierra Leone in 1895. The crops they introduced and distributed provided the basis of a viable agricultural industry in these West African colonies.

As a result of intensive exploration Africa was being apportioned among the European powers, all determined to develop the resources of their new possessions. The Consul-General at Zanzibar, Sir John Kirk, had at his own expense formed a garden where he grew plants, some acquired from Kew. His efforts with a local rubber vine were rewarded with rubber sales in the Zanzibar market worth £200,000 in 1891. Sir Harry H. Johnston, Commissioner in British Central Africa, also took a personal interest in his territory's vegetation. It soon became apparent to colonial officials that an intensive floristic survey was urgently required. The *Flora Capensis*, recording the vegetation of southern Africa, and the *Flora of Tropical Africa* had been undertaken by Kew in 1859 and 1868 respectively. Only three volumes of each series had been published before coming to a halt, still far from completion. Thiselton-Dyer received a sharp reminder from the Foreign Office of his responsibilities in 1891 when there was no sign of his resurrecting them.

> Lord Salisbury is of opinion that a proper knowledge of the Flora of Tropical Africa would do much to aid the development of the territories over which this country has recently acquired our influence, and he would therefore suggest that the completion of the work in question should at once be carried out.[9]

Administration was a constant distraction for Thiselton-Dyer who, perhaps somewhat reluctantly, assumed editorial responsibility for the two floras; four volumes of the *Flora Capensis* took him well into retirement, and although he also saw four volumes of the *Flora of Tropical Africa* through the press, his successors, Sir David Prain and Sir Arthur Hill, finished it.

James Britten, never missing a chance of exposing what he perceived were Kew's shortcomings, blamed Thiselton-Dyer for the delay in publishing the *Flora of Tropical Africa* in an editorial note in the *Journal of Botany* for January 1901. His entirely erroneous and mischievous statement that "the work remained in abeyance during thirty-one years, for the greater part of which – i.e. since 1872 – it was in the hands of the present editor who issued the first instalment of the continuation in 1896" so incensed Thiselton-Dyer that he began libel proceedings, forcing a public apology from Britten in two issues of the *Gardeners' Chronicle* and the *Journal of Botany*.

With work on the African flora resumed, the seventh and final volume of the *Flora of British India* rolled off the press in 1897. Its author, Sir Joseph Hooker, now in his 81st year, had been engaged on it for more than 30 years but, like his son-in-law, he too had found that official duties impeded progress. A century later, this classic tropical flora is still the only one to cover the entire Indian sub-continent. The death in 1896 of Henry Trimen, Director of the Botanic Garden at Peradeniya, had left his *Handbook to the Flora of Ceylon* unfinished. Hooker accepted an invitation to complete it, but a task which he had anticipated accomplishing in 18 months took him to 1900. Kew is currently participating in a revision of this flora. Hooker's researches and observations on the Indian flora and its affinities with other floristic regions, conducted over half a century, were succinctly presented in an essay in the *Imperial Gazetteer of India* (1907).

In 1884 Hooker had joined the directors of the botanic gardens at Calcutta and Saharanpur at the India Office in London to consider a strategy for reforming botanical study in India. There existed no central control or direction of botanical research at colleges or at botanical gardens which were administered locally. When Thiselton-Dyer in 1885 advised the Madras Government on the establishment of a botanical department for its Presidency, he took the opportunity to advocate cooperation with similar departments elsewhere in India. This pioneering work by Kew helped the formation in 1887 of the Botanical Survey of India with authority to investigate the flora of the sub-continent and to co-ordinate all botanical research. In 1891 George King became the first Director of the Botanical Survey of India and, with Thiselton-Dyer's prompting, his post as Superintendent of the Calcutta Botanic Garden was redesignated 'Director'.

The first tentative links between Kew and India had been made when Sir Joseph Banks advised the East India Company about the economic plants they should seek and cultivate. It was again through Banks that Kew began an association with the West Indies, probably the closest of any of the relationships formed with the Colonies. They were the first tropical and sub-tropical territories occupied by the British: Bermuda (1612), St Kitts (1624), Barbados (1627), Nevis (1628), Antigua and Montserrat (1632), Jamaica (1655). After some years of subsistence farming, the settlers grew tobacco and cotton as cash crops but by the mid-eighteenth century sugar cane, originally imported from Brazil, ranked as their staple product, the English market taking all that the islands could produce. Such was the wealth generated by the sugar trade that Britain judged the West Indies a more valuable possession than Canada. The Amsterdam Botanic Garden sent young Arabian coffee plants to Dutch Surinam in 1718; ten years later coffee seeds were sown in Jamaica. From its foundation in 1754 the Society for the Encouragement of Arts, Manufactures and Commerce (now the Royal Society of Arts) actively encouraged the development of the resources of the British Colonies which at that time principally meant North America and the West Indies. In 1758 the Society offered a gold medal to the first person to despatch mango seeds to the West Indies, adding cinnamon and other crops in 1760, and at the same time recommending that land should be allotted in the colonies for gardens and nurseries for the making of experiments in raising such rare and useful plants as are not the spontaneous growth of the kingdom or said colonies.

General Robert Melville, Governor of the Windward Islands and a member of the Society, while touring St Vincent in 1765, responded positively to a proposal by George Young, surgeon of the British garrison there, for a botanical garden. Some 20 acres of land near Kingston provided a site for such a garden, the first in the West Indies, with Young as its superintendent. In 1773 its healthy cinnamon and nutmeg trees earned Young the Society's gold medal.

Its only rival, the private garden of Hinton East, Receiver-General of Jamaica, successfully acclimatised mango, yam, jack tree, date palm and all sorts of ornamental plants. A Bill supported by East in the Jamaica House of Assembly led to the establishment of a 'Tropical Garden' at Bath in 1779. East had been corresponding with Sir Joseph Banks in London as early as 1784, sending seeds to him and the Royal Gardens at Kew. He questioned Sir Joseph about the value of breadfruit as an alternative food to bananas. Everyone now turned to Sir Joseph Banks, whose "knowledge, philanthropy, patriotism & influence are superior to all others", to organise an expedition to the Pacific. His first attempt with his nominee, Captain Bligh on HMS *Bounty*, failed as everybody knows. The second voyage, again directed by Sir Joseph, despatched Bligh on HMS *Providence* in 1791. This time the venture succeeded and breadfruit and other plants collected en route in the Pacific, Timor and St Helena were delivered to the Botanic Garden at St Vincent, now managed by Alexander Anderson, and to several gardens in Jamaica. Some had been set aside for Kew and to these were added generous gatherings of the West Indian flora. In 1794 the committee responsible for the public gardens in Jamaica reported complete satisfaction with everything Bligh had brought them and offered Sir Joseph "whatever the island affords in the botanical line that may be agreeable to you".[10]

The breadfruit transfer greatly enhanced the reputation of Sir Joseph Banks who generously advised all who approached him: Thomas Dancer, for instance, who consulted him about the restoration of the now run-down Botanic Garden at Bath in Jamaica, and Alexander Anderson who gratefully received "some hints . . . towards aiding him in his plan of giving an account of the rise & progress of the Botanical Gardens" at St Vincent.[11] And Kew was a fortunate beneficiary in this climate of goodwill. "I have a great many plants I wish to send to Kew which I am pretty sure it has not yet obtained," Anderson informed Banks.[12]

Anderson had been given charge of the St Vincent Botanic Garden when the island was liberated in 1785 from French occupation. Prompted by Banks, the Secretary for War, Sir George Yonge, requested plants from the Calcutta Botanic Garden to restock St Vincent's. Sir Joseph saw this overture as the beginning of mutually beneficial exchanges between the two institutions.

View of the Botanic Garden at St Vincent from the Superintendent's house. Lithograph of a drawing by Revd Lansdown Guilding in his *Account of Botanic Garden in Island of St Vincent*, 1825.

To exchange between the East & West Indies the productions of nature usefull for the support of mankind, that are at present confined to one or other of them, to increase by adding this variety, the real quantity of the produce of both countrys, & by that means their population, furnishing at the same time to the inhabitants new resources against the dreadful effects of hurricanes & droughts . . . are the more immediate objects of his Majesty's present institutions.[13]

The jack fruit, clove, custard apple and a superior variety of mango were promised by Calcutta, which in return got West Indian mahogany (*Swietenia mahagoni*) and other potentially useful Caribbean plants.

After Banks's death in 1820, Kew's connections with the West Indies weakened. William J. Hooker, then Professor of Botany at Glasgow, received plants from the Revd Lansdown Guilding, Chaplain on St Vincent, for depicting in *Curtis's Botanical Magazine*. When he became Director at Kew in 1841, the institution soon resumed its contacts with the West Indies. It was a Kew plant collector, William Purdie, who filled the post of Government Botanist and Superintendent of the Botanic Garden in Trinidad, and a Kew gardener, Nathaniel Wilson, who took charge of the Bath Garden in Jamaica. Wilson, who had arrived with a gift of seeds from Kew in 1861, later received cinchona seeds "through the liberality of the British Government and recommendation of Sir W.J. Hooker at Kew".

With official permission to initiate a series of colonial floras, Sir William selected the West Indies as his first choice since its flora was well represented in the Kew Herbarium and he could rely on the cooperation of the islands' Governors and botanical gardens. A. Grisebach's *Flora of the British West Indian Islands* (1854–64) set a commendable standard for other floras in the series to emulate.

Following his father as Director, Joseph Hooker accelerated the flow of plants from Kew. Castleton, a new garden established in Jamaica, received its share of Kew's bounty.

> The garden has thus been enriched by the importation of new and important species with a rapidity perhaps unsurpassed in colonial gardens. Such indeed has been the enormous demand upon the royal gardens [at] Kew, from which nearly all plants have come, that the resources of that great establishment have been drained, with respect to tropical species, in a manner never before experienced.[14]

Alarmed by rumours of Hooker's threatened resignation during the Ayrton scandal in 1872, the Committee of West India Planters and Merchants expressed its concern to the Prime Minister.

> On behalf of the Committee I am desired to bring to your notice the great benefits which have resulted from valuable suggestions and the kind assistance to those colonies [of the West Indies] at all times by Dr Hooker.[15]

Of all the botanical and horticultural appointments made in the West Indies during the latter half of the nineteenth century, that of Daniel Morris had the greatest impact on the agricultural economy of the region. A graduate of Trinity College Dublin, he spent two profitable years at Peradeniya in Ceylon studying the fungus *Hemileia vastatrix* that was threatening the survival of the coffee plantations. In 1879 he accepted the post of Director of the Public Gardens and Plantations in Jamaica, where he continued to broaden his knowledge of tropical economic botany. Having reminded the Government of Jamaica that the aims of his department included the promotion of agriculture, the introduction and distribution of plants of economic importance, and practical advice in setting up new industries, Morris put into operation such a radical reorganisation that Kew predicted his department becoming "the headquarters of botanical enterprise" in the Caribbean.

Kew had prophesied a new era of prosperity for the islands if they reduced their dependency on the sugar trade, encouraged a greater diversification of crops, and targeted the United States market.[16] Kew supplied a variety of crops for trial cultivation: Indian tea, tobacco from Cuba and Manila, Chinese dwarf bananas, superior onion seed from Europe, mangoes from Bombay, mangosteens from the Moluccas and, inevitably, cinchona and rubber. Seeds of Liberian coffee, believed to be immune to the blight that attacked *Coffea arabica*, were sent to the West Indies in 1874, only two years after specimens had been raised at Kew. Ten years later Sir Joseph Hooker received a present of three boxes of Liberian coffee from grateful Dominica. So highly did Daniel Morris rate the commercial prospects of Liberian coffee that he published a lengthy account of its cultivation.[17]

Regardless of this introduction of alternative crops – cocoa in Trinidad and Grenada, limes, fruit and coffee in Dominica, dye-woods, bananas and pimento in Jamaica – sugar still controlled the economy of the West Indies. But the combination of the repeal of duties on foreign sugar and competition from European sugar beet had a devastating effect to which the West Indies responded by building new factories, improving methods of pest control and experimenting with other varieties such as elephant sugar cane from Cochin-China which reached Jamaica via Kew in 1880.

The Royal (West Indian Finance) Commission, appointed in 1883, recommended the formation of plant committees on the smaller islands as channels of communication with Daniel Morris's department in Jamaica, the source of professional advice and the supply of seeds. Sir Joseph Hooker's response to the Colonial Office on this proposal was that

> Plants, however, cannot be grown or propagated in a committee room, and botanical stations on a cheap basis are an essential condition for doing anything in an effective way.[18]

He endorsed Morris's perception of botanic stations as inexpensive units on a few acres of ground requisitioned for trials of new crops and their distribution. Thiselton-Dyer, now Assistant Director at Kew, thought they should also publish cultivation notes for the benefit of growers. In 1885 the Colonial Office approved the formation of these botanic stations linked to Jamaica, each island making a financial contribution to the cooperative venture. Barbados and Grenada were the first to establish them, St Lucia followed in 1889, St Vincent converted its original and now abandoned Botanic Garden in 1890 and Dominica joined in 1891, the year in which the Governor of the Leeward Islands approved the appointment of a Superintendent of Agriculture responsible for the botanic stations on Antigua, Dominica, Montserrat and St Kitts.

Their programme and performance were as variable as the crops they grew. St Vincent sought Kew's advice on the cultivation and processing of arrowroot; Barbados concentrated on raising sugar canes from seed; Grenada published a series of slim *Bulletins of Miscellaneous Information* (obviously inspired by Kew's more ambitious periodical); Dominica by 1897 had distributed about 165,000 economic plants. During their infancy Kew nurtured them, providing plants, advice and sometimes qualified staff, but Thiselton-Dyer was disappointed by a lack of positive commitment to the project.

> I must confess that I have sometimes entertained the idea that neither the Home nor the Local Governments have very much at heart the success of the Botanical Stations scheme for developing the material prosperity of the West Indian Colonies.[19]

Herbarium and office in
the Botanic Garden,
Trinidad in 1890.

Some settlers on Antigua wanted their station closed or converted into a public
park. Many planters were reluctant to abandon traditional crops and practices.
In 1807 Alexander Anderson of the St Vincent Botanic Garden had warned
the Society of Arts in London that planters feared that any new crops might
imperil the future of sugar production. It was an attitude still prevalent 80
years later.

Daniel Morris monitored the progress of the botanic gardens and stations
from Kew where he went to be Assistant Director in 1886. Early in 1890,
Thiselton-Dyer, unhappy about their management, got the Colonial Office's
permission for Morris to inspect them and to confer with the Botanical Departments in Trinidad and Jamaica "to solve finally the innumerable petty administrative difficulties which have hindered the effective working of the scheme".[20]
Morris left in November 1890, escorting several Wardian cases of Gambia
plants (*Uncaria gambir*) from the Straits Settlements, used in tanning and
potentially a new commercial undertaking for the West Indies. He was back
again in 1895, this time in the Bahamas, and in 1896 acted as an Expert
Adviser on Agricultural and Botanical Questions to the West India Royal
Commission.

This Royal Commission was commanded to report on prevailing conditions
in the sugar-producing islands "and to suggest such measures as appear to you
best calculated to restore and maintain the prosperity of those Colonies and of
their inhabitants". Its conclusions, published in the autumn of 1897, had a
substantial appendix – Morris's "Report on the agricultural resources and
requirements of British Guiana and the West India islands".[21] The Commission, aware of the dangers of a monoculture, urged prudent investment in

other profitable crops. While acknowledging the valuable services rendered by the botanical establishments in Jamaica and Trinidad and the botanic stations, the Commission advocated a single body to co-ordinate these disparate efforts. They proposed a department of agriculture responsible for promoting existing and new industries and for supervising the botanic gardens and botanic stations – the genesis, in fact, of the Imperial Department of Agriculture for the West Indies. In announcing the Government's approval of this recommendation in the House of Commons on 2 August 1898, the Secretary of State for the Colonies declared that

> this establishment should be placed under the direction of Dr Morris, Assistant Director at Kew, who is marked out, as I think anyone who knows anything of Kew will admit, by special qualifications for an important position of this kind.

Morris's departure was viewed with some dismay by Thiselton-Dyer, who had come to rely on his exceptional experience of tropical agriculture. During the ten years Morris filled the post of Imperial Commissioner for Agriculture, cane sugar was improved by hybridisation, sea cotton introduced, and lime and cocoa cultivation developed, the *West Indian Bulletin* appeared quarterly, and seven intercolonial conferences were convened. His staff included a mycologist and an entomologist to advise on plant diseases. Morris's success prompted the creation of similar departments of agriculture in India, Uganda, Southern Rhodesia, Northern Nigeria, Malaya and elsewhere, often manned by staff trained in the West Indies.

When Joseph Chamberlain announced Morris's appointment in the House of Commons, he took the opportunity of acknowledging his department's indebtedness to Kew.

> I do not think it is too much to say that at the present time there are several of our important Colonies which owe whatever prosperity they possess to the knowledge and experience of, and the assistance given by, the authorities at Kew Gardens. Thousands of letters pass every year between the authorities at Kew and the Colonies, and they are able to place at the service of these colonies not only the best advice and experience, but seeds and samples of economic plants capable of cultivation in the colonies.

His predecessor, Lord Ripon, had been equally complimentary in 1895.

> I find that much goodwill has always existed among our friends at Kew, but we must bear in mind that one of the great works which have been undertaken by Kew is to educate the Colonies to recognise the nature of their various natural products, and the advantage of introducing new products. [22]

Kew's relations with the Colonial Office, strengthened by common objectives, were probably more intimate than with any other government department, an association officially recognised in 1902 when Thiselton-Dyer was formally appointed botanical adviser to the Secretary of State for the Colonies. But this privileged relationship was gradually passing to a usurper. The Imperial Institute of the United Kingdom, of the Colonies and India and the Isles of the British Seas, constituted by Royal Charter in 1888, was not only a museum of empire, situated in South Kensington, but also had scientific and commercial goals including the development of natural resources – in 1902 it was described as the 'Kew for chemistry'. Sir David Prain and Sir Arthur Hill, who succeeded Sir William Thiselton-Dyer as botanical adviser to the Colonial Office, continued to promulgate Kew's mission overseas, but as the Colonies became self-governing and had their own agricultural experts and research facilities, their dependency on Kew correspondingly declined.

Sir William
Thiselton-Dyer
(1843–1928).

CHAPTER NINETEEN

"The botanical metropolis of the world"

*Kew Bulletin – D. Prain succeeds
Thiselton-Dyer – dispute with student
gardeners – forestry museum – changes in
opening hours – Suffragettes attack Kew –
Japanese Gateway – Flagstaff – A. Hill as
Director – Empire Marketing Board –
Bedgebury Pinetum – new glasshouses – Kew
in wartime – death of Hill*

In APRIL 1887 both the Botanical Department in Jamaica and the Botanic Garden in Trinidad started a *Bulletin*, obviously emulating Kew's *Bulletin of Miscellaneous Information* launched three months earlier. Thiselton-Dyer had begun editing Kew's annual report in 1876, turning it into a useful compendium of information on commercial enterprises in the Colonies. When it ceased to appear after 1882, the excuse given was the "protracted illness" of the Curator, according to an unconvincing answer to a Parliamentary Question in the House of Commons on 17 March 1887. But in fact even before he became Director, Thiselton-Dyer had contemplated a more ambitious successor to the annual report, and one that would appear more frequently:

> I am anxious to submit to the Bd. a proposal which I have for some time had under consideration for systematically putting on record for the information of the public generally and of our colonial and Indian correspondents the more important facts and documents relating to the commercial and economic subjects which from time to time pass through our hands at Kew.[1]

The first issue of the *Bulletin of Miscellaneous Information*, soon abbreviated to *Kew Bulletin*, appeared in January 1887. From the outset it was a channel of information on economic plants and products. Since only 250 copies were printed, it was specifically aimed at civil servants, colonial administrators, colonial botanical establishments, and commerce generally. In its first year it abstracted reports on fibre plants, crops in Mauritius, an indigenous Abyssinian

Lilium Henryi

PLATE 27 *Lilium henryi*. Lithograph of watercolour by Lilian Snelling (*Genus Lilium*. Supplement by A. Grove, 1933, plate 2). Miss Snelling drew her specimen at Kew in 1923. The lily had been discovered by Augustine Henry in the I'Chang gorges on the Yangtze River in 1888; he sent a few bulbs to Kew where it flowered in 1889. It was later collected in bulk by E. H. Wilson for the Veitch nursery.

PLATE 29   The Bamboo Garden.   *Photograph by A. McRobb*

PLATE 28   *(opposite)*   The Pagoda in a morning mist.   *Photograph by M. Svanderlik*

PLATE 30 Hercules and the Serpent draped in icicles. *Photograph by M. Svanderlik*

cereal and a potentially useful Cape boxwood for wood engravers, and speculated on the source of a mysterious 'oil of Ben', reputed to have been used by
clock and watch makers. Diseases which threatened important economic crops
always received prominence: cocoa in Trinidad, sugar cane in Barbados, palms
in British Honduras, bananas in Fiji, vanilla in the Seychelles, potatoes in
India, wheat in Cyprus, and melons and apples in the Cape. An occasional
feature under the title of 'Decades Kewenses' described plants new to science.
In 1891 a new series of miscellaneous notes recorded staff appointments at
Kew, and acquisitions to the Gardens, Herbarium and Museums.

Difficulties in the mechanics of its publication, blamed by the Director on
the "scarcely concealed hostility" of the Stationery Office, elicited an assurance
in Parliament on 19 May 1892 that the *Bulletin* would continue. By the close
of the century, however, it had reverted to an irregular appearance which
Thiselton-Dyer still attributed to the Stationery Office, but one wonders
whether the absence of an Assistant Director after Daniel Morris's departure
to the West Indies in 1898 was the real reason. From 1902 to 1905 only a
series of appendices (seed list, library accessions, new garden plants and staff
list) was published, giving rise to James Britten's sour comment that the *Bulletin*
had "succumbed to a form of appendicitis".[2]

Britten's persistent attacks, although merely an irritant, nevertheless added
to the stress of the Director's final years in office. "Dyer told me last Monday
that he would retire next year," Sir Joseph Hooker confided to Sir Inglis
Palgrave, "and that he was overdone, and suffering from gout and confusion
in the head after work."[3] Ill-health and exhaustion may have precipitated his
retirement which came in December 1905. One senses that he was glad to
depart.[4] Towards the end of his directorship he admitted that "I believe I have
the reputation among my staff of being not too amiable but perfectly just, *i.e.*
according to my lights."[5] Beatrix Potter, an impartial observer, made this
shrewd assessment of him: "I should imagine he is a short-tempered, clever
man with a very good opinion of his Establishment, and jealous of outsiders."[6]

Always conscious of the dignity of his office, Thiselton-Dyer challenged
opposition, enjoyed confrontation, and since self-doubt never appeared to
trouble him, conceded defeat reluctantly. He was direct in speech, incisive in
style, intolerant of incompetence, but while generally considered a martinet,
there were those who had experienced his kindness, consideration and generosity. Nevertheless, whatever the flaws and contradictions in his personality, he
made substantial contributions to Kew's progress and reputation. He reorgan-
ised its administration, neglected for years by his father-in-law.

> It has been my business to acquire a direct personal control over the
> whole establishment and in all details.[7]

And no detail was too trivial for his personal attention: from the organisation
of office files to the proper care of garden tools. He wrote surprisingly little,
"the disappointment of my life", he confessed to David Prain. However, he
had chosen to concentrate on colonial affairs which undoubtedly gave him

Sir David Prain (1857–1944). Portrait in crayon and watercolour by A. de Biden Footner, 1929. *Kew Library.*

immense satisfaction. With his respect for authority and ceremonial and love of dressing up, he would have been in his element as a colonial governor. Being Director at Kew had given him a tantalising hint of what he had missed. As son-in-law of Sir Joseph Hooker, he brought the dynastic rule of the Hookers to a fitting conclusion. Unlike Hooker, who in retirement still visited the Herbarium or had specimens sent to him at Sunningdale, Thiselton-Dyer rarely came back, preferring to reminisce by letter from distant Gloucestershire with his successor, David Prain, whose appointment met with Hooker's approbation.

> I knew Prain well, a charming cultivated man and literally the only one known to me fit for the directorship at Kew, as an economic and scientific botanist with experience of garden work and official duties.[8]

David Prain had joined the Indian Medical Service in 1885 and, after duty with a regiment and in military hospitals, obtained a posting in 1887 more to his liking, that of Curator of the Herbarium in the Botanic Garden at Calcutta. When he became its Director, he continued his research on the Indian flora and, on behalf of the Government of Bengal, investigated the cultivation of hemp, wheat, pulses and indigo. Thiselton-Dyer welcomed this product of Empire somewhat grandiloquently: "India has recognised its debt by supplying Kew with a director."[9]

Of all the unfinished business Prain had inherited from his predecessor, that of the status and wages of the gardeners demanded immediate attention. He revealed some sympathy with their grievances by telling the Board of Agriculture and Fisheries in January 1906 that it was unreasonable to expect young gardeners to buy their obligatory blue serge clothes, recommending that their first issue be provided at government expense. Foremen and labourers received modest pay awards in 1906 but the wages of sub-foremen and gardeners remained unchanged. When the matter was discussed in the House of Commons on 23 May 1907, the Board of Agriculture refused to consider any increase in their wages, claiming that they were really apprentices, receiving free tuition. "The most valuable part of their wages is the experience in scientific horticulture which they gain." The outraged gardeners protested to the President of the Board of Agriculture on 28 May:

> . . . we as Gardeners and not as students decline to attend any lectures given by officials or others in the Royal Botanic Gardens in lieu of wages. We desire a legitimate wage and shorter hours, leaving us free to attend lectures where we wish.[10]

Prain supported the official policy of the Board which in 1906 had made some adjustments to the terms of engagement for gardeners: the minimum age limit reduced from 20 to 19, four instead of five years pre-entry training required, attendance at lectures compulsory, and a 'subsistence allowance' substituted for 'wages'. He did concede, however, their demand for lectures in official working hours but this was not implemented until 1908. One of the gardeners' spokesmen in the House of Commons, the MP for Sunderland, quoted with evident disgust from an Official Return in a debate in Parliament on 11 June 1908.

> The man whom a Dependency or Colony wishes to avoid is the man whose ambition is limited to the receipt of his pay . . . It is because the conditions under which admission to Kew is offered are such as to exclude those whose main incentive to work is immediate comfort, and whose ambitions are satisfied with a colourless career; and [it is] because these conditions do not discourage men of higher and more reflective type of which the Empire stands in need, that the reputation of Kew is so high among those who indent upon her for men.

One can only speculate on the identity of the official who wrote this astonishing piece of special pleading. It has resonances of Thiselton-Dyer but Prain, too, shared the same outlook. For him, Kew was an "Imperial School of Horticulture" which offered training facilities to a privileged few.

Yet he had the welfare of his gardening staff at heart. In 1908 'The Gables' was built as a residence for four foremen on the site of the stable block of Cambridge Cottage. After the First World War he tried to acquire hostel accommodation for young gardeners. Plans were actually prepared for a building to house 60 of them on a site off Ferry Lane behind the Herbarium.[11] The conversion of 'Courtlands' and 'Gloucester House' in the Kew Road was contemplated and even Cambridge Cottage itself became a possibility. Prain intended emptying Cambridge Cottage by transferring the timber exhibits there to Kew Palace, a proposition promptly vetoed by the Lord Chamberlain's department.

Cambridge Cottage, transferred to Kew's custody upon the death of the Duke of Cambridge in 1904, had been scheduled as a forestry museum by Sir William Thiselton-Dyer. Six rooms on the ground and first floors of the central block and east wing, adapted for a display of British timbers and their utilisation, tree diseases and forestry equipment, was opened to the public in 1910. Later on specimen timbers from the Orangery were added, making it a general wood museum.

When Richmond Town Council petitioned in 1907 for an extension of opening hours during the summer months, Prain at first resisted as stubbornly as his predecessors, disdainfully reminding the councillors that "the Royal Botanic Gardens are primarily a scientific institution maintained for the benefit of the whole Empire". The first weakening of his resolve appeared in 1912 when the opening time was advanced from noon to 10 o'clock from May to October; final capitulation followed in January 1921, with the public being admitted daily at 10 a.m. throughout the year, excluding Christmas Day.

Prain does not appear to have been particularly innovative, but rather a defender of the *status quo*. Following a successful experiment with official guides at the British Museum in 1911, the proposal to employ them at Kew met with little enthusiasm from the Director who argued that a key plan and notices of items of interest were sufficient. Reluctantly he agreed to a trial period during the summer months, with the appointment in 1913 of S.T. Dunn as guide. Initially the charges for tours were 2s 6d in the morning and 1s in the afternoon, reduced in 1914 to 6d and 3d respectively. With little public support, the experiment was abandoned in 1920.

The need to generate revenue probably lay behind these high charges and certainly occasioned the imposition of an entrance fee in 1916: a penny on Mondays, Wednesdays, Thursdays and Saturdays; Tuesdays and Fridays were reserved for 'students' who paid 6d for the privilege of a day's uninterrupted 'special study'.

Kew Gardens never lacked publicity whether in Parliament or in the national and provincial press. There was always something newsworthy – rebellious gardeners or limited public access or charges for admittance or the formidable boundary wall – but in 1913 it achieved the notoriety of banner headlines. "Mad women raid Kew Gardens" shrieked the *Daily Express*. "Orchids at Kew ruined" lamented the *Standard*. In the early hours of Saturday, 8 February, suffragettes invaded the Gardens, smashed the glass in three of the orchid houses, broke pots and deliberately destroyed plants. "An attack on plants is as cowardly and cruel as one upon domestic animals or those in captivity," complained the *Gardeners' Magazine*. Clues to the identity of the perpetrators of this "outrage" were "an envelope bearing the inscription 'Votes for women' in an uneducated hand, some feminine fingerprints, a handkerchief and a bag". Another attack 12 days later burnt down the Refreshment Pavilion and placards proclaiming women's suffrage were found in the vicinity. The destruction of the Pavilion was no great loss. It had been crudely fabricated in 1888 "a tiled roof, framed in the cheapest possible manner, supported on rough wooden posts with wooden divisions forming three separate rooms inside". A temporary replacement was erected in 1914 while architects discussed the style of its successor: an elegant structure with classical pillars and a dome, or something akin to the Orangery, or a feature whimsically eastern to partner the Pagoda. What eventually emerged in 1920 was a stark, functional building with no architectural pretensions whatsoever, designed by the Ministry of Works.

The erection of the temporary tea pavilion in 1914 required the renewal of lawns and the replacement of shrubs in the vicinity, probably supervised by W.J. Bean, the Assistant Curator in charge of the Arboretum. It seems that Prain seldom visited the Gardens, content to leave their management to his Curator, William Watson, and his Assistant Director, Arthur Hill. Cambridge Cottage garden, a random grouping of trees and shrubs within brick walls, presented an entirely new landscaping project. Two entrances were made in its southern boundary wall, much of the planting thinned out but the lilacs and two ancient yews retained, paths altered, the conservatory removed, and a geometrical medicinal herb garden laid out in the main area of grass. While much of this work was being carried out during the winter of 1907/8, the semi-circular hedge of yew behind the Palm House was replaced by holly, and two gravelled walks leading to the Pagoda and Syon Vistas were grassed.

Pagoda Vista with the Temperate House, Palm House and Refreshment Pavilion. The planting of young trees had been completed in the winter of 1908–9. Sixteen large trees had been moved to the Vista from other parts of the Gardens with the assistance of Barron's tree transplanter.

William Barron's tree transplanter, purchased in 1866, was replaced by a new model in 1936. It was known as 'the devil' by the Arboretum men who in one winter moved 60 trees varying in weight between two and seven tons. Their roots were protected in balls of earth, and few trees were lost.

William Watson had originally wanted the site for the herb garden in Cambridge Cottage grounds for a new aquatic garden, conceived in an informal manner. This was one of the few occasions when the Director did not defer to his Curator, preferring a formal design and a location just east of the Rock Garden. Built in 1909, supposedly inspired by the sunken garden at Hampton Court, it comprised seven tanks or pools, the central one exclusively for water lilies.

The search for a suitable setting arose again in 1910 when Kew was offered a splendid replica of a sixteenth-century Buddhist gateway in Kyoto, made specially for the Japan-British Exhibition at the White City in London. The Bamboo Garden had been favoured as a location but after a tour of the grounds with some officials of the Kyoto Exhibitors' Association, 'Mossy' Hill near the Pagoda was chosen. Sir William Chambers's 'Mosque' – the derivation of 'Mossy' Hill – had once stood on this gentle undulation, an appropriate prede-cessor for the Chokushi-Mon (Gate of the Imperial Messenger), re-erected there in 1911. Being a delicate wooden structure, it needed repair in 1936 and again in 1957, and now stands shrouded in plastic sheeting awaiting yet another restoration. When the Japanese poet, Kyoshi Takahama, visited Kew in 1936, the beauty and serenity of the Gardens inspired him to compose a haiku:

'Even sparrows
Freed from all fear of man
England in spring.'

Japanese Gateway, made of Hinoki wood, *Retinospora obtusa*, with cedar bark shingles on the roof. The carvings depict episodes in the relationship between Kosekko and his pupil, Choryo.

Palm House terrace in
Edwardian days.

In 1979 this haiku – a Japanese lyric poem of 17 syllables – was carved on a
block of granite and placed near the Japanese Gateway in the poet's honour.

The grounds at Kew partly reflect the tastes and skills of successive directors
and curators and their staff. Nesfield's layout had never been popular with the
gardeners, and gradually his designs had been modified – a simplification of
a flower-bed or the removal of a topiaried shrub or the turfing of a gravelled
path – and this process of piecemeal erosion continued under Prain. The
transformation of Nesfield's garden behind the Palm House has already been
noted. When visitors were seen to avoid his gravelled path through Syon Vista,
it was covered with soil and sown with grass seed. But as the grass never grew
well on such a foundation, invariably turning brown in dry summers, the
gravel was eventually dug up in 1913. In the same year Nesfield's Deodar
cedars along the Broad Walk – mostly pathetic specimens – gave way to Atlantic
cedars.

But Nature as much as man dictates the character of a garden – plants die
and storms devastate a landscape. A violent storm on 28 March 1916 destroyed
a number of Kew's historic and notable trees: a Minorca box (*Buxus balearica*),
one of the finest in the country, and a rare bay willow (*Salix pentandra*). The
sole survivor of the 'Seven Sisters' elms, once a prominent Kew landmark,
also succumbed.

The First World War upset routine maintenance. Women gardeners who
replaced conscripted men had expanded to a work force of more than 30 by
1917. The Palace lawn was ploughed in 1918, preparatory to planting it with

Palm House terrace planted with onions during 1914–8 war.

potatoes. Whatever had survived of Nesfield's garden in front of the Palm House disappeared under beds of onions, the flower plots along the Broad Walk yielded to cabbages, lettuces and root vegetables. The flowers which reappeared on the Palm House Terrace in 1920 were arranged in less complicated patterns.

Fifty-eight years after its erection in 1861, the flagstaff, its life prolonged by patching and preservatives, gave way to another, 55 feet taller, a gift of the Provincial Government of British Columbia.

Another gift, also from North America, was soon regretted. In 1907 the former Chairman of the Chiswick District Council had persuaded Prain that some grey squirrels would be a desirable addition to Kew's fauna. At that time the Gardens had definite zoological proclivities: exotic waterfowl, storks, pelicans and even a few penguins. The Duke of Bedford, who had imported some American grey squirrels, gladly presented two pairs. The native red squirrel, just about hanging on at Kew, was soon ousted by his American cousins which multiplied so fast that their numbers had to be controlled by shooting.

Towards the end of Sir Joseph Hooker's management, Chinese introductions had just begun to make a visual impact at Kew. In 1885, the year of Hooker's retirement, Augustine Henry, an official in the Chinese Customs Service, offered to collect plants for Kew. Most of his despatches were destined for the Herbarium, but the Gardens received his bulbs of the late-summer-flowering *Lilium henryi* PLATE 27. Although Henry made numerous discoveries, gardeners had to wait for the consignments of professional collectors like Wilson, Forrest, Farrer, Purdom and Kingdon Ward. E.H. Wilson, formerly a student gardener at Kew, had through Thiselton-Dyer's recommendation obtained the post of collector in China for the Veitch nursery. An outstanding collector, his introduction of *Lilium regale* gave him the most satisfaction. George Forrest concentrated mainly on rhododendrons and primulas. From 1914 to 1916 Reginald Farrer teamed up with William Purdom, one of the leaders of the students' revolt at Kew and now a seasoned collector in China. Frank Kingdon Ward started his long collecting career in 1909, exploring Burma and Tibet as well as China. With the imminent closure of Veitch's Coombe Wood nursery, Kew purchased from their stock in 1913 some 250 Chinese trees and shrubs. Wilson's rhododendrons found a home on the side of King William's Temple but some trees in the Arboretum had to be replanted to accommodate the Chinese intruders. Occasionally an entire genus was moved to a new site in order to preserve family kinship since 'promiscuous planting' was contrary to the Gardens' taxonomic display.

Kew acquired some Chinese plants from subscribers to Wilson's expeditions – enthusiastic gardeners like J.C. Williams of Caerhays in Cornwall and Ellen Willmott of Warley Place in Essex. There appears to be some evidence that in 1920 Miss Willmott had contemplated handing over her Warley garden to Kew.[12]

Though this never happened, the relations between Ellen Willmott and Kew were mutually beneficial – she presented plants and in return had her collections identified. In 1904 Thiselton-Dyer asked her for financial assistance in publishing a supplement to G.A. Pritzel's *Iconum botanicarum index locupletissimus*. Pritzel is remembered as the industrious German librarian who compiled a list of over 100,000 references to illustrations of flowering plants, ferns and fern allies he had located in books and periodicals published since the time of Linnaeus. The fruits of his labour were published in 1855, with a supplement in 1866. Prain corresponded with Miss Willmott about the project and W. Botting Hemsley, recently retired as Keeper of the Herbarium, agreed to collaborate in the new work. At some stage Miss Willmott ceased to be involved but in 1917 the Royal Horticultural Society undertook the revision of Pritzel, and the following year Otto Stapf was appointed editor with the resources of the Kew Herbarium and Library, of which he was Keeper, at his disposal. The first volume of the *Index Londinensis* appeared in 1929, was completed in 1931 with the sixth volume, and supplemented by a further two volumes in 1941. Many botanists in Europe and the United States supplied references, often from obscure publications, to make this work invaluable for locating published illustrations since 1753 (the date of Linnaeus's *Species Plantarum*). It follows the arrangement of the *Index Kewensis* with species listed under an alphabetical sequence of genera; the *Index Kewensis*'s format of bibliographical citations was also adopted.

Prain, as editor of the *Index Kewensis*, made an important departure from its original aims with the fourth supplement in 1913. He abandoned Kew's practice of making taxonomic judgements; in future all validly published generic and specific names were to be included without heeding synonymy, a rule still operating today.

As a botanist who enjoyed taxonomy, Prain probably welcomed the opportunity to take on the editorship of the *Flora of Tropical Africa* and *Flora Capensis* when Sir William Thiselton-Dyer was no longer able to continue. He also contributed articles to the *Kew Bulletin* which he had resurrected in 1906. It had been in abeyance since 1901 and its re-emergence incited James Britten to note in the *Journal of Botany* that "the new Director has been clearing out the pigeon-holes of his predecessor". The Treasury had tried to axe the *Bulletin* in 1892 on the grounds of expense, and in 1917 its existence was threatened by the Stationery Office, which was required to defer the publication of all inessential books and periodicals. This was not an unreasonable act with wartime paper shortages, but when the *Kew Bulletin*'s suspension was announced *The Times* called it a "false economy", *Nature* strongly protested, and the British Science Guild argued that by "suspending the publication of the *Bulletin*, the link connecting Kew with the whole of the botanic stations of the Empire is broken". Faced with such opposition, the Stationery Office rescinded its suspension order, provided the editor exercised "due regard to economy" in his selection of articles. Those who had supported Kew in this crisis drew attention to its international activities. Lord Bryce called Kew "the botanical metropolis of the world", with the *Kew Bulletin* offering "the greatest possible service to those who are endeavouring to develop the vegetable resources for many parts of the Dominions".[13] With that sort of reputation to uphold, Prain had a survey of plants of economic value to the Dominions and Colonies quickly put together and published in the *Kew Bulletin*.[14]

However, with the expansion of the work of the Imperial Institute, the formation of the Commonwealth Agricultural Bureaux, the Empire Cotton Growing Committee and overseas departments of agriculture, Kew's imperial role gradually diminished. It still advised the Colonial Office, and for many years Prain chaired the Advisory Council for Plant and Animal Products of the Imperial Institute. After the war fewer openings abroad became available to Kew gardeners, who now sought advancement in this country in private and public gardens and in nurseries.

Miss Matilda Smith, appointed botanical artist after W.H. Fitch's defection, retired in 1921. She was followed by Gerald Atkinson, who remained in the post until 1959 in the dual role of artist and photographer.

The renting of three or four acres of land near Kew Gardens for an "orchard" and a "kitchen garden" to study plant diseases had been proposed by Mr Middleton of the Board of Agriculture and Fisheries in November 1911. He believed that "the extension of the purview of Kew to fruit and vegetables would be very popular with gardeners",[15] a prospect which received a lukewarm response from Prain, irritated by Middleton's reference to Kew as "a horticultural institution". Independently of Kew, land was found near Ruskin Avenue, more or less where the Public Record Office now stands, experimental plots were planted and investigations carried out until 1921. In the summer of 1913 the Board of Agriculture and Fisheries purchased Chestnut and Gumley Cottages on Kew Green for mycological research on plant diseases. These two Georgian dwellings, merged into one building, provided facilities for the pathological research formerly done in conjunction with the staff of the Jodrell Laboratory. With the transfer of the Board of Agriculture's laboratory to Harpenden in Hertfordshire in 1919, the Imperial Bureau of Mycology became the next occupant. It had been founded to participate in the eradication of crop diseases in the Empire under the aegis of the Colonial Office. When this accommodation became full it moved in 1930 to a new building just behind the Kew Herbarium.

David Prain, who had been knighted in 1912, retired in February 1922. After the high profile Kew had enjoyed under Sir William Thiselton-Dyer, the institution had needed a respite, time to recover from autocratic rule. Through his urbanity and skills in conciliation, Prain brought stability, eased staff tensions and delegated duties, at the same time consolidating the achievements of his predecessors.

Sir Arthur William Hill (1875–1941).

There seemed little doubt that his Assistant Director, Arthur Hill, would automatically succeed him. When the Treasury had re-instituted the post of Assistant Director in 1907 Hill, then a young lecturer at Cambridge, had got the job. It was "the combination of readiness to travel, and the willingness and ability to undertake office work and drudgery that led me to ask him to come to Kew", Prain revealed years later.[16]

As Director, Hill was beholden to the Empire Marketing Board for developments that he could not have pursued without its financial support: extra temporary staff, the incorporation into the Herbarium of a vast accumulation of dried specimens, more overseas travel, and the installation of a quarantine house.

The Empire Marketing Board, formed in 1926 to foster trade within the Empire, had as one of its principal goals the scientific and economic investigation of Imperial resources. To this end it granted Kew £4,000 in 1927 to appoint an Economic Botanist and to send its staff overseas on botanical missions.

Hugh Charles Sampson, with a background of agricultural experience in the Transvaal and India, made his first tour as Economic Botanist in 1927, visiting British Guiana and the West Indies. Before his retirement in 1938, he had been invited to British Honduras, West Africa, the Bahamas and Kenya, advising on agricultural problems, and collecting a few plants for Kew's Herbarium as well. When a Conference of Colonial Directors of Agriculture, held in London in 1931, recommended that Kew should be responsible for "the dissemination of information on economic plants", Sampson responded by compiling a useful inventory.[17]

Kew undertook not only the dissemination of information but also provided a vital service in the transit of plants to Dominion and colonial plant breeders. To avoid diseases being transported with plants going to new habitats, a quarantine house was erected at Kew at the Board's expense where, once plants had been cleared of infection, cuttings were propagated and despatched. For more than 50 years Kew offered this service to growers and botanists overseas seeking disease-resistant plants. These included species of bananas collected in the Far East, the Pacific and East Africa destined for the West Indies, *Cassava* and

Plants being placed in a Wardian case for despatch overseas. Being an expensive method of transport, it was confined to those which could not be sent by any other means. It remained in use until the early 1960s. Plants are now often sent by air in polythene bags.

Passiflora species from South America for East Africa and special clones of cocoa from Trinidad for West Africa. All were subjected to Kew's quarantine procedures.

With another grant from the Empire Marketing Board, Kew collaborated with the Imperial Institute in 1928 in an exercise to grow *Aleurites fordii* and *A. montana*, the source of tung oil, in the Colonies. At that time China, the sole source of supply of this quick-drying varnish oil, was in a state of great unrest. Many of the trials failed but eventually viable plantations were established in South Africa and Burma.

Financed by the Empire Marketing Board, Kew's botanists went on expeditions to the African continent and British Guiana, but the member of

Moving the Chilean wine or honey palm, *Jubaea chilensis*, some 20 feet from the side of the Temperate House to the middle in 1938. Its leaves were then touching the sloping roof. The total weight of the tree and the ball of soil was almost 54 tons. It is possibly the largest tree under glass anywhere in the world.

staff who travelled most frequently and extensively was the Director himself. While at Cambridge, Hill had seen Iceland and explored the Andes; as deputy to Sir David Prain, who conscientiously remained in his office, he represented Kew on official visits to the Dominions and Colonies. Now, thanks to the Board's generosity, he could really indulge in style his fondness for travel: he attended conferences in the United States and the West Indies; a visit to a botanic garden in the Cameroons gave him an excuse to see Nigeria; a winter was agreeably spent in South and East Africa; a tour of Australia and New Zealand took in Java, Malaya and Ceylon en route. He evidently enjoyed his ambassadorial role – attending official functions, lecturing, and establishing or renewing Kew's links with the Empire.

The Empire Marketing Board came to an end in 1933 but the Director persuaded the Treasury to make permanent most of the temporary posts created with the Board's funding. When Sampson retired in September 1938, Sir Geoffrey Evans, Principal of the Imperial College of Agriculture in Trinidad, filled the post of Economic Botanist.

T.F. Chipp's service in the Gold Coast and Singapore presumably recommended him as an excellent choice as deputy in 1922 to Hill, the last of Kew's directors motivated by an imperial mission. Hill encouraged his staff to establish working relations with colonial institutions. In 1929 C.E. Hubbard at Kew made a temporary exchange with W.D. Francis of the Botanic Garden in Brisbane. After his visit to Australia and New Zealand in 1927–8, Hill recommended that both countries send liaison officers to Kew. No action followed but the Third Imperial Botanical Conference in August 1935 resolved to appoint liaison officers for Australia, New Zealand and Canada at Kew. Only Australia implemented the resolution by the secondment of C.A. Gardner, Government Botanist at Perth, in 1936. Today tours of duty at Kew are regularly undertaken by botanists from Australia, South Africa and India.

"Broadly, it may be said that Kew is devoted in the first instance to Imperial and economic interests and research." That was one of the observations of the Royal Commission on National Museums and Galleries, appointed in 1927 to report on the statutory requirements, organisation and accommodation of institutions in London and Edinburgh.[18] This inspection included Kew Gardens and, inevitably, the unresolved matter of its relations with the Natural History Museum surfaced once more. At times the Commission favoured a transfer of the Natural History Museum's herbarium to Kew, but since it failed to reach a unanimous decision, its final report circumspectly noted that "each herbarium is an indispensable wheel in the machinery of the Institution to which it is attached".[19]

The Commission sympathetically received Hill's case for an extension to his Herbarium. The need was accepted by the Office of Works and a third wing was ready for occupation by the autumn of 1932.

Without the resources of its Herbarium, the publication of colonial floras would have suffered. The last part of the *Flora Capensis* appeared in 1925 and of the *Flora of Tropical Africa* in 1937. The first volume of the *Flora of West Tropical Africa* came out in 1927 and, complementing it, the *Useful Plants of West Tropical Africa* in 1937.

The Director gave as much attention to the domestic scene at Kew as he did to international affairs. William Watson, Curator for the whole of Sir David Prain's administration, retired in the year Hill became Director. For some years Hill depended on the advice of his successor, W.J. Bean, one of the Gardens' ablest Curators. A knowledgeable arboriculturist who had been in charge of the Arboretum for many years, Bean's recommendations on the formation of a new pinetum were valued by the Director.

Industrial expansion along Brentford's riverside had for years been a major source of local air pollution which adversely affected conifers, especially Spruce and Silver Fir. In 1920 Sir David Prain had seriously considered another site

National Pinetum,
Bedgebury.

where the soil and climate were more suitable for Kew's conifers but constraints
on public expenditure blocked the scheme. Aware of Kew's predicament, the
Forestry Commission suggested to Hill in May 1922 a joint venture in the
planting of a new pinetum on the Commission's land. Bean had rejected several
sites before approving Bedgebury Park on the Sussex/Kent border, 50 acres
of which were acquired by the Forestry Commission in 1924. The National
Pinetum, as it was called, jointly managed by the Forestry Commission and
Kew, had as its principal objective "the growing of a natural collection of
conifers and taxads for the purpose of study by botanists, horticulturists,
foresters and the public generally". The Commission was primarily interested
in forest plots, created in 1929 on an additional 40 acres. The responsibility
for developing the remaining 50 acres as an arboretum or public amenity fell
to Kew, whose tree expert, William Dallimore, started planting in 1926. He
designed long grassy glades or rides, leaving a cluster of old trees, including
deciduous ones, and a thicket of rhododendrons around Marshall's Lake.
Hardy species and varieties of conifers crowned the high ground, others defined
the lines of avenues, or were huddled in groups, never as solitary specimens,
and all were placed with a sensitive regard to shape and colour. The main axes
were subsequently named Dallimore Avenue and Hill Avenue after the two
men who had done so much to establish the pinetum.

Within weeks of his appointment as Director at Kew in September 1943,
Professor Edward Salisbury recommended the closure of the pinetum, preferring the formation of small experimental plots throughout the country in order
to test conifers under different ecological conditions.[20] No one would claim
Bedgebury to be an ideal site: it suffers from soil deficiencies and frost pockets
in winter. The well-known forester, Alan Mitchell, is on record as having
said that had it not been a requirement that the pinetum should be close to
London, better sites could have been found in Scotland. Kew continued to

serve on Bedgebury's management committee until 1965 when it acquired Wakehurst Place in Sussex.

Hill always maintained an interest in all landscaping activities at Kew, personally approving all bedding schemes and preserving existing views and open spaces. When the trees in Cedar Vista were felled because constant clipping had severely mutilated them, he sanctioned the plan to widen the vista by some 40 feet. An oblique view of Syon House across the Thames was created by the removal of shrubbery at the north end of Rhododendron Dell. A judicious thinning of the *Lonicera* collection opened up a prospect of the Azalea Garden, and additional paths in the Queen's Cottage grounds disclosed new aspects of the Cottage.

The Director wanted a replacement for the Temple of the Sun, demolished during a storm in 1916. When post-war austerity frustrated the Board of Agriculture's intention to rebuild it, Hill let it be known that Kew would consider suitable substitutes. Mount Clare in Roehampton just beat him in taking possession of a neglected Georgian temple in the grounds of nearby Manresa House. An option on a garden feature at Amisfield in Lothian had to be declined as being too large for Kew. He even tried to persuade Lord Rothschild to part with Nell Gwyn's Temple at Tring. A modest octagonal shelter of Western Red Cedar wood, erected in the grounds of Cambridge Cottage, was his only success.

He had better luck with glasshouses: a new rhododendron house in 1925/6; a larger Economic House in 1930; and in 1936 a South African Succulent House for small desert plants from the Karoo region. Its designer, Duncan Tucker, also supervised the building of the nearby Sherman Hoyt Cactus House.

Mrs Hoyt of Pasadena had displayed a collection of the desert flora of southern California against a painted desert landscape at the Chelsea Flower Show in 1929. Not only did she present Kew with the plants and the panorama but she also generously offered to build a special house for them. It was attached to the north end of the T-Range and opened to the public in 1932. The panorama, which survived the destruction of the T-Range complex during the 1980s, now flanks one of the entrances to the Princess of Wales Conservatory.

Other glasshouses were renovated, adapted or extended. In 1925 the Temperate House acquired an annexe on the west side of its northern octagon. This teak structure, providing space for tender Chinese and Himalayan rhododendrons, was taken down in 1977.

Hill, now Sir Arthur, died from injuries received in a riding accident in November 1941. Through his correspondence, his lectures and, above all, his overseas tours, he personified Kew's paternalism. He was a competent and benevolent administrator, wholeheartedly dedicated to the Royal Botanic Gardens; they were, according to Dr John Ramsbottom, "almost a religion" with him. For the next two years the Economic Botanist, Sir Geoffrey Evans, replaced him as acting Director.

Before his untimely death Sir Arthur Hill had witnessed some of the disruption at Kew after the outbreak of World War II. Irreplaceable specimens

from the Herbarium and rare and valuable books and paintings from the Library found sanctuary in the comparative safety of Oxford and Gloucestershire. Qualified women gardeners gradually replaced men called up for military service. Air-raid shelters and allotments disfigured Kew Green. The lawns in front of Kew Palace were ploughed and planted with potatoes and other root crops, as in the First World War. Demonstration vegetable plots offered practical advice to the public.

Now that the country was denied access to vital commodities in continental Europe and the Far East, government departments sought Kew's assistance in finding alternative sources and substitutes for important vegetable products. Kew was represented by Dr Ronald Melville on the Vegetable Food and Drugs Committee, which considered means of intensifying food production and of maximising the use of available medicinal plants such as Belladonna, Henbane, Stramonium, Digitalis, Squill and Cascara.

In July 1940 Kew agreed to cultivate pharmaceutical plants which, requiring a long period of maturation, made them uneconomic for the commercial grower – *Atropa belladonna*, for instance, which could not be cropped under two years. In the autumn of 1941 a plot at Kew yielded a ton of fresh leaves and tops of Belladonna for processing by a commercial drug house. The National Federation of Women's Institutes gathered Digitalis seed and troops of boy scouts dug up *Colchicum* corms for planting at Kew.

Deprived of imported citrus fruit, children got vitamin C in their diet from rose-hip syrup. Melville helped to identify the numerous species, varieties and hybrids of native British roses whose hips were then analysed for their vitamin

Demonstration vegetable plot, 1939–45 war.

C content (it was discovered that the richest hips were produced by the roses of northern England and Scotland). Between 1941 and 1945 nearly 2,000 tons of hips were harvested by volunteers and some 10 million bottles of National Rose-hip Syrup processed.

The Jodrell Laboratory examined potential rubber-yielding plants, including species of Russian and British *Taraxacum* (dandelions). Kew's scientists, in collaboration with the National Physical and Chemical Laboratories, confirmed that the exceptional strength of the bast fibre of nettles (*Urtica dioica*) might be suitable for reinforcing plastics used in aircraft construction.

Even the Pagoda played its part in the war effort. The Royal Aircraft Establishment dropped "small bomb shapes" through holes cut in each of its floors to examine "the flight of these models during a vertical drop of 100 feet or more".

CHAPTER TWENTY

Towards the future

*E. Salisbury as Director – Australian House
– quarantine services – G. Taylor as Director
– Wakehurst Place – Queen's Garden –
Herbarium and Library – J.
Heslop-Harrison as Director – Orangery
converted – J.P.M. Brenan as Director –
Aiton House – A. Bell as Director – Banks
Centre – Princess of Wales Conservatory –
Palm House rebuilt – Board of Trustees –
Professor G. Prance as Director – Kew
Foundation – Friends of Kew – Jodrell
Laboratory enlarged – plant conservation –
St Helena*

KEW HAD ESCAPED relatively unscathed during the bombing of London. A total of some 30 high explosives had fallen within the Gardens, breaking glass in the Temperate House, Palm House, North Gallery and the Museum facing the Pond, and severely damaging the Tropical Water Lily House. With the end of the war, rehabilitation slowly got under way. Plants lost through flying glass were replaced, and books and herbarium specimens were returned from temporary storage.

The responsibility for supervising this restoration fell upon Edward Salisbury, formerly Quain Professor of Botany at London University, who became Director in September 1943. An eminent ecologist, it is not surprising that in 1944 his first exercise was the extension of Kew's British flora by creating a habitat for calcicolous species on a chalk garden on the east side of the ice-house. Limestone was used in 1949 to build a Clematis Wall near Berberis Dell, replacing the unsightly display of species and varieties on wooden supports near the Victoria Gate. The removal of limestone from the Rock Garden and the substitution of Sussex sandstone, started in 1929, continued whenever finance and manpower permitted.

Nature itself imposed changes on the landscape. The annual reports for the years immediately after the war lament the death through age and decay

Australian House; now, in
1994–5, being
transformed into an
Evolution House.

A design for a replacement
Palm House by E.
Bedford of the Ministry of
Works, incorporating the
Coronation arches.

of many mature and historic trees, some calculated to be about 200 years old. In 1950 a group of very large beech trees was felled; in 1953 two Cedars of Lebanon near the Japanese Gateway, their life-span having been prolonged by tree surgery, eventually died; the following year an elm near Cumberland Gate, one of the largest in the Gardens, was removed. Their departure provided space for replanting with specimens of greater botanical interest.

The Australian Government commemorated the Director's visit to the Antipodes in 1949 with the gift of a glasshouse. The prefabricated frame of aluminium alloy of the Australian House guaranteed resistance to corrosion and the elimination of constant repainting. Its steeply pitched roof gave the optimum angle through which light could be transmitted during winter. Although the largest house erected at Kew after the completion of the Temperate House, it could still only display a mere sampling of the Australian flora. The transfer of pot plants to this new house in 1952 made it possible to rearrange the South African plants in the Temperate House. Aluminium was also used in a replacement succulent house, completed in 1956.

The lack of adequate maintenance during the war had rendered the Palm House unsafe for the public, who were excluded from 1952 onwards. An engineer's report recommended its entire replacement, and at one time serious consideration was given to the idea of adapting the arches that had spanned The Mall during the Queen's coronation as the main structural members of a new house. Fortunately, the Ministry of Works decided to save this outstanding example of Victorian engineering; the basement was strengthened, corrosion treated, glazing bars stripped, repaired and realigned wherever necessary, and new glass installed.

Sir Geoffrey Evans, who had temporarily taken charge of Kew after Sir Arthur Hill's death, reverted to his post of Economic Botanist in 1943. He participated in a programme of introducing food crops from Peru and Bolivia into East Africa, and advised the Colonies on varieties of crops with superior yield, quality and disease resistance. Kew had supplied limited quarantine facilities for the transfer of plants within the Commonwealth since 1927, but now increased demands from plant breeders justified a properly equipped, intermediate quarantine station. Funded by a Colonial Development and Welfare grant, Kew built a new quarantine house in its Melon Yard in 1951. Bananas, rubber and especially cocoa were propagated and despatched to colonial research stations. When Evans retired at the end of 1953 he was replaced by T.A. Russell, formerly of the Colonial Agricultural Service.

In 1947 the Colonial Research Committee convened a group of distinguished biologists to consider the feasibility of a Colonial Biological Survey. They recommended help be given to Kew in its compilation of a *Flora of Tropical East Africa* (*i.e.* Kenya, Tanganyika and Uganda), just started before the outbreak of war. The first parts appeared in 1952, a work which still continues in collaboration with the East African Herbarium and universities in the region. In 1951 Dr R. Keay began his revision of the *Flora of West Tropical Africa*.

Salisbury retired in September 1956, exhausted and frustrated by the diffi-

Sir Edward Salisbury
(1886–1978).

culties of restoring Kew Gardens to normality under conditions of post-war austerity and financial constraint. It has to be said that his failure to involve staff in discussion and decision-making alienated them, causing resentment and lowering morale. Professor A.R. Clapham made this assessment of him:

> During his period of office Salisbury's achievements, considerable as they were, nevertheless fell below what had been expected of him. This was partly through lack of relevant experience and partly through reasons of temperament and judgement.[1]

His successor, Dr George Taylor, Keeper of the Department of Botany at the Natural History Museum and a prominent member of the Royal Horticultural Society, was known to many of the scientific and gardening staff at Kew. He found them, he said later, "in a state of revolt",[2] and he learned that the Ministry of Agriculture had under consideration the formation of a committee to report on issues specifically affecting Kew. Asked to suggest members of this committee, Taylor nominated Sir Eric Ashby, Vice-Chancellor of Queen's University Belfast, as Chairman. The Director made the most of his opportunities to promote his own vision of Kew's future, and the recommendations of the Ashby Visiting Group in 1957 endorsed, as we shall see, many of his proposals.

The bicentenary celebrations, scheduled for 1959 with the likelihood of a royal visit, provided the Director with an excuse to concentrate on improvements in the grounds. Snowdrops, primroses, narcissi and Martagon lilies were planted in abundance around the Queen's Cottage. In the same year (1958) the collection of roses rambling rather chaotically over tree trunks which looked like "a petrified forest" gave way to about 20,000 ericas, the nucleus of a new Heath Garden near the Pagoda. A Rose Pergola bisecting the Herbaceous Ground made the old one flanking the Rock Garden redundant.

With the aim of restoring Chambers's Orangery to its original function and elegance, the Director had its timber exhibits removed, the narrow gallery and staircase dismantled, and the two Francavilla statues, associated with Frederick Prince of Wales, placed among tubs of citrus fruit. Burton's and Turner's Palm House was renovated in time for the official reopening in June 1959. Too late for the royal visit by the Queen and the Duke of Edinburgh, a new flagstaff, the third to be donated, was erected on 5 November.

The momentum of change never slackened during the 1960s: a major reconstruction of the Rock Garden was finished in 1968; azaleas were planted on the south-facing slope of the Japanese Gateway; a depressing mass of laurel on the east side of the lake was removed, and rhododendrons, camellias and magnolias displaced other dismal groupings of evergreens. An incongruous element in the landscape, the Clematis Wall – now supporting a dispirited display of blooms – disappeared in this campaign to restore colour and vitality to the grounds.

But these improvements, however commendable, did not address Kew's urgent need for more space to accommodate a broader selection of the world's

Sir George Taylor (1904–1993).

flora. In 1949 Salisbury had attempted a partial solution when he entered into negotiations with the Duke of Northumberland for land at Syon House to be used as an arboretum and nursery. Nothing had been settled when Taylor took over. Under certain conditions, the 15 acres of Tresco Gardens in the Isles of Scilly became available to Kew in 1957, but this renowned garden, although it would have been an asset, would not have solved Kew's problems and the offer was declined. A large estate was required, preferably already developed, not too far from London, with a different climate and soil to extend the diversity of cultivated plants. It so happened that during the deliberations of the Ashby Visiting Group, the Director learned of the possibility of Kew being

offered the management of a large garden in Sussex. The Group, persuaded
by his arguments for a satellite garden, recommended that this solution to
Kew's congestion be given serious consideration.

The future of Wakehurst Place, an estate on the Sussex Weald, was at that
time being discussed by its owner, Sir Henry Price, with the National Trust.
Gerald Loder, later Lord Wakehurst, had acquired it in 1902 and, with his
incomparable head gardener, Alfred Coates, had created an outstanding woodland garden. Their planting of trees and shrubs from China, Australasia and
elsewhere in a semi-naturalistic manner had met with the approval of their
hypercritical neighbour, William Robinson. Shortly after Lord Wakehurst's
death in 1936, Sir Henry Price, an affluent tailor, bought the estate and made
some alterations, but kept its best features. He bequeathed it to the National
Trust which passed its management to Kew in 1965.

This mature garden of nearly 500 acres fulfilled Kew's requirements: soils
ranging from clay to a light silt-loam, a higher rainfall, and sheltered frost-free
places PLATE 31. Kew's major collections of *Rhododendron, Acer, Nothofagus,
Betula* and certain other genera have since been moved to this more hospitable
habitat. Its fine collection of conifers allowed Kew to withdraw from the joint

Formal walled gardens at Wakehurst Place.

Rock Walk in Bloomer's Valley, Wakehurst Place. Roots of trees cling spectacularly to massive sandstone outcrops.
Photograph by A. McRobb.

stewardship of the Bedgebury Pinetum. Unlike Kew, where trees and shrubs are grouped in taxonomic order, Wakehurst Place observes a geographical sequence: the Westwood Valley has been allocated to the Asian collection, trees of the southern hemisphere congregate in Coates Wood, the North American collection flourishes in Horsebridge Wood, and areas around Bloomers Valley will be planted with specimens from Circumboreal, Mediterranean and Irano-Turanian regions. This is natural landscaping at its best; the only concession to formality is the walled garden near the grey-stone Elizabethan mansion, now maintained as a memorial to Sir Henry Price.

While the future of Wakehurst Place was being negotiated, work had begun on a new conservatory, a Filmy Fern House, and a new garden at Kew. During the royal visit in the bicentenary year, the Duke of Edinburgh's uncomplimentary remarks on the squalid appearance of the ground behind Kew Palace prompted thoughts of remedial action. Earlier attempts had been made to develop the site: Samuel Fortrey, the first owner of the Dutch House, had "Orchards, Gardens, Plots" on some seven acres there[3]; Levett Blackborne, the owner during the latter half of the eighteenth century, had a "close of meadow behind the house planted with elms". Since the land was subject to flooding by the Thames, it remained a tree-lined paddock for most of the time. In 1898 the 2nd Duke of Cambridge had a plot there to store his bulbs after they had been lifted from his Cambridge Cottage garden. In 1933 Sir Arthur Hill undertook to make something of the site but nothing appears to have happened. When the Duke of Edinburgh peered out of the rear windows of the Palace he saw semi-derelict allotments and trees surmounting a tip formed of ash from the glasshouse boilers. With the raising of the tow-path, flooding was no longer a threat, and a garden far more ambitious than any that had preceded it now became feasible.

The site – barely an acre sandwiched between the Palace and a high boundary wall – presented a challenge to its designers. Kew departed from its normal landscaping practices by creating a garden contemporary with the Palace, a garden that demonstrated seventeenth-century fashions in plants and garden ornaments. This ambitious project nearly came to grief soon after its conception through economy measures imposed by the Ministry of Works. Stanley Smith, a friend of Sir George Taylor (he had been knighted in 1962), saved it from drastic modification with a generous offer to meet half the costs provided the work was completed before the Director's retirement in 1971.

It was never intended to be a garden appropriate for that particular house, being obviously too elaborate and too crowded for such a confined space. It deliberately set out to be a pastiche of Stuart garden design – a geometric parterre, pleached alley, clipped box, mount, statues and a fountain. From the outset a decision had been taken to reinstate the Palace's arcaded loggia and steps which had been removed during one of its earlier renovations. A watercolour by Paul Sandby, now in the Royal collections, enabled the Ministry of Work's architect, Peter Holland, to install an accurate reconstruction, conveniently forming an architectural link with the parterre beyond.

The parterre itself has two focal points: an Italian wellhead and also a

Paul Sandby's drawing of 1776 reveals that the riverside of Kew Palace was no more than grass and trees. In G.E. Papandiek's lithograph of *c.* 1820 some beds of shrubs had been planted but the garden was still vestigial.

fountain, a bronze copy of Verrochio's 'Boy with a dolphin', gently spraying jets of water into a pool below. The northern vista is closed by a curved clipped hedge, or exedra, around which are placed five terms, all that have survived intact out of the eight carved for Frederick Prince of Wales. To the west, the sunken garden, lined on three sides by a pleached walk of laburnum, contains annuals, biennials, and bulbs as examples of plants grown for their culinary or reputed medical uses or as decoration. A seventeenth-century gate pier, restored and gilded, thrusts up out of a density of vegetation. A hornbeam walk lies to the east of the parterre with a statue of a rather effeminate young man with a flute at one end and a copy of a Hampton Court stone urn at the other. The garden at Pitmedden in Scotland supplied about 1,500 box cuttings to clothe the mount constructed from an accumulation of cinders and garden rubbish. It serves as a podium for a decorative wrought iron rotunda of open trellis work surmounted by a dome with a gilded finial, a platform from which to view the patterned parterre below or the Thames beyond the wall.

The designers of this garden copied or adopted features in existing period gardens. The gazebo is indebted to Packwood House at Hockley Heath; the iron frame of the laburnum walk closely follows the design of the one at Greys Court at Henley; and the brick-paved path leading from the courtyard to the gazebo replicates the design of a path at Barrington Court at Ilminster.

Garden labels quote relevant extracts from John Gerard's *Herball* (1597 and 1633) and John Parkinson's *Paradisi in sole* (1629), the two principal works consulted for guidance on contemporary plants. Rosemary, sage and lavender are used as bedding in the parterre; the periwinkle provides ground cover; fennel, thyme, borage and other herbs intermingle with hypericum, dianthus, candytuft and other familiar flowers. The garden is probably at its best in spring, but even when there is little in flower, its geometry, defined by paving and clipped evergreen shrubs, invites inspection.

The Queen's Garden seen from the rotunda on the mound.

The sunken garden with part of the laburnum walk in the Queen's Garden.

When the Queen formally opened the Queen's Garden, as it is now called, in May 1969 she also officiated at the opening of a new wing to the Herbarium which included accommodation for the library. A new wing has been built about every 30 years, and the latest one with four floors marks a departure from the 'well' design of its predecessors. For the first time books, manuscripts and paintings have been brought together from their scattered locations in numerous rooms and corridors. The quality as well as the extent of its stock justify its reputation as one of the world's great botanical libraries.

The Herbarium, with about 6,000,000 specimens, ranks as one of the world's largest. The need for additional Herbarium space had been recognised by the Ashby Visiting Group, which was also concerned about the duplication of activities at Kew and the Natural History Museum. In 1951 both institutions had agreed on spheres of interest in certain groups of plants and floristic regions. In 1959 the Treasury appointed a small committee under the chairmanship of Wilfred Morton to consider the division of responsibilities between Kew and South Kensington. Agreement was reached in the allocation of geographical areas for taxonomic research on vascular plants: the Natural History Museum would deal with Europe, North-West Africa, North and Central America, West Indies, and the Polar regions; Kew would concentrate on Asia, the rest of Africa, Madagascar and the Mascarene Islands, South America and Australasia (including Polynesia). Kew transferred its collections of mosses, algae and lichens to the Natural History Museum while retaining responsibility for fungi and Gymnosperms. The Natural History Museum responded by sending its fungi collections to Kew. This arrangement with some pragmatic adjustment is still in operation.

The building of an additional wing to the Herbarium complied with a recommendation of Ashby's committee which also approved another of the Director's objectives: a radical improvement in the facilities of the Jodrell Laboratory. Taylor had made a convincing case for a cytogeneticist to work with the taxonomists, and for a physiologist to investigate such problems as the germination of seeds and the difficulty of striking cuttings. In 1963 the Jodrell Laboratory, patently too small and inadequately equipped, was demolished and replaced two years later by a much enlarged building on two floors with laboratories for plant anatomy, physiology and cytology, and a lecture theatre for public and scientific meetings and for Kew's student gardeners.

When Taylor arrived at Kew, student gardeners were examined in scientific and horticultural subjects during a two-year studentship for the award of the Kew Certificate. The Ashby Group accepted his proposal that their training should be slanted towards the National Diploma in Horticulture awarded by the Royal Horticultural Society. In 1963 a three-year diploma course based on a more intensive syllabus was launched, and the total intake of students increased from 40 to 60.

The education of the public as well as that of students came under scrutiny. A start was made with the appointment of a guide-lecturer in 1960. The heterogeneous collections in the museums, badly displayed and poorly labelled, were obvious targets for overhaul. With a new botanical exhibition gallery in the Natural History Museum and the building of the Commonwealth Institute, the successor to the Imperial Institute, Kew was in danger of losing any justification for a museums service. Its four museums were reduced to two by removing timber exhibits from the Orangery and closing the original museum near the Rock Garden. A thorough weeding of exhibits followed: ethnographical objects like a totem pole from British Columbia and a native canoe from Sierra Leone went to the British Museum; the Horniman Museum received specimens, and the Forest Products Research Laboratory accepted samples of timber. What

The main Library in the Herbarium wing, officially opened by Her Majesty The Queen in May 1969.

The rebuilt Jodrell Laboratory, opened in 1965. To the left of it is part of the Aquatic Garden, formed in 1909.

remained was displayed or stored in the museum facing the Pond or in Cambridge Cottage, now known as the Wood Museum.

In the midst of all these reforms a rumour circulated in 1969 that Kew was about to be closed and its scientific staff dispersed to other institutions. The radio and national press reported it, some Conservatives vowed to oppose it, the Labour Government vehemently denied it, and a petition is said to have been presented to the Queen. No one ever discovered the genesis of this absurd report, but the demonstration of genuine concern revealed the high regard in which Kew was held.

Sir George Taylor retired on 31 May 1971, content, as he confessed in the unpublished fragment of his autobiography, that "practically all the proposals which I had put before the Ashby Committee had been implemented".[4] His ability to establish good working relations with civil servants, his membership of numerous societies, frequently as chairman or officer, and his circle of influential friends, all aided his mission to transform Kew into a dynamic institution. He introduced features in the Gardens to remind the public of Kew's long history and its royal associations: a period garden at Kew Palace, the restoration of Princess Augusta's Orangery, the acquisition of a Coade stone copy of the Medici Vase made for George IV, and the installation of the statue of Hercules and the Serpent from Windsor Castle. Before he retired he rescued from probable extinction *Curtis's Botanical Magazine*, which had been edited by Kew staff since 1841. The magazine, never free from financial crises, had gained a respite in 1921 when its copyright was assigned to the Royal Horticultural Society, which then subsidised its publication for nearly half a century. In 1966 the Bentham Moxon Trustees at Kew helped with artists' fees and a few years later acquired the copyright. It still survives under the Trustees' imprint as the oldest illustrated magazine in the world, and with a broadening of its editorial policy may enjoy many more years.

Professor John Heslop-Harrison returned to this country from the United States, where he had held a chair of botany in the Institute of Plant Development at the University of Wisconsin, to become the next Director. He had a wing of his official residence at 49 The Green converted into a cell physiology laboratory and, with the installation of scanning electron microscopes there and in the Jodrell Laboratory, new fields of investigation were open to Kew's scientists. Extra accommodation in the Jodrell Laboratory was found by transferring the Physiology Section during 1974–5 to the mansion at Wakehurst Place.

For some years it had been apparent that watering the citrus fruit in the Orangery hastened the spread of dry rot in the building. After the ceiling had partly collapsed, the Director decided in late 1971 that the plants would have to be moved. It now became an orientation centre – an awful expression – a place for educational displays, exhibitions, and a book and gift stall.

As a distinguished experimental botanist, Professor Heslop-Harrison had a personal interest in advancing scientific research at Kew. His democratic approach to administration, involving the staff in consultative groups, marked a welcomed departure from the institution's traditional autocratic rule. He

Professor John
Heslop-Harrison.

RIGHT:
Professor John Patrick
Micklethwait Brenan
(1917–1985).

resigned in August 1976, and after a brief spell in the United States, returned to Britain to take up the post of Royal Society Professor at the Welsh Plant Breeding Institute.

J.P.M. Brenan, Deputy Director and Keeper of the Herbarium, through a natural humility, somewhat reluctantly assumed the responsibilities and challenges of Director in September. His term of office is particularly notable for the fruition of a number of building projects, some of which had already been mooted.

Aiton House, the first project to be completed, was opened in June 1977 on part of the site of George III's Castellated Palace. It provides accommodation for the Curator and his Deputy, the technical propagation unit, plant records and the planning and information unit, its name appropriately commemorating Kew's first Curator, William Aiton.

A new Quarantine House was in operation by the end of 1979, with improved facilities for inspecting the international traffic in tropical crops for breeding purposes. The site of the former Quarantine House had been earmarked for another Alpine House not far from its predecessor, a conventional timber building with a pitched roof, top and side ventilation, and small panes of glass. This new house had been considered by Taylor and given priority by Heslop-Harrison. Begun in 1976, completed in 1978, and officially opened in April 1981, the glasshouse represents a revolutionary concept in the cultivation of alpines: maximum light obtained by the use of a minimal structural frame. It is an eye-catching pyramid with cantilevered sides over a moat serving both as a gutter and as a reservoir for rainwater. Automated ventilators ascend the sloping roof in three louvred flights. The house does not dwarf its 2,000 diminutive inhabitants displayed in simulated habitats amongst Sussex sandstone: plants requiring moisture are placed around a small pond and waterfall; a corner has been landscaped for those from dry regions; peat provides acid, and limestone chippings alkaline conditions; Arctic plants stand on a refrigerated bench; others that can tolerate the English climate are planted outside on flanking rockwork.

PLATE 32 *Gustavia augusta.* Watercolour by Margaret Mee (1904–89) of the Amazon jungle, completed in 1985. The egret provides scale to a composition that includes aroids, bromeliads and orchids. This drawing is part of a collection of Margaret Mee's work in the Kew library.

PLATE 31 *(previous page)* The Tudor mansion at Wakehurst Place in Sussex. *Photograph by M. Svanderlik*

PLATE 33 *Johannesteijsmannia magnifica.* Watercolour by Mary Grierson, at one time employed by Kew, based upon colour slides and herbarium specimens. This spectacular palm of the undergrowth of the tropical forest in peninsular Malaysia is found in very few localities and is under threat through forest clearance and illegal collection of its seed. *Royal Botanic Gardens, Kew*

PLATE 34 *(overleaf)* A confusing reflection of girders on the glass of the Temperate House taken at night in July 1992. *Photograph by Ms S. K. Ellison*

The Temperate House had never been an entirely satisfactory building: its wide timber sashes reduced the passage of light, and inefficient heating in the south wing limited the choice of plants. It had suffered structural damage during the last war and, even after repairs, still leaked. Safety requirements demanded the removal of most of the stone urns in 1952. A surveyor's report in 1972 noted widespread wrought iron corrosion and deterioration in the masonry. When glass and metal started falling in 1973, the building was closed to the public. Restoration began in 1977 and, being a Grade I listed building, its architectural integrity had as far as possible to be observed.

Most of the plants were removed to temporary accommodation, but the Chilean wine palm and seven other trees, too large to move, remained, protected by polythene sheeting and portable heaters. It took three and a half years to complete renovations and improvements. Timber sashes, altered during repairs in 1936–7, were replaced by a modular pattern of aluminium glazing bars, thereby increasing light transmission and allowing the installation of semi-automatic ventilators. The incongruous teak annexe for rhododendrons, added in 1925, vanished. A new boiler house in the neighbouring Stable Yard replaced the one beneath the Temperate House. The firm of Chilstone in Kent replicated urns for the roof cornice, based on the design of one on the Victoria Gate. Decorative motifs of swags of fruit were remoulded, and one of the statues had its missing arm restored.

In landscaping its interior, the Curator's staff rejected the formal, unimaginative layout of beds, side benches and tarmac paths. Pools of water were introduced, a feature was made of a cascade over huge boulders of Sussex sandstone, and a few statues now peer coyly from the vegetation. Plants are arranged geographically, progressing from the Mediterranean in the south

The Alpine House, opened in 1981, has electronic environmental control and photoperiod lighting. *Kew photograph.*

wing, through South Africa, temperate Central and South America, Australasia and the islands of the Pacific to Asia in the north wing. Now that Wakehurst Place can grow rather tender plants like the Coral Vine (*Berberidopsis corallina*) from Chile, and the Grande and Arboreum series of rhododendrons, the space they once occupied has been freed for rhododendrons from the mountainous regions of South East Asia. This rejuvenated building, completed in the autumn of 1980, was reopened by Her Majesty The Queen in May 1982.

During the 1970s the Curator, John Simmons, submitted a brief for a large glasshouse able to simulate a number of climatic zones. The first draft of this ambitious project was ready in 1979, but it would be Brenan's successor who would officiate at its opening.

In 1977 the Standing Commission on Museums and Galleries approved the siting of a new Museum between the Herbarium and Kew Palace.

The last addition to this growing complex of buildings before Professor Brenan retired in October 1981 was a new Administration Block, a redevelopment of the Old Director's Office on Kew Green, opened in May 1981.

As the incoming Director, Professor E.A. Bell of London University and Head of Plant Sciences at King's College, maintained this building momentum. In 1982 he judged entries for the new museum near Kew Palace – the same location, in fact, that had been tentatively proposed for an earlier museum in 1855. Work on it – the future Sir Joseph Banks Centre for Economic Botany – began in 1985, and MCP Landscape Consultants together with the Curator and his staff skilfully transformed the five-acre site into a new attractive feature for the Gardens: two interconnecting lakes, a cascade, lawns and a sinuous path over gentle undulations, a sympathetic setting for the single-storey building whose curved glass roof recalls the elegance of the Victorian conservatory.

Professor E. Arthur Bell.

In the summer of 1984 work started on the construction of a large glasshouse – bigger in area than the Temperate House – to replace the T-Range and the Ferneries, a cluster of 26 houses constantly in need of repair. The architect, Gordon Wilson, was briefed to design a building that required minimum maintenance, reduced energy losses, and offered a compatible environment to plants from different tropical habitats. His solution represents a radical departure from the curvilinear lines of the Palm House and the classicism of the Temperate House. Factors that influenced his design included the latest technology, horticultural requirements, and the need for aesthetic conformity with the Gardens' historic landscapes. Ten climatic zones and several micro-areas are controlled by an integrated automated system, the southern section for the flora of the dry tropics, and the northern for the moist tropics. Here are displayed some of Kew's specialist collections: begonias, bromeliads, ferns, insectivorous plants, orchids, succulents, and rarities like the *Welwitschia* from the Namib Desert of South West Africa, bedded in heated soil. Climbers and epiphytes conceal structural columns, balconies and handrails support other plants. Everything is shown in as naturalistic a manner as possible – ferns cling to a dripping rock face, a cascade dramatises the setting for the South American water lily. Paths at various levels bring visitors close to the vegetation. Unlike the Palm House and the Temperate House where the structure

Temperate House, reopened in 1982 after extensive restoration. *Kew photograph.*

of the building is a conscious presence, here one is aware only of the plants.

This low-lying building, much of it below ground, has been treated by the architect as a 'glazed hill', as a backdrop to the Rock Garden with planting, paving, water and a gentle cascade, reminiscent of a Persian or Mughal chadar, to link it with the surrounding gardens. Earth excavated during its construction has added another man-made mound close to the Broad Walk. The Princess of Wales Conservatory, which was opened by Princess Diana on 28 July 1987, commemorates Princess Augusta whose architect, Sir William Chambers, designed in 1761 the Great Stove, in its day one of the country's largest glasshouses.

In the very year that contractors started on the Princess of Wales Conservatory, the Palm House had to be closed when the structure again became unsafe. Surveys carried out between 1980 and 1983 revealed alarming evidence of extensive corrosion to the metalwork despite the restoration in the 1950s and subsequent maintenance. Constant humidity within the house contributed to this progressive deterioration, but there were also faults inherent in the design of the building, and materials of adequate quality and strength had not always been used. Following the daunting decision to dismantle it, the Curator and his staff had to face the problem of storing the plants. Palms too big to move were reluctantly felled and climbers cut down; other plants were discarded after propagules had been taken; temporary accommodation was found in a new glasshouse near the Lower Nursery and in the Aroid House; and duplicates were given to other botanic gardens.

Scrupulous attention to accuracy and authenticity guided the restoration of the Palm House, but some modern materials were used for more reliable maintenance: durable stainless steel glazing bars replaced badly corroded

The demolition of an untidy sprawl of glasshouses to make way for the Princess of Wales Conservatory. *Kew photograph*.

Princess of Wales Conservatory and part of the Rock Garden. *Photograph by A. McRobb*.

Encephalartos altensteinii was one of the first plants to be returned to the Palm House after its latest restoration. The oldest of all the glasshouse plants at Kew, this cycad had been introduced from South Africa by Francis Masson in 1775.

wrought iron ones, mild steel strengthened the main arches, and ductile iron was manifestly preferable to brittle cast iron; special paints protected the metalwork; toughened glass was set in silicon rubber mastic. An atomised humidification system was installed.

The original planting of palms and other trees in boxes had been resisted by the first Curator, John Smith, who eventually persuaded William Hooker to let him replant some of them in beds. This obviously superior method of cultivation is now followed throughout the Palm House except in the apsidal ends where benches with potted plants have been retained as a historical feature. Although a new heating system has been installed, it has been possible to retain the cast iron floor grilles that covered the former hot-water pipes. Paths have been widened for the greater convenience of the public. A phytogeographical arrangement of plants has displaced the former partly systematic order; in 1868 Joseph Hooker had introduced the concept by rearranging the pot plants on the benches into geographical areas. The presentation is now one of mixed communities in a tropical rainforest, the south wing having the flora of tropical Africa, the north wing that of Asia, Australasia and the Pacific, with tropical South America and the tallest palms occupying the centre transept. A Marine Display in the former basement boiler room is an innovation proposed by the Curator. Nineteen tanks in a circular gallery grow plants of a British rocky shore and salt marsh, a mangrove swamp and a coral reef, and all with some of their marine inhabitants.

It could be said that the most important event in Professor Bell's directorship arose from the promulgation of the National Heritage Act in 1983.

The apses of the Palm
House have retained their
original bench ends with
pots of plants to
demonstrate the original
method of display.
Photograph by R. Desmond.

The Library established a
Paper Conservation Unit
for the preservation of
drawings and documents
in 1987.

Professor Ghillean Prance as President of the Kew Guild during its centenary year, 1993. On the wall behind him is a portrait of an earlier Director, Sir Joseph Hooker.

Under its authority, responsibility for Kew Gardens passed from the Ministry of Agriculture, Fisheries and Food on 1 April 1984 to a Board of twelve Trustees. Such an arrangement had been proposed before. In 1838 the Office of Woods and Forests had recommended that Kew should be managed by trustees representative of London's educational, scientific and horticultural communities;[5] both the Hookers had contemplated a board of trustees or visitors; the Ashby Visiting Group in 1959 had favoured an independent grant-aided institution with trustees. Now for the first time, Kew's functions were broadly defined in an Act of Parliament.[6] The responsibility for advising the Trustees rests with the Director currently Professor Sir Ghillean T. Prance, who succeeded Professor Bell on 1 September 1988.

Professor Prance (knighted in 1995), who studied botany at Oxford, has returned to his native land after 25 years at the New York Botanical Garden where he distinguished himself as the world's leading authority on the flora of Brazil's Amazon forests PLATE 32. The destruction of the rainforests has made him a passionate environmentalist, determined that Kew should play its part in global conservation.

However, faced with progressive reduction in government funding, securing adequate financial resources to maintain the Gardens and its research programmes is an over-riding preoccupation. In March 1990 Kew set up a Foundation with the sole object of raising funds for projects not covered by grant aid and self-generated money. The Foundation now raises £2,000,000 a year towards Kew projects. In June of the same year the Friends of the Royal Botanic Gardens, Kew was launched not only to bring in extra income but also as an effective means of advertising Kew and involving the public in some of its activities. Fees for courses and lectures on a wide range of botanical and horticultural subjects, and entrance charges all contribute revenue.

Kew fortunately received a government grant for an extension to the Jodrell Laboratory, completed in 1993, trebling its size. Up to the demolition of the original building in 1963, research had been confined largely to plant anatomy; rebuilding in 1965 added laboratories for plant physiology and cytogenetics. A few years later the Agricultural Research Council set up a biochemistry unit in adjacent accommodation. The latest enlargement of the Jodrell Laboratory has brought its scattered units together under one roof (excepting the Seed Bank which remains at Wakehurst Place): anatomy, cytogenetics, molecular systematics, biochemistry and biological interactions. Biological interactions investigates secondary compounds of plants – chemicals which might be employed in the treatment of AIDS and diabetes (the Medical Research Council has funded a programme seeking new chemical agents with anti-viral properties), in the protection of crops from insects and pathogens, and in the improvement of biological control agents.

Research on plant chromosomes by the cytogenetics section and the work of the newly-formed molecular systematics section enlarge the taxonomists' understanding of plant evolution and classification. Taxonomy is now a discipline within systematic biology which

> Provides the philosophical and nomenclatural framework for all biological research and for all users of organisms. Agriculture, forestry, land utilisation and conservation programmes all depend upon an accurate and stable way of naming, relating and identifying organisms whether they be useful crops or pests.[7]

The millions of dried specimens filed in Kew's Herbarium provide vital data for the correct identification of plants. Additions are constantly being made as more species are discovered by professional and amateur collectors. The classifying and naming of this material is an essential prerequisite to the preparation of regional floras and monographic reviews of taxa. The publication of floras,

Kew botanists Mark
Coode and Sara Oldridge
in a long-boat on the
Tutong River, Brunei
Darussalam in 1990.

initiated by Sir William Hooker as the first stage in utilising the plant resources
of the Empire, is still a major activity at Kew. The completion of the *Flora
of Tropical East Africa* and *Flora Zambesiaca* is in sight, and a revision of
Useful Plants of West Tropical Africa is in progress. In addition to its own floras[8],
Kew is collaborating with other institutions: with the National Herbarium at
Peradeniya on a new flora of Sri Lanka, with the Sugar Industry Research
Institute in Mauritius and ORSTOM in Paris on the *Flore des Mascareignes*,
and with the Forestry Department of Brunei on a check list of their forest
vegetation. Its taxonomists also contribute to other floras, of which *Flora of the
Guianas*, *Flora Malesiana*, *Flora Neotropica* and *Flora of Turkey* are examples.

There is probably not a time when a member of the Kew staff is not overseas collecting plants for the Herbarium or for cultivation in the Gardens, or gathering seeds for the Seed Bank, or advising other botanical gardens, or participating in conferences and setting up exhibitions. Kew used to depend on professional and private plant collectors but now its botanists and horticulturists make their own discoveries, at the same time gaining first-hand experience of ecosystems. Sir Arthur Hill began the practice of using botanical staff as collectors with grants from the Empire Marketing Board. A great deal of the world's flora has now been explored and sampled by Kew: Africa, partly for historical reasons, has attracted most expeditions, but Brazil now comes a close second; the palms, legumes and orchids of Madagascar's fast disappearing flora provide another mission; Irian Jaya (the Indonesian half of New Guinea), which unlike Madagascar still retains much of its forest cover, is receiving attention; relays of botanists visit Brunei. One could go on reciting countries or regions – even oceanic islands – duly inspected by Kew and the motivation is not always collecting *per se* but conservation.

Population growth and commercial greed are primarily responsible for the alarming destruction of the world's flora. It has been predicted that a quarter of our plant species may be extinct by about 2050, many of potential value as medicine and food. There have always been lone voices condemning this irresponsible eradication of our natural resources. The German scientist and explorer, Baron von Humboldt, witnessed with apprehension the destruction of cinchona trees in South America in 1799; half a century later Richard Spruce reported the wanton felling of rubber trees in Amazonia. In 1850 the British Association for the Advancement of Science appointed a committee "to consider the probable effects in an economic and physical point of view of the destruction of tropical forests". Its Secretary, H.F.C. Cleghorn, a pioneer conservationist in India, had campaigned against the slashing and burning of forests for temporary pasture by Indian farmers and peasants. His committee deplored the loss of plants of value to science through the indiscriminate elimination of large tracts of forest land. George Gardner, a former pupil of Sir William Hooker at Glasgow University, expressed similar disquiet when he arrived in Ceylon in 1844 to take charge of the botanical garden there.

> Botanists of future time will look in vain for many of the species which their predecessors had recorded in the annals of science as natives of the island.[9]

When Sir Joseph Hooker botanised in Sikkim its flora was still intact; 30 years later great swathes of it had been lost to cinchona (introduced, ironically, with the help of Kew) and tea plantations. When Sir Arthur Hill visited South Africa in the winter of 1930–1, he advised the Government to establish botanical reserves for the native flora under threat.

The grounds of the Queen's Cottage formed Kew's first 'nature reserve', created by royal command. Much of it consisted of exotics which have gradually been replaced by native species, while dead trees and fallen branches have been left as breeding grounds for insect life. Kew's wild flowers and butterflies have benefited from the introduction of carefully controlled grass cutting programmes. Kew is also concerned to protect peat habitats by using alternatives to peat in potting compost.

The construction by the Southern Water Authority during 1976–8 of a reservoir which extends into the Wakehurst Place estate presented an opportunity, as proposed by the Curator, to create a botanical reserve on the 150 acres (60 hectares) of ground around this sheet of water. Much of this area had already been designated a Site of Special Scientific Interest by English Nature for plants native to the woodlands, meadows and wetlands of South East England, and for the unique bryophytes on the sandstone rock formations.

Kew first became involved in international conservation programmes when Sir Peter Scott invited Ronald Melville, after his retirement from Kew, to be a member of the Survival Service Commission of the International Union for the Conservation of Nature and Natural Resources (IUCN). Aided by the resources of Kew, he compiled a Red Data Book of threatened flowering plants and gymnosperms, published in 1970. Melville set alarm bells ringing when he estimated that 20,000 species were in danger. The first United Nations conference on the Human Environment was held in Stockholm in 1972. Then in 1973 Washington was the venue for the first meeting of the Convention on International Trade in Endangered Species of Wild Fauna and Flora (CITES). The United Kingdom signed a treaty which acknowledged the need to control international trade in animals and plants likely to be at risk through commercial exploitation. Under the terms of this Convention Kew was designated the UK Scientific Authority for plants in 1976. In 1974 Professor Heslop-Harrison had set up a Threatened Plants Committee at Kew with G.Ll. Lucas as Secretary, who was also responsible for Kew's own Conservation Unit to advise on CITES, etc. In 1975 NATO's Eco-Sciences Panel sponsored an international conference at Kew on 'The functions of living plant collections in conservation-orientated research and public education'. Two of its resolutions – the long-term conservation of seeds and the propagation of rare and endangered species – have received special attention at Kew. Another international conference – 'The practical role of botanic gardens in the conservation of rare and threatened plants' – was convened at Kew in 1978.

Over the past two decades the living collections have been greatly enriched with authentic natural source material, much of which emanates from Kew's own collecting missions as well as exchanges. This has included access to previously restricted regions such as western China and Madagascar. Kew Gardens now cultivates roughly 40,000 plant taxa – about 10% of the world's population of seed plants and ferns, some of them near extinction in the wild. There are palms and cycads in the Palm House under threat in their native habitats, the Princess of Wales Conservatory offers sanctuary to endangered succulents from Madagascar, and endemics from the Canary Islands and some rarities from St Helena are successfully grown in the Temperate House. The micropropagation unit at Kew, by employing the latest techniques, rears plants which often do not respond to conventional methods. This work is increasingly in support of collaborative species recovery programmes with further progeny given to other botanical gardens and institutions engaged in conservation.

Plants have always been grown from seed at Kew, and surplus seed exchanged with other gardens. The seeds were kept in the Arboretum cold store until 1968 when a new seed unit offered improved storage facilities. This was the genesis of the present Seed Bank, which also conducts research on the storage and germination of seeds. The viability of seeds can be considerably prolonged – perhaps for centuries – by reducing their moisture content and housing them in a deep freeze temperature of minus 20 degrees centigrade. Rather than collect seed from cultivated plants, Kew now concentrates on wild plants offering greater genetic purity. Two Kew collectors harvest seeds, especially from arid regions where the vegetation has a precarious existence, and from other plants, particularly those of potential economic value which are in danger. The Seed Bank also handles seed for use in developing countries on behalf of the International Board for Plant Genetic Resources. Some 4,000 species, including a sampling of the British flora, are now stored in the Seed Bank, an impressive figure but nevertheless representing only 1% of the world's flowering plants.

The germination of terrestrial orchid seeds can be frustratingly difficult since they depend on a symbiotic relationship with particular fungi. With financial assistance from Sir Robert and Lady Sainsbury in 1983, followed by a generous endowment of £1,000,000, Kew is now engaged in a project of propagating threatened British and European orchid species.

Botanic gardens are crucial components in conservation, and since 1988 Kew has been advising on the restoration of the neglected botanic garden at Limbe in the Republic of Cameroon, in association with the management of two neighbouring forest reserves with perhaps the richest rainforest flora in Africa.

While Kew plays a prominent role in programmes to save the tropical forests of the Old and New World, it also has a particular interest in the flora of small islands, which through their isolation are home to many endemics. After the war Kew joined expeditions to Socotra, the Seychelles and Aldabra, which between them have an estimated population of over 300 endemic species, a large number on the verge of extinction. Easter Island's *Sophora toromiro* survives only in European and Chilean botanic gardens, and Kew is part of a European initiative to reintroduce it to its native habitat. *Helianthemum bystropogophyllum* has been returned to the Canary Islands; the micropropagation unit has increased the number of Mauritius's campanula, *Nesocodon mauritianus*; the rare *Lactoris fernandeziana* has been successfully propagated for Juan Fernandez island; it is hoped to re-establish a thriving population of *Ramosmania heterophylla* on Rodrigues; and the seed Kew is storing from the few survivors of *Impatiens gordonii* will in due course be sown in the Seychelles.

St Helena, an island which has had relations with Kew since the second half of the eighteenth century, is a microcosmic example of what has happened elsewhere in the world: settlement followed by the devastation of virgin vegetation, the introduction of new and often aggressive plants, and consequently a deteriorating environment. The Portuguese, who discovered it in 1502, began the erosion of its thickly-wooded slopes, felling trees for firewood, and introducing goats which soon overran the island. The Dutch briefly occupied it before

OPPOSITE:
Some examples of island flora that have become extinct or are in danger.
(a) St Helena. *Nesiota elliptica*. Extinct; survives only in cultivation.
(b) Hawaii. *Alsinidendron trinerve*. Only six plants known to have survived. Kew has returned seed as part of a reintroduction programme.
(c) Rodrigues. *Gastonia rodriguesiana*. Fewer than 15 trees left on the island.
(d) St Helena. *Trochetiopsis erythroxylon*. Extinct.

A

B

C

D

the British East India Company in 1659 made it a station for water and fresh supplies for their East Indiamen. The endemic gumwood provided fuel for their kitchens, and the cabbage tree and redwood timber for their huts. Governor Roberts, who tried to restore order out of anarchy on the island in 1708, attempted in vain to halt the disappearance of the vanishing woodlands. Captain Cook's *Endeavour* called at St Helena in May 1771, giving the young Joseph Banks a chance to botanise and to appraise the island's capabilities for growing "most if not all the vegetables of Europe together with the fruits of the Indies". But while he eagerly collected specimens for his herbarium, he appeared indifferent to the fate of its flora. His subsequent dealings with St Helena were influenced by this perception of it as suitable for growing certain crops, and as a temporary convalescent home for live plants transported on long sea voyages. In March 1787, the Governor invited Banks to send Francis Masson "to explore our Cabbage Tree Lands (as we call them). I think he may find some plants worthy of notice."[10] Another Governor, Robert Brooke, assured Banks that Henry Porteous, the island's gardener, had been instructed not to deviate from Banks's directions on the plants he should cultivate.

James Wiles, custodian of the precious breadfruit on HMS *Providence*, also fulfilled the precise instructions given to him by Banks when the ship reached St Helena in 1793.

> On your arrival at St Helena, the Commanding Officer will, in obedience to this instruction, direct you to deliver to the Governor & Council there such proportion of bread fruit & other usefull trees & plants as can be spar'd with propriety, for the use of the East India Company, which will be replaced by such trees & plants as the Governor & Council may think fit to send from the Company's Garden for the use of the West India Islands & of his Majesties Botanic Garden at Kew . . . You are to fill it [*i.e.* any spare space on the ship] with such wild plants of the Island as are not yet introduced into the English gardens.[11]

Six plants from St Helena were taken on board *Providence* for Kew. His Majesty's Royal Garden often received gifts of plants from the island's governors; some were exotics left there by passing ships, others came from the island's diminishing flora. In 1791 Governor Brooke notified Banks of the despatch of "four more chests of trees and plants . . . they are mostly natives of the island and reared in the manner you directed".[12]

Ferns still flourished under a canopy of purple and white cabbage trees when William Burchell was resident schoolmaster and acting botanist from 1805 to 1810.[13] But he sadly noted in his journal that

> the decayed remains of trees and shrubs which must formerly have covered all these hills . . . the soldiers and inhabitants have been suffered barbarian-like to cut down the trees with a wanton waste.[14]

St Helena passed from the control of the East India Company to the British Government in 1834, and for a time a moratorium was imposed on any agricultural and horticultural development. Joseph Hooker, surgeon and naturalist with the *Erebus* and *Terror* expedition, visited it in 1839 and 1843, seeking out some of the endemics in its "wonderfully curious little flora". His father, Sir William, received requests for plants and trees, especially pines, from the island's Governor. In 1868 Joseph Hooker, now Director, nominated Joseph Chalmers, a gardener on Ascension Island, to take charge of St Helena's new cinchona plantation. Hooker also suggested trials with tobacco, but his sending cocoa was an optimistic gesture. Chalmers, to his credit, tried to limit the encroachment of his plantations on the woodlands, and Hooker, although enthusiastic about these new crops, recognised the need to protect the indigenous flora.

> It is most desirable also that some of the extremely interesting native plants, which have become very scarce, should be preserved from extinction, viz: *Solidago rotundifolia* [*Psiadia rotundifolia*, now presumed extinct] and *Commidendron robusta* [*Commidendrum robustum*, once a very common tree, now reduced to a few specimens].[15]

His Kew annual report for 1867 noted the receipt of "seeds and plants of species now all but extinct in that interesting island".

When Daniel Morris, then Director of Public Gardens and Plantations in Jamaica, went to St Helena in 1883 to see what steps should be taken to reform its agriculture, he advised the abandonment of the cinchona plantation and suggested more suitable crops. He also urged the appointment of a Horticultural and Agricultural officer, a post not filled until 1928. His report to the Colonial Office pointed out that the island's climate had been adversely affected by the loss of its forests.

Until recent years, Kew maintained intermittent relations with St Helena: in 1933 it was asked to advise on the eradication of *Opuntia*, which had been introduced with Banks's encouragement; in 1938 the Governor wanted to know whether Rhodes grass (*Chloris gayana*) could be a viable crop for the islanders to grow. The beginnings of a closer relationship followed the visit in 1983 of a Kew garden supervisor to set up facilities for propagating the endemic flora. In 1990 the island was revisited by another Kew horticulturist who proposed a rescue programme for surviving endemics, in particular the St Helena olive (*Nesiota elliptica*), represented by a lone plant. Less than 1% of the forest that had once completely blanketed the island now remains; the clearings made by generations of settlers are eroded patches; water supplies are precarious. In 1991 the Overseas Development Administration funded a Sustainable Environment and Development Strategy for St Helena with Kew as a participant advising on habitat conservation and propagating some of the rarer endemics. It is, according to the Secretary General of Earth Summit, "probably the first true plan for sustainable development anywhere in the world".

The Earth Summit meeting, a United Nations Conference on Environment and Development, held at Rio de Janeiro in June 1992, and attended by 110 Heads of State and over 1,400 non-governmental organisations, unfortunately failed to get unanimous resolutions on climate, biodiversity, forest conservation and sustainable development. Kew was also represented at this meeting by the Deputy Director. Although there was agreement on certain strategies for improving the environment, much still depends on the cooperative efforts of individual nations and organisations. In committing the United Kingdom to the Biodiversity Convention, the Prime Minister announced the Darwin Initiative for the Survival of Species. As a result, the expertise of Kew and six other organisations in this country is now being tapped to assist the conservation and sustainable use of the planet's resources. The forest regeneration project in the Republic of Cameroon is just one instance of Kew's current activities which fall within the remit of the Darwin Initiative.

Sir Joseph Banks, the two Hookers, and Sir William Thiselton-Dyer vigorously promoted the agricultural and horticultural uses of the world's flora, and economic botany still ranks high among Kew's priorities. The seeds of potential crops for deserts and semi-arid lands, the home of some 20% of the world's population, are stored in the Seed Bank. With funding from OXFAM in 1981 and, more recently, from the Clothworkers Foundation, Kew has compiled a database of plants that can support the inhabitants of dry regions. SEPASAL (Survey of Economic Plants for Arid and Semi-Arid Lands) has now documented some 6,000 species of under-used plants. In 1984 Kew hosted an international conference on economic plants for arid lands.

A current venture aims to manage an area in North East Brazil before the pressures from 50 million people – about 30% of the country's population – degrade the present semi-arid scrubland into desert. Since the 1980s Kew's botanists have been working with their Brazilian colleagues on schemes to protect the ecosystem and in 1992 launched 'Plantas do Nordeste: local plants for local people'. Taxonomists, economic botanists and information officers are collaborating on programmes to improve forage crops, to create fuelwood plantations, and to encourage greater use of native medicinal plants.

Economic and ethnobotanical activities operate from the Sir Joseph Banks Centre for Economic Botany where some of Sir William Hooker's collections, and later accessions of plant extracts and artefacts, are now made more accessible through a computerised catalogue.

Sir William's first museum, a converted fruit store, was adapted again in 1990, re-emerging as the Education Centre of the School of Horticulture.

Another building of the Hooker era, the tropical Water Lily House, was extensively restored not only to display ornamental aquatics and climbers, especially cucurbits, but also economically significant plants such as jute, khushus, lemon-grass, rice and taro.

In 1994 Kew acquired an additional building, previously occupied by the International Mycological Institute (formerly the Commonwealth Mycological Institute), to which the Herbarium's 700,000 specimens have been transferred.

Changes still go on in the Gardens where the landscaping has never been static – new flower and shrub beds, a grass garden, or radical redesigning. The Rock Garden, never an entirely satisfactory concept, has been an irresistible challenge to gardeners. Boulders have been gradually replaced, contours

altered, and now a water and bog feature with a bridging cascade has immensely improved its northern area, bringing it into a more sympathetic bonding with the Princess of Wales Conservatory. Coming to completion near the Conservatory is a Secluded Garden. The primeval cycads, which were never returned to the Palm House, will find a more dramatic setting in the Australian House, now being converted to present the story of plant evolution. Huge artificial rocks, simulated lava flow, bubbling mud and appropriate noises will, it is hoped, create a convincing ambience for the living relatives of ancestral plants.

During its 276 years Kew has been in the possession of the Royal Family, and been the responsibility of the Lord Steward's department, the Office of Woods and Forests, the Office of Works, the Ministry of Agriculture, Fisheries and Food, and now operates under a Board of Trustees. It has survived serious crises with the assistance of friends and supporters which it has never lacked; its objectives may have changed or been modified over the years, but successive directors have always believed in its primary role as a scientific institution. Through its collections, its research, its training facilities, its publications and its heritage, this premier botanic garden is well-equipped to execute its succinctly-defined mission:

> To enable better management of the earth's environment by increasing knowledge and understanding of the plant kingdom – the basis of all life on earth.[16]

Sir Joseph Banks Centre for Economic Botany, opened in 1990. The architect's brief was a building that would minimise energy consumption, reduce maintenance costs, and not clash with Kew Palace.

APPENDICES

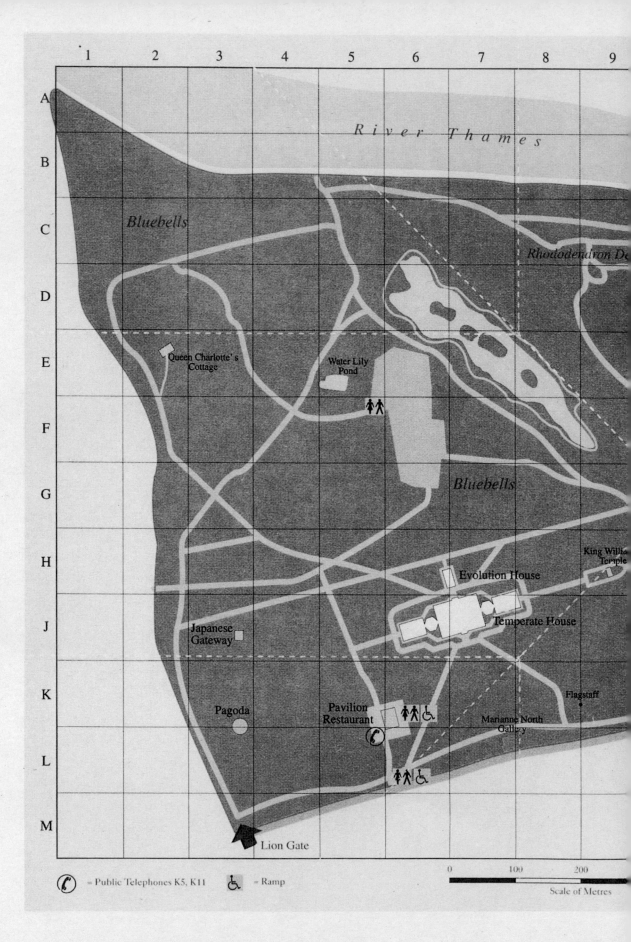

	1	2	3	4	5	6	7	8	9

A

B

River Thames

C

Bluebells

Rhododendron De

D

E

Queen Charlotte's Cottage

Water Lily Pond

F

G

Bluebells

H

King Willia Temple

Evolution House

J

Japanese Gateway

Temperate House

K

Pagoda

Pavilion Restaurant

Flagstaff

Marianne North Gallery

L

M

Lion Gate

(C) = Public Telephones K5, K11   ♿ = Ramp

0   100   200

Scale of Metres

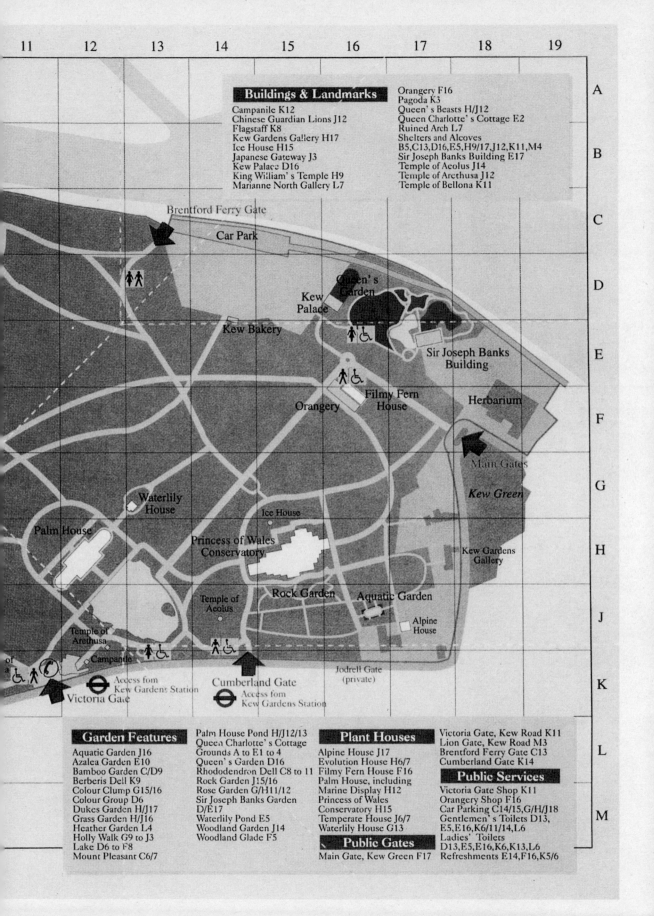

Outline map of Kew Gardens, 1995.

SOME BASIC FACTS

Acreage
Kew: *c.* 330 acres (132 hectares).
Wakehurst Place: *c.* 498 acres (*c.* 200 hectares).

Living Collections
The world's largest documented botanical collection consisting of about 40,000 plant taxa representing about 25,000 species which is 10% of the total number of the world's species of seed plants and ferns. The total number of accessions grown at Kew and Wakehurst Place is over 90,000 species.

Seed Bank
The world's largest seed bank containing more than 4,000 species.

Herbarium
The world's biggest collection with some 6,000,000 specimens and over 700,000 specimens of fungi. Included in this collection are 270,000 type specimens, representing a quarter of the world's named plants.

Economic Plant Collections
Some 80,000 items including plant products, associated implements and artefacts.

Library
One of the world's most important botanical collections with more than 750,000 items including
    120,000 books
      3,000 periodical titles
    250,000 letters
    175,000 drawings and prints

Buildings
39 listed buildings including:

Grade I Ancient Monuments
Kew Palace (1631)
Queen Charlotte's Cottage (*c.* 1771)

Grade I
Orangery (Chambers, 1757–61)
Pagoda (Chambers, 1761–2)
Palm House (Burton and Turner, 1844–8)
Temperate House (Burton, 1861–2; 1898–9)

Grade II
Outbuildings of former White House (1730s, etc.)
Temple of Arethusa (Chambers, 1758; rebuilt and moved)
Ruined Arch (Chambers, 1759)
Temple of Bellona (Chambers, 1760; rebuilt and moved)
Temple of Eolus (Chambers, pre-1763; rebuilt by Burton, 1845)
Former Aroid House (Nash and Wyatville, 1836)
King William's Temple (Wyatville, 1837)
Main Gates on Kew Green (Burton, 1845)
Campanile (Burton, 1847)
Water Lily House (Turner?, 1852, rebuilt)
Museum near Pond (Burton, 1856–7)
Lion Gate and Lodge (mid-19th century)
Unicorn Gate (mid-19th century)
Victoria Gate (1868; moved in 1889)
Marianne North Gallery (Fergusson, 1882)
Temperate House Lodge (Nesfield, 1867)
Japanese Gateway (1910; re-erected at Kew 1911)

Glasshouses
Princess of Wales Conservatory, covering 4,490 square metres, is the largest botanical glasshouse in the world.

The former Aroid House, also known as the Architectural Conservatory, is the oldest glasshouse in the Gardens.

Flagstaff

Erected in 1958, 225 feet long, it is entered in the *Guinness Book of Records* as the tallest in the world.

Largest Tree

Quercus castaneifolia (north west of the Water Lily House) measures about 110 feet (34m) with a trunk of 22 feet (6.9m). One of the finest specimens anywhere.

Kew and Wakehurst between them have 164 specimens of trees that are the largest of their species in the British Isles.

Oldest Tree

This is difficult to identify but a possible candidate may be an eighteenth-century sweet chestnut, *Castanea sativa*, on the south side of the Lake.

Soil

River sand and gravel, some 70 feet deep, over London Clay; consequently it has to be constantly improved by the addition of organic matter.

CHRONOLOGY

This chronology presents significant and interesting facts, quotations and publications.

1631 Dutch House (now Kew Palace) built for Samuel Fortrey, possibly on the site of Sir Hugh Portman's house: three storeys; 70 feet long and 50 feet wide; the cellars may have been retained from the previous house. (*See* official guide by John Charlton for detailed description)

1678 "Hence I went to my worthy friend Sir Henry Capel [at Kew] brother to the Earle of Essex: it is an old timber house, but his garden has the choicest fruit of any plantation in England, as he is the most industrious and understanding in it." John Evelyn's diary for 27 August.

Evelyn returned again on 30 October 1683: "He had repair'd his house, roof'd his hall with a kinde of cupola . . . The two greene houses for oranges and mirtles communicating with the roomes below, are very well contriv'd. There is a cupola made with pole-work betweene two elmes at the end of a walk, which being cover'd by plashing the trees to them, is very pretty: for the rest there are too many fir trees in the garden."

1691 "Sir Henry Capell's garden at Kew has as curious [ever]greens, and is as well kept as any about London . . . His terrace walk, bare in the middle, and grass on either side, with a hedge of rue on one side next a low wall, and a row of dwarf trees on the other, shews very fine, and so do from thence his yew hedges with

trees of the same at equal distance, kept in pretty shape with tonsure . . ." (John Gibson. *Archaeologia* vol. 12, 1809, p. 185)

It was Capel's house that Frederick Prince of Wales was to acquire in 1731 and have remodelled by William Kent.

1697 Dutch House acquired by Sir Richard Levett from Samuel Fortrey's grandson.

1707 Duke of Ormonde acquired a former hunting lodge in the Deer Park, making improvements.

1718 Duke of Ormonde went into exile in 1715, and his Richmond property was leased to the Prince and Princess of Wales as a summer residence. It became known as Richmond Lodge and Gardens. Painting by Jan Griffier of Syon House, *c.* 1710, shows part of the south front of Richmond Lodge. Plan of its rooms (now in the Royal Library) is reproduced in H.M. Colvin: *History of the King's Works*, vol. 5, 1976, p. 219.

1724 "From hence we came to Richmond, the delightful retreat of their Royal Highnesses, the Prince and Princess of Wales, and where they have spent the fine season every summer for some years." (Daniel Defoe: *Tour thro' the Whole Island of Great Britain 1724–6*)

1727 The coronation of the Prince and Princess of Wales as George II and Queen Caroline.

1728 Charles Bridgeman appointed Royal Gardener to George II and Queen Caroline, but he had already started relandscaping Richmond Gardens by mid-1720s. Richmond Gardens extended as far as the hamlet of Kew where the Queen acquired the lease of three houses including the Dutch House.

1729 Work began on extending the Duke of Ormonde's riverside terrace to Kew (9 October 1729. Works 4/4. P.R.O.); completed in 1734 (9 April 1734. Works 4/6. P.R.O.)

William Kent was engaged to design garden buildings.

1730 "The King is about to purchase the Bowling Green at Richmond to add to the grounds to the Royal Gardens [i.e. Richmond Gardens]" (*The Craftsman* 25 April 1730)

1731 Frederick Prince of Wales leased the Capel dwelling at Kew from Lady Elizabeth St André.

Hermitage, designed by W. Kent, erected in Richmond Gardens at a cost of £1,114. (March 1731. Works 5/58. P.R.O.)

1732 W. Kent who had been engaged by Prince Frederick as architect to enlarge the Capel house received his first payment in September 1732. A floor plan of this plain classical building is given by Sir William Chambers in his *Plans and Elevations . . . of . . . Buildings at Kew* 1763.

1733 Mount raised in Richmond Gardens to obtain a view of the Thames.
". . . the King's Gardens [at Richmond] which are prodigiously fine & contain above 400 acres of ground – all this ground is laid out in walks, wildernesses and flower plots, there are likewise some fields of corn interspersed which is chiefly done for the benefite of the game . . ." (May 1733 Sir John Clerk: "Travels at home". Scottish Record Office)

1734 Richard Butt, Kew nurseryman, supplied plants to Prince Frederick in 1734 and 1735. (Sir George Lee's papers. Buckingham Record Office)

1735 In 1734/5 eight terms or herms were carved for Prince Frederick.

Andrew Telf, the mason, was "setting up the pedestalls & marble statues at East End of Grass Walk" at Kew. (Duchy of Cornwall Accounts (D.C.O.))

Painter's account for work done at Kew: "24 figures in the Garden, 3 times done". (D.C.O.)

Merlin's Cave, designed by W. Kent, built in Richmond Gardens. Elms were planted on the mound near the Cave.

1736 Marriage of Frederick Prince of Wales to Princess Augusta of Saxe-Gotha on 8 May.

1737 Queen Caroline died on 20 November.

"seat on mount walk", "oke paling by the mount walk" at Kew. (D.C.O.)

"I passed a fortnight at Kew very agreeably, the Prince of Wales lives there quite in private, without form; passes his time wholly with his family who[m] he treats in so obliging a manner and with such an easy

familiarity as makes the attendance very agreeable." (Lady A. Irwin to Lord Carlisle, 19 July 1737. Historical Manuscripts Commission, series 42, 15th report, p. 186)

1738 On the death of Charles Bridgeman on 17 July, Thomas Greening, father and son, were appointed head gardeners at Richmond Lodge.

1740 "Mr Scutcheon & Mr Penley, foremen to the late Mr Bridgeman, did attend & were examined as to what they knew of works specified in the several bills. The first bill for feeding Tygers in the Royal paddock Garden was read to them, neither of which knew anything of the contents, it having been done in his late Majesty's reign." (22 April 1740. Work 6, vol.16, f.109. P.R.O.). A report in *The Craftsman* (29 May 1731) confirms that Queen Caroline had a menagerie.

1749 The Prince of Wales expanded his estate at Kew southwards with purchases in 1749 and 1750. (Crest 2/1245. P.R.O.; Sir George Lee's papers. Buckingham Record Office)

House of Confucius, designed by Joseph Goupy, probably erected in 1749. Consisted of a room and two closets with small saloon above; walls decorated with chinoiserie designs and episodes in the life of Confucius.

1750 Lord Bute appointed Lord of the Bedchamber in the Prince's household.

1751 The antiquarian, George Vertue, was commissioned by the Prince of Wales to make drawings of some 20 philosophers to adorn a mount.

The Prince of Wales died on 20 March. William Chambers designed

a mausoleum for him but it was never built.

1752 According to Sir George Lee's papers, Princess Augusta's garden consisted of 152 acres, either purchased or rented.

Princess Augusta's household accounts refer to "pheasantry in the Garden", "new greenhouse", "old greenhouse", "kitchen garden", "antelope ground", and for 17 June: "In moving lead figure and pedistal from Wilderness, Kew Lane to the Green House". Francavilla statues, ordered from Italy by the late Prince, arrive and remain in their packing cases.

1753 Robert Greening replaced the head gardener, John Dillman, whose duties were confined to the kitchen garden, orangery and melon ground. In June Greening gave an estimate of the work to be done: "To sinking two parts of the field which is to be added to the lawn, so as to give a view of the water from the [White] House".
"To planting the Mount and adjacent ground will take up an assortment of 14,400 plants".

Further dredging of the lake to be done. (Sir George Lee's papers. Buckingham Record Office)

Account of Solomon Brown, bricklayer: "To building & paving a drain from the end of the Bason to & cross the Horse Pond with rock [?] work to D° & laid in Terras according to a plan delivered". (Princess Augusta's household accounts. D.C.O.)

"To digging out the sand in that part of the circumducting walk from the Pond to ye Chinese Arch & laying it with gravel." (Princess Augusta's household accounts. D.C.O.)

1755 According to Sir George Lee's papers 110 acres of the Pleasure Ground on the Kew estate had been "walled & paled".

1756 "Six new large orange tubs". (Princess Augusta's household accounts. D.C.O.)

1757 Possibly the year in which John Hill met Lord Bute who subsequently employed him at Kew.

William Chambers engaged to instruct the Prince of Wales in architecture and to design garden buildings at Kew. He designed a bridge for the House of Confucius, which was moved in 1758 to the east end of the lake, and also the Gallery of Antiques (a screen of eight Palladian windows with statuary); work began on building the Orangery.

Robert Greening submits plans for converting the nursery into a wilderness.

1758 "I have been informed by Mr [Lancelot] Brown who called on me this morning that Greening who had the care of her Roy¹ Highness Garden at Kew dyed yesterday. He at the same time expressed how happy should he be if he was appointed to succeed him in that office." (Lord Brooke to Earl of Guildford. 12 March 1758. North MS 87109. Bodleian Library.) In the event, John Haverfield succeeded Robert Greening on 5 April 1758.

Chambers's Alhambra, Temple of Arethusa and Temple of Pan built.

1759 William Aiton recruited from Chelsea Physic Garden to manage the small "Physick Garden" at Kew.

"The terrace in the Royal Gardens of Richmond is generally allowed to be one of the finest walks in England . . . but the pleasure and beauty of this fine scene is wholly frustrated by the neglect of watering the road by the side of it . . . clouds of dust." (*London Chronicle* 31 May 1759)

Chambers's Temple of Victory and Ruined Arch built; the latter was

constructed with bricks and stone-faced. The interior of the Temple of Victory which commemorated the battle of Minden was embellished with stucco decoration. Much later, medallions of naval heroes were added: Duncan, Howe, Nelson, Rodney and Vincent.

J.H. Müntz's Gothic Cathedral probably added in 1759.

1760 Accession of George III on 25 October.

Chambers's Menagerie, Exotic Garden, Temple of Bellona and Theatre of Augusta constructed. The interior of the Doric Temple of Bellona displayed garlands and medallions recording the names of British and Hanoverian regiments serving in the Seven Years' War.

1761 George III married Princess Sophie Charlotte of Mecklenburg-Strelitz on 8 September.

Chambers's Pagoda (completed in 1762), Mosque and Temple of the Sun, and Great Stove built.

J. Smeaton's water pump installed.

1762 John Haverfield, senior and junior, take over the management of Richmond Gardens.

According to Peter Collinson, "This spring, 1762, all the Duke of Argyle's rare trees and shrubs were removed to the Princess of Wales's garden at Kew" by his nephew, Lord Bute. (*Transactions of Linnean Society*, 1811, p. 275)

"The house [at Kew] is not large, nor is the garden, in comparison with others, but there is much to see in it in the way of summerhouses, some of which are still being built. There are close upon twenty such little buildings." (Kielmansegge (1902), p. 77)

1763 Chambers's Temple of Eolus and Temple of Solitude completed by

this time but there is uncertainty whether the Temple of Peace was ever finished although several contemporary works describe it.

W. Chambers. *Plans, Elevations, Sections, and perspective Views of the Gardens and Buildings at Kew in Surry.*

1764 George III and his family began the first of their summer sojourns at Richmond Lodge, usually from June to October.

The young Joseph Banks probably met William Aiton for the first time.

Plan of "Richmond Garden with proposed alterations" prepared by Lancelot Brown for the King.

1765 "Mr Kirby [Clerk of Works] mentioned that great alteration & improvement are making in Richmond Gardens by Mr Launcelot Browne." (Minutes and Proceedings of Board of Works, 11 October 1765. Works 4/13. P.R.O.)

"Paving the seat [i.e. Temple of Eolus] on the Laurel Mount". (Princess Augusta's household accounts. D.C.O.)

1766 Merlin's Cave demolished in Brown's relandscaping of Richmond Gardens.

1767 The ha-ha along the Thames was constructed in 1767 according to Janet Dunbar (*Prospect of Richmond* 1966, p. 98). Presumably it had been in existence some time by 1792 when it was repaired – "repairing the ha-ha wall next the River". (Works 5/81. Richmond Garden, Midsummer quarter 1792. P.R.O.) It was repaired again in 1810. (Works 5/99. P.R.O.)

1768 The Royal Archives have several detailed invoices of plants purchased for Princess Augusta's garden from nurserymen such as

John Cree, John Busch, Lee and Kennedy, James Gordon, William Malcolm, William Watson. (R.A. 55512, 55544–6, 55557, 55584, 55624–5)

John Hill: *Hortus Kewensis*: 2nd edition, 1769. Lists 3,400 species in cultivation at Kew. "The unfortunate *Hortus Kewensis* of Dr Hill has long since been consigned to its merited obscurity." (*Monthly Review*, 1790, no.1, p. 44)

1769 Princess Augusta "lives altogether in town, and never lies a night at Kew; all she sees of those delightful gardens are Tuesday and Saturday mornings when the King and Queen breakfast with her." (Lady Mary Coke: *Journals* 1889–96, vol.3, p. 103)

1770 "Mr Warren [carpenter] is desired to repair the seat [Temple of Eolus] which is now so much decay'd & stands upon the Laurel Mount". (Chambers's letter-books. 15 May 1770. Add MS 41133, f.14. B.L.)

1771 Joseph Banks and Daniel Solander, on the successful conclusion of the *Endeavour* voyage, presented to George III on 10 August at Kew.

Prince of Wales and Prince Frederick move into the Dutch House.

1772 Princess Augusta died on 8 February. George III and his family move to White House in May.

An estimate for the demolition of Richmond Lodge. (3 July R.A. 15981)

The gardener, Francis Masson, sent to the Cape in South Africa to collect plants.

1773 By 1773 Joseph Banks was exercising some authority at Kew.

Brown still engaged in transforming

Richmond Gardens; Hollow Walk (now Rhododendron Dell) formed.

John Hill: *Twenty-five new Plants, rais'd in the Royal Garden at Kew*

"Catalogue of trees & shrubs in his Majesty's botanic collection at Kew planted in the year 1773." This manuscript record lists 177 genera and 791 species and varieties. (Kew archives)

1774 "I sent in October last . . . a large collection of seeds of plants & trees collected chiefly in Abyssinia: they were directed to Mr Eaton [Aiton] at the King's Garden at Kew." (James Bruce). (D.T.C. vol.1, ff.67–8)

1775 The cycad, *Encephalartos altensteinii*, brought back from S. Africa by F. Masson. It still survives in the Palm House where it is labelled as "the oldest pot plant in the world".

1776 David Nelson joined Capt. J. Cook's third voyage to collect plants, seeds and insects for Banks. F. Masson goes on a collecting expedition in the Canaries and Azores.

John Greg of Dominica offers plants to Kew including a clove: "I wish this box to ornament Kew." (Add MS 33977, ff.61 2. B.L.)

1777 George III began to make greater use of Windsor as a country residence.

1779 F. Masson collects plants in the West Indies.

1781 Leasehold of Levett Blackborne's estate, including Dutch House, conveyed to Queen Charlotte.

1782 Daniel Solander died on 13 May.

1783 Kew had a "matchless

collection of plants". (W. Curtis: *Catalogue of the . . . Plants cultivated in the London Botanic Garden*, 1783, p. 16)

F. Masson sent to Spain, Portugal and Tangier to collect plants.

1784 On the death of John Haverfield, senior, he was succeeded as head gardener at Kew by William Aiton.

Roofs of the Pagoda covered with slates.

"Painting [wooden] bridge between the two [Royal] Gardens over the [Love] Lane". (Works 5/73. P.R.O.)

1785 Anton Hove joins HMS *Nautilus*'s voyage to West Africa as collector for Kew.

An Act of Parliament authorised the closure of Love Lane but Richmond Gardens and Kew Garden were not physically united until 1802.

By about 1785 Chambers's Menagerie had been converted into a lawn.

1786 F. Masson returns to S. Africa on another collecting expedition.

Painting "the letters on the flower sticks [i.e. plant labels] at . . . 3d per dozen, three dotts to be allowed equal to a letter". (7 April. Works 4/17. P.R.O.)

1787 "The gardens of Kew undoubtedly are, upon the whole, extremely pleasing, but it has been thought, and not without reason, that with regard to the ornaments and buildings therein, a fondness for Turkish and Chinese chequer work has too much prevailed, in preference to the more beautiful models of Grecian and Roman architecture." (*New Display of the Beauties of England*, vol.2, 1787, p. 283)

Anton Hove collects plants in India.

1788 George III's first serious attack of porphyria confined him to Kew on 29 November "for the benefit of air and exercise".

A new hothouse built for Masson's African introductions; as it eventually filled with Australian plants, it became known as the Botany Bay House.

L'Héritier's *Sertum Anglicum* (1788) included engravings of plants that had been drawn at Kew.

1789 "This year the Paeonia Moutan was at first placed in the Stove. One day in the depth of a severe winter as Mr Bauer was walking around the Melon Ground in company with Mr Aiton he observed something red peeping through the snow, he poked it out with his stick saying 'there is something alive'. Mr Aiton exclaimed 'It is my Moutan and took possession of it.'" (Sir James E. Home. "Robert Brown letters", vol. 1, f. 214. Natural History Museum)

W. Aiton *Hortus Kewensis*. 3 vols.

1790 Sir Joseph Banks appoints Francis Bauer as flower painter at Kew.

"Report speaks in the most exaggerated terms of numberless plants now raising in the gardens at Kew, which never flourished there before; many of them totally unknown". (*Monthly Review* January 1790, p. 50)

M. Meen: *Exotic Plants from the Royal Gardens at Kew.*

1791 Archibald Menzies joined HMS *Discovery* as surgeon and naturalist and collected plants for Banks on the voyage.

David Burton who went to Australia as Superintendent of Convicts agreed to collect plants for Banks.

1792 George L. Staunton on

Macartney mission to China promised to collect plants for Banks.

"I am concerned for the fate of the breadfruit plants, as well as the box with others, for the King's Garden at Kew. Mr Aiton writes me that they had all perished, but that the seeds were in good condition." (W. Roxburgh in India to Banks. 17 August 1792. Add MS 33979, ff. 171–3. B.L.)

Hothouse, 60 feet long, built for Cape and Australian plants.

". . . building a shed for the Kangerous at the [Queen's] Cottage" (Works 5/81 and 5/92. P.R.O.)

1793 William Aiton, who died on 2 February, was succeeded by his son, William Townsend Aiton.

HMS *Providence*, under Captain Bligh, arrived home with plants from the Pacific and West Indies for Kew.

Christopher Smith, formerly plant collector on *Providence*, appointed "Botanical Gardener" at Calcutta Botanic Garden, where Roxburgh had become the new Superintendent. Until his retirement in 1813, Roxburgh despatched many plants for Kew.

1794 George III purchased Revd Thomas Methold's house and gardens of about 3–5 acres.

1795 W.T. Aiton also took charge of Richmond Gardens when John Haverfield junior resigned.

Large consignment of plants sent from Kew to Maria Feodorovna, Grand Duchess of Russia.

1796 W.T. Aiton's contract (LS10/ 4. P.R.O.) specifies the acreage at Kew: Kitchen Garden (12 acres), Pleasure Ground (33 acres), Lawns (55 acres), Great Lawn (41 acres), Lake (7 acres).

Large consignment of plants from Calcutta Botanic Garden.

Delineation of Exotick Plants cultivated in the Royal Garden at Kew (with illustrations of ericas by F. Bauer)

1798 Masson in N. America collecting plants for Kew.

1799 "repairing paving in 'Temple of Solitude'" (Works 5/88. P.R.O.) Large order for plants placed with Messrs Dickson of Edinburgh.

1800 George Caley employed by Banks as a collector in Australia.

1801 Botanic Garden placed upon the establishment of the Royal Household, coming thereby under the authority of the Lord Steward.

George III again ill and moved to Kew for treatment.

Building of the Castellated Palace, designed by James Wyatt, started.

HMS *Investigator* on a survey of the Australian coast had Peter Good, a Kew gardener, on board.

1802 While the demolition of the White House was proceeding, the Royal Family moved into the Dutch House.

Boundary walls between Richmond Gardens and Kew Garden demolished and the two estates united.

1803 "Among other things that were curious to me, there is the first Hydrangia that was brought to this country from which all the rest . . . have been originally taken. It has been at Kew, between twenty and thirty years, and is so thick that I counted more than eighty flowers." (F. Horner to his sister 10 September)

Kew gardener, William Kerr, despatched to Canton to collect Chinese plants.

Span house, 40 feet long, built for small Cape and Australian plants.

"Taking down & removing the Temples of Bellona & Arethusa by His Majesty's Command" . . . "repairing the colums" of Temple of Bellona. (Michaelmas quarter. Works 5/92. P.R.O.)

1804 "Altering the flues & pits of the old Cape House [at Kew] for Guyana plants" seized from a French ship by a British privateer. (Works 5/93. P.R.O.)

The dispersal of the animals in the Menagerie adjacent to the Queen's Cottage began.

Plants or seeds received from Kerr in China, George Caley and Robert Brown in Australia.

1805 Seeds or plants were received among numerous consignments from James Wiles at the Botanic Garden in Jamaica, W. Roxburgh in Calcutta, J.F. von Jacquin at the Imperial Gardens, Vienna, and from the *Investigator* voyage.

Kew also made despatches – several, for example, to W. Kerr at Canton.

1806 William J. Hooker introduced to Sir Joseph Banks by Dawson Turner.

W.T. Aiton requested permission to rebuild the dilapidated pinery which supplied fruit for the Royal table. (13 August. LS 10/4, f.57. P.R.O.)

The Menagerie having been dispersed, Aiton reported that the 3 acre paddock there had been converted, as instructed, into a flower garden. (20 December. LS 10/4, f.64. P.R.O.)

American plants bought at John Lyon's sale, 12–15 July (*see* 'Kew Record Book 1804–1826', Kew archives). Several large consignments of plants were purchased from Loddiges nursery at Hackney.

1807 Gothic Cathedral demolished. (Works 5/96. P.R.O.)

1808 Mango ripened at Kew in the autumn. (*Transactions of Horticultural Society*, vol. 1, 1809, p. 151)

1810 Jonas Dryander died on 19 October.

W.T. Aiton: *Hortus Kewensis*, 2nd edition, 1810–13. 5 vols.

1811 Prince of Wales appointed Regent on 8 February.

1812 W. Kerr arrives in Ceylon to take charge of its botanical garden.

1813 26 acres of "land cropp'd with oats and barley from Stafford Walk to Pagoda" and another 23 acres "cropp'd with turnips and buck wheat". (LS 10/4, ff.90–1. P.R.O.)

Hailstorm on 28 June broke 13,000 panes of glass.

House of Confucius repainted, paying particular attention to its decorative features. (Works 5/102. P.R.O.)

1814 W. Kerr died in Ceylon.

The Kew gardeners, A. Cunningham and J. Bowie, sail for Brazil to await transport for Australia and S. Africa where they were to collect plants.

Hortus Kewensis Epitome.

1816 Alexander Moon, a Kew gardener, replaced W. Kerr in Ceylon.

Lord Amherst's mission to China included the Kew gardener, James Hooper. David Lockhart, a Kew gardener, joined Captain Tuckey's exploration of the Congo river.

Arrival of seeds collected by Cunningham and Bowie in Brazil.

1817 Cunningham and Bowie arrive in Australia and S. Africa respectively and seeds are received

from them. G. Caley sent seeds from St. Vincent. After his appointment as Superintendent of the Botanic Garden in Calcutta, N. Wallich despatched the first of many consignments of plants to Kew.

1818 On 11 July in the Queen's drawing-room in Kew Palace, the Duke of Clarence married Princess Adelaide and the Duke of Kent remarried the Dowager Princess of Leiningen.

Queen Charlotte died in Kew Palace on 17 November.

Wallich sent seeds of the Himalayan *Rhododendron arboreum*; the seedlings were lost through over-heating but another lot in 1820 survived.

1819 Prince Regent began negotiations for the purchase of Hunter House (now the Herbarium).

Sir Joseph Banks probably made his last visit to Kew when he saw the male cone produced by *Encephalartos altensteinii*, the cycad introduced by Masson in 1775.

1820 On the death of George III on 29 January, the Prince Regent succeeded him as George IV.

Death of Banks on 19 June.

Purchase of Hunter House for George IV was finalised during April/May.

G.E. Papendiek: *Kew Gardens: a series of twenty-four drawings on stone*

1821 Plants and seeds were still being received from Cunningham and Bowie, and David Lockhart in Trinidad was also a donor.

1822 John Smith was transferred from the Royal Forcing Garden at Kensington Palace to Kew.

Herbaceous collection in about one acre in the Botanic Garden was rearranged.

"Neither this [Kew] nor any other of the royal gardens are at all kept in order as they ought to be, not on account of the want of skill in the royal gardeners but from the want of support from their employers." (J.C. Loudon: *Encyclopaedia of Gardening* 1822)

1823 Hunter House which had been purchased for George IV was sold to the Commissioners of Woods and Forests in July.

John Smith promoted foreman of the hothouses and propagating department.

James Bowie's contract as plant collector in S. Africa terminated.

Plants received from A. Moon in Ceylon, orchids from D. Lockhart in Trinidad, also donations from Horticultural Society of London.

1824 About this time the Cape, Botany Bay and Palm Houses had glazed frames, called bulb borders, along their lengths – some 234 feet long and 5 feet wide – heated by flues from these houses, and filled primarily with the collections of Masson and Bowie. (Smith (1880), pp. 312–3)

1825 Western end of Kew Green enclosed with railings and two lodges. New entrance to the Botanic Garden. (Works 5/109. P.R.O.)

Low lean-to house, 40 feet long, was remodelled with two compartments for S. African succulent plants.

A cutting from a willow near Napoleon's grave on St. Helena was planted.

1826 "Kew . . . unquestionably possesses the most complete collection of exotic plants in Europe . . . but is in general rather neglected. Yew-trees are found here of the height of our firs, and very fine specimens of holly and evergreen oaks; but in other respects the old Queen's plantations are not very tasteful." (9 December. Prince Pückler-Muskau)

Tender aquatics moved to oval tanks in a renovated lean-to house. (J. Smith (1880) pp. 247–8). Old lean-to house, formerly a vinery, repaired for the heath collection.

Palms purchased from Loddiges nursery.

1827 W.T. Aiton appointed Director-General of His Majesty's Gardens.

George IV ordered demolition of Castellated Palace.

1828 Grass collection in concentric circular beds, 100 feet in diameter, replanted by J. Smith.

Palm House (60 feet long, 16 feet wide and 15 feet high at the back) had its height raised by 4 feet.

"With respect to the pleasure-ground, I have no high opinion of it. The trees are in a miserable state, and badly distributed. I met with a little piece of water [*i.e.* the remains of the lake], which I think would have been better omitted." (Jacob Rinz. *Gardener's Magazine*, vol.5, 1829, pp. 380–1)

1830 On the death of George IV on 26 June, the Duke of Clarence succeeded him as William IV.

A. Cunningham, Kew's only plant collector abroad, recalled.

1831 Enclosure of part of Kew Green revoked and the lodges and railings removed.

Duke and Duchess of Cumberland assigned Hunter House as a residence, and the Duke given *c.* 400 acres of meadow in Deer Park and Richmond Gardens.

Aiton's request for land from the Pleasure Grounds for an arboretum refused. (19 November. T1/3903. P.R.O.)

1832 The Pagoda "was formerly open to the public, but the mischievous habit so common to Englishmen, of cutting their names on every part of the woodwork, has caused it to be closed." G.W. Bonner: *The Picturesque Companion to Richmond and its Vicinity*, *c.* 1832. It would appear that there was at one time limited public access to the Pagoda. "The building is not now open to the general public although in former years visitors were permitted to ascend the winding staircase and view the gardens around." (*Gardeners' Chronicle* 28 March 1908, p. 202)

1834 The extent of the private lawn in front of Kew Palace defined by a ha-ha and railings. (29 April. Works 16/589. P.R.O.)

Jeffry Wyatville designed a Palm House which was never built.

1835 George Barclay joined HMS *Sulphur* as a plant collector for Kew.

A. Cunningham appointed Colonial Botanist in New South Wales.

1836 One of John Nash's conservatories at Buckingham Palace re-erected at Kew. This Architectural Conservatory measured 80 feet in length, 40 feet in width and 26 feet in height.

Bust of victor of battle of Minden, Ferdinand of Saxe-Gotha, brother of Princess Augusta, put into Temple of Bellona, and the arms of Earl Waldegrave who fought at Minden emblazoned on the walls. Subsequently the Temple of Bellona was frequently called the Temple of Minden.

Lean-to house erected against back wall of a propagating pit (35 feet long and 14 feet wide) mainly for ferns.

1837 Queen Victoria succeeded William IV who died on 20 June.

Duke of Cumberland ceased to occupy Hunter House on his accession to the throne of Hanover.

King William's Temple, originally called the Pantheon, or Temple of Military Fame, built to a design by J. Wyatville.

John Armstrong emigrated to N.W. Australia, promising to send plants to Kew.

"The whole space of the botanic garden occupies nearly three acres [this obviously excluded the arboretum]: it is surrounded with a fifty feet wall, and divided into small beds, or patches, for the cultivation of rare vegetable productions." (*Horticultural Journal* October 1837, p. 97)

". . . it is so thoroughly disgusting to see the bad arrangements, the ignorant attendants, the slovenly state of the place and the abominable state of the plants." (*Gardeners' Gazette* 14 October 1837, p. 648)

1838 The Treasury instituted an investigation of all the royal gardens, and Kew in particular, under the chairmanship of John Lindley.

1839 "Aiton was very friendly to me & very open . . . a new [glass] house has just been constructed which cost £6,000." (Sir William Hooker to Dawson Turner. 29 January)

1840 The Treasury approved the transfer of Kew from the Lord Steward's Department to the Office of Woods and Forests on 25 June. This was shortly followed by an announcement that Sir William Hooker had been appointed Director.

Kew's flower painter, Francis Bauer, died on 11 December.

E. & G.N. Driver's plan of Kew, 1840, shows 15 acres immediately north of the Pagoda in use as arable land.

J. Lindley *et al. Report upon the Present Condition of the Botanical Garden at Kew, with Recommendations for its Future Administration.* House of Commons.

1841 W.T. Aiton relinquished control of Botanic Garden on 26 March to Sir William Hooker.

Lord Lincoln appointed First Commissioner of Woods and Forests to which Kew had been transferred.

John Smith appointed Curator in December.

Four acres between Orangery and Architectural Conservatory added in November.

Heath House doubled and joined to span-roof glasshouse. Most of the orange trees in the Orangery sent to Kensington Palace.

Receipt of plants and seeds collected by J. Armstrong en route to Australia and from G. Barclay on HMS *Sulphur*; plants presented by Duke of Bedford, James Bateman, William Borrer, John Clowes, A.B. Lambert, John Parkinson; botanic gardens at Birmingham, Hull, Liverpool, Manchester and Horticultural Society of London were also donors; purchases made from nurseries such as Backhouse of Darlington, Jackson of Kingston, Low of Clapton, Veitch of Exeter and Whitley and Osborn of Fulham. Purchases and exchanges significantly increased under Hooker.

1842 Large window inserted in both ends of the Orangery. Double propagating pits altered and new ones built; fern house enlarged.

1843 Expenditure on Kew given in a Parliamentary return, 22 March: £1,191 (1838–9), £1,681 (1839–40), £2,078 (1840–1), £2,887 (1841–2), £2,563 (1842–3); repairs to buildings were an additional charge.

An additional 45 acres extended the Botanic Garden to just beyond the Pond. Wire fence erected to separate Botanic Garden from the Pleasure Grounds.

Joseph Burke collected plants in N. America on behalf of Kew and Lord Derby. William Purdie engaged to collect plants in West Indies and S. America for Kew and Duke of Northumberland.

Decision taken to demolish the House of Confucius and Temple of Pan, both in a ruinous state. (Office of Woods and Forests. 26 July 1843 (Works 1/26, f.277), 19 March 1844 (Works 1/26, f.494. P.R.O.)

Lean-to house (60 feet long and 17 feet wide) doubled to make a span-roofed house for the Duke of Bedford's orchids. A dry stove and a span greenhouse, both 40 feet long, were joined to provide accommodation for heaths and cacti.

Plants collected by Joseph Hooker in the Falkland Islands despatched in two glazed cases to Kew.

1844 King of Hanover surrendered 44 acres in the Pleasure Grounds.

With the appointment of Decimus Burton, Richard Turner and W.A. Nesfield, work began on planning the landscaping of the grounds and building the new Palm House.

The old Cape House was doubled to form a span-roofed structure and Burton added two wings for the finest Australian and New Zealand plants. (Plan in *Gardeners' Chronicle* 1 February 1845, p. 67). Other glasshouses were also renovated and enlarged.

Queen Elizabeth's Elm blown down with only a fragment of the trunk remaining; this relic still survived in 1891. Tradition said this historic tree had been planted by Queen Mary I and that Queen Elizabeth had sat under it. A fragment of the dismembered tree was made into a kitchen table for Osborne House.

Kew's annual reports published; the first one retrospective to 1841.

1845 When W.T. Aiton retired, the Pleasure Grounds of 178 acres and the Old Deer Park of *c.* 350 acres passed to Hooker's management on 9 July. The Old Deer Park was let for grazing to a tenant farmer. The King of Hanover surrendered another 80 acres.

Broad Walk begun. Mound near Pond cleared of trees and shrubs and grassed. Ice-house mound replanted. Deodar Avenue planted. Open rustic trellis replaced tall wooden fence around the Queen's Cottage to reveal it to the public.

Sir William noted in 1845 that 14 plant houses, excluding pits and frames, were in use at Kew. Existing houses were adapted or replaced. Plan of seed pit at Kew in *Gardeners' Chronicle* 27 December 1845, p. 872.

Main Gates, designed by D. Burton. Burton replaced wooden Temple of Eolus with stone replica but without the revolving seat. Burton also recommended that Pagoda should be restored to Chambers's original design. (10 July. Works 16/590. P.R.O.)

Some Metropolitan policemen were employed to patrol the grounds; after the Crimean War, army pensioners were appointed gate keepers and park constables.

A large quantity of trees were purchased from Lawson of Edinburgh, presumably for the

Arboretum being planned by Nesfield.

Curtis's Botanical Magazine (new series) illustrated many plants grown at Kew – an editorial policy that has continued to the present day.

1846 Lord Morpeth followed Lord Lincoln as First Commissioner of Woods and Forests.

Royal Kitchen Garden of 14 acres added, to be converted into the

Herbaceous Ground. Crab Mound formed with surplus soil from Broad Walk. Old Deer Park boundary wall/fence replaced by ha-ha. Work started on enlarging and deepening the Pond.

Stone pine, *Pinus pinea*, planted near Cambridge Cottage. It lost a large limb in a snowstorm in January 1926; gradually tilted and since the spring of 1941 has been supported by a prop.

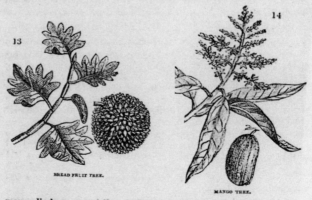

A page from the first guide to Kew Gardens, published in 1847. Illustrated by W.H. Fitch.

Greenhouse (110 feet long by 40 feet wide and 17 feet high) erected. Orchid House (90 feet long and 32 feet wide) built. Ladies cloakroom built on Broad Walk near Kew Palace.

Revd John Clowes's orchid collection necessitated the building of a new house. Plants received from Joseph Burke. Huge *Echinocactus platyacanthus* sent by F. Staines in Mexico. *Amherstia nobilis* presented by East India Company. Seeds of *Victoria amazonica* purchased but none survived.

1847 Berthold Seemann joined HMS *Herald* and collected plants for Kew during the voyage.

Site of former Castellated Palace laid out as a lawn with shrubbery; called Queen Elizabeth's lawn after the ancient elm (*see* 1844). American Garden behind the Palm House formed. The Dell, formerly Hollow Walk, replanted with shrubbery.

Campanile, 107 feet high, built. Brentford Ferry Gate opened.

Wardian cases were now in regular use.

Kew Gardens, or a Popular Guide to the Royal Botanic Gardens of Kew (this reached its 30th edition in 1885)

1848 Joseph Hooker in northern India collecting plants for Kew.

Final eight acres in the Pleasure Grounds held by King of Hanover released.

Palm House finished. Decimus Burton was the architect; Richard Turner as engineer provided the ironwork; and Grissell & Peto were the contractors. Turner designed 60 wrought iron half arches to span wide spaces with 16 cast iron pillars for additional support; he also advised on heating and ventilation. Twelve Burbidge and Healy boilers were placed in two vaults beneath the floor; perforated cast iron plates covered coils of hot water pipes. Scraping of the metalwork during the last restoration revealed that the exterior had been painted a deep blue-green; green-tinted sheet glass used. With the completion of the Palm House, the old palm house and Chambers's Great Stove were to be demolished, and their contents transferred to the new building.

Nesfield implemented his plans for the parterres around the Palm House; planting of the Arboretum under way.

Museum opened to the public in an old fruit store adapted by Burton. John Smith's cottage in the Botanic Garden converted by Burton into an office for the Director and a reading room for gardeners.

Plants received from Kew's collectors: W. Purdie and B. Seemann, and the first batch of seeds from Hooker in India.

1849 W.T. Aiton died on 9 October.

"Let us keep straight along this broad gravel-path [Broad Walk]; the turf on each side laid out, you see, in beds of evergreens, flowers and flowering shrubs: that chimney shaft [campanile] is an unsightly termination, is it not? I wish it could change places with the fine old trees in its rear." (*Florist* 1849, p. 285)

Victoria amazonica, in a tank in a propagating house, produced its first flower bud on 21 November, but did not fully open.

Number of glasshouses increased to 21.

Treasury allowed a small annual grant for the purchase of books.

[W.J. Hooker] *Rhododendrons of Sikkim Himalaya* (1849–51)

Between 1843–9, Kew's plant collectors sent 92 Wardian cases of plants as well as seeds and dried specimens.

1850 Arboretum had 2,325 species and 1,156 varieties of trees and shrubs.

Victoria amazonica now in full bloom. *Plectocomia elongata* var. *philippinensis* was the first palm to reach the roof of the Palm House.

As well as being a recipient of plants and seeds, Kew was also a generous donor sending consignments to the British colonies and European botanical gardens, and maintaining a mutually beneficial exchange with British nurserymen.

1851 In the reorganisation of the Office of Woods and Forests, Kew was transferred to the new department of Works and Public Buildings. Responsibility for managing the Old Deer Park remained with Woods and Forests. Lord Seymour appointed First Commissioner of Works and Public Buildings.

13½ acres were taken from the Old Deer Park and added to Queen's Cottage grounds in order to tidy the boundaries of the two properties. (Works 19/868. P.R.O.)

On 15 December Methold House on Kew Green became the Director's official residence.

Kitchen Garden and Paddock, belonging to late King of Hanover, reverted to the Crown and was added to Kew Gardens.

Nursery of about 4 acres established in the Pleasure Grounds. Syon Vista with gravelled walk formed; completed in 1852; planted in 1854.

According to 1851 annual report a lodge for the foreman of Pleasure Grounds was built at Pagoda (Lion) Gate; however, in January 1863 permission was given to erect another Lodge at Lion Gate. The first one had been built from salvaged bricks.

Temperate Fern House built.

16 Wardian cases of economic plants despatched to the colonies; seeds of *Victoria amazonica* also sent.

Botanical items from the Great Exhibition in Hyde Park donated to the Museum.

1852 Lord John Manners and Sir William Molesworth were successively First Commissioner of Works.

William Milne contracted by Kew to collect plants while with HMS *Herald* in the Far East.

About this time the American Garden behind the Palm House was completed. Private garden for Queen Victoria laid out at Queen's Cottage.

Some of the tallest palms in the Palm House were sunk into beds in the floor to give them a few more years' growth before they touched the roof. The celebrated double coconut, *Lodoisia sechellarum*, presented by Mauritius Botanic Garden. Water Lily House for *Victoria amazonica* built.

Ground floor of Hunter House made available to Director as temporary accommodation for herbarium and library. W.A. Bromfield's herbarium and library presented.

1853 Allan A. Black appointed curator of the Herbarium.

About 4,500 herbaceous plants were now in cultivation. Garden of medicinal plants made in the former Kitchen Garden and Paddock of the late King of Hanover. Simple fountain installed in the Pond.

Hooker now using his annual report to campaign for a Temperate House.

James Niven: *Catalogue of the Hardy Herbaceous Plants in the Royal Gardens of Kew*

1854 Many trees and shrubs identified with iron tallies.

Part of the metal floor in the Palm House removed and 3 beds, each 8 feet wide, made on each side of the centre passage to accommodate vigorously growing trees. *Amherstia nobilis* presented by Mrs Lawrence.

George Bentham presented his herbarium and botanical library. About 1,000 drawings by Indian artists, done for J.F. Cathcart, presented by his sister (a few of them are reproduced in J.D. Hooker's *Illustrations of Himalayan Plants*, 1855). 2,188 drawings of Kew plants, executed mainly by G. Bond and T. Duncanson, presented by W.A. Smith. Extensive correspondence of N. Wallich during 28 years as Superintendent of Calcutta Botanic Garden presented.

1855 Sir Benjamin Hall appointed First Commissioner of Works.

Joseph D. Hooker appointed Assistant Director on 5 June.

Four-acre nursery established in the grounds of Kew Palace to supply trees for the London parks. Succulent House built: a span-roof building, 200 feet long, 30 feet wide and 15 feet high; formerly the succulents had been crowded into two lean-to houses.

W.J. Hooker: *Museum of Economic Botany, or a Popular Guide to the Museum of the Royal Botanic Gardens of Kew*. This guide was frequently revised.

1856 "I was much struck with the bad state of the flower-beds and the untidy appearance of the grass and the extremely bad condition of the walks in the Gardens, and I wish . . . that you will inform Mr Smith

[Curator] that I cannot suffer such things in future." (Sir Benjamin Hall to the Director, 19 July). Following this reprimand, many more flower-beds were planted and by 1859 they numbered about 400.

Work started on creating a lake in the Arboretum. Queen's Cottage grounds "much improved and adorned with trees and shrubs". There were 504 species of ferns in cultivation. (J. Smith (1880), pp. 322–3)

Old Botany Bay House demolished.

D. Burton designed a new museum facing the Pond: a plain brick building about 112 feet long and 36 feet wide on three floors. The Curator's son, Alexander Smith, was put in charge of the museums.

1857 Charles Barter on Niger expedition under W.B. Baikie began sending dried specimens.

Garden patrols, principal gate-keeper and museum attendants put into uniform.

Vases and pedestals added to the Gardens. The firm of J.M. Blashfield at Poplar supplied Kew with terra-cotta vases about this period.

1858 Lord John Manners returned as First Commissioner of Works.

J.R. Jackson appointed Curator of the museums.

Indian collections of W. Griffith, A. Falconer and J.W. Helfer, and Roxburgh drawings acquired from East India Company.

1859 H. Fitzroy appointed First Commissioner of Works.

Gustav Mann replaced the late Charles Barter as a plant collector in West Africa. Robert Cross joined C.R. Markham on expedition to South America to collect cinchona.

Before a drinking fountain was installed in 1859, the public could get "a delicious glass of pure spring water" from an attendant at the Museum on the Pond (*Gardeners' Chronicle*, 16 August 1859, p. 289). The *Builder* (27 August) regretted the absence of "a place of refreshment".

Decimus Burton designed the Temperate House which was under construction in 1859.

Lectures given to young gardeners.

W.H. Harvey and O.W. Sonder: *Flora Capensis* (not completed until 1925). A. Grisebach: *Flora of the British West Indies* (1859–64; the first of a series of colonial floras)

1860 W.F. Cowper appointed First Commissioner of Works.

"Before Kew was altered to the present style it was the best example one had in England of our free-and-easy mode of dealing with flat surfaces by means of judicious planting and easy winding walks. The change from this has added some of the best features of the geometric style . . . and Kew Gardens are now in the mixed style." (Donald Beaton, *Cottage Gardener*, 18 September)

Mount Pleasant formed with surplus earth excavated for the gravel path in Syon Vista.

Forcing house for germinating cinchona seeds erected at the expense of the India Office.

1861 Richard Oldham sent to the Far East to collect plants.

Majolica fountain, made by Minton and Co., erected near the old Museum. (Works 16/922. P.R.O.)

Flagstaff raised on mound formerly occupied by the Temple of Victory. A gift of Edward Stamp, a timber merchant, it was a spar of Douglas fir, 159 feet in length. He had

presented a shorter one in 1859 but it broke while being erected.

Small house for tropical tree ferns built. Sir William Chambers's Great Stove demolished and its contents transferred to the new Temperate House. The wisteria which once covered the eastern end of the stove still survives on an iron frame.

G. Bentham: *Flora Hongkongensis*

1862 Belt of trees and shrubs planted along the riverside from Brentford Ferry Gate to Syon Vista to hide the view of the new docks and railway terminus in Brentford.

The centre block and two octagons of the Temperate House finished. Centre block: 212 feet 6 inches by 137 feet 6 inches and nearly 60 feet at its highest point; octagons: 54 feet in diameter and 25 feet high. Roof of wrought iron and columns of cast iron.

In 1862/3 plants were transferred to the Temperate House from the Palm House which now concentrated on tropical plants, from the Architectural Conservatory which was filled with aroids and tropical large-leaved climbers, and from the Orangery which was converted into a museum of colonial timbers.

Plants received from G. Mann in West Africa and R. Oldham in the Far East.

Allan Cunningham's Australian specimens and manuscripts presented. William Borrer's British herbarium and Mrs A.E. Griffiths's herbarium of marine algae acquired. Bust of Sir William Hooker by T. Woolner presented by H. Christy.

1863 Daniel Oliver appointed Keeper of the Herbarium and Library.

The sculptor, John Bell (1812–95), made a fountain for Kew with the

figure of a child raising a shell to its lips.

Collection of timbers, cabinet and furniture woods from Great Exhibition of 1862 installed in Orangery.

Large collection of plants and seeds sent to St Helena at the request of the Governor.

Sir William Hooker presented the Colonial Office with a scheme for a series of colonial floras which had been earlier approved in principle.

G. Bentham: *Flora australiensis*, 1863–78

J.A. Grant's plant collections, made during J.H. Speke's expedition to Central Africa added.

1864 The curator, John Smith, retired; replaced by another John Smith, formerly gardener to Duke of Northumberland.

Some 13,000 species in cultivation; more than 3,000 species and varieties of trees and shrubs planted in the Arboretum. In December the Director considered gravelling Pagoda Vista because of worn grass.

Side arches of the Ruined Arch opened and its surrounds planted with ferns, ivies, etc.

J.D. Hooker: *Handbook of the New Zealand flora*

P.B. Ayres's herbarium of Mauritius plants and F. Boott's specimens and drawings relating to carices presented.

1865 Sir William Hooker died on 12 August, and succeeded by his son, Joseph Dalton Hooker, as Director on 1 November. Post of Assistant Director abolished. Post of Curator of the Pleasure Grounds also abolished and his duties in 1866 combined with those of the Curator of the Botanic Garden.

Much of the collections in the Palm

House repotted and rearranged.

J.D. Hooker: *Catalogue of the Plants distributed at the Royal Gardens, Kew . . . from the Herbaria of Griffith, Falconer and Helfer*

John Lindley's orchid herbarium purchased. W.J. Burchell's herbarium of South African, South American and St Helena plants presented.

1866 In 1859 the Gardens employed one policeman and three liveried constables; in 1866 the number was increased and "deserving labourers" appointed.

The Water Lily House, built in 1852, converted into an Economic Plant House.

The Gardens purchased one of W. Barron's machines for transplanting mature trees.

Sir William Hooker's herbarium, library and correspondence purchased.

1867 Rock garden of Reigate sandstone built on the north side of the ice-house.

Severe snows in January destroyed many tender pines and cypresses. Napoleon's willow chopped down.

A chimney in each of the wings in the Palm House replaced the flue which extended the length of the tunnel to the Campanile.

Temperate House Lodge, designed by Eden Nesfield, built (plans in Victoria and Albert Museum).

1868 A.H. Layard appointed First Commissioner of Works.

Queen's Gate erected, and Cumberland Gate opened.

Thorn Avenue made of *Crataegus* and *Pyrus* species. Mound near the Pond replanted. Avenue cut through the riverside plantation of trees created in 1862. Hedge, to serve as a wind-break, planted along the top of the ha-ha bordering the

Old Deer Park. 802 species and varieties of ferns and 48 of fern allies in cultivation.

T-Range of glasshouses, designed by Henry Ormson, built during 1868–9. The centre portion, 53 feet by 40 feet, was allocated to tender palms and a tank for *Victoria amazonica*; there were also two lateral wings, 70 feet long, 24 feet wide and 12 feet high, and a back wing, 170 feet long, 20 feet wide and 12 feet high.

F. Oliver: *Flora of Tropical Africa* (completed in 1937)

Herbarium and papers of M.J. Gay presented by J.D. Hooker.

1869 A.S. Ayrton appointed First Commissioner of Works.

Original layout of Herbaceous Ground replanned as a series of straight beds. Work began on shaping the gravel pit which over the next five or six years was transformed into Berberis Dell. King William's Temple mound replanted and the building itself enclosed by a fence to keep out vandals.

Two orchid houses erected.

1870 Rose Walk flanking Herbaceous Ground created. Palm House terrace relevelled for the first time since its construction in 1846; "its angles have been filled with large beds of laurels and rhododendrons and the whole terrace bordered with ivy" (Annual Report). The landscaping of the Lake almost finished. Kew reluctantly indulged in carpet bedding.

1871 Cedar Vista, composed of Atlantic cedars, planted. Douglas firs and evergreen oaks planted along Syon Vista. Rose garden laid out on north side of King William's Temple. Pinetum on south side of the Lake nearly finished – contained some 1,200 specimens, some

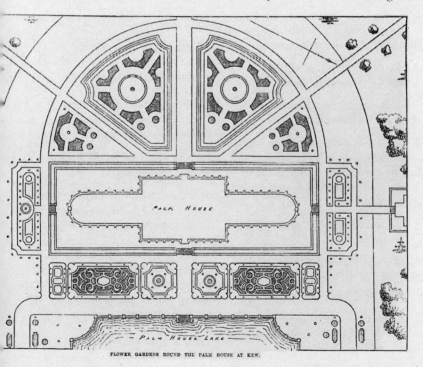

W.A. Nesfield's parterres around the Palm House in 1872. *Garden*, 17 August 1872, p. 151.

transferred from Sir William Hooker's pinetum.

Student gardeners form the Kew Mutual Improvement Society.

1872 Acacia Avenue formed. Extensive replanting of the American Garden on west side of Palm House.

Isleworth Ferry Gate opened.

J.D. Hooker: *Flora of British India* (1872–97)

W.A. Leighton's lichen herbarium presented.

1873 New walk from Isleworth Ferry Gate to the Pagoda. Shallow brick tank for aquatics *c.* 70 feet by 15 feet, sited at northern end of Herbaceous Ground. W. Wilson Saunders presented his "unrivalled collections of South African, etc. bulbs".

Manuscript copy of W. Roxburgh's *Flora Indica* presented by his son.

1874 About 20,000 species in cultivation. Holly Walk (which for part of its route followed Love Lane) formed; also Isleworth Vista lined with Irish yews. Ice-house rock garden enlarged and hardy fernery added. Two-acre nursery of rare ornamental trees established near Kew Palace as a reference collection for the London parks.

1875 William Thiselton-Dyer appointed Assistant Director on 12 June.

John Stuart Mill's herbarium presented.

1876 "During the past year the assemblage on Kew Green on all public holidays of large crowds of persons, who do not appear to visit the gardens, has become a difficulty of considerable magnitude. Men with articles of different kinds for sale, music, games, etc. have been, of course, attracted by the

concourse, and the whole green has had on these occasions the aspect of a fair." (Annual Report).

H.A. Wickham's consignment of rubber seeds reach Kew on 14 June, and in August seedlings were despatched to Ceylon.

1877 The Director toured western North America and returned with specimens for the Herbarium and Museum.

"The flower gardening at Kew this year consists of a loud shout of colour along the main walk backed by the never-ending kidney-shaped Rhododendron beds. The flower garden in front of the great Palm-house has the effect of a dowdy carpet bag not much toned down by time." (*Garden*, 11 August)

A wind-break of trees east of the Queen's Cottage formed to protect the Pinetum. More planting of trees along the riverside to hide the docks at Brentford.

E.G. Rivers designed and installed six tubular boilers in the Palm House to replace the original twelve.

Kew Gardens Public Rights Defence Association formed.

Wing added to the Herbarium at the rear of Hunter House.

Jodrell Laboratory built on site of an old propagating pit: a brick building, 40 feet long and 20 feet wide; James Elder of Hayes was the contractor.

J.G. Baker: *Flora of Mauritius and the Seychelles*.

Indian herbarium of C.B. Clarke and tropical African plants of G. Schweinfurth presented.

1878 W.H. Fitch ceased to contribute plates to *Curtis's Botanical Magazine*.

Plantation of trees on riverside, created in 1862, thinned and a good

collection of oaks and elms exposed. Many new gravelled paths were created and a steam roller was purchased for the purpose.

W.J. Burchell's drawings of St Helena presented by his sister. Revd M.J. Berkeley's mycological herbarium presented, and N.J. Dalzell's Indian herbarium purchased.

1879 C.B. Clarke, Indian Civil Service, seconded to Kew for three years to assist Sir Joseph Hooker in compiling *Flora of British India*.

"It would, we think, be a great mistake to practise bedding out of any kind on a large scale at Kew flower-beds, let that be left for the parks and pleasure gardens proper." (*Garden*, 1 November)

Work still progressing on the taxonomic grouping of trees and shrubs in the Arboretum. Violent hailstorm on 3 August – some hailstones five inches in circumference – shattered nearly 39,000 panes of glass.

Brick tank, 80 feet by 20 feet, replaced the smaller one installed in 1873 for hardy aquatics.

1880 Sweet Chestnut Avenue formed. 3,669 trees and shrubs supplied to London parks.

John Smith: *Records of the Royal Botanic Gardens, Kew*

When the India Office dispersed the remnants of the India Museum, it gave Kew its "botanical and ethno-economical collections", an 'Iron Room' to sort the collections, and the services of Mordecai Cooke for a limited period. The herbaria of Bishop Goodenough, W. Munro and W.P. Schimper presented.

1881 The flower-beds in front of the Palm House, now water-logged, were replaced largely by grass plots. G.C. Joad bequeathed his large alpine collection which made a new rock garden a priority.

Extension built to museum facing the Pond to accommodate the collections from the India Museum. It consisted of a wing, 50 feet by 20 feet, with a new staircase; the India Office paid for its construction. (Works 16 29/5. P.R.O.)

British herbarium of H.C. Watson and F. Curry's mycological herbarium presented.

1882 Rock Garden designed by Thiselton-Dyer and stocked with G.C. Joad's alpines. Mound, surmounted by Temple of Eolus, converted into a wild garden, planted with daffodils and other bulbs. American Garden thinned out, enlarged, and made into the Azalea Garden. Gravel path in Syon Vista covered with soil and sown with grass seed. Nursery, established in 1855 to supply trees for London parks, closed down.

Marianne North Gallery opened to public in July. Designed by James Fergusson: red brick with Bath stone dressing, low-pitched, zinc-covered roof, iron verandah; large room, 50 feet by 25 feet.

Kew Volunteer Fire Brigade given accommodation in the Melon Yard with entrance in boundary wall in Kew Road near Descanso House; closed down when incorporated into Richmond Fire Brigade. (Works 16 471. P.R.O.)

W. Botting Hemsley: *Gallery of Marianne North's Paintings of Plants and their Homes.* (6th edition, 1914)

Herbaria of W.A. Leighton, R. Baron and Botanical Record Club acquired.

1883 India Office paid the salary of an additional assistant in the Herbarium.

Police establishment reduced by one sergeant and four constables and

replaced by five park constables appointed by the Office of Works.

The screw pine, *Pandanus odoratissimus*, in the Palm House fruited for the first time.

Small studio added to Marianne North Gallery. Two light iron galleries with spiral staircases erected in Timber Museum in the Orangery.

Official Guide to the Museums of Economic Botany. No. 1. Dicotyledons and Gymnosperms. (4th edition, 1930)

1884 Iris Garden planted in lawn opposite Cambridge Cottage garden under the supervision of the distinguished iris grower, Michael Foster.

Old Water Lily House demolished.

Herbarium and drawings of J.T. Moggridge presented.

1885 Sir Joseph Hooker who retired in November was succeeded by his son-in-law, William Thiselton-Dyer.

Printed seed lists for exchange purposes started.

M.C. Cooke's fungi herbarium and part of T. Moore's fern herbarium purchased. J.R. Forster's herbarium acquired.

1886 George Nicholson succeeded John Smith as Curator.

Official Guide to the Museums of Economic Botany No. 3. Timbers. (3rd edition, 1927)

1887 Former rockery near the ice-house adapted for a collection of hardy ferns, some presented by E.J. Lowe and others from the bequest of late W.C. Carbonell.

On 20 October *Gardeners' Chronicle* suggested that the Pagoda be converted into a "tea-house".

Kew's first Alpine House built at the north end of the Herbaceous Ground.

F. Sander, nurseryman of St Albans, presented a collection of orchids.

Bulletin of Miscellaneous Information, Royal Botanic Gardens, Kew. (Kew's first regular periodical)

Herbaria of W.C. Carbonell, J.E. Leefe and A.F. Oldfield presented.

1888 The Gardens' first Refreshment Pavilion opened on 1 August.

The five busts in King William's Temple removed to Buckingham Palace. ('Pleasure Grounds, 1845–1911', ff.70–1. Kew archives)

1889 The giant aroid, *Amorphophallus titanum*, flowered in June – the first time in Europe and outside its native Sumatra. Subsequent flowerings at Kew – June 1901 and July 1926, for example – always attracted crowds. It was reported that visitors never stayed long to admire it, repelled by its offensive odour.

Victoria Gate opened on 27 May.

M.J. Berkeley's fungi herbarium and J. Ball's herbarium and library presented.

1890 J.G. Baker appointed Keeper of Herbarium and Library.

1891 Garden of hardy bamboos formed in a disused gravel pit; completed by April 1892. "The garden is in the form of a shallow depression with sloping banks."

Alpine House enlarged to 40 feet in length, 9 feet in width and 8 feet six inches in height.

More than 1,600 drawings done for the early volumes of *Curtis's Botanical Magazine* purchased.

T.R.A. Brigg's herbarium presented.

1892 D.H. Scott made Honorary Keeper of the Jodrell Laboratory in October.

Monkey Puzzle tree, *Araucaria araucana*, grown from seed brought back from S. America by Archibald Menzies in the 1790s, died.

The old 'Cape House', built in 1792 and enlarged by Decimus Burton in 1844, was partially rebuilt, its wooden roof replaced by an iron one and lantern ventilators added. Temperate Fern House, adapted from a house built in 1803, was replaced by light iron span-roof house, 60 feet long, 23 feet wide and 13 feet high. Presentation of J.C. Forster's collection of filmy ferns in 1888 necessitated the building of Filmy Fern House, 50 feet by 14 feet, on north side of large fernery.

First volume of *Index Kewensis* appeared.

Selected books from Daniel Hanbury's library presented.

1893 Old Oak Avenue formed; extended in 1906–7.

Kew Guild of present and former members of staff formed, and an annual *Journal* published.

Collection of Japanese artefacts presented by J.H. Veitch. Conrad Dressler's bust of Marianne North presented by Mrs A. Symonds.

1894 Some 20,000 species and varieties in cultivation at Kew; of this total about 3,000 were hardy trees and shrubs.

The Director had recommended that the practice of using green glass in certain houses should be discontinued; he now proposed extending this prohibition to all houses. (28 November. Kew 93. Kew archives)

Palm House heating system renewed with the pipes carried up to the level of the lantern.

Hand-list of Trees and Shrubs grown in the Arboretum Part 1. Polypetalae. (The first of a series of lists of Kew's collections of living plants)

Wilson Forster's portrait of Daniel Oliver and Mrs Janet Hutton's collection of flower drawings presented.

1895 Joseph Chamberlain appointed Secretary of State for the Colonies.

4½ acres of Palace Meadow in front of Kew Palace added. Wire fence which had divided the Botanic Garden and the Pleasure Grounds since 1843 was removed in April.

c. 6,000 herbaceous plants, 1,116 species of ferns and 97 species of fern allies in cultivation. Flagstaff mound planted with bulbous and herbaceous plants. Sunken Rose Garden created near Pagoda; planted with species roses. Mount Pleasant lowered. Path in Queen's Cottage grounds lengthened. "Ice on Lake strong enough to walk & wheel barrows on, etc.; at work on islands." (W.J. Bean: "Arboretum records". Kew archives)

"Relief house", 56 feet long, 23 feet wide and 13 feet 7 inches high, erected in Melon Yard for plants temporarily removed from their houses under repair.

Hand-list of Herbaceous Plants (3rd edition, 1925); *Hand-list of Ferns and Fern Allies* (2nd edition, 1906); *Official Guide to the Museums of Economic Botany No. 3, Monocotyledons and Cryptogams* (2nd edition, 1928)

1896 227 species and 340 varieties of conifers in cultivation.

First women gardeners recruited. Gravel on Palm House terrace reduced to a path and the remainder

turfed. "Widened the Riverside Avenue which starts at Brentford Gate. Thickened up the belt of shrubs between it & the ha-ha." (W.J. Bean. "Arboretum records". Kew archives)

A pelican was presented to the Gardens with two more the following year.

Hand-list of Coniferae (4th edition, 1938 edition included Bedgebury Pinetum). *Hand-list of Orchids* (3rd edition, 1962)

1897 Lily Pond for tender aquatics made off Cedar Vista.

Nepenthes House built as an extension to the T-Range: 70 feet long, 12 feet wide and 9 feet 6 inches high. South Wing of Temperate House opened in July: iron columns with sashes and frames of wood. With the bankruptcy of the contractor, modifications were made to the design.

Hand-list of Tender Monocotyledons excluding Orchideae. (2nd edition, 1915).

1898 Kew Palace passed to the Office of Works and, with the Queen's permission, was opened to the public.

Queen's Cottage and 37 acres of ground transferred from Lord Steward's department. (28 January 1898. Works 19/453. P.R.O.) Pictures and furniture in the Cottage removed to Windsor Castle. The grounds were opened to the public in May 1899.

Part of the T-Range redesigned for a better display of plants. Orchid houses, erected in 1869, remodelled; "The mixed construction of wood and iron (or rather rolled steel) now generally employed at Kew was adopted." (*Kew Bulletin* 1898, p. 337).

D. Morris: *Report on the Economic Resources of the West Indies* (*Kew*

Bulletin, Additional Series 1);
*Selected Papers from the Kew Bulletin:
vol.1. Vegetable Fibres (Kew
Bulletin, Additional Series 2)*

Presentation of R.H. Beddome's
mosses and lichens and C.S.P.
Parish's orchid drawings. E.
Bristow's portrait of Mrs Aiton
presented.

1899 William Botting Hemsley
appointed Keeper of Herbarium and
Library.

Bluebells and narcissi planted in
Queen's Cottage grounds.

Encephalartos woodsii presented by
Natal Botanic Garden. It has been
described as the rarest plant in
cultivation at Kew, possibly the last
surviving specimen in the world.

Head gardener at Government
House in the Falkland Islands,
formerly a Kew gardener, presented
three penguins.

North Wing of Temperate House
opened in May. Its contractors were
Mackenzie and Moncur of
Edinburgh. The use of rolled steel
made it a lighter structure than the
South Wing. Each wing is 112 feet
6 inches by 62 feet 6 inches. A
number of houses in the Melon
Ground or Yard were repaired. The
Melon Yard takes its name from its
use in the Royal Kitchen Garden.
During the 1890s and later it was the
main propagating department.

Hand-list of Tender Dicotyledons
(2nd edition, 1931); Library
catalogue (*Kew Bulletin. Additional
Series 3*; Supplement 1919)

R.H. Beddome's herbarium of
mosses and hepaticae presented.
Revd L. Guilding's drawings of
St Vincent flora purchased.

1900 *List of published Names of
Plants introduced to Cultivation, 1876
to 1896. (Kew Bulletin. Additional
series 4)*

J.E. Winterbottom's herbarium of
Himalayan plants presented.

1901 William Watson appointed
Curator on 1 August. J.M. Hillier
appointed Keeper of the Museums.

Rose Pergola put up between Rock
Garden and Herbaceous Ground.

Small forcing house, built in 1860
for cinchona, demolished.

1902 Director appointed botanical
adviser to the Colonial Office.

Piece of ground between Brentford
Gate and former Kew Palace lawn
added to the Gardens.

About 8,000 herbaceous plants and
4,500 hardy trees and shrubs in
cultivation. "Commenced remaking
and replanting Rhodo beds along the
Broad Walk, putting as far as
possible one variety in each bed."
(W. Dallimore. "Arboretum
records". Kew archives)

Storks had been kept since 1890; in
1902 a pair of young storks – said to
be the first bred in this country –
were reared at Kew.

Another wing added to the
Herbarium.

John Day's orchid drawings
presented.

1903 On 1 April Kew Gardens was
transferred from the Office of Works
to Board of Agriculture and
Fisheries.

246 species and 451 varieties of
conifers in cultivation. About 80
Austrian pines planted in the
Herbarium ground to form a screen
"to block out ugly buildings at
Brentford". Simplified flower-beds
on Palm House terrace.

R.C.A. Prior's herbarium of S.
African and West Indian plants
presented.

1904 Following the death of 2nd

Duke of Cambridge on 17 March,
Cambridge Cottage and its garden
were added to Kew Gardens.

Succulent House replaced by a
lighter and more elegant structure:
200 feet long, 30 feet wide and 20
feet high.

Stone seat installed in Arboretum in
memory of Miss Cassell,
Superintendent of the College for
Working Women (*Kew Bulletin*
1909, p. 160). It was destroyed
during an air-raid in February
1944. (*Journal of Kew Guild* 1944,
p. 398)

Electric light installed in the
Herbarium.

1905 Sir William Thiselton-Dyer
retired on 15 December; succeeded
as Director by David Prain in
January 1906.

Reorganisation of the late Duke of
Cambridge's garden under way. Kew
Palace stables demolished.

1906 Semi-circular yew hedge on
west side of Palm House being
replaced by holly; two openings
were made in this holly hedge in
1907–8 – one connecting with the
path to King William's Temple, the
other to the path leading to the
Azalea Garden. Trees near the
riverside in Queen's Cottage
grounds removed in order to plant
species, varieties and hybrids of
willows and related trees. Two
entrances made in southern
boundary wall of Cambridge Cottage
garden.

J.D. Milner: *Catalogue of Portraits
of Botanists exhibited in the Museums
of the Royal Botanic Gardens; Wild
Fauna and Flora of the Royal Botanic
Gardens, Kew (Kew Bulletin.
Additional series 5); Selected Papers
from the Kew Bulletin: vol.3. Rubber
(Kew Bulletin. Additional series 6)*

Bronze cast of medallion of Sir
Joseph Hooker presented by the

artist, Frank Bowcher. Mrs H. Jameson's herbarium of Cape plants presented.

1907 Arthur William Hill appointed Assistant Director.

Medicinal garden laid out in Cambridge Cottage grounds.

Two pairs of grey squirrels presented by Duke of Bedford.

1908 Otto Stapf appointed Keeper of Herbarium and Library in December.

16 large trees from various parts of the Gardens transplanted to Pagoda Vista to complete its avenue of trees.

The Gables built on Kew Green as residences for foremen.

The first Underground Railway poster of Kew Gardens issued. W.J. Bean: *The Royal Botanic Gardens, Kew: Historical and Descriptive;* C.B. Clarke: *New Genera and Species of Cyperaceae (Kew Bulletin. Additional series 8)*

1909 L.A. Boodle appointed Assistant Keeper in charge of Jodrell Laboratory.

Governments of West African colonies employed an assistant in Herbarium to help in compilation of *Flora of West Tropical Africa.*

Collection of Japanese cherries planted on slight mound between Palm House and King William's Temple. Shrubs and some trees removed from Cambridge Cottage grounds and paths made.

New aquatic tank replaced the rather primitive brick construction of 1873. It consisted of seven tanks: a central one 72 feet by 30 feet; four corner tanks, each one 26 feet square; and two side tanks, each 38 feet by 9 feet.

1910 Part of Cambridge Cottage converted into a museum of British forestry and timbers.

1911 E.H. Wilson's Chinese rhododendrons planted in Rhododendron Dell.

Japanese Gateway, 'Chokushi-Mon', presented to Kew in 1910 after the closure of Japan-British Exhibition, erected.

1912 More replanting along the riverside during 1912 and 1913 to improve the effectiveness of screening industrial development in Brentford; elms replaced by Holm oaks, pines, laurels, etc.

Popular Official Guide published and frequently reissued; S.T. Dunn and W.J. Tutcher *Flora of Kwangtung and Hong Kong (Kew Bulletin. Additional series* 10)

22 drawings of Chinese conifers executed under supervision of Robert Fortune acquired.

1913 T.A. Sprague and J. Hutchinson collect plants in the Canary Islands.

Official guide lecturer appointed in April 1913 to 1915 and 1919–20. (*Gardeners' Chronicle* 1919, p. 228)

Suffragettes destroyed plants in three orchid houses on 8 February; they returned on 20 February and burned down Refreshment Pavilion. Two were apprehended: Olive Wharry and Ida Inkle.

Work on removing all the gravel and sand in the walk in Syon Vista and replacing with soil finally completed in December after seven or eight years intermittent labour. Deodar cedars along Broad Walk mostly replaced by Atlantic cedars. Rock Garden partly rebuilt with Cheddar limestone. Vine pergola, 500 feet long, erected near Refreshment Pavilion. Two more of famous Seven Sisters elms removed, leaving only three. Collection of orchids presented by Sir George Holford.

Marble bust of Sir Joseph Hooker from clay model by Pennacchini acquired. Wallich herbarium transferred from Linnean Society.

1914 150 trees, mainly oak, poplar, birch, Douglas fir, planted in Queen's Cottage grounds. 24 of E.H. Wilson's Chinese prunus added to the collection near Holly Walk. Orchid collection of late Sir Trevor Lawrence presented. Three horses taken by the Army in August.

Temporary Refreshment Pavilion erected.

E. Crossland's collection of fungi and drawings purchased.

1915 Women gardeners replace men on military service.

24 Chinese oaks collected by E.H. Wilson planted.

1916 Violent storm on 28 March caused ancient Cedar of Lebanon to fall on the Temple of the Sun, wrecking it. The storm also blew down the last of the Seven Sisters elms, reputedly planted by the daughters of George III; young elms were planted as replacements on the lawn just north of Water Lily House. (Engraving in *Gardeners' Chronicle* 15 September 1883, p. 336)

Royal Naval balloon landed on Palace lawn.

67 flower drawings by eminent artists presented by widow of Sir Arthur Church.

1917 Letters from Charles Darwin to Sir William Thiselton-Dyer presented.

1918 Palace lawn ploughed up in January and 27 tons of potatoes lifted in August.

1919 In 1911 female shoots had been grafted on to the male *Ginkgo* *biloba* which produced four fruits in autumn of 1919.

Replacement flagstaff, 214 feet tall, presented by the Government of British Columbia, raised on 18 October.

Official Guide to the Museums of Economic Botany. No. 4. British Forestry

About 600 watercolour flower drawings by Indian artists, said to have come from Claude Martin's collection, were presented.

1920 New Refreshment Pavilion, designed by R.D. Allison of the Ministry of Works, opened; described by Pevsner as "a nice straightforward piece of building".

1,000 watercolours of Swiss flowers by M.L. Moxon presented.

1921 Matilda Smith retired in July as botanical artist after 23 years in the post.

Palace lawn, after having been planted with potatoes, was reseeded with grass. Collection of witch hazels planted on side of King William's Temple. Prolonged drought severely affected rhododendrons and conifers.

War memorial unveiled in Temple of Arethusa on 25 May; bronze tablet designed by Sir Robert Lorimer.

1922 Sir David Prain retired as Director in February, and was succeeded by the Assistant Director, Arthur Hill. T.F. Chipp appointed Assistant Director. W.J. Bean followed W. Watson as Curator. A.D. Cotton appointed Keeper of Herbarium and Library. Gerald Atkinson appointed botanical artist.

A geometrical arrangement of beds of medicinal and economic plants made in Cambridge Cottage garden. Kew given permission to plant trees and shrubs on land adjacent to

Brentford Dock. (1/Riverside/3. Kew archives)

J.H. Holland: *Useful Plants of Nigeria, including Plants suitable for Cultivation in West Africa and other Tropical Dependencies of the British Empire(Kew Bulletin. Additional series 9)*

Bequest of W. Hancock's herbarium of Chinese plants. Drawings of Indian plants by Mrs King, S. African plants by Mrs F.G. Crossman, and orchid drawings by H.G. Reichenbach acquired.

1923 Semi-circular parterre on west side of the Palm House converted into a rose garden with some 6,000 rose bushes in 113 beds, each with a single variety. The "American plants" formerly grown there had not done well on this low-lying damp site. Cedar Vista widened by some 40 feet during 1923–4. Queen Mary, "using an ordinary garden spade", planted a *Ginkgo biloba* on the site of the Temple of the Sun. Terrace with brick wall and steps built on south side of King William's Temple. "Greenfly more abundant than anyone can remember."

165 flower drawings, done for *Curtis's Botanical Magazine* between 1830 and 1834, purchased.

1924 6,300 species and varieties of trees and shrubs in cultivation. Last remnant of gravel walk in Syon Vista near the Palm House grassed. Experimental plot near the Herbarium for plants used in genetical studies. National Pinetum formed at Bedgebury in conjunction with Forestry Commission.

Kiosk built near Orangery for the sale of guides and postcards.

Royal Botanic Gardens, Kew: Illustrated Guide

1925 288 species and 398 varieties

of conifers in cultivation. 20,000 snowdrop bulbs planted under horse chestnuts near Broad Walk.

Teak annexe for tender Chinese rhododendrons added to west side of Temperate House: 60 feet long, average width 40 feet. Heated span-roof house erected in Arboretum nursery.

Detailed annual report now included in *Kew Bulletin*; *Hand-list of Hardy Monocotyledons*; *Hand-list of Rock Garden Plants* (edition 4, 1934)

34 paintings by Margaret Meen presented.

1926 William Dallimore appointed Keeper of the Museums.

With the development of railway terminal in Brentford, the Riverside Avenue was extended.

1927 Hugh Charles Sampson appointed Economic Botanist.

Flower-beds on Palm House terrace to display a more varied selection of plants. Very heavy fall of snow did considerable damage to conifers and oaks.

Span-roof quarantine house for bananas erected in Melon Yard, 34 feet long, 20 feet wide and 16 feet high: architect W. Duncan Tucker.

J. Hutchinson and J.M. Dalziel: *Flora of West Tropical Africa* (completed in 2 vols in 1936; 2nd edition in 3 vols, 1954–72)

1928 One of the two 'horseshoe' shaped paths, halfway along the Broad Walk removed; Nesfield had introduced them partly to save a fine Turkey oak on the east side of the Walk.

When one of the six horses in use died, a motor lorry replaced it.

Annexe to T-Range for insectivorous plants replaced by a larger structure.

Fire station in the Kew Road near Descanso House closed and converted into store rooms for the Gardens.

Single-seater aircraft crashed in flames near west end of Syon Vista.

1929 Thomas William Taylor appointed Curator.

A programme to substitute Sussex sandstone for limestone in the Rock Garden initiated.

1930 Charles Russell Metcalfe appointed Assistant Keeper at the Jodrell Laboratory (regraded Keeper in 1946).

Imperial Bureau of Mycology moved from 17–19 Kew Green to a plot behind the Herbarium.

Northern end of Rock Garden extended.

Economic House demolished and replaced by a larger structure during 1930–31.

Mrs Ellis Rowan's paintings of New Guinea plants presented.

1931 John S.L. Gilmour appointed Assistant Director.

New gasometer, 160 feet high, being built in Brentford, dominating the northern skyline of Kew Gardens. The gas company having agreed that the Director should have a say in the choice of colour to paint it, an innocuous pale grey was chosen. (1/SMO/2. Kew archives)

1932 John Coutts appointed Curator.

Removal of clipped golden yews in rose garden behind the Palm House.

Sherman Hoyt Cactus House, designed by J.H. Markham of Ministry of Works, built; canvas panorama (now in Princess of Wales Conservatory) painted by Perry McNeedy.

Third wing, designed by J.H. Markham, added to the Herbarium.

Collection of wood specimens and paintings of the Indian flora bequeathed by Alfred Hay.

1933 Shelter of Western Red Cedar, designed by J.H. Markham, placed in Cambridge Cottage grounds. Red brick piers of two gates to Cambridge Cottage grounds, erected in 1906, were rebuilt with old brick to harmonise with the wall.

Wing walls of Ruined Arch rebuilt and the two recesses stripped of the ivy which had completely concealed them.

1934 Artist's studio and photographic darkroom added to the Jodrell Laboratory.

1935 Cherries planted along path from the Rose Garden to the Temperate House.

Aquatic tank near the Jodrell Laboratory repaired, refaced and central tank raised to give greater depth for water lilies.

Paintings by Lilian Snelling for H.J. Elwes's *Genus Lilium* acquired.

1936 John Hutchinson appointed Keeper of the Museums.

Tree-transplanting machine replaced by an improved model.

South African Succulent House, designed by Duncan Tucker: 67 feet long and 13 feet wide, built of teak.

Japanese Gateway restored.

H.C. Sampson: *Cultivated Crop Plants of the British Empire and the Anglo-Egyptian Sudan* (*Kew Bulletin, Additional series 12*)

1937 William M. Campbell appointed Curator.

Circular bed at Pond end of Broad Walk enlarged to some 20 feet in

diameter. Last of the Shire horses replaced by Suffolk Punch, making a team of five.

J.M. Dalziel: *Useful Plants of West Tropical Africa* (2nd edition, 1985, by H.M. Burkill, in progress)

1938 Sir Geoffrey Evans appointed Economic Botanist.

Tulip trees, *Liriodendrum tulipifera*, replaced most cedars on Broad Walk. Chilean Honey Palm, *Jubaea chilensis*, moved from side to middle of Temperate House.

Alpine House enlarged slightly.

Alterations to Marianne North Gallery: protective railing which projected into the Gallery reduced in size; flooring of small red, black and buff square tiles replaced by brown Ruboleum covering.

Aircraft made a forced landing near Palm House on 5 January.

1939 Alpine House rebuilt, widened and lengthened.

1941 Sir Arthur Hill killed in riding accident on 3 November; Sir Geoffrey Evans became Acting Director.

The Gardens participated in the war effort with demonstration plots of vegetables and by serving on the Vegetable Food and Drugs Committee.

1943 Edward J. Salisbury appointed Director in September.

1944 Chalk Garden created on ice-house mound.

1946 William B. Turrill appointed Keeper of Herbarium and Library.

1948 Norman L. Bor appointed Assistant Director. Frank N. Howes appointed Keeper of the Museums.

Only two horses still at work in the grounds; motor mowers and lorries were gradually replacing them.

1949 Clematis Wall and mound created near Berberis Dell.

1950 Chilean wine palm which had been raised from seed in 1843 flowered and fruited for the first time in the Temperate House. 13 very large beech trees removed, "the majority of them 180 to 200 years old".

1951 New quarantine house erected: 70 feet by 20 feet.

1952 Palm House, being unsafe, was closed to the public. Australian House opened in March. Designed by S.L. Rothwell of Ministry of Works; constructed of aluminium sections by Crittall Manufacturing Company: 93 feet long, 52 feet wide and 32 feet high at its apex; hot-water pipes connected with Temperate House boilers.

Flora of Tropical East Africa (in progress)

1953 T.A. Russell replaced Sir Geoffrey Evans as Economic Botanist.

Two large Cedars of Lebanon near the Japanese Gateway, reputed to have been planted in 1761, were removed.

1954 Four houses in the T-Range rebuilt in teak.

1955 Larger teak house replaced the quarantine house built in 1927. One of the ferneries rebuilt in teak with improved natural lighting.

1956 Sir Edward Salisbury retired in September and was succeeded as Director by George Taylor.

Portland stone replicas of the Queen's Beasts placed in front of the Palm House.

1957 Charles E. Hubbard appointed Keeper of Herbarium and Library.

Ashby Visiting Group appointed by Ministry of Agriculture to report on future of Kew Gardens.

Orangery closed in May.

Japanese Gateway restored.

1958 Rose Garden near the Pagoda converted into a Heath Garden with the planting of 20,000 ericas. Snowdrops, primroses, narcissi and Martagon lilies planted in Queen's Cottage grounds. Pair of stone Chinese Lions sited near the Pond.

Aluminium-alloy Succulent House replaced the house demolished in 1955. Orchid house built.

Dispersal of some ethnographical items from the museums to other institutions. Cambridge Cottage now the 'Wood Museum'.

J.B. Gillett: *Indigofera (Microcharis) in Tropical Africa with the Related Genera Cyamopsis and Rhynchotropis (Kew Bulletin, Additional series, 1)*

Flower and fungi drawings by F. Clarke acquired.

1959 Bicentenary year of Kew Gardens, taking the formation of the Physic Garden in 1759 as its foundation date.

Post of Secretary replaced that of Assistant Director in September.

Morton Committee convened to report on the division of responsibilities between Kew and the Natural History Museum.

Brick rose pergola built in the Herbaceous Ground.

Renovated Palm House reopened to the public. Orangery restored to its original function with tubs of citrus fruit (interior galleries put there when it was a Timber Museum dismantled).

New flagstaff erected in November.

1960 Lewis Stenning appointed Curator. Mary Grierson appointed botanical artist. Post of guide/lecturer created.

Flora Zambesiaca (in progress)

1961 A new remote boiler house in the Shaft Yard was built and fitted with oil-fired heating for the Palm House. An automatic watering system was also installed; the two chimney stacks removed a few years later.

Horses no longer in use in the Gardens.

1962 The last Wardian case delivered plants from Fiji; these cases had been in use at Kew for some 120 years.

Temperate Fern House rebuilt.

'Iron Room', an austere lecture room for several generations of students, was demolished.

1963 Three-year diploma course for student gardeners introduced.

Grass Garden moved from Herbaceous Ground to east side of T-Range.

Conservatory House rebuilt. Jodrell Laboratory demolished.

Statue of Hercules and the Serpent installed in the Pond.

1965 J.P.M. Brenan appointed Keeper of Herbarium and Library and Deputy Director.

Management of Wakehurst Place, acquired by the National Trust, transferred to Kew. Kew withdrew from joint management of Bedgebury Pinetum.

Filmy Fern House built on north side of Orangery: aluminium-alloy glazing bars; plants displayed on tufa rock behind glass to ensure correct growing conditions. Water

Lily House restored after extensive wartime damage – "roof glazing pattern and ventilators . . . altered in 1965".

Cafeteria near Kew Palace opened.

New Jodrell Laboratory, designed by C.G. Pinfold of Ministry of Public Building and Works with additional laboratories, a lecture theatre and classrooms for student gardeners.

Lady Douie's large collection of drawings of European and Indian plants presented.

1966 Richard L. Shaw appointed Curator.

Doryanthes palmeri flowered in Australian House in March – the first time in 25 years.

E. Guest and others: *Flora of Iraq*

1967 Rosemary Angel appointed officer-in-charge of the museums.

Larger beds for herbaceous plants laid out in Cambridge Cottage garden (also known as the Duke's Garden).

Agave atrovirens var. *laussima* flowered and was displayed in the Orangery.

1968 Seed unit installed in Stable Yard where seeds were preserved under refrigerated conditions.

Flower drawings by Mrs Read Brown presented.

1969 Keith Jones appointed Keeper of the Jodrell Laboratory.

Queen's Garden officially opened by Her Majesty The Queen on 14 May; she also opened an additional Herbarium wing which incorporated the Library, previously dispersed throughout the building.

Clematis Wall removed.

W.D. Clayton: *Revision of the Genus*

Hyparrhenia (*Kew Bulletin. Additional series 2*)

Acquisition of drawings by Ann Lee and Barbara Everard.

1970 The dangerous state of the Temperate House necessitated its closure.

Palynology unit formed in the Herbarium.

R.W.G. Dennis: *Fungus flora of Venezuela and Adjacent Countries* (*Kew Bulletin. Additional series 3*)

1971 Following the retirement of Sir George Taylor on 31 May, Professor J Heslop-Harrison was appointed Director on 1 June.

Kew Club for social gatherings by members of the staff formed.

Kew Record of Taxonomic Literature (still published)

1972 John Simmons appointed Curator.

Orangery, unsuitable for growing citrus fruit, was converted into an exhibition area and a bookshop in May.

Cell physiology laboratory established in former servants' wing of the Director's residence on Kew Green.

Correspondence and records of the publisher Lovell Reeve, who published a number of works emanating from Kew, acquired.

1973 Move of Physiology Section of Jodrell Laboratory to the mansion at Wakehurst Place completed. With it went the seed unit where it now operates a Seed Bank.

Old Filmy Fern House demolished.

1974 Conservation Unit formed.

Buddleja collection sited near the Water Lily House. "This is part of the policy of using small collections

of summer flowering shrubs in the north end of the Gardens to extend the colour interest and release areas in the south of the Gardens for the expansion of other collections." (*Journal of Kew Guild* 1974, p. 333). Herb collection in Cambridge Cottage grounds replaced by an educational display of bulbous monocots and the borders reserved for variegated plants. Micropropagation unit formed; building completed in 1976. Two new glasshouses erected in the Melon Yard following the demolition of old houses and sheds.

1975 New circular bed formed at the north end of Broad Walk; the vase in its centre used to stand in front of the Orangery, and the tiled surround was unearthed from the Cottage grounds. Four large granite blocks, formerly part of old London Bridge, made into a "feeding pier" for waterfowl on the Lake. Formal walled garden at Wakehurst Place dedicated to Sir Henry Price who had bequeathed the estate to the National Trust.

Quarantine House demolished to provide a site for a new Alpine House.

Gumley Cottage converted into student accommodation.

H.K. Airy Shaw: *Euphorbiaceae of Borneo* (*Kew Bulletin. Additional series 4*)

1976 Following the resignation of Professor Heslop-Harrison in August, J.P.M. Brenan became Director. Peter S. Green appointed Keeper of the Herbarium and Deputy Director.

Kew designated the Scientific Authority for Plants for the UK under the terms of the CITES treaty in 1973.

G.E. Wickens: *Flora of Jebel Marra (Sudan Republic) and its Geographical Affinities* (*Kew Bulletin. Additional* series 5); *Flore des Mascareignes: La Réunion, Maurice, Rodrigues* (in progress)

1977 Ice-house cleared, path made, planting improved and temporarily opened to the public.

Aiton House, which houses Curator's office and other sections, opened on 23 June. Designed by Manning, Clamp and Partners in collaboration with the Department of the Environment. Restoration of the Temperate House started.

Palm House promenade concert on 15 July to commemorate the Silver Jubilee of Her Majesty The Queen was the first of the open-air concerts which have become an annual event.

D.N. Pegler: *A Preliminary Agaric Flora of East Africa* (*Kew Bulletin. Additional series 6*); R.D. Meikle: *Flora of Cyprus* (vol. 2, 1985)

1978 Attractive summer bedding schemes along the Broad Walk of ornamental vegetables and herbs. Some visitors, however, objected to lettuce, red cabbage and flowering French beans. *Amherstia nobilis* from India made one of its rare flowering appearances.

Queen's Cottage renovated and Hogarth's prints, which had once adorned the ground floor room, were returned from Windsor Castle where they had been found in boxes in the Round Tower. Wooden ticket office built.

1979 4,419 species and 1,562 cultivars in the arboreta at Kew and Wakehurst Place.

Haiku poem by Kyoshi Takahama on block of granite placed at base of Japanese Gateway.

New quarantine house built.

1980 Annual Kew Medal award founded for meritorious service by members of staff.

Loder Valley Reserve at Wakehurst Place opened on 9 July by Sir Giles Loder. It commemorates Gerald Loder, 1st Lord Wakehurst, who developed the Wakehurst estate at the beginning of this century.

Discovery of cellar and soakaways of outbuildings of Cambridge Cottage. (*Journal of Kew Guild*, 1980, pp. 915–8)

With the closure of the Kew Observatory in the Old Deer Park, the Meteorological Office set up a small climatalogical station in the Gardens.

F.R. Fosberg and S.A. Renvoize: *Flora of Aldabra and Neighbouring Islands* (*Kew Bulletin. Additional series 7*); H.K. Airy Shaw: *Euphorbiaceae of New Guinea* (*Kew Bulletin. Additional series 8*)

1981 Professor J.P.M. Brenan retired as Director on 31 October; succeeded as Director by Professor E.A. Bell.

Area around King William's Temple replanted with shrubs mainly from Mediterranean region.

New Alpine House opened in April; designed by Simon Woolf of R.H. Partnership and Property Services Agency: 46 feet square pyramid reaching about 27 feet at its apex; galvanised steel frame and aluminium glazing bars.

New administrative block on Kew Green opened on 1 May, an extension of the former Director's office incorporating Ted Pooley's cottage and yard.

Survey of Economic Plants for Arid and Semi-Arid Lands (SEPASAL) database developed and maintained at Kew.

1982 Transfer of the grass collection (more than 600 species) to the former Iris Garden near Cambridge Cottage garden completed.

Ice-house renovated, lighting installed and opened to the public.

Temperate House, under restoration from 1978, reopened by Her Majesty The Queen on 13 May. Architects were Manning, Clamp and Partners; engineers were Posford, Pavry and Partners.

1983 Sainsbury Orchid Conservation Project launched.

Demolition of T-Range and associated glasshouses begun.

D.N. Pegler: *Agaric Flora of the Lesser Antilles* (*Kew Bulletin. Additional series 9*); D.N. Pegler: *Genus Lentinus: a World Monograph* (*Kew Bulletin. Additional series 10*)

Pharmaceutical Society's collection of crude drugs transferred to Kew.

1984 Responsibility for Kew Gardens was vested in a Board of Trustees under the authority of the National Heritage Act, 1983 on 1 April 1984.
"So far as practicable and subject to the provisions of this Act, the Board shall –

(a) carry out investigation and research into the science of plants and related subjects, and disseminate the results of the investigation and research,

(b) provide advice, instruction and education in relation to those aspects of the science of plants with which the Board are for the time being in fact concerned,

(c) provide other services (including quarantine) in relation to plants,

(d) care for their collections of plants, preserved plant material, other objects relating to plants, books and records,

(e) keep the collections as national reference collections, secure that they are available to persons for the purposes of study, and add to and adapt them as scientific needs and the Board's resources allow, and

(f) afford to members of the public opportunities to enter any land occupied or managed by the Board, for the purpose of gaining knowledge and enjoyment from the Board's collections . . ."

G.Ll. Lucas was appointed Keeper of the Herbarium and Deputy Director.

35 English oaks, *Quercus robur*, planted in Queen's Cottage grounds to launch 'Beautiful Britain 1984'.

Palm House closed in September for major repairs. About a thousand plants were moved out, many to a commercial-type glasshouse near the Tea Bar.

Kew magazine; continuation of *Curtis's Botanical Magazine*. G.R. Proctor: *Flora of the Cayman Islands* (*Kew Bulletin. Additional series 11*); P. Ives editor: *Catalogue of the Living Plants Collections, Part 1: Non-flowering Plants*; S.A. Renvoize: *Grasses of Bahia*

1985 History of the modern rose demonstrated in a garden on the south side of the Palm House.

Sir David Attenborough buried a time capsule in the foundations of the Princess of Wales Conservatory under construction. (R. Desmond and F.N. Hepper: *A Century of Kew Plantsmen*, 1993, p. 70, for details).

N.P. Taylor: *Genus Echinocereus* (Kew magazine monograph). G.E. Wickens and others, editors: *Plants for Arid Lands*

1986 *Protea cynaroides* from the Cape produced five flowers in the restored Temperate House. The last time it flowered at Kew had been in 1826.

Kew no longer acted as an Intermediate Quarantine Station.

R.W.G. Dennis: *Fungi of the Hebrides*; J.W. Grimes and B.S. Parris: *Index of Thelypteridaceae*; R.M. Harley and N.A. Simmons: *Florula of Mucugê*; D.N. Pegler: *Agaric Flora of Sri Lanka* (*Kew Bulletin. Additional series 12*); W.D. Clayton and S.A. Renvoize: *Genera Graminum – Grasses of the World* (*Kew Bulletin. Additional Series 13*)

1987 Professor Michael D. Bennett appointed Keeper of the Jodrell Laboratory in October.

5 acres of ground between Kew Palace and the Banks Building were being landscaped. Continuing development of Queen's Cottage grounds by re-establishing it with native species. Crocus bulbs, a gift of *Reader's Digest* to celebrate its 50th anniversary, were planted in the lawns on either side of the path from the Victoria Gate to King William's Temple: they amounted to 1.6 million bulbs, representing the current membership of *Reader's Digest*. Although biological control was tried at Kew during the 1870s, it was not until 1987 that a programme of biological pest control was introduced in the glasshouses.

Princess of Wales Conservatory opened by Diana, Princess of Wales on 28 July. Architect: Gordon Wilson of Property Services Agency; contractor: Kier Southern Ltd. Steel structure, braced laterally by hollow Vierendeel tubes on the exterior;

aluminium glazing bars. In 1991 the Institution of Mechanical Engineers awarded it its Heritage Hallmark Award as the world's most advanced energy-efficient conservatory.

Interior glass screen in Filmy Fern House removed to improve the effectiveness of the display.

Violent storm on 15/16 October destroyed many trees at Kew and Wakehurst Place.

Jodrell Laboratory awarded a grant by the Medical Research Council to participate in the Directed AIDS Research programme.

Paper Conservation Unit set up in the Library to treat and repair books, archives, drawings and prints.

Kew Index (the first of annual issues, replacing the former cumulated supplements of the *Index Kewensis*); P. Cribb: *Genus Paphiopedilum* (Kew Magazine Monograph); G.P. Lewis: *Legumes of Bahia*

1988 Professor E.A. Bell retired on 30 June; succeeded by Professor G. Prance on 1 September.

Art Gallery in Cambridge Cottage opened. Tea bar near Brentford Gate replaced.

P. Cribb and I. Butterfield: *Genus Pleione* (Kew Magazine Monograph); C. Grey-Wilson: *Genus Cyclamen* (Kew Magazine Monograph); J. Stewart: *Orchids* (Kew Gardening Guides); F.R. Fosberg and S.A. Renvoize: *Flora of Aldabra and Neighbouring Islands* (*Kew Bulletin. Additional series 7*); H.K. Airy Shaw: *Euphorbiaceae of New Guinea* (*Kew Bulletin. Additional series 8*); S.A. Renvoize: *Hatschbach's Paraná Grasses*

1989 First full-time seed collector appointed.

Japanese-style garden on roof building in Herbarium quadrangle. Designed by Peter Riddington with pavilions, rocks and pebbles.

Orangery Tea Room opened in July.

D. Bridson and L. Forman: *Herbarium Handbook* (and subsequent revisions); C.S. Chao: *Guide to Bamboos grown in Britain*; B.A. Lewis and P. Cribb: *Orchids of Vanuatu*; G.P. Lewis and P.E. Owen: *Legumes of the Ilha de Maracá*; J.M. Lock: *Legumes of Africa: a Check-list*; B. Mathew: *Genus Lewisia* (Kew Magazine Monograph); S.G. Oldridge and others: *Wild Mushroom and Toadstool Poisoning*; S.A. Robertson: *Flowering plants of Seychelles*; P. Taylor: *Genus Utricularia* (*Kew Bulletin. Additional series 14*); *Catalogue of Living Plant Collections. Part 2: Orchids*

1990 Kew Foundation formed in March. Friends of Kew formed in June.

His Royal Highness The Prince of Wales gave a lecture on 6 February on the conservation of rainforests.

Bed of roses planted in June commemorating Edith Holden, the author of *Country Diary of an Edwardian Lady*.

Her Royal Highness The Princess Royal opened the Sir Joseph Banks Centre for Economic Botany on 20 March. Architects: Manning, Clamp and Partners in conjunction with Property Services Agency.

After extensive restoration, the Palm House was officially reopened by Her Majesty The Queen Mother on 6 November. Consultant architect: H. Clamp; contractor: Balfour Beatty Ltd. Mild steel used for strengthening the main arches and stainless steel for the glazing bars; new heating system installed; humidification system using compresssed air.

School of Horticulture opened by His Royal Highness The Duke of Gloucester on 7 September. The building had originally housed Sir William Hooker's first museum in 1848.

Six-part television series on Kew Gardens on Channel 4 of TV beginning 9 November.

Kew's 150th anniversary of its being taken over by the State in 1840 commemorated by four postage stamps (20p, 29p, 34p, 37p). They featured trees and buildings at Kew, designed by Paul Leith.

D.K. Abbiw: *Useful plants of Ghana*; L. Ponsonby: *Marianne North at Kew Gardens*

1991 Border of hardy ferns planted behind the Orangery.

Agave americana flowered in Princess of Wales Conservatory, where glass panes were removed for the tall flower stem to continue growing.

Marine display in basement of the Palm House opened by Her Royal Highness Princess Margaret on 10 July.

Kew Bakery rebuilt in late 1991 after a fire in June 1990.

T.A. Cope and H.A. Hosni: *Key to Egyptian Grasses*; V. Eggli and N.P. Taylor, editors: *IOS Index of Names of Cactaceae*; R.J. Johns: *Pteridophytes of Tropical East Africa: a Preliminary Check-list of the Species*; A.J.M. Leeuwenberg: *Revision of Tabernaemontana. 1. Old world Species*; B.A. Lewis and P. Cribb: *Orchids of the Solomon Islands and Bougainville*; J.M. Lock and K. Simpson: *Legumes of West Asia: a Check-list*; T.D. Pennington: *Genera of Sapotaceae* (published with New York Botanical Garden); N.P. Taylor: *Genus Melocactus (Cactaceae) in Central and South America*, (reprinted from *Bradleya*)

1992 Dr Charles H. Stirton appointed Deputy Director (Science) and John Lavin Deputy Director (Operations).

Gradual redevelopment of Cambridge Cottage garden emphasising large herbaceous plantings. Magnolia Walk, sponsored by the Batsford Foundation Arboretum; first plantings made near the Main Gates. *Alluaudia ascendens* flowered for the first time since its introduction to Kew in 1957.

Lord Howe reopened restored Water Lily House on 3 August. Its alloy bars installed in its last restoration in 1965 replaced by stainless steel.

Mrs Anita Roddick opened the Victoria Gate Visitor Centre and Shop on 19 March. Extension added to the Pavilion Restaurant.

R.K. Brummitt: *Vascular Plant Families and Genera*; R.K. Brummitt and C.E. Powell, editors: *Authors of Plant Names*; J. Dransfield: *Rattans of Sarawak* (published with Sarawak Forest Department); S. Dransfield: *Bamboos of Sabah* (with Forestry Department, Sabah); I. Friis: *Forests and Forest Trees of Northeast Tropical Africa* (*Kew Bulletin. Additional series 15*); R.M. Harley and T. Reynolds, editors: *Advances in Labiate Science*; S. Hollis and R.K. Brummitt: *World Geographical Scheme for Recording Plant Distributions* (with Hunt Institute for Botanical Documentation); D. Hunt, editor: *CITES Cactaceae Checklist* (with International Organisation for Succulent Plant Study); C.M. Pannell: *Taxonomic Monograph of the Genus Aglaia Lour (Meliaceae)* (*Kew Bulletin. Additional series 16*); B.S. Parris and others: *Plants of Mount Kinabalu. 1: Ferns and Fern Allies*; D.A. Simpson: *Revision of Genus Mapania*. J. Stewart: *Orchids of Kew*

1993 Kew Enterprises started trading on 1 April.

Replanting of the Cherry Walk, to be phased over three years, sponsored by Sakura Bank of Japan. Bed of daffodils planted on Kew Green in commemoration of centenary of Kew Guild. Coconut in the Palm House fruited for the first time.

All 20 volumes of *Index Kewensis* recorded on CD-ROM.

P. Boyce: *Genus Arum* (Kew Magazine Monograph); D.N. Pegler and others, editors: *Fungi of Europe: Investigation, Recording and Conservation;* D.N. Pegler and others: *British Truffles. A Revision of British Hypogeous Fungi*; M.G. Pimenov and M.V. Leonov: *Genera of the Umbelliferae*; M. Thulin: *Flora of Somalia* vol. 1; J. Wood and others: *Plants of Mount Kinabalu. 2. Orchids*

1994 Mycological herbarium moved into one of the former buildings of the International Mycological Institute off Ferry Lane, Kew Green.

His Royal Highness The Duke of Kent opened the new North American section of the Rock Garden on 27 April. Bee Garden near Queen's Garden established. Secluded Garden near the Princess of Wales Conservatory nearing completion. Philadelphus collection being planted near the Pagoda. Australian House being converted into an Evolution House. *Ensete superbum* flowered for the first time.

Her Majesty The Queen opened an extension to the Jodrell Laboratory on 18 March.

Kew commissioned its first TV commercial.

E.F. Anderson and others: *Threatened Cacti of Mexico*; R.E.R. Banks and others, editors: *Sir Joseph Banks: a Global Perspective*; U. Eggli and N. Taylor, editors: *List of Names of Succulent Plants (other than Cacti) from Repertorium Plantarum Succulentarum (1950–1992)*; F.N. Hepper and I. Friis: *Plants of Pehr Forsskal's "Flora Aegyptiaco-Arabica"* (published with Botanical Museum, Copenhagen); M.H. Kurmann and J.A. Doyle: *Ultrastructure of Fossil Spores and Pollen*; A.J.M. Leeuwenberg: *Revision of Tabernaemontana. 2. New World species and Stemmadenia*; J.M. Lock and J. Heald: *Legumes of Indo-China: a Check-list*; C. Stapleton: *Bamboos of Bhutan: an Illustrated Guide*; C. Stapleton: *Bamboos of Nepal: an Illustrated Guide*; J.J. Wood and P. Cribb: *Checklist of the Orchids of Borneo*

BENTHAM-MOXON TRUST

The botanist, George Bentham (1800–1884) bequeathed to nominated trustees at Kew his residuary estate "in the preparing and publishing botanical works or in the purchase of books or specimens for the botanical establishment at Kew aforesaid or in such other manner as they might consider best for the promotion of botanical science . . ."

The following selection of purchases, publications and assistance in the employment of staff indicates the extent of Kew's indebtedness to the Trust.

1887 Collection of life-size drawings of bromeliads belonging to the late Edouard Morren of Liège.

1890 £24 to Augustine Henry for expenses in collecting specimens in Western China.

From 1890 the Trust supported the publication of *Icones Plantarum*, an occasional journal founded in 1837 by Sir William Hooker. Its original purpose was to describe and illustrate rare and interesting plants in his personal collections. After his death *Icones Plantarum* selected items from the Kew Herbarium. It ceased publication in 1990.

1894 Mycological works from the library of M.C. Cooke.

1898 Collection of letters of Sir Joseph Banks. Books included: Petrus de Crescentiis: *Opus Ruralium Comodorum*, 1471; Leonhard Fuchs: *De Historia Stirpium*, 1551; John Worlidge: *Systema Horticulturae* 1682; Francis Bacon: *Sylva Sylvarum*, 1627.

1900 Two volumes of flower drawings by Revd Lansdown Guilding of St Vincent.

1903 *Hortus Sanitatus* c. 1490.

1904 *Hortus Sanitatus*, 1496.

1908 Conrad von Megenberg: *Buch der Natur*, 1482.

1909 Pierre J. Redouté: *Choix des Plus Belles Fleurs*, 1827–33.

A page from a manuscript herbal and bestiary of about 1370. An early purchase of the Bentham-Moxon Trust.

One of the drawings from the Tankerville collection, purchased by the Bentham-Moxon Trust in 1932: *Musschia aurea*, a plant collected in Madeira by Francis Masson in 1777.

1911 M.C. Cooke's fungi drawings.

1912 South American herbarium of Edouard André.

1914 Pliny: *Historia Naturalis Libri xxxvii*, 1507.

1917 Orchid drawings by Mrs Janet Ross.

1918 Lady Barkly's orchid drawings.

1920 16 herbals published between 1529 and 1744.

1925 Drawings of Ngamiland plants by Mrs Lugard.

1926 34 drawings by Miss S.A. Drake and others for J. Lindley's *Sertum Orchidaceum*, 1838.

1927 B. Carrichter: *Kräutterbuch*, 1601.

Grant towards the publication of W.B. Turrill's *Plant life of Balkan Peninsula* 1929.

1928 E.J. Wallis's collection of Kew photographs (5,000 negatives and 1,800 mounted photographs).

1930 A. Bennett's collection of Potamogetons.

1931 Over 1,000 specimens of Chinese plants from W.Y. Chun.

Miss W.J. Barker employed in South Africa preparing specimens and drawings of the native flora for Kew.

N.E. Brown's drawing of Mesembryanthemums.

Residue of the estate of Miss Margaret Louisa Moxon (her drawings of British and Swiss plants had already been presented) and Alfred Edward Moxon "to be applied and administered by the Bentham Trustees along with the Bentham Fund for furthering the study of Botany, the publishing of works on flowers and botany, and the making of grants for travelling botanists, and for the purpose of botanical literary collections and books on botany for use at Kew Gardens, and for any other purpose for which the Bentham Fund might be used and applied." The two funds were amalgamated as the Bentham-Moxon Fund.

1932 Tankerville collection of 648 flower drawings including some by G.D. Ehret and also many by Margaret Meen who drew plants at Kew.

Financial assistance towards the cost of producing illustrations for D. Prain and I.H. Burkill's *Account of genus Dioscorea*, 1936–8.

Grant to E. Marsden Jones towards the cost of his transplant and plant breeding experiments.

1933 47 drawings of plants *c.* 1760 by Simon Taylor who drew at Kew and elsewhere.

1933–4 Temporary assistants employed in the Herbarium.

1935 Janet R. Perkins Trust. This and subsequent trusts and funds were kept separate but administered by the Bentham-Moxon Trust.

1936 Published J.D. Snowden's *Cultivated Races of Sorghum*.

1937 W.B. Grove's collection of Coelomycetes.

1938 S. Garside's collection of South African plants.

1958 H.K. Airy Shaw was engaged to revise J. Willis's classic *Dictionary of Flowering Plants and Ferns*. It was published by the Trust in 1966.

1960 Financial assistance towards the collecting trips of Paul Furse in Iran.

William Dallimore Memorial Trust Fund.

1961 Employed two assistants to prepare supplement to *Index Londinensis*.

1964 George Conrad Johnson Memorial Fund.

1965 Frank Kingdon Ward Trust. N.Y. Sandwith Bequest.

1966 394 flower drawings by Indian artists done for B. Heyne (1770–1819).

1967 Published *Darwin and Henslow: the Growth of an Idea: Letters 1831–1860*, edited by N. Barlow.

Financial assistance for a wrought iron superstructure to Venetian well head in Queen's Garden.

Landsman's Bookshop Prize Fund.

1968 Stanley Smith Gift Fund. Ernest Thornton-Smith Travel Scholarship.

1969 Purchase of important collection of Sir Joseph Hooker's correspondence.

Miss Doris Ruth Butcher Bequest.

1970 The Trust had for some years helped to pay the salaries of artists employed by *Curtis's Botanical Magazine*. When the Royal Horticultural Society announced that it was no longer able to publish it, the Bentham-Moxon Trust acquired the copyright and assumed responsibility for publishing it.

The Bentham-Moxon Trust organised the erection in May 1974 of a commemorative stone on George Bentham's grave in Brompton Cemetery.

1971 B.A. Krukoff Fund for Study of African Botany.

1974 Commemorative stone placed on George Bentham's grave in Brompton Cemetery.

1977 Published vol. 1 of R.D. Meikle's *Flora of Cyprus*; vol. 2, 1985.

1978 Donald Dring Memorial Prize Fund.

1979 Published *Survival or Extinction*, edited by H. Synge and H. Townsend.

1980 Henry Idris Matthews Memorial Trust Scholarship Fund.

1982 Kazimierz and Winifred E.A. Kaminski Memorial Fund. John Gilbert Prize.

1983 Supported collecting trips to Thailand and Sabah.

1984 W.M. Curtis Bequest.

1985 Supported expedition by Kew assistants to Malaysia.

Herbert Kenneth Airy Shaw Memorial Fund.

1986 Pat Brenan Memorial Fund.

1987 Mrs Marjorie Hurley Bequest.

1989 Sainsbury Orchid Trust. Eleanor Constance Bor Trust.

1991 W.R. Marshall Bequest.

The Bentham-Moxon Trust which is a registered charity now administers more than 20 separate trusts and grant-supported schemes.

ADMISSION OF THE PUBLIC

Times of Opening

The public were admitted to the two contiguous royal estates on different days: Kew on Thursday but later altered to Monday, and Richmond on Sunday from June to September. George III once declared he never liked Richmond Gardens on Mondays – "they seem so dirty" after the public invasion on Sundays.

> Kew now became quite gay [in 1776], the public being admitted to the Richmond Garden on Sundays, and to Kew Gardens on Thursdays. The Green on these days was covered with carriages, more than £300 being taken at the [Kew] bridge . . . Parties came up by water, too, with bands of music, to the Ait opposite the Prince of Wales's house [Kew Palace]. (Mrs Papendiek: *Journals*, 1887, vol.1, p. 77.)

London Gazette (7 July 1787) announced that the garden at Kew would in future be open on Mondays. *The Times* (25 June 1800) stated that Kew would not be accessible to the public on Mondays. On 23 June 1801 *The Times* reported that both Richmond and Kew Gardens would be open on Sundays during the summer.

> . . . we entered [Kew] by a gate which we found open, and proceeded through a thickly planted shrubbery to the interior . . . we were met by a game-keeper, who told us that strangers were not now permitted to visit the gardens and that we must therefore return as we have come. (Anonymous diary of a visitor from Birmingham in 1817.)

> The Gardens are open to the public on Sunday from Midsummer till Michaelmas. The hours of admission are from ten in the morning till sunset; and the entrance is in the upper side of Kew Green, whilst another is in Kew Road. (John Evans: *Richmond and its Vicinity*, 1824, p. 219.)

> Kew-gardens to the opening of which there had been some interruption, are now regularly opened every Sunday evening till half-past nine o'clock. (*The Times*, 20 July 1825.)

According to John Smith, the source of much of our knowledge of Kew's domestic affairs, William IV withdrew the privilege of public access to the Royal Gardens ('Unpublished memoirs of Kew Gardens', p. 339). After the King's death, W.T. Aiton allowed visitors but their treatment by his staff provoked complaints. John Lindley said that

> You entered unwelcome, you rambled about suspected, and you were let out with manifest gladness shown at your departure. (*Gardeners' Chronicle*, 24 July 1847, p. 486.)

The distinguished Indian botanist, William Griffith, as a student had experienced this barely concealed hostility.

> Such was the principle of exclusion on which such Gardens were then conducted, that on one occasion he [Griffith] was induced to conceal a plant in his hat, that had been presented to him by one of the foremen, not so much for the purpose of carrying [it] off successfully, as with a view of protecting the liberal benefactor who gave it him for examination,

from the consequences of so great an offence. (J. McClelland: *Journal of Agricultural and Horticultural Society of India*, 1845, part 1, pp. 5–6.)

In 1839 the small Botanic Garden was open daily from one to six or to dusk in winter, excluding Sundays, throughout the year; Richmond Gardens opened on Thursdays and Sundays (*The Times* 10 July 1839). This restricted access to Richmond Gardens did not bother the *Florists Journal* (1840, p. 76) which considered them "a scene of mere amusement, and not of information". No entrance fee was charged, provided visitors were "decently dressed" and they accepted being escorted by a gardener.

Sir William Hooker made these hours of admission more flexible when he arrived at Kew in 1841 by allowing "respectable individuals" in as early as eight in the morning, and, furthermore, they were free to wander without an escort. Omnibuses to Kew Bridge provided the only means of public transport.

In 1851 the Pleasure Grounds – the area outside the Botanic Garden then being transformed into an arboretum – opened daily from mid-May to mid-September, from one o'clock to six o'clock in the afternoon. In 1853, at the insistence of the First Commissioner of Works, both the Botanic Garden and Pleasure Grounds opened on Sundays, mid-May to mid-September, from two o'clock. In 1854 closing time was extended to seven o'clock or sunset. In 1859 the Office of Works accepted a memorial from the inhabitants of Isleworth, Hounslow and Brentford that the Pleasure Grounds be opened in October as well. In 1864 the times of opening of the Pleasure Grounds, now the Arboretum, were the same as those of the Botanic Garden, that is, every day of the year except Christmas Day: weekdays, one o'clock to sunset; Sundays, two o'clock to sunset.

The public agitation for opening Kew Gardens in the mornings began in the 1870s. The first concession was opening at ten o'clock on the four Bank Holidays in the year from 1877. Kew's rather spurious excuses for not admitting the general public in the mornings were contemptuously dismissed by William Robinson.

> To those who know anything of botanic gardens and their management, the 'scientific purpose' objection to their being thrown open early would be amusing if it were not so untrue and so puerile. (*The Parks and Gardens of Paris*, edition 3, 1883, p. 64.)

The public were refused admittance if they did not fulfil the regulation that "visitors must be decently dressed." In 1884 the local newspaper complained that boatmen "and those who follow the pleasures of the river" strolled about Kew Gardens "in their deshabille, and not infrequently to the extent of exposing their lower limbs". Sir Joseph Hooker accordingly instructed the Curator:

> The above practice has also attracted the notice of the Board. While persons properly clothed from head to foot in boating costume are not to be excluded, men with bare legs or breasts or others unbecomingly dressed are not to be admitted.

Such attire was "inadmissible and in bad taste in a national scientific establishment". ('Kew gates 1845–1913', f.16. Kew archives.)

In February 1883 the Treasury advised Sir Joseph Hooker to resist changes to the times of opening from "local residents whose interests may often conflict with those of the public". However from 1 April 1883 the Gardens were open from noon on weekdays and from one o'clock on Sundays.

In 1898 the Gardens opened from ten o'clock during June to September, excluding Sundays, an experiment opposed by Kew but imposed by the Office of Works. The experiment, however, was not repeated the following year. Queen's Cottage grounds were opened to the public, June to September, but the Cottage itself remained closed.

Students, artists and photographers got special permission to visit the Gardens in the mornings from six a.m., April to October, and from six-thirty a.m., March and November, and from seven a.m., January, February and December.

In 1912 opening time was advanced to ten o'clock from mid-May to mid-October; closing time was eight p.m. or an hour before sunset. From January 1921, the Gardens were open every day, except Christmas Day, from ten a.m. Following pressure from the MP for Acton, closing time in summer was nine p.m. from 1923, but reverted to eight p.m. in 1932. In January 1987 opening time was changed to nine-thirty a.m. and closing time between four p.m. and six p.m., depending on the time of sunset.

Entrance Charges

1916 1d. admission fee to the Gardens imposed at the suggestion of the Committee on Public Retrenchment in the interests of national economy. Tuesday and Friday were 'student' days for study, sketching and photography: 6d. all day or five shillings for permit. Sundays were free.

1924 1d. fee abolished on 21 April but the charges on 'student' days were maintained.

1926 1d. fee reintroduced on 1 January following a recommendation of the Select Committee on Estimates.

1929 1d. fee abolished on 5 August.

1931 1d. fee reimposed on 5 October.

1938 Standing Commission on Museums and Galleries recommended the abolition of 6d. fee on 'student' days.

1951 Fee increased to 3d. on 1 April; 6d. 'student' fee on Tuesday and Friday finally abolished.

1971 With the introduction of decimal coinage 3d. became 1p. in the new currency.

1980 Fee increased to 10p.

1983 ·· ·· ·· 15p.

1985 ·· ·· ·· 25p.

1987 ·· ·· ·· 50p.

1989 ·· ·· ·· £1.00; concessionary rates introduced.

1990 ·· ·· ·· £3.00 from 1 November.

1992 ·· ·· ·· £3.30.

1993 ·· ·· ·· £3.50.

1994 ·· ·· ·· £4.00.

Attendance Figures

No figures before the appointment of Sir William Hooker in 1841 exist.

1841	9,174	
1842	11,400	
1843	13,492	
1844	15,114	
1845	28,139	
1846	46,573	London and South Western Railway to Richmond opened.
1847	38,951	

1848	91,708	Palm House opened.
1849	137,865	
1850	179,627	
1851	327,900	Great Exhibition, Hyde Park. Arboretum opened daily during summer.
1852	231,010	
1853	331,210	Gardens opened on Sundays in summer. Kew Bridge railway station opened.
1854	339,164	
1855	318,818	Very bad weather.
1856	344,140	
1857	361,978	
1858	405,376	
1859	384,698	
1860	425,314	The considerable increase in flower-beds during the next decade attracted crowds to Kew.
1861	480,070	
1862	550,132	International Exhibition in London.
1863	401,061	
1864	473,307	Arboretum open every day of the year except Christmas Day.
1865	529,241	
1866	488,765	
1867	494,909	
1868	502,369	
1869	630,594	Kew Gardens station opened.
1870	586,835	
1871	577,084	
1872	553,249	
1873	683,870	
1874	699,426	
1875	678,002	
1876	596,865	
1877	687,972	Gardens opened at 10 a.m. on Bank Holidays.
1878	725,422	
1879	569,134	Dismal weather.
1880	723,681	
1881	836,676	
1882	1,244,167	
1892	1,354,157	
1901	1,460,169	Coronation of Edward VII.
1906	2,339,492	
1909	3,360,221	
1911	3,704,606	Coronation of George V.
1912	3,815,427	Gardens opened at 10 a.m., June to September.
1914	4,082,011	
1915	4,300,330	
1916	713,922	Admission charged for the first time; turnstiles installed.
1920	1,131,771	
1921	1,236,308	Gardens opened daily from 10 a.m.
1924	1,535,855	Admission fee abolished except on Tuesday and Friday.
1926	1,162,547	Admission fee reimposed.
1929	1,169,900	Admission fee abolished except Tuesday and Friday.
1930	1,544,212	
1931	1,389,184	Admission fee reintroduced.

1941	825,372	World War II.
1942	1,230,241	
1943	1,401,001	
1946	1,567,954	
1951	1,205,957	
1960	1,082,271	
1981	953,153	
1991	749,139	
1994	988,801	

The high attendance figures from 1882 must be treated with caution. Their reliability was questioned by the Royal Commission on National Museums and Galleries when it interrogated Sir Arthur Hill on 21 October 1927:

> We have a figure in some of the Treasury papers which gives for 1914 four million visitors to Kew; 1915 4,300,000; are those figures authentic? [Sir Arthur Hill]: No. I cannot say that they are. The way in which the numbers were concocted in old days before we had turnstiles was very casual. The constables at the gates were supposed to count the visitors who came in. Probably on weekdays, when numbers were few, their countings were fairly accurate, but on Sundays, Banks Holidays and crowded days I rather gather from what I have learned that they counted the number passing through the gate for half an hour and multiplied that number by the number of hours that the gardens were open. We always regarded the figures sent up with considerable suspicion. Since the turnstiles have been placed I have attempted to estimate the number of visitors. (Minutes of Evidence, p. 17, para. 110.)

GATES AND PUBLIC ENTRANCES

Eighteenth century records inform us that the public could enter Richmond Gardens from an entrance near Richmond Green and depart from a gate near Kew Green. There was also a gate in the boundary wall of Princess Augusta's garden near the Pagoda.

By the early 1820s the main entrance to the Botanic Garden, according to John Smith's unpublished memoirs (p. 23), was by a small wooden door in the arboretum wall which stretched from the last house on the south side of Kew Green (55 The Green) to the stables slightly east of Kew Palace. Its site would have been just beyond the present ticket office at the Main Gates where a path turns to the left. Another gate a little farther on in the same wall led to the Pleasure Grounds. (See map on p. 413)

In 1825 railings and two lodges carrying the royal supporters, the lion and the unicorn, flanked a gate on Kew Green when George IV ordered the enclosure of its western extremity (railings across the Green mark the site). Because this annexation thus blocked access to the Botanic Garden by the gate in the arboretum wall, a new public entrance was made in 1825 near the present administrative offices on the south side of Kew Green (a plaque marks its approximate site): "three small doorways have been approved by His Majesty, also to be affixed in situations of the most convenient access into the Royal Gardens on public days of admission." (W.T. Aiton to Colonel Stephenson. 2 April 1825. Works 1/13, ff.267–8. P.R.O.)

> A new public entrance has lately been made to these gardens; it is perhaps more convenient for the public, as its door may be approached with a carriage, but the walk, proceeding along a narrow crooked passage, has nothing of the magnificence, variety, and beauty of the portion of the pleasure ground which bursts on the view, on entering the old door-way. (*Gardener's Magazine*, vol. 1, 1826, p. 223).

The entrance is clearly shown on E. & G.N. Driver's map of 1840 (copy in Kew archives).

This inconspicuous and unpretentious entrance fell far short of Sir William Hooker's concept of what was appropriate for the former Royal Gardens, now undergoing transformation into a national institution. After he moved into the late Sir George Quentin's house on the Green in 1852 (49 The Green), the Director closed the gates that had served as the principal entrance for some 27 years. Under Decimus Burton's supervision, Kew now boasted an entrance that proclaimed its elevated status. These Main Gates, as they are now called, opened in April 1846, and consist of double gates for carriages, flanked on either side by single gates for pedestrians. The elaborate wrought iron work in a Jacobean style had been made by Walker of Rotherham (Works 1/27, ff.343, 428; 1/29, f.190, P.R.O.). The incorporation of the royal coat of arms with the letter 'V' confirmed the Director's determination to maintain Kew's royal links. The floral and fruit motifs on the Portland stone piers had been carved by John Henning the younger (1801–1857) who had earlier carved the classical reliefs on Decimus Burton's triple screen at Hyde Park Corner. John Smith said that the curved iron railings on both sides had been fashioned from the fence that had once enclosed part of Kew Green; he also claimed that the vases on top of the outer stone pillars had come to Kew during William IV's reign. Burton failed to get approval to build a small lodge near the Main Gates.

Burton also designed a gate "plain in style but well adapted in accordance with the old Palace" (1847 Annual Report) as a riverside entrance to Kew Palace. He had intended mounting the gate piers with the lion and unicorn. Instead we have two badly eroded

The Main Gates on Kew Green, designed by Decimus Burton, were opened in 1846.

Decimus Burton's gate to Kew Palace. Map reference D14 (pp. 354–5).

heraldic figures said to have once stood on the gates to one of the houses owned by Sir Richard Levett: the Dutch House or the Queen's House. They have been described as "dogs" but they are so worn that it is impossible to tell. The gate, after having been in store for some years, was restored and rehung in 1985 on a new site near the Kew Bakery.

In 1847 the public were able to enter the Gardens by the Brentford Ferry Gate.

In September 1849 Mr Jesse, presumably an official of the Office of Woods and Forests, recommended the purchase for £7 from Mr Long of Richmond (a builder?) of the statue of the unicorn, one of the two royal supporters that had stood on the lodges on Kew Green. The lion presumably already stood above the Lion Gate shown on E. & G.N. Driver's plan of 1840 close to where Cumberland Gate now stands. It was moved to the Pagoda Gate, then renamed the Lion Gate, and the Unicorn gave its name to the gate which was an entrance to the Pleasure Grounds, but by the 1880s was used only by staff. The public could also get into the Pleasure Grounds by the Pagoda Gate and the Brentford Ferry Gate; in 1872 the Isleworth Ferry Gate with its wooden drawbridge over the river boundary ditch added another entrance. Once in the Pleasure Grounds, visitors could enter the Botanic Garden by one of the small gates in its wire fence.

The Unicorn and Lion (Pagoda) Gates as the only entrances in the long boundary wall in the Kew Road were patently inadequate for the increasing volume of visitors. In anticipation of a railway station at Kew, double gates with single side gates were erected in 1868 opposite the Temperate House. These cast iron gates, called the Queen's Gate, were made by Coalbrookdale Ironworks; the Portland stone piers bore the crown and 'VR' within roundels. Cumberland Gate, a pair of gates between brick piers, opened in 1868 just over 300 yards north of Queen's Gate. Erected near the site of old stables which had been demolished the previous year, its name commemorates the Duke of Cumberland, later King of Hanover. Queen's Gate, despite public demand, never opened but was re-erected in 1889 opposite Lichfield Road, an event necessitated by the new site for the railway station, and renamed Victoria Gate.

Other gates:
Confusingly, there is another Queen's Gate. It is adjacent to the Lion Gate and leads to the golf course in the Old Deer Park. For many years it was a private entrance to the Queen's Cottage.

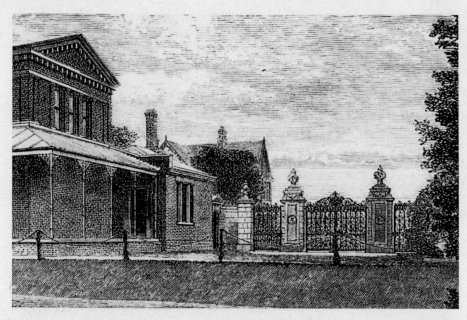

The Queen's Gate stood next to the Marianne North Gallery before being moved and re-erected as the Victoria Gate. *Gardeners' Chronicle*, 3 December 1881, p. 725.

Staff can get into Kew Gardens along the path from this Queen's Gate through Oxhouse or Oxenhouse Gate in the southern boundary of the Gardens. It takes its name from a shed and yard for oxen that stood in the Old Deer Park. E. & G.N. Driver's map of 1840 and W.T. Warren's map of 1843 indicate its position. (E.G. Driver's letter of 18 September 1844. Class 16, piece 31/10. P.R.O.)

King's Steps (King's Stair) Gate is shown as an entrance to the Queen's Cottage grounds over the ha-ha (indicated on 1894–6 25 inch Ordnance Survey map).

Edward's Gate is shown on Thomas Chawner's plan of 1837 (1839 copy in Surrey Record Office and Kew) in the boundary wall just north of the present Victoria Gate. Named possibly after Edward, Duke of Kent; nearby there was a path called Duke of Kent's Ride.

Princesses Gate was an entrance to Kew Palace lawn across the boundary ditch; shown on 1863–6 25 inch Ordnance Survey map.

Jodrell Gate opened in 1965 when the new Jodrell Laboratory was finished.

The royal supporters, the lion and the unicorn, which once stood on lodges on Kew Green, now surmount two gates on the Kew Road.

The Unicorn Gate (map reference K11 (pp. 354–5)) is not a public entrance.

BOUNDARY WALLS

The Gardens' boundary wall dominating the Kew Road is almost as memorable a landmark as the Pagoda. Richmond Gardens had a brick wall running along its boundary with Love Lane (J.J. Boydell: *History of the Principal Rivers of Great Britain* vol. 2, 1796, p. 37). In 1755 some 110 acres of Princess Augusta's estate had been "walled & paled in for Pleasure Garden" (Sir George Lee's papers). John Smith stated that the original belt of trees extending from the Temple of Eolus to the Old Deer Park had been planted against the Kew Road boundary wall. Since we know that this periphery of trees was in existence by the early 1760s, the wall itself must be older – it may be the one referred to in 1755. Sir Richard Phillips in his *A Morning's Walk from London to Kew* (1817) described an amazing panorama of the ships of the British navy, executed in chalk by a disabled seaman on the entire 1½ mile length of the boundary wall.

The long gardens of the houses on the south side of Kew Green were all enclosed by walls which joined a boundary wall running west from about the site of the present Cumberland Gate to the ice-house; from there the boundary wall went north past the east side of the Orangery to meet another boundary wall from Kew Green. The configuration of all these walls is clearly indicated on the large-scale version of the 1771 map in Richmond Museum. William IV started the process of thinning them by ordering the removal of the west and north boundary walls of the arboretum. Sir William Hooker continued their demolition. Soon after his appointment the wall separating the original Botanic Garden from the Pleasure Grounds disappeared; as private gardens were incorporated their boundary walls were pulled down. One that has survived bounded the western side of the Royal Kitchen Garden and now divides the Herbaceous Ground from the Rock Garden. Three feet of pierced brickwork were added to it in 1870. Unfortunately a gale blew down 60 feet of it in February 1965, but it had been rebuilt by February 1968.

Not everyone welcomed the destruction of these walls. A contributor of the *Gardeners' Chronicle* (29 August 1874, p. 271) declared that "the old walls had an especial attraction for me, as they sheltered many things I had never met with in the gardens and nurseries I had worked in or visited." Sir Joseph Hooker also regretted the disappearance of at least one of them.

> The removal of a similar wall on the south side of the grounds (facing the Deer Park) . . . affords an example of what would happen if the East Wall were removed [he is referring to the Kew Road boundary wall]. This was unquestionably a great advantage as it exposed beautiful views, and there was no collection there – but the trees and shrubs near it were blasted . . . ('Kew-walls, fences, etc.' 1/ADM/25. Kew archives)

Boundary stone with the initials of Christopher Appleby.

Land that had been leased to Queen Caroline or to her son Frederick Prince of Wales was demarcated not by walls but by unobtrusive square stones, some of which still survive in the Gardens. One such stone on Mount Pleasant near the Lake has 'RTM 1728' carved on it, the letter T being larger than the other two. These initials are those of Robert and Mary Thoroton* who granted to Queen Caroline in June 1728 the leases on the Dutch House and on the freehold land at Kew, the southern boundary of which passed through the spot now called Mount Pleasant. At the other end of the Gardens, near the Rock Garden, another stone proclaims 'CA 1743'. 'CA' was Christopher Appleby the owner of Cambridge Cottage; according to Burrell's and Richardson's map of Richmond for 1771 Ann Appleby still owned the land immediately south of the Cottage.

Other stones, some since moved, denote the boundary between Kew and Richmond which passes through the Temperate House, the Lake and Mount Pleasant. Their original positions are shown on E. & G.N. Driver's plan of 1840, *c.* 1865 25 inch Ordnance Survey map, and Works 16 1610. P.R.O.

* I am indebted to John Cloake for these identifications.

TREES

The aerial view of Kew Gardens, taken in 1992, and reproduced as endpapers to this book, shows how extensive is its tree cover. Despite the considerable loss of trees during the memorable storms in 1987 and 1990, there are still more than 2,000 species and varieties in cultivation. In order to maintain the landscape and to expand the number of species, more trees and shrubs are planted every year at Kew and Wakehurst Place. The two gardens between them can claim to have one of the finest collections of temperate species and hardy exotics in the world.

Sophora japonica. One of James Gordon's Chinese introductions which he obtained from France in 1753. Planted in Princess Augusta's garden. Map reference G16 (pp. 354–5). *Photograph by A. McRobb.*

Corsican pine, *Pinus nigra* var. *maritima.* Brought back from the south of France in 1814 by R.A. Salisbury. A specimen was planted at Kew a few years later. It fortunately survived a lightning strike in 1992. Map reference F17 (pp 354–5). *Photograph by A. McRobb.*

The planting of trees began in earnest when Charles Bridgeman landscaped Queen Caroline's Richmond Gardens. Most of his trees have died through age, disease or storm but a sweet chestnut, *Castanea sativa*, near the Lake, could be one of his; a Tulip Tree, *Liriodendron tulipifera*, close to the Azalea Garden, is perhaps another survivor of Queen Caroline's garden. We can only ascertain with any certainty the age of a tree when it is felled and its growth rings are counted. In March 1914 a storm uprooted one of the largest Tulip Trees in the country (an engraving of it appears in the *Gardeners' Chronicle* 23 August 1890, p. 219). Its annual growth rings indicated it was about 150 years old and thus had been planted in Richmond Gardens in the early years of the reign of George III, possibly by 'Capability' Brown. One of the largest red oaks (*Quercus rubra*) in Britain, which stood in the southern part of Kew Gardens, was blown down in March 1916. Its annual rings revealed that it had been planted about 1746 when Frederick Prince of Wales was beginning to take a renewed interest in his Kew estate. Another tree contemporary

Lucombe oak, semi-deciduous hybrid between the evergreen cork oak, *Quercus suber*, and the deciduous Turkey oak, *Quercus cerris*. Raised by an Exeter nurseryman, William Lucombe, about 1763. W.A. Nesfield moved Kew's specimen some time after February 1846 when it stood in the way of his proposed Syon Vista. Map reference G10 (pp. 354–5). *Photograph by A. McRobb.*

The Caspian Sea Chestnut-leaved oak, *Quercus castaneifolia*. Raised from seed imported in 1843, it was planted at Kew in 1846, and has now reached 110 feet, making it the largest tree in terms of volume at Kew. Map reference G13 (pp. 354–5). *Photograph by A. McRobb.*

A storm on 28 March 1916 brought a Cedar of Lebanon down on the Temple of the Sun, one of Sir William Chambers's most graceful garden buildings.

One of the many trees destroyed at Wakehurst Place during the violent storm on 16 October 1987.

with Frederick, an elm in the northern end of the Garden and lost in 1929, had been raised about 1738. This tree and others of the same period had been planted on small mounds, presumably to relieve the monotony of the flat terrain.

The great storm which swept across South East England during the early hours of 16 October 1987 destroyed or severely damaged about 10 per cent of Kew's 9,000 trees, including historic ones such as the *Zelkova carpinifolia* near the Herbarium and the Turkey oak, *Quercus cerris*, close to the Pond, both more than 200 years old; rare and first introductions were also casualties. Wakehurst Place, which took the full force of the storm, was devastated: between 15,000 and 20,000 trees were blown down or irreparably damaged. Remedial action such as replanting and landscaping has done much to lessen the effects of this catastrophe.

The Maidenhair Tree, *Ginkgo biloba*. Its date of introduction from China is uncertain but James Gordon, a nurseryman in Mile End, had it in stock in 1758. Kew's specimen may have been among those transferred from the Duke of Argyll's garden in 1762. John Smith said it had been originally trained against a wall "like a fruit tree". Map reference F16 (pp. 354–5). Near it is a North American Locust Tree, *Robinia pseudoacacia*, perhaps another of the Duke of Argyll's stock. *Photograph by A. McRobb.*

SCULPTURE

The oldest examples at Kew are the five terms standing in front of the curved hedge in the Queen's Garden, and the stone plinth with cherub heads in the central block of the Temperate House. The assumption that they are associated with Frederick Prince of Wales is considered in Chapter 2.

Another piece with royal connections is the Medici vase near the Wisteria Cage. It was made of an artificial stone akin to stoneware by Mrs Eleanor Coade and her relatives, John Seely and William and Thomas John Croggon, at their kilns in Lambeth on the site of the present Royal Festival Hall. Coade stone, noted for its durability and its ability to withstand weathering, was ideal for architectural and garden ornaments. Two of the Coade factory's most popular items were the Borghese and Medici vases, reproduced from copies in Robert Adam's possession. A fine example of the Borghese vase stands in the Temple of Flora at Stourhead. When John Nash ordered a Medici vase in 1825 for the Royal Lodge at Windsor he chose one of their stock pieces made in 1817; it was mounted on a specially designed pedestal with bas-reliefs and George IV's cipher on each of the four panels. When Sir George Taylor saw it in store in the works yard at Hampton Court during the 1960s, he arranged for it to come to Kew (QE 1074. Kew archives). The figures on the vase (the original is in the Uffizi Gallery in Florence) have been identified as being Greek leaders consulting the oracle at Delphi before departing for Troy, but this is purely conjecture. Unfortunately this splendid example of Coade stoneware has been crudely repaired with cement and deserves better restoration. Croggon, who supervised its manufacture, was also responsible for another Medici vase, dated 1826, which stands in

One of eight terms commissioned by Frederick, Prince of Wales. Now stands in the Queen's Garden. Map reference D16 (pp. 354–5). *Photograph by A. McRobb.*

Coade stone Medici Vase. Map reference G16 (pp. 354–5). *Photograph by A. McRobb.*

RIGHT: Sundial erected by William IV to commemorate an astronomical investigation made in the White House in 1725. The stepped base has been removed. Map reference E15 (pp. 354–5).

the circular bed at the Pond end of the Broad Walk. It had been moved from Whitehall Gardens about the time the buildings there were demolished before the war, and first placed in Lancaster House and then in Kensington Palace Orangery. In 1957 the London Museum agreed to its being on permanent loan to Kew where it was installed in 1958 (Works 16 1748. P.R.O.)

William IV agreed to the transfer of a sundial to Kew in 1832. It was one of a pair that had stood in the Privy Garden at Hampton Court, its marble pedestal bearing the cipher of William III. It was placed on a stepped base on Kew Palace lawn to mark the approximate site of an important astronomical observation, made there in 1725. It had taken place in the predecessor to the White House, where Samuel Molyneux, Secretary to the Prince of Wales (the future George II), lived with his wife, Lady Elizabeth Capel. An amateur astronomer, Molyneux had set up a 24-foot telescope in a room on the south side of the house (its possible location is discussed in the *Miscellaneous Works and Correspondence of the Rev. James Bradley*, 1832, p.xiv). There his friend James Bradley (1693–1762), Professor of Astronomy at Oxford "made the first observations which led to his two great discoveries: the aberration of light and the nutation of the earth's axis". In 1930 the ugly steps and railings around the sundial were replaced by an octagonal Portland stone base flush with the grass. In 1963 the bronze sundial itself made by the great English clock and watch maker Thomas Tompion, was replaced by a replica. The original can be seen in Kew Palace. The pedestal was returned to Hampton Court in 1995 and a copy substituted.

Another sundial, not on display at the moment, was erected in the medicinal garden in the grounds of Cambridge Cottage in 1929. Designed by Professor C. Vernon Boys, it consists of five cubes of bronze screwed together in the form of a Greek cross upon a

Sundial designed by C. Vernon Boys. At present in store.

trapezoid base; 'the dial carries on its upper surface a statement giving the position of Kew and directions for adding or subtracting a certain number of minutes to or from the time indicated by the dial to obtain Greenwich mean time . . .' (*Kew Bulletin* 1930, p. 13). It is mounted on one of the balusters of the old Kew Bridge, which was replaced in 1903.

Two statues were added to the Gardens in 1929: a bronze figure leaning on a spade by Arthur G. Atkinson (the plaster cast was exhibited in the Royal Academy in 1890) entitled "Out in the Fields"; and "A Sower", executed by Sir Hamo Thornycroft in 1886, a gift of the Royal Academy. Its pedestal was designed by Sir Edwin Lutyens and A. Drury.

The Lion of England. One of the Queen's Beasts in front of the Palm House.

"Out in the Fields" by A.G. Atkinson. Map reference J15 (pp. 354–5).

RIGHT: "A Sower", by Sir Hamo Thornycroft. Map reference H17 (pp. 354–5).

One of the two Chinese lions facing the Pond. Map reference J12 (pp. 354–5). *Photograph by A. McRobb.*

The Queen's Beasts, ten heraldic figures displaying the armorial bearings of Her Majesty The Queen's forebears, were positioned along the eastern front of the Palm House in 1956. They are Portland stone replicas by James Woodford of plaster models he had made to stand outside the western annexe to Westminster Abbey on the occasion of the Queen's Coronation.

A pair of Chinese lions facing the Pond – probably eighteenth century but possibly dating to the Ming dynasty, (1368–1644) – were presented in 1958 by Sir John Ramsden of Bulstrode Park, Bucks. He also gave Kew the Venetian well-head now in the Queen's Garden.

In 1963 the crude pipe which had served as a fountain in the Pond was replaced by a fine bronze composition of Hercules wrestling with Achelous in the form of a snake (PLATE 30). It is a splendid sight when all 20 jets are in operation. It had stood on the East Terrace at Windsor Castle before being moved to storage at Hampton Court. The original plaster cast by François Joseph Bosio had been exhibited in the Paris Salon in 1814; in 1825 a bronze cast was placed in the Jardin des Tuileries where it can still be seen. The copy, now at Kew, had been cast by C. Crozatier for George IV in 1826.

A less grand but charming fountain, occupies a pool in the Queen's Garden: a bronze cast of Andrea del Verrocchio's "Boy with a Dolphin", a replica of a copy in the Victoria and Albert Museum. The original is in the Palazzo Vecchio in Florence.

Kew is again indebted to the Victoria and Albert Museum for the loan of lead statues of a "Shepherd" and "Shepherdess" by John Cheere (1709–1787). These can be seen in the central block of the Temperate House.

LEFT: Copy of Verrocchio's "Boy with a Dolphin" in the Queen's Garden. Map reference D16 (pp. 354–5).

RIGHT: "Shepherdess" by John Cheere in the Temperate House. *Photograph by A. McRobb.*

ORNAMENTAL IRONWORK

It was a memorable occasion for all who were privileged to see the renovated Palm House, bright with fresh white paint, before the plants were returned: the vertical thrust of columns, the curving lines of the glazing bars, the arched ribs, and the spirals of the stairs. Although Richard Turner's natural exuberance had been curbed by Decimus Burton, he still managed to express his fondness for decoration with plant forms on stairs and gallery rails, and a generous scattering of sunflower heads. "I can report," wrote the gardener and journalist Arthur Hellyer, "that the Palm House is more lovely than I have ever known it. It is the first time I have ever seen it without plants, and, though as a gardener I hate to admit this, it is even more beautiful without them."

The Temperate House, denied any contribution from Turner, is more restrained. There are token florets on the ironwork, but far more striking are the fortuitous patterns of curved braces and the vista along the length of the catwalk.

The fine Victorian ironwork on the entrance gates, especially the Main Gates, and modern examples in the Queen's Garden, should not be overlooked.

Interior of the restored Palm House below, and spiral staircase on far right. *Photographs by A. McRobb.*

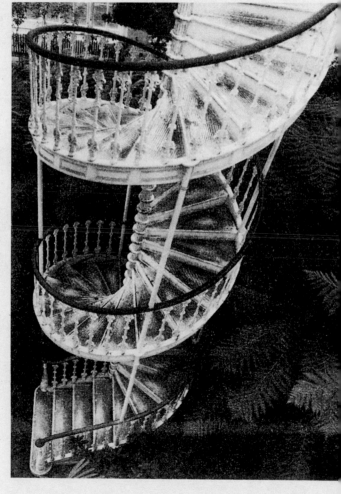

A view of one of the wings from the gallery in the Palm House.

BELOW LEFT: The gallery in the Temperate House.

WATER SUPPLIES

While William Chambers was building his garden temples, John Smeaton designed an engine to raise water from a well near the Temple of Confucius. Smeaton (1724–94) was a civil engineer who planned harbours, constructed canals and bridges and built the first Eddystone lighthouse. His machine, concealed in shrubbery, with a 30 foot canopy to shade the two horses that powered it, was in operation by 1761. It supplied water to the lake and to the ornamental basins in Princess Augusta's garden.

John Smeaton's machine for raising water from a well in Princess Augusta's garden. W. Chambers: *Plans . . . of the Gardens . . . at Kew*, 1763.

Lawns were usually watered from horse-drawn barrels on wheels, equipped with a single spout or sprinkler heads. The contract appointing John Haverfield as head gardener at Richmond Gardens in 1762 listed among the equipment in his charge "water barrows" and "watering tubs and watering pots". The renewal of W.T. Aiton's contract in 1796 mentioned a "water engine" pulled by a couple of horses. Aiton had two watermen in 1838, one of whom described his duties. "The watering was done by cans from a cask filled from a pump at the back of the office and wheeled about. Pumping was considered very hard work and ale was allowed twice a day – at 12 a.m. and 4 p.m." (*Journal of Kew Guild*, 1903, p. 131.)

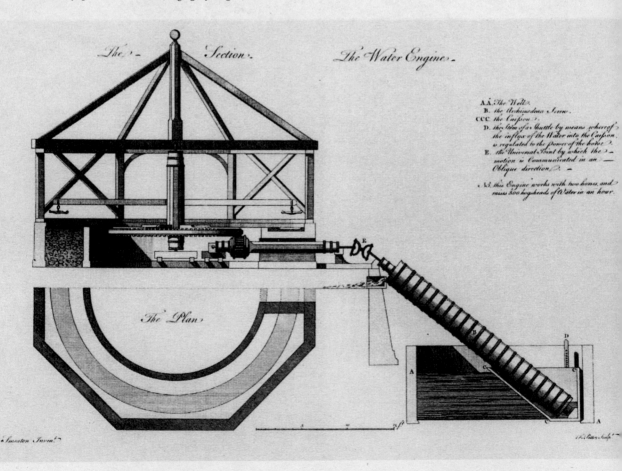

Smeaton's water engine in the 1820s. Lithograph by George Papendiek.

In his report to the Board of Woods and Forests in June 1846, Decimus Burton thought a "deep spring" the most economical and reliable source of water for the Gardens. He recommended that a well should be sunk either in one of the boiler rooms beneath the Palm House or near the proposed smoke shaft "and that a steam engine and pumps should be erected from which water would be thrown into the ornamental Pond, and to such parts of the garden to supply basins, or stand pipes required for the use of the gardeners." (Kew 207. Kew archives.)

In 1850 the Board of Woods and Forests, having decided to replace Smeaton's water engine, sought the advice of the engineers, Messrs Easton and Amos. James Easton recommended the installation of a steam engine and pump at the base of Burton's Campanile with a suction pipe to the Thames near the Brentford Ferry. This would extract river water, which would be pumped into a cast iron tank, capable of holding 4,500 gallons, at the top of the tower. A pipe from this tank would conduct water under pressure to water cocks throughout the grounds. Unfortunately the suction pipe also deposited fine particles of alluvia on the leaves of plants. James Simpson and Company tried to solve this problem about 1855 by sinking a well in the gravel near the river and laying a new pipe to a storage tank, a cemented underground reservoir near the Temple of Eolus on the top of the Mound, 25 feet square and 9 feet deep. (*Gardeners' Chronicle*, 1 September 1855, p. 583.)

In 1861 Sir William Hooker's new Lake was ready to be filled with river water through a culvert. In 1864 an engine house near the Arboretum nursery was built to pump water from the Lake for use in the Gardens. Adequate filter beds to clean this water had not yet been constructed by 1867 when James Simpson reported that river water, properly filtered, was superior to well water "highly charged with organic and mineral matters". He urged that the Lake be deepened, cleared of unnecessary vegetation, and permanent filter beds constructed. ('Kew water supply, 1844–1925', f.17. Kew archives). With Office of Works approval in June 1868, Simpson was engaged to construct filter beds near the Temperate House. The filtered water was pumped to a covered reservoir in Richmond Park with a capacity of 250,000 gallons, and returned under pressure through a water main to the Gardens. Sir Henry Barkly learnt from his friend, Sir Joseph Hooker, that he was "now introducing water mains and stand cocks all over the grounds". This reservoir also supplied official residences at Kew and the public drinking fountains. One is not surprised that visitors sampling this water in the drinking fountains complained about its quality. In 1876 the Treasury allowed mains to be laid by the Southwark and Vauxhall Company to the residences, museums and drinking fountains in Kew Gardens.

The erratic pattern of the Thames's tides – often very high in winter and extremely low in summer – made it difficult to keep the Lake filled. In 1887 a larger culvert was substituted and in the early 1890s the storage capacity of the Lake was increased by excavating "the vast accumulation of mud, in great part London sewage". In 1895 an additional water engine was installed, and by 1897 a new system of service pipes had been laid throughout the grounds. Also in 1897 another main was laid from the Richmond Park reservoir. New filter beds were built in 1899–1900. Electric pumps added a small quantity of water to the Lake when tides were too low for gravity feed. Yet all these improvements could not meet Kew's water demands – in 1905 it asked for the storage capacity of the Richmond Park reservoir to be trebled.

The pumping station (its position mid-way between the Lake and the Temperate House is clearly shown on early key plans of the Gardens) was near enough to the recently created Water Lily Pond for condensed water from its engine to be discharged into it. It was an opportunistic method of providing sufficient warmth to grow tender aquatics in the open.

During the severe drought of 1921, when the rainfall was less than half of normal, losses of hardy plants were high but, unexpectedly, even greater losses were suffered in the glasshouses. An analysis of the Lake water conducted by W.B. Turrill (*Kew Bulletin* 1922, pp. 13–5) disclosed an exceptionally high level of salt. Turrill reasoned that a reduced flow of fresh water from the higher reaches of the Thames allowed salt tidal water to come further upriver. The brackish Lake was emptied and refilled several times and this dependency on it was reduced by building large rainwater tanks. The Palm House, the Temperate House and the fern houses already had them and during 1922 and 1923 they were added to other glasshouses.

In 1931 Kew Gardens began getting its water through Richmond Borough mains and a few years later the pumping station machinery was dismantled and its buildings converted into accommodation for carts and motor transport; in 1963 the disused filter beds were finally broken up. In 1973, with the Metropolitan Water Board assuming responsibility for Kew's water supply, river water from the Richmond Park reservoir was replaced by treated water from the Petersham pumping station. Heavily chlorinated and alkaline water, unsuitable for some glasshouse plants, can fortunately be treated.

Mist spraying of the Gardens had been introduced in the early 1930s; in 1961 overhead automatic watering was installed in the Palm House, a more sophisticated version of the overhead watering which was used shortly after the Palm House was first opened.

The Palm House now has two 50,000-gallon rainwater tanks sunk in the ground on the Pond side; the rainwater storage facilities in the restored Temperate House have been increased by 60 per cent; the Princess of Wales Conservatory holds its rainwater in 60,000-gallon capacity underground tanks; and the new Alpine House also has its own reservoir. These underground tanks now provide 70 per cent of the water requirements of the larger glasshouses.

FRANCIS BAUER'S FLOWER PAINTINGS

The Hookers, father and son, suspected that the Bauer drawings had been removed from Kew with dubious authority. In a letter to his successor, Sir William Thiselton-Dyer, Sir Joseph Hooker wrote that "Presumably it was Aiton inspired by Brown who engineered the transfer of the Bauer drawings to the Brit. Mus. Bot. Dept." (19 November 1904. 'Hooker letters to Dyer, 1870–1909, f.179. Kew archives). This confusion about the legality of the transfer is understandable.

As his employer, Sir Joseph Banks was the legitimate owner of Bauer's official work, never disputed by the artist and confirmed in writing on 31 August 1813:

> I the undersigned do hereby declare that all the sketches and drawings which I have made since the year 1790 (as well as those which I have already delivered, as those which I have still under my care in my house) are the property of, and belong to the Right Honble Sir Joseph Banks, K.B., I having during that time received from him an annual salary.
>
> I likewise declare that all the sketches and drawings which I shall make hereafter, as long as I shall continue to be engaged for that purpose by the Right Honble Sir Joseph Banks, K.B., and any of them which might be found in my house at the time of my death (should that event happen during the continuance of my so being engaged) shall be the property of, and belong to the Right Honble Sir Joseph Banks . . . (Add. MS 52281, f.107. B.L.)

In a codicil to his will (21 January 1820. Prob. 10/box 4514. P.R.O.) Sir Joseph Banks bequeathed his library, herbarium, manuscripts, drawings, etc. in 32 Soho Square and the neighbouring house in 17 Dean Street to the Trustees of the British Museum but gave Robert Brown a life interest in them. Bauer received an annuity of £300 provided he continued to reside on Kew Green and drew the plants in the Royal Gardens, the said drawings to "be added to the collection now in his hands and which shall revert to me or to my representatives at the time of his death". In the eventuality of any doubts arising about the interpretation of the bequest, Brown and Bauer were empowered to adjudicate "in a manner so as to be most favorable to them".

A second codicil to the will made on 7 March 1820 bequeathed all the drawings Bauer had made of Kew plants to the King. Sir Joseph reminded his executors that the post of botanical artist at Kew had been created at his own expense.

> I did this under the hope amounting almost to an expectation that the truths of my opinion [that Kew needed the services of a botanical artist] would in due time become manifest and that the charge of maintaining Mr Bauer would then be transferred from me and placed on the establishment of the Garden . . . In case, however, of it being deemed inexpedient by His Majesty's advisers to make this small addition to the Royal Establishment of the Gardens, it is my will . . . that my heirs . . . continue . . . to pay Mr Bauer . . . the sum of three hundred pounds during the term of his natural life, provided that he continues to make drawings and sketches . . . and deposits the same in the hands of William Townsend Aiton Esquire, and his successors, to be added to those before me beqeathed to the Royal Establishment of the Botanic Gardens at Kew. (Both codicils are given in

full in House of Commons Botanical Work Committee return 1901,
pp. 101–2).

This arrangement was confirmed by W.T. Aiton in a letter he wrote to Allan Cunningham.

> He has left his Library, Herbarium and his etceteras of curiosities to his
> worthy librarian, your friend and mine, Mr Robert Brown, & finally to
> the British Museum. His collection of drawings by Mr Bauer to His
> Majesty to be deposited in my hands for the establishment at Kew Gardens.

In consultation with Bauer, now an octogenarian, Brown drafted a memorandum regarding the disposal of his drawings at Kew in view of "the altered circumstances of the establishment to which they were beqeathed" (25 May [1840]. 'R. Brown letters', vol. 1, f.36. Natural History Museum). Bauer's will of 15 August 1840 (Prob. 10/box 5968. P.R.O.) reflected their accord by bequeathing his eight portfolios of finished drawings and sketches at Kew to the British Museum. His decision was influenced by Kew being

> no longer on the same footing as it was when that bequest was made,
> namely the private Botanic Garden of the King; and as there seems but
> little probability of the views of Sir Joseph Banks earnestly expressed by
> him in his bequest being carried into effect, namely the appointment of a
> resident draughtsman.

Bauer felt justified in seeking Queen Victoria's permission that the Kew drawings should be added to those already in the British Museum where they would be more accessible to botanists. Bearing in mind the uncertainty about Kew's future at that time, it was a reasonable decision and one which the Crown accepted. In May 1841, five months after Francis Bauer's death, the eight portfolios were formally handed over to Robert Brown at the British Museum.

When Bauer's anatomical drawings, commissioned by Sir Everard Home, were declined by the College of Surgeons, the King of Hanover, formerly the Duke of Cumberland, purchased and presented them in November 1841 and January 1842 to the University of Göttingen together with miscellaneous sketches and notes, amounting to 20 volumes.

Memorial to Francis
Bauer in St Anne's
Church, Kew Green, by
Sir Richard Westmacott,
1840.

KEW GREEN

A number of the houses around the Green have been residences of the Royal Family and court officials or of administrators, botanists and gardeners employed by the Royal Botanic Gardens. The tombs and memorials of some of them can be seen at St Anne's Church which, after its consecration in 1714, enjoyed royal patronage for many years. In 1770 George III commissioned his Clerk of Works, Joshua Kirby, to enlarge the building; in 1822 George IV presented the organ which had been intended for the Castellated Palace; in 1837 William IV instructed Sir Jeffry Wyatville to make a further enlargement, adding the Doric porch on the west front. In 1851 the 2nd Duke of Cambridge had a mausoleum built at the East end for his late father – he and the Duchess were transferred to Windsor in 1930 by command of Queen Mary. The tombs of the Haverfield and Aiton families, Joshua Kirby and Sir William and Sir Joseph Hooker are still decipherable in the churchyard. Inside the church itself plaques commemorate Lady Dorothy Capel, the Aitons, Francis Bauer and both Hookers. (See plan on p. 418)

Houses around the Green having connections with the Gardens
SOUTH SIDE:

Descanso House, 199 Kew Road (near the corner of Kew Green) GRADE II

When W.T. Aiton indicated his wish to relinquish control of the small Botanic Garden in November 1840, he wanted to remain in the house in which his family had lived for 80 years. Descanso House was thus probably built about 1760–1 as a residence for William Aiton. W.T. Aiton continued to stay there until his death in 1849.

In 1846 Decimus Burton had recommended that, in due course, the house should be converted into offices. Sir William Hooker sought permission in 1849 to rent it as a residence and as a repository for his herbarium. This request was refused by the Office of Works which let the house on an annual tenancy from 1852. In 1898 the Treasury agreed to convert it into an "Institute for the gardeners, offices for the Curator and his clerks and as a residence for the Assistant Curator." In 1935 part of its front garden was lost during the widening of the Kew Road.

The name 'Descanso House', was given to it by George Willison, a tenant from 1888 to 1892. Once a merchant in Brazil, he gave it the Portuguese name for "a resting place".

Gumley Cottage, 17–19 The Green GRADE II

The early 18th-century houses, Gumley and Chestnut Cottages, were purchased in 1913 by the Ministry of Agriculture and Fisheries and converted into a plant pathology laboratory. When the laboratory moved to Harpenden in 1919, the building was occupied by the Imperial Bureau of Mycology which, in 1930, moved to a new building, designed by J.H. Markham, in Ferry Lane off the Green. Gumley Cottage, then the residence of senior Kew staff, was severely damaged by fire in July 1951. Since conversion in 1974–5, it has provided hostel accommodation for a limited number of students.

23 The Green

The botanical artist, Francis Bauer, was a tenant there from 1804 to 1837; previously,

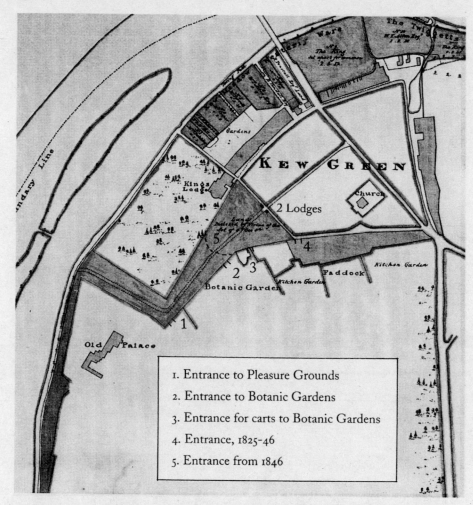

1824 map of Kew Green, with the shaded part indicating the extent of George IV's enclosure. This enclosure blocked the public entrance to the Pleasure Grounds(1) and the Botanic Garden(2). At that time carts entered through gate(3). From 1825 the public were admitted at a new entrance(4) until 1846 when Decimus Burton's Main Gates were erected.

from at least 1795, he had lived in Denmark House, 356–8 Kew Road at the junction with Mortlake Road,. In 1837 he moved to Eglantine Cottage, 20 Waterloo Place, Kew Green; the house was demolished in 1939/40 and the site is now 74 The Green.

31 The Green (formerly Llewellyn House)

Home of the botanical artist W.H. Fitch, who had previously lived at Hope Cottage (1866) and 6 Priory Park Terrace (1883), both off Kew Green.

King's Cottage or Church House, 33 The Green   GRADE II

According to Miss O. Hedley, this dwelling had been built by Christopher Appleby, who also owned other property on the Green. It became "A house built for Lord Bute to study in, where none of his family resides" (caption to an engraving in the *Political Register* 1767). It was settled on the Duke of Cumberland by George III in 1806. It has been stated that it was known as Bute House until the early 1840s, when the Secretary of the Office of Woods and Forests changed it to Church House.

Cambridge Cottage, 37 The Green   GRADE II

Houses have stood on this site from at least the early seventeenth century. Christopher Appleby, a barrister of the Middle Temple, "built and furnished" a house there (his will of 25 August 1739); his widow, Ann, who inherited the property leased it to Lord Bute in November 1758. In 1772 George III acquired it and placed his two sons – the Princes Edward and William – there. In 1806 he settled it on the Duke

Kew Green with the road
leading to the Palace.
Mid-1760s. Drawn by
George Bickham.

A VIEW of LORD BUTE's ERECTIONS at Kew, with som

Engraved for the Polit

of Cambridge who made it his permanent residence in 1838. George Robinson, the Clerk of Works, provided extra bedrooms and a drawing-room in a new East wing in 1839–40. Space for this expansion was found when the adjacent barracks were no longer needed. In 1840–1 a Portland stone Doric *porte-cochère* was added, and in 1843 a verandah was built on the garden front. When his mother died in 1889, the 2nd Duke of Cambridge occupied it until his own death in 1904. It was then incorporated into Kew Gardens and the central block and east wing converted into a forestry museum in 1910.

The Gables, 39–45 The Green

Four houses for gardening staff were built in 1908 on the site of the stables of Cambridge Cottage. The gateway was preserved and the original ornamental gables copied in the modern pair of houses.

Administration offices, 47 The Green

In the early 1840s the workshops and offices of the Clerk of Works were here, with the 1825 entrance to the Botanic Garden next to it. In 1931–2, J.H. Markham of the Office of Works designed a new Director's Office to replace a cottage which had been the Curator's residence before being converted into library facilities for gardeners and a Director's office. The present administrative block was opened in 1981.

South side of Kew Green. *Political Register*, May 1767.
A The White House or Kew House, the residence of the Dowager Princess of Wales.
C Lord Boston's house (now 55 The Green).
D General Graeme's house.
E Pennel Hawkins's house.
F Mr Howlett's house.
G Hell House where Princess Augusta's servants were quartered.
The Director's residence, 49 The Green, occupies the site of E-G.
The two buildings with prominent gables are now part of 39–45 The Green.
H Cambridge Cottage where Lord Bute's family lived.
I Lord Bute's study (now 33 The Green).

Director's official residence, 49 The Green   GRADE II

The site of several early eighteenth century houses – the properties of Mr Howlett
and of the King's Surgeon, Pennell Hawkins, and the sinister-sounding Hell House.
The last-named was described in the *Political Register* for 1767 as "A house for the
P[rincess] D[owager] of Wales's servants". It had been used as a charity school by
Lady Dorothy Capel, and was acquired by Frederick Prince of Wales in 1731. It was
demolished by 1814.

In August 1794 George III acquired the property from the Revd Thomas Methold
together with its gardens and orchards (Crest 2/1245. P.R.O.; indicated as nos. 733–
8 on 1771 map of the Gardens). In 1802 the Earl of Cardigan was a tenant. In 1846,
Methold's garden (about 3–5 acres) was added to the Botanic Garden. Sir George
Quentin, Riding Master to the Queen, lived here; after his death it became the official
residence of Sir William Hooker on 15 December 1851. The stables were then pulled
down and the 1825 entrance to the Botanic Garden closed. It has been subjected to
alterations and improvements which still continue; in 1972, for instance, a cell physiology laboratory was briefly established in the former servants' wing.

Royal Cottage, 51 The Green   GRADE II

This much altered house, created out of two Georgian dwellings, was, until recent
years, a 'grace and favour' residence. It was once the home of Mrs Charlotte Papendiek,
assistant keeper of the Wardrobe and reader to Queen Charlotte. Her published
journals are the source of much anecdotal information on the Royal Family at Kew
and Richmond.

53 The Green   GRADE II

A modest Georgian house, once the residence of George Robinson, who made alterations to Cambridge Cottage and submitted a design for a new Palm House at Kew.
When he was transferred to Hampton Court in 1845, the house became John Smith's
official residence. Other Curators including George Nicholson and W.J. Bean have
lived here.

55 The Green   GRADE II

The house is shown on John Rocque's 1734 map as being the property of Frederick
Prince of Wales. It has an attractive Corinthian door and is very early eighteenth
century or perhaps late seventeenth century. The *Political Register* for 1767 stated it
was then occupied by Lord Boston, Chamberlain to Princess Augusta. As a youth, the
Duke of Cambridge lived there. T.W. Phillips, Secretary of the Office of Woods and
Forests, rented it. Joseph Hooker, following his appointment as Assistant Director in
1855, moved into the house at the end of October or the beginning of November.
Having enlarged three rooms on the west side at his own expense, further payment
of rent was waived for him. Two small back rooms were added in 1922. After Joseph
Hooker moved to the Director's residence in 1865, this house became the official
residence of Keepers of the Herbarium.

EAST SIDE:

Haverfield House, 24 The Green   GRADE II

The most impressive house on this side of the Green, it was the home of several
generations of Haverfields.

NORTH SIDE:

Herbarium

A substantial Georgian House. John Rocque's map of 1734 declared it to be another property belonging to Frederick Prince of Wales. This 1734 map shows two buildings close together – the present Hunter and Hanover Houses – but in shape they differ from the two figured on the 1852 map of the Gardens. John Cloake has established the sequence of owners from Arthur Manley in 1650, but Frederick's name does not appear. Christopher Appleby was the owner from 1710, succeeded by his widow, Ann, until 1765 when Peter Theobald, a local philanthropist, acquired the property. Theobald purchased the house immediately adjacent to his on the west side, once the home of the painter Peter Lely, demolished it in 1773–4 and added the site to his garden. Theobald's widow remained there after his death in 1778; when she died about 1796, John Maud briefly owned the house until 1800, when Robert Hunter, a London merchant, took possession. When Hunter died in 1812 the house passed to his son, also Robert, who sold it to George IV in 1820. The grounds of Hunter House which stretched towards Kew Palace had a coach house, stables and glasshouses and a kitchen garden and orchard on the riverside – altogether about six acres. In addition to Hunter House, George IV also acquired two houses immediately to the east, owned by the Orton family and at one time the home of the miniature painter Jeremiah Meyer, and also part of Brentford Ait. Hunter House now became the King's Lodge but George IV's ownership was brief: it was sold to the Commissioners of Woods and Forests in July 1823 for £18,250. Its history after that date has already been told in Chapter 12: occupancy by the Duke of Cumberland, later King of Hanover, and its transfer in 1852 to Kew Gardens.

When the Duke of Cumberland lived there a porter's lodge intruded on the Green, causing a sharp bend in the road. It was demolished in 1881 and its railings brought nearer to the Herbarium.

Jeremiah Meyer, as has already been stated, was the tenant of what is now called Hanover House on the corner of Ferry Lane (formerly Meyer's Alley), consisting of two dwellings. Following the King of Hanover's death in 1851, officials of the Office of Works lived in it until J.G. Baker, First Assistant in the Herbarium, was installed there in late 1878; O. Stapf and W.B. Hemsley followed and then successive Keepers of the Museums, concluding with F.N. Howes. In 1898 part of the building was demolished to reduce fire risks to the adjacent Herbarium.

The rear of Hunter House which was demolished in the 1870s when the first wing to the Herbarium was built.

Kew Green at one time tapered to a point beyond Hunter House, taking the road to Kew Palace and the White House. Some of the outlying buildings of the White House survived its demolition in 1802. The house nearest to the sundial features on John Rocque's map of 1734 and its stepped gables suggest a kinship to the predecessor of the White House. Miss Hodgson, custodian of the Queen's Cottage, occupied it in 1898 when Sir William Thiselton-Dyer suggested it might be turned into a foreman's residence after she terminated her tenancy. The large building with prominent pediments west of it was the kitchen block. It does not feature on Rocque's 1734 map but appears on Chambers's 1763 map. A detailed description of the Georgian kitchen and its equipment is given by John Charlton in *Kew Palace*, 1956, pp. 8–9. A laundry block which stood just south of the kitchen is indicated on the key plan to the Gardens for 1904 but not on the next one in 1909. Part of the service wing on the west side of Kew Palace was demolished in the 1880s; the remaining buildings in this wing became a works depôt in 1899.

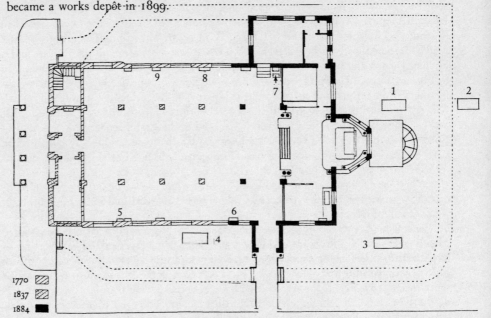

St Anne's Church, Kew Green. Graves and memorials with associations with Kew Gardens.
1 John Haverfield (c. 1694–1784).
2. Sir William and Sir Joseph Hooker.
3. William Aiton (1731–1793).
4. John Joshua Kirby (1716–1774).
5. William Townsend Aiton (1766–1849).
6. Lady Dorothy Capel (–1721).
7. Sir William J. Hooker (1785–1865).
8. Sir Joseph D. Hooker (1817–1911).
9. Francis Bauer (1758–1840).

William Aiton's tomb where his son, William Townsend Aiton, was also buried. Part of the inscription was recut on the occasion of the centenary of the Kew Guild in 1993. The full inscription is given in the *Kew Bulletin*, 1910, pp. 307–9.

A Wedgwood plaque commemorating Sir William Jackson Hooker. The white jasper profile was modelled by the Pre-Raphaelite sculptor, Thomas Woolner. R.F. Palgrave, a cousin of Sir Joseph Hooker, designed fronds of fern on green jasper – ferns were Hooker's principal interest. They include *Asplenium*, *Blechnum* and *Polypodium* species. A copy of this plaque can be seen in the Victoria and Albert Museum.

A Wedgwood plaque commemorating Sir Joseph Dalton Hooker. Portrait in white relief on green ground by Frank Bowcher, a sculptor and medallist. The surrounding plants, all associated with Hooker's career, were modelled by Matilda Smith, his cousin and botanical artist at Kew.
Clockwise from upper right panel:
Nephenthes albo-marginata (pitcher plants and Malayan flora).
Rhododendron thomsonii (Indian flora).
Celmisia vernicosa (Flora Antarctica).
Cinchona calisaya (Introduction of cinchona into India).
Aristolochia mannii (African flora).

BRENTFORD AND LOT'S AITS

Aits or eyots – small islands in the River Thames – were frequently osier beds. There were more than 20 in the 1740s between Chiswick and Sunbury, now fewer than 10 remain. The cluster of three near Kew Bridge is fortunately not disfigured by buildings like the best known of all the aits – Eel Pie Island.

Before the river banks were built up, the Thames was shallower, the Brentford or Mattingshawe Aits were not covered at high tide as they often are now, and a restaurant of sorts was built on one of them. William Hickey recorded in his memoirs that in 1780 he "dined upon the island off the town of Brentford, where there is a house famous for dressing pitchcooked eels, and also for stewing the same fish". The Prince of Wales had an assignation there in 1779 with the actress Mary ('Perdita') Robinson. Robert Hunter who lived opposite the ait in Kew denounced its "House of entertainment, which has long been a harbour for men and women of the worst description, where riotous and indecent scenes were often exhibited during the summer months on Sundays". He leased the island in 1811 so that poplars might be planted to spare Kew Palace the sight of industrial Brentford. Jeremy Bentham wrote to Sir Samuel Bentham in mid-September 1820 that "Aiton has just bought [the ait] for the King [from Hunter's son], and a little house is to be built upon it." The surveyors who were negotiating the purchase of Hunter House on behalf of the Commissioners of Woods and Forests in 1823 assessed the aits as a desirable acquisition, the means of effectively screening the "inferior buildings in Brentford".

In his correspondence and annual reports during the 1870s, Sir Joseph Hooker expressed concern that the frequent inundation of the Brentford Aits, the consequent loss of soil and uprooting of trees would lead eventually to their disappearance. In 1892 Kew became a parish of Richmond which commendably took immediate steps to preserve the amenities of the northern reaches of its newly expanded borough. In 1893 the Council approved the purchase of the islands (some 4½ acres), and Sir William Thiselton-Dyer lost no time in congratulating Richmond Corporation on its prudent move, offering, at the same time, Kew's assistance in conserving them. And, true to his word, he directed their replanting in 1894. He vigorously opposed the attempt by the Thames Steam Tug and Lighterage Company to extend its barge-building works on Lot's Eyot in 1899. He reminded the participants in this affair that the Office of Works had agreed "that the preservation of the amenities of Kew is primarily an imperial and not a local concern".

When ship-builders tried again to develop Lot's Eyot in 1925, Sir Arthur Hill voiced concern. They agreed to convey a strip of land, planted with trees, to the Brentford Council which would maintain it. The islands were replanted in 1927 and again between 1962– 4. Besides providing a desirable screen, these small islands have now assumed a new role, that of a habitat for wildlife, in particular birds such as herons, cormorants and kingfishers.

W·H· FITCH DEL·

A·E·SMITH Sc

The ait near Kew bridge.
Wood engraving of a
drawing by W.H. Fitch.
Gardeners' Chronicle, 21
October 1882, p. 529.

BRIEF BIOGRAPHIES

This is a selection of monarchs, statesmen, patrons, administrators, scientists, gardeners, plant collectors, architects and artists who have contributed to the development of Kew Gardens.

Many of the entries are based on those appearing in my *Dictionary of British and Irish Botanists and Horticulturists*, published by Taylor and Francis in 1994. The entries give date and place of birth and death, qualifications, honours and awards, brief details of career, important publications and a few biographical references for further information.

ABBREVIATIONS USED:

b.	born	FLS	Fellow of Linnean	R.H.S.	Royal Horticultural Society
d.	died		Society	R.S.	Royal Society
D.N.B.	Dictionary of National	FRS	Fellow of Royal Society	VMH	Victoria Medal of Honour
	Biography	LS	Linnean Society	VMM	Veitch Memorial Medal

ROYALTY

CAROLINE, H.M. Queen
(1683–1737)

b. Anspach 1 March 1683 *d.* London 20 Nov. 1737

Daughter of Margrave of Brandenburg-Anspach. Married George Augustus, Prince of Hanover, 1705. Accompanied him to England as Princess of Wales, 1714; her husband became George II in 1727. Knowledgeable gardener. Engaged Charles Bridgeman and William Kent to design and embellish the gardens at Kensington Palace and Richmond Lodge.

References: *D.N.B.* P. Quennell: *Caroline of England*, 1939. D. Jacques: *Georgian Gardens*, 1983, *passim*. R. Strong: *Royal Gardens*, 1992, pp. 37–47.

FREDERICK LOUIS, Prince of Wales (1707–1751)

b. Hanover 6 Jan. 1707 *d.* London 20 March 1751

Created Prince of Wales, 1729. Married Princess Augusta, 1736. Leased Kew estate in 1731 but little work done on the garden until the last few years of his life. The antagonism between him and Queen Caroline was discussed in Lord Hervey's Memoirs, 1727–37.

References: *D.N.B.* G. Young: *Poor Fred: the People's Prince*, 1937. A. Edwards: *Frederick Louis, Prince of Wales, 1707–1751*, 1947. S. Jones: *Frederick, Prince of Wales: a Patron of the Rococo*. In C. Hind, editor: *Rococo in England: a Symposium*, 1986, pp. 106–12. K. Rorschach: *Frederick, Prince of Wales (1707–51), as Collector and Patron* (*Walpole Society*, vol.55, 1989/90, pp. 1–76). K. Rorschach: 'Frederick Prince of Wales: taste, politics and power' (*Apollo*, vol.34, 1991, pp. 239–45).

AUGUSTA, Princess of Wales
(1719–1792)

b. Saxe Gotha 30 Nov. 1719 *d.* London 8 Feb. 1772

Princess of Saxe Gotha. Married Frederick Prince of Wales, 1736. She implemented her late husband's 'Plan' for the garden at Kew with advice from Lord Bute. William Chambers added garden buildings.

References: O. Hedley: *Queen Charlotte*, 1975, *passim*. W. Blunt: *In for a Penny: Prospect of Kew Gardens*, 1978, *passim*.

O. Hedley: 'Augusta, Princess of Wales at Kew' (*Country Life*, 30 July 1981, pp. 401–5). R. King: *Royal Kew*, 1985, *passim*. R. Strong: *Royal Gardens*, 1992, pp. 49–63.

GEORGE III (1738–1820)

b. London 4 June 1738 *d.* Windsor 29 Jan. 1820

King, 1760. Married Princess Charlotte, 1761. Although more interested in architecture and agriculture, he supported Sir Joseph Banks in his development of the gardens at Kew. "Since the King has forborne the violent exercise which used to contribute so much to His Majesty's amusement, he has taken more interest in the Botanic Garden . . ." (Sir Joseph Banks to Lord Hawkesbury, 13 Sept. 1803.)

References: *D.N.B.* J. Brooke: *King George III*, 1972. J. Roberts: *Royal Artists*, 1987. H.B. Carter: *Sir Joseph Banks, 1743–1820*, 1988.

SOPHIE CHARLOTTE, H.M. Queen (1744–1818)

b. Mirow, Mecklenburg-Strelitz 19 May 1744 *d.* Kew 17 Nov. 1818

Princess of Mecklenburg-Strelitz. Married George III, 1761. She and her daughters were instructed in botany by Sir J.E. Smith and in flower painting by F. Bauer. Kew was one of her favourite residences until she acquired Frogmore where she created her own garden.

References: *D.N.B.* J.L. Gilbert: 'Queen Charlotte as a botanist' (*Gardeners' Chronicle* 1 Nov. 1968, pp. 21–2). O. Hedley: *Queen Charlotte*, 1975. J. Roberts: *Royal Artists*, 1987. R. Strong: *Royal Gardens*, 1992, pp. 65–76.

GEORGE IV (1762–1830)

b. London 12 Aug. 1762 *d*. Windsor 25 June 1830

Prince Regent, 1811. King, 1820. He made Kew's head gardener, W.T. Aiton, responsible for all the royal gardens and, as a result, Kew itself suffered from neglect. "George IV allowed no alteration to be made in the establishment [at Kew], influenced perhaps more by regard for his father's memory than by any personal pleasure that he himself derived from it." (W.J. Hooker to Dawson Turner, 1840).

References: *D.N.B.* J. Richardson: *George IV*, 1966. C. Hibbert: *George IV: Regent and King*, 1975.

WILLIAM IV (1765–1837)

b. London 21 Aug. 1765 *d*. Windsor 20 June 1837

Duke of Clarence, 1789. King, 1830. "The King is said to be attached to Kew, and to be desirous of improving both the botanic garden and the pleasure ground." (*Gardener's Magazine* vol.7, 1831, p. 687). "Although Mr Aiton was much respected by the family of George III and George IV, but on account of his obtaining the charge of Hampton Court Gardens for his brother John, he lost the friendship of the Duke of Clarence who was then living at Bushy and had applied for the situation for his gardener. On the Duke of Clarence becoming King in 1830, he at once appointed different gardeners to take charge of the various Royal Gardens. Mr Aiton was allowed to retain charge of the Botanic and Pleasure Grounds at Kew [and Buckingham Palace gardens]. The King also blamed him for being the adviser of George IV in enclosing Kew Green."

(J. Smith's unpublished memoirs of Kew Gardens, ff.487–8).

References: *D.N.B.*

ERNEST AUGUSTUS, Duke of Cumberland (1771–1851)

b. Kew 5 June 1771 *d*. Herrenhausen 18 Nov. 1851

Son of George III. Duke of Cumberland, 1799. King of Hanover, 1837. Had property on Kew Green including Hunter House and much of the Pleasure Grounds as a game preserve. He did not approve of Sir W.J. Hooker's changes to Kew.

References: *D.N.B.* H.v. Thal: *Ernest Augustus Duke of Cumberland and King of Hanover: a Brief Survey of the Man and his Time*, 1936. R. Fulford: *Royal Dukes*, 1973, pp. 205–51.

CAMBRIDGE family

The Duchess of Cambridge who lived in Cambridge Cottage on the Green sometimes resented the changes the two Hookers were making at Kew. Sir William also had to cope diplomatically with the gardening enthusiasms of her daughter, Princess Mary Adelaide. An entry in her diary for 29 Sept. 1860 declared: "I have of late taken to gardening and Sir William Hooker has in a measure made over the arrangement and colouring of the flower-beds to me . . ." Sir William Hooker dedicated *Rhododendrons of Sikkim-Himalaya* to the Princess.

References: C.K. Cooke: *A Memoir of Her Royal Highness Princess Mary Adelaide, Duchess of Teck. Based on her Private Diaries and Letters*, 1900. 2 vols.

PATRONS AND BENEFACTORS

Eighteenth century

BANKS, Sir Joseph (1743–1820)

b. London 13 Feb. 1743 *d*. Isleworth, Middx 19 June 1820

MA Oxon. 1763. FRS 1766. FLS 1788. Baronet 1781. To Newfoundland, 1766–7. With Capt. Cook on HMS *Endeavour* 1768–71. Iceland, 1772. President, Royal Society, 1778–1820. A wealthy man who used his considerable influence and

entrepreneurial skills to advance the Royal Gardens at Kew. Without his guidance Kew might never have evolved into an institution of international repute.

References: *D.N.B.* E. Smith: *Life of Sir Joseph Banks*, 1911. H.C. Cameron: *Sir Joseph Banks*, 1952. W.R. Dawson: *Banks Letters*, 1958. H.B. Carter: Sir Joseph Banks and the Plant Collection from Kew sent to the Empress Catherine II of Russia (*Bulletin Natural History Museum, Historical Series*, vol.4, 1974). P.O'Brian: *Joseph Banks*, 1987. H.B. Carter: *Sir Joseph Banks: a Guide to Biographical and Bibliographical Sources*, 1987. H.B. Carter: *Sir Joseph Banks, 1743–1820*, 1988. R.E.R. Banks and others, editors: *Sir Joseph Banks: a Global Perspective*, 1994.

DRYANDER, Jonas Carl (1748–1810)

b. Gothenburg, Sweden March 1748 *d*. London 19 Oct. 1810

MA Lund 1776. FLS 1788. To England 1777, where he succeeded Solander as Banks's librarian. H.B. Carter calculated that between 1777 and 1809 he paid 161 visits to Kew, mainly to identify new plants. Edited *Hortus Kewensis* 1789 and part of 2nd edition 1810–13.

References: *D.N.B.* H.B. Carter: *Sir Joseph Banks*, 1988, *passim*.

HALES, Revd Stephen (1677–1761)

b. Bekesbourne, Kent 17 Sept. 1677 *d*. Teddington, Middx 4 Jan. 1761

BA Cantab. 1700. DD Oxon. 1733. FRS 1718. Perpetual curate, Teddington, 1709. Friend of Frederick Prince of Wales whose widow made him Clerk of the Closet. Plant physiologist who advised on the heating of the Great Stove at Kew.

References: *D.N.B.* A.E. Clark-Kennedy: *Stephen Hales*, 1929. D.G.C. Allan and R.E. Schofield: *Stephen Hales: Scientist and Philanthropist*, 1980.

STUART, John, 3rd Earl of Bute (1713–1792)

b. Edinburgh 25 May 1713 *d*. London 10 March 1792

Friend of Frederick Prince of Wales. A competent botanist and horticulturist who advised Princess Augusta on the development of her estate at Kew. Created notable gardens at Luton Hoo, Beds. and Highcliffe, Hants. Had a fine botanical library. Published *Botanical Tables*, 1785.

References: D.N.B. J.A. Lovat-Fraser: *John Stuart, Earl of Bute*, 1912. A.M. Coats: *Lord Bute*, 1975. K.W. Schweizer: *Lord Bute: Essays in Reinterpretation*, 1988.

SOLANDER, Daniel Carlsson (1733–1782)

b. Pitea, Norrland, Sweden 19 Feb. 1733 *d.* London 13 May 1782

MD Uppsala. DCL Oxon. 1771. FRS 1764. To England 1760. Assistant Librarian, British Museum, 1763; Keeper, Natural History Department, 1773. With Banks on Cook's *Endeavour* voyage, 1768–71 and to Iceland, 1772. Librarian to Banks, 1771–82. Assisted Aiton in the early planning of *Hortus Kewensis*.

References: D.N.B. *Transactions American Philosophical Society* vol.58, part 8, 1968, pp. 1–67.

Nineteenth century

BROWN, Robert (1773–1858)

b. Montrose, Angus 21 Dec. 1773 *d.* London 10 June 1858

Studied medicine at Edinburgh. DCL Oxon. 1832. FRS 1811. FLS 1822. Naturalist on HMS *Investigator*'s voyage to Australia, 1801–5. Librarian to Sir Joseph Banks who bequeathed him a life-interest in his collections which were left to British Museum. Keeper of Botany at British Museum. It was Banks's wish that he continued to be botanical consultant at Kew. "In my time no specimens were sent to London for names. Mr Brown then frequently visited Kew and I occasionally obtained names from him but more in a conversational way than for any real purpose." (J. Smith to W. Carruthers 22 Sept. 1873). Brown contributed to *Hortus Kewensis*, 1810–13.

References: D.N.B. D. Mabberley: *Jupiter Botanicus: R. Brown of British Museum*, 1985. H.B. Carter: *Sir Joseph Banks*, 1988, *passim*.

LINDLEY, John (1799–1865)

b. Catton, Norfolk 5 Feb. 1799 *d.* Turnham Green, Middx 1 Nov. 1865

PhD Munich 1832. FRS 1828. FLS 1820. Assistant Secretary, Horticultural Society of London, 1827; Secretary, 1858–63. Professor of Botany, University College London, 1829–60. Prolific author and editor. His report on the future of the Royal Gardens at Kew in 1838 did much to save them from dismemberment. At one time he appeared the most promising candidate for the post of director at Kew.

References: D.N.B. F.W. Oliver: *Makers of British Botany* 1913, pp. 164–77. J.R. Green: *History of Botany in U.K.* 1914, pp. 336–53. *Orchids from Botanical Register*, 1991 (biographical introduction by W.T. Stearn).

RUSSELL, John, 6th Duke of Bedford (1766–1839)

b. 6 July 1766 *d.* Rothiemurchus, Perthshire 20 Oct. 1839

FLS 1816. MP Tavistock, 1788–1802. Had notable garden at Woburn Abbey. Especially interested in agriculture. Actively supported Sir W. Hooker in his candidature for the director's post at Kew.

References: D.N.B. W. Hooker: *Copy of Letter . . . to D. Turner . . . on the Death of the late Duke of Bedford*, 1840.

First Commissioners of Woods and Forests:

In March 1840 Kew was transferred from the Lord Steward's Department to the Office of Woods and Forests. It was fortunate for Sir W. Hooker that very shortly after his appointment Lord Lincoln became First Commissioner.

NEWCASTLE-UNDER-LYME 5th Duke of, Henry Pelham Fiennes Pelham-Clinton, styled Earl of Lincoln (1811–1864)

b. London 22 May 1811 *d.* Clumber Park, Notts. 18 Oct. 1864

BA Oxon. 1832. 5th Duke, 1851. Politician. During his term of office as First Commissioner of Woods and Forests, 1841–6 "Hooker received much encouragement and was allowed complete freedom of action by Lincoln, who was a

sympathetic patron of natural science" (*Curtis's Botanical Magazine*, dedication to vol.71, 1845). While he was First Commissioner, Decimus Burton, Richard Turner and W.A. Nesfield were engaged to implement Hooker's reforms.

References: D.N.B.

CARLISLE, 7th Earl of, George William Frederick Howard (1802–1864)

b. London 18 April 1802 *d.* Castle Howard, Yorks. 5 Dec. 1864

BA Oxon. 1823. Known by courtesy title of Lord Morpeth. MP Morpeth 1826. Politician. First Commissioner of Woods and Forests, 1846–50. ". . . his relations with Sir William Hooker were of a most happy nature. Any suggested improvement in the Gardens had his support . . ." (*Curtis's Botanical Magazine*, dedication to vol.74, 1848.)

References: D.N.B.

Some other statesmen particularly sympathetic to Kew:

BALL, John (1818–1889)

b. Dublin 20 Aug. 1818 *d.* London 21 Oct. 1889

MA Dublin 1847. FRS 1868. FLS 1856. MP Co. Carlow, 1852. Under-Secretary for the Colonies, 1855. Supported Kew's proposal for the series of colonial floras. Collected plants with J.D. Hooker in Morocco, 1871. "Mr Ball was Parliamentary Under-Secretary of State for the Colonies and member for Limerick some twenty years ago, and a most powerful friend of Kew – indeed we owe to the establishment of the Herbarium on a proper footing to his personal efforts with the Ministry in 1865." (J.D. Hooker to G. Maw. 14 March 1871.)

References: D.N.B. *Curtis's Botanical Magazine*, dedication to vol.90, 1884.

DUFF, Sir Mountstuart Elphinstone Grant (1829–1906)

b. Eden, Aberdeenshire 21 Feb. 1829 *d.* London 12 Jan. 1906

MA Oxon. 1853. FRS 1881. FLS 1872.

MP, Elgin 1857–81. Under-Secretary of State for India, 1868–74, and for the Colonies, 1880–1. Supported Sir Joseph Hooker's projected *Flora of British India* while he was at the India Office.

References: D.N.B. Curtis's Botanical Magazine, dedication to vol.117, 1891.

MITFORD, Algernon Bertram Freeman-, 1st Baron Redesdale
(1837–1916)

b. London 24 Feb. 1837 *d.* Batsford Park, Glos. 17 Aug. 1916

FLS 1896. VMH 1904. Diplomatic service; in Japan, 1866–70. Secretary, Office of Works, 1874–86 when he took a particular interest in the public parks, and supported Sir Joseph Hooker's improvements at Kew. Created a fine arboretum at Batsford Park where he cultivated bamboos. Wrote *Bamboo Garden*, 1896.

References: D.N.B. Curtis's Botanical Magazine, dedication to vol.123, 1897. Lord Redesdale's autobiography: *Memories*, 1915; *Further Memories*, 1917.

CHAMBERLAIN, Joseph
(1836–1914)

b. London 8 July 1836 *d.* Birmingham 2 July 1914

MP. Secretary of State for the Colonies, 1895–1903. As a keen gardener he promoted Kew's interests. He urged the Chancellor of the Exchequer to sanction the completion of the Temperate House in the 1890s. As Colonial Secretary he enlisted Kew's services; the Director was appointed his Botanical Adviser in 1902.

References: D.N.B. Kew Bulletin, 1914, pp. 233–6.

Colonial Governors:

BARKLY, Sir Henry (1815–1898)

b. Montegle, Ross-shire 24 Feb. 1815 *d.* London 21 Oct. 1898

KCB 1853. GCMG 1874. FRS 1864. Governor, British Guiana, 1849–53; Jamaica, 1853–56; Victoria, 1856–63; Mauritius, 1863–70; Cape, 1870–77. In correspondence with Kew from 1852. Sent

many living plants to Kew together with flower drawings by his wife and daughter.

References: D.N.B. Kew Bulletin, 1898, pp. 335–6.

MACGREGOR, Sir William
(1846–1919)

b. Towie, Aberdeenshire 20 Oct. 1846 *d.* Aberdeen 3 July 1919

MB Aberdeen 1872. KCMG 1889. Medical officer in Colonial Service in Seychelles, 1873; Mauritius 1874; Fiji, 1875. Administrator, British New Guinea, 1888–89. Governor, Lagos, 1901–4; Newfoundland, 1904–9; Queensland, 1909–14. Sent plants to Kew from New Guinea, Lagos and Labrador.

References: D.N.B. Kew Bulletin, 1907, pp. 76–8; 1908, pp. 135–7; 1920, pp. 31–2. *Flora Malesiana*, vol.1, 1950, pp. 337–42.

RAWSON, Sir Rawson William
(1812–1899)

b. London 8 Sept. 1812 *d.* London 20 Nov. 1899

KCMG 1875. Government Secretary, Canada, 1842. Treasurer, Mauritius, 1844. Colonial Secretary, Cape of Good Hope, 1854–64. Governor, Bahamas, 1864; Jamaica, 1865; Windward Islands, 1869–75. "One of many Colonial Governors who have done good service to Kew and botanical science" (*Kew Bulletin*, 1899, p. 221). Encouraged the Colonial Parliament in Cape Town to support the publication of *Flora Capensis*.

THURN, Sir Ferdinand im
(*c.* 1852–1932)

d. Prestonpans, E. Lothian 8 Oct. 1932

MA Oxon. KCMG 1905. KBE 1918. Curator of Museums, Georgetown, British Guiana, 1877–82. Magistrate, Pomeroon, 1882–91. Government Agent, N.W. District, 1891–99. Colonial Secretary, Ceylon, 1901. Governor, Fiji, 1904–10. Sent plants to Kew.

References: Kew Bulletin, 1932, pp. 461–2.

THURSTON, Sir John Bates
(1836–1897)

b. London 31 Jan. 1836 *d.* Suva, Fiji Feb. 1897

Sheep farmer, N.S.W. *c.*1859–62. British Consul, Levaka, 1867–69; Fiji, 1869. Deputy Governor, Fiji, 1882; Governor, 1887–97. In correspondence with Kew for some 20 years; sent plants from Solomon Islands and Fiji.

References: D.N.B. Kew Bulletin, 1897, p. 169.

DIRECTORS

HOOKER, Sir William Jackson
(1785–1865)

b. Norwich, Norfolk 6 July 1785 *d.* Kew 12 Aug 1865

FRS 1812. FLS 1806. Knight of Hanover, 1836. Professor of Botany, Glasgow 1820. Director, Kew 1841–65. Prolific author.

References: D.N.B. J.D. Hooker: 'Sketch of the Life and Labours of Sir William Jackson Hooker' (*Annals of Botany*, vol.16, 1902, pp.ix-ccxxi). F.W. Oliver, editor: *Makers of British Botany*, 1913, pp. 126–50. M. Allan: *The Hookers of Kew, 1785–1911*, 1967.

HOOKER, Sir Joseph Dalton
(1817–1911)

b. Halesworth, Suffolk 30 June 1817 *d.* Sunningdale, Berks 10 Dec. 1911

MD Glasgow 1839. FRS 1847. FLS 1842. CB 1869. KCSI 1877. GCSI 1897. OM 1907. Surgeon and naturalist on HMS *Erebus* 1839–43. Collected plants in Sikkim and Nepal, 1848–51. Assistant Director, Kew, 1855; Director, 1865–85. Pioneer phytogeographer and friend of C. Darwin. Publications include *Botany of Antarctic Voyage*, 1844–60 and *Flora of British India*, 1872–97.

Sir Joseph Hooker never enjoyed administration and viewed the admission of the public to the Gardens as a necessary evil. He would have been far happier in charge of a purely scientific institution or as an independent researcher within the scientific community.

References: *D.N.B. Kew Bulletin*, 1912, pp. 1–34, 439–40. *Proceedings of R.S. part B.*, vol.85, 1912, pp.i–xxxv. L. Huxley: *Life and Letters of Sir Joseph Hooker*, 1918. W.B. Turrill: *Joseph Dalton Hooker*, 1963. M. Allan: *The Hookers of Kew*, *1785–1911*, 1967.

THISELTON-DYER, Sir William Turner (1843–1928)

b. Westminster 28 July 1843
d. Witcombe, Glos. 23 Dec. 1928

BA Oxon. 1867. BSc London 1870. FRS 1880. FLS 1872. KCMG 1899. Professor of Natural History, Royal Agricultural College, Cirencester, 1868. Professor of Botany, Royal College of Science, Dublin, 1870. Professor of Botany, RHS, 1872. Assistant Director, Kew, 1875; Director, 1885–1905. Although Thiselton-Dyer always used the compound form of his name which acquired a hyphen about 1890–1 (*Journal of Botany*, 1892, p. 89), his contemporaries frequently addressed him as Dyer.

References: *D.N.B. Kew Bulletin*, 1929, 32, pp. 65–75. *Nature*, 9 Feb. 1929, pp. 212–5. *Proceedings of R.S. part B.*, vol.106, 1930, pp.xxiii–xxix. C.C. Gillispie: *Dictionary of Scientific Biography*, vol.13, 1976, pp. 311–4.

PRAIN, Sir David (1857–1944)

b. Fettercairn, Kincardineshire 11 July 1857 *d.* Whyteleafe, Surrey 16 March 1944

MA Aberdeen 1878. MB Edinburgh 1882. FRS 1905. FLS 1888. VMH 1912. VMM 1933. CMG 1912. Indian Medical Service, 1884–7. Curator of Herbarium, Royal Botanic Garden, Calcutta, 1887; Director, 1898–1905. Director, Kew, 1905–22.

References: *D.N.B. Proceedings of L.S.*, 1943–4, pp. 223–9. *Obituary Notices of Fellows of R.S.*, vol.4, 1944, pp. 747–70.

HILL, Sir Arthur William (1875–1941)

b. Watford 11 Oct. 1875 *d.* Kew 3 Nov. 1941

BSc Cantab. 1897. DSc 1919. FRS 1920. FLS 1908. KCMG 1931. VMH 1934. VMM 1936. Demonstrator and then lecturer in botany, Cambridge,

1899–1907. Assistant Director, Kew, 1907–22; Director 1922–41.

References: *D.N.B. Nature*, 22 Nov. 1941, pp. 619–23. *Journal of Kew Guild*, 1942, pp. 129–39. *Obituary notices of Fellows of R.S.*, vol.4, 1942, pp. 87–100.

EVANS, Sir Geoffrey (1883–1963)

b. Walmersley, Lancs. 26 June 1883
d. Mayfield, Sussex 16 Aug. 1963

MA Cantab. Indian Agricultural Service, 1902–23. Director of Cotton Culture, Queensland, 1923–26. Principal, Imperial College of Tropical Agriculture, Trinidad, 1926–38. Economic botanist, Kew, 1938–53; Acting Director, 1941–3.

References: *Kew Bulletin*, 1954, pp. 32–3. *Nature*, vol.200, 1963, pp. 214–5.

SALISBURY, Sir Edward James (1886–1978)

b. Harpenden, Herts. 16 April 1886
d. Felpham, Sussex 10 Nov. 1978

DSc London 1913. FRS 1933. FLS 1909. VMH 1952. VMM 1935. Senior lecturer in Botany, East London College, 1914. Senior lecturer, University College London, 1919; Reader in Plant Pathology, 1924; Professor of Botany, 1929–43. Director, Kew, 1943–56. Prolific author.

References: *D.N.B. Nature*, vol.178, 1956, pp. 454–5; vol.279, 1979, pp. 357–8. *Journal of Kew Guild*, 1978, pp. 714–6. *Biographical Memoirs of R.S.*, vol.26, 1980, pp. 503–41.

TAYLOR, Sir George (1904–1993)

b. Edinburgh 15 Feb. 1904 *d.* Dunbar, Lothian 12 Nov. 1993

BSc Edinburgh 1926. Knighted 1962. FRS 1968. VMH 1956. Assistant Keeper of Botany, British Museum (Natural History) *c.* 1928; Deputy Keeper, 1945–50; Keeper, 1950–6. Director, Kew, 1956–71.

References: *Independent*, 16 Nov. 1993. *Garden*, Feb. 1994, pp. 74–5. *Taxon*, Aug. 1994, pp. 510–2.

HESLOP-HARRISON, John (1920–)

b. Middlesbrough 10 Feb. 1920

Durham University. FRS 1970. Lecturer in Agricultural Botany, Durham. Lecturer in Botany, Queen's University Belfast, 1946–50. Reader in Botany, University College London, 1950. Professor of Botany, Queen's University Belfast, 1954–60. Professor of Botany, Birmingham, 1960–7. Professor of Botany, Institute of Plant Development, Wisconsin University, 1967. Director, Kew, 1971–6.

BRENAN, John Patrick Mickelthwait (1917–1985)

b. Chislehurst, Kent 19 June 1917
d. 26 Sept. 1985

MA Oxon 1940. FLS 1952. VMH 1978. Imperial Forestry Institute, Oxford. Kew Herbarium, 1948; head of African section, 1959. Keeper of Herbarium and Deputy Director, 1965; Director, 1976–81.

References: *Journal of Kew Guild*, 1972, pp. 108–9; 1981, pp. 19–23. *The Times*, 28 Sept. 1985. *Kew Bulletin*, 1987, pp. 287–96.

BELL, Ernest Arthur (1926–)

b. 20 June 1926

King's College, Newcastle. PhD Trinity College Dublin. Biochemist, King's College London. Professor of Botany, University of Texas, 1968. Professor of Biology, London University and head of Department of Plant Sciences, King's College, 1972–81. Director, Kew, 1981–8.

PRANCE, Sir Ghillean Tolmie (1937–)

b. Suffolk 13 July 1937

D.Phil. Oxon. 1963. FRS 1993. Knighted 1995. New York Botanical Garden, 1963; B.A. Krukoff Curator of Amazonian Botany, 1968–75; Director of research, 1975–81; Vice-President, 1977–81; Senior Vice-President, 1981–8. Director, Kew, 1988–

References: C. Langmead: *A Passion for Plants: the Biography of Professor G.T. Prance*, 1995.

BOTANISTS

BAKER, John Gilbert (1834–1920)

b. Guisborough, Yorks. 13 Jan. 1834
d. Kew 16 Aug. 1920

FRS 1878. FLS 1866. VMH 1897.
Curator and Secretary of Thirsk Botanical
Exchange Club. First Assistant in Kew
Herbarium, 1866. Lectured to Kew's
student gardeners from 1874. Keeper of
Herbarium and Library, 1890–9.
Authority on ferns. Contributed text to
A. Parsons's illustrations in *Genus Rosa*,
1910–4. *Curtis's Botanical Magazine*
dedicated to him in 1880.

References: *Kew Bulletin*, 1899, pp. 17–8;
1920, pp. 319–20. *Proceedings of R.S. part
B.*, 1921, pp.xxiv-xxx. C.C. Gillispie:
Dictionary of Scientific Biography, vol.1,
1970, pp. 412–3.

BENTHAM, George (1800–1884)

b. Stoke, Devon 22 Sept. 1800 *d.* London
10 Sept. 1884

LLD Cantab. 1874. FRS 1862. FLS
1826. Secretary, Horticultural Society of
London, 1829–40. Bentham was never on
the Kew staff but from 1854, when he
presented his herbarium and library, he
worked at Kew almost daily for the next 30
years. He compiled two of Kew's colonial
floras (Hong Kong and Australia) and with
J.D. Hooker the *Genera Plantarum*,
1862–83, a scheme of classifying plants
still in use at Kew. His legacy formed the
basis of the Bentham-Moxon Trust.

References: *D.N.B. Proceedings of L.S.*,
1884–5, pp. 90–104. *Annals of Botany*
vol.12, 1898, pp.ix-xxx. B.D. Jackson:
George Bentham, 1906. C.C. Gillispie:
Dictionary of Scientific Biography, vol.1,
1970, pp. 614–5.

BOODLE, Leonard Alfred
(1865–1941)

b. St John's Wood 5 May 1865
d. 22 Aug. 1941

ARCS. FLS 1888. Demonstrator at Royal
College of Science for 7 years under
D.H. Scott whose assistant he became when
Scott was Hon. Keeper of Jodrell
Laboratory at Kew, 1892. Assistant Keeper
of Jodrell Laboratory, 1909–30. Published
papers on vascular system of Pteridophytes.

An exceptionally modest and self-effacing
man.

References: *Kew Bulletin*, 1930, p. 328;
1941, pp. 236–9. *Journal of Kew Guild*,
1941, pp. 82–3.

BOR, Norman Loftus (1893–1972)

b. Ireland 4 May 1893 *d.* London 22
Dec. 1972

MA Dublin 1927. DSc Edinburgh. FLS
1931. Indian Forest Service, 1921–42.
Assistant Director, Kew, 1948–59.
Authority on taxonomy of grasses.

References: *Kew Bulletin*, 1975, pp. 1–10.

CHIPP, Thomas Ford (1886–1931)

b. Gloucester 1 Jan. 1886 *d.* Kew 28 June
1931

BSc London 1909. FLS 1912. Gardener,
Syon House, 1904–6; Kew, 1906.
Assistant, Kew Herbarium, 1906–8.
Assistant Conservator of Forests, Gold
Coast, 1910–4. Assistant Director,
Singapore Botanic Garden, 1919–20.
Deputy Conservator of Forests, Gold
Coast, 1921–2. Assistant Director, Kew,
1922–31.

References: *Journal of Kew Guild*, 1931,
pp. 81–2. *Kew Bulletin*, 1931, pp. 397–8,
pp. 433–40.

COOKE, Mordecai Cubitt
(1825–1914)

b. Horning, Norfolk 12 July 1825
d. Southsea, Hants 12 Nov. 1914

ALS 1877. VMH 1902. India Museum
at India Office, 1862. In charge of Lower
Cryptogams in Kew Herbarium,
1880–92. Prolific author.

References: *D.N.B. Missing Persons*, 1993,
pp. 150–1. M.P. English: *Mordecai Cubitt
Cooke*, 1987.

COTTON, Arthur Disbrowe
(1879–1962)

b. London 15 Jan. 1879. *d.* Hertford 27
Dec. 1962

FLS 1902. VMM 1935. VMH 1943.
Assistant, Kew Herbarium, 1904–22;
Keeper, 1922–46. Mycologist. Wrote
good account of Cambridge Cottage in
Journal of Kew Guild, 1942, supplement,
pp. 1–22.

References: *Journal of Kew Guild*, 1941,
pp. 4–6, 1963, pp. 335–6.

GILMOUR, John Scott Lennox
(1906–1986)

b. London 28 Sept. 1906 *d.* Cambridge 3
June 1986

MA Cantab. FLS 1932. VMH 1957.
VMM 1966. Assistant Director, Kew,
1931–46. Director, RHS, Wisley,
1946–51. Director, Cambridge Botanic
Garden, 1951–73.

References: *Journal of Kew Guild*, 1981,
pp. 582–3. *Garden*, 1987, pp. 452–5.

HEMSLEY, William Botting
(1843–1924)

b. East Hoathly, Sussex 29 Dec. 1843
d. Broadstairs, Kent 7 Oct. 1924

FRS 1889. FLS 1896. VMH 1909. Kew
gardener, 1860. Assistant, Kew
Herbarium, 1865–7; 1883; Keeper,
1899–1908. An autodidact who was fluent
in several languages and a taxonomist and
bibliographer. Frequent contributor to
Gardeners' Chronicle. Compiled catalogue
of paintings in Marianne North Gallery in
1882. *Curtis's Botanical Magazine*
dedicated to him in 1898.

References: *Kew Bulletin*, 1909, pp. 22–3;
1924, pp. 389–92. *Journal of Kew Guild*,
1925, pp. 331–7; 1943, pp. 292–4.
Proceedings of R.S. part B., 1925, pp.i-ix.

HILL, Sir John (1714–1775)

b. Peterborough, Northants. 1714
d. London 21 Nov. 1775

MD St Andrews 1750. Knight of Order
of Vasa 1774. Apothecary, botanist,
journalist. Prolific author. His precise
connections with Kew are obscure but it
would appear that under the patronage of
Lord Bute he obtained plants for the Royal
Garden and provided botanical expertise.
His *Hortus Kewensis* 1768 is the first
published record of the plants at Kew.
*Twenty-five New Plants, rais'd in the Royal
Garden at Kew*, 1773.

References: *D.N.B. Short Account of the
Life, Writings and Character of the late Sir
John Hill*, 1779. F.W. Oliver, editor:
Makers of British Botany, 1913,
pp. 84–107. L.L. Woodruff: 'The
versatile Sir John Hill' (*American*

Naturalist, vol.60, 1926, pp. 417–42).
G.S. Rousseau: 'The much-maligned
doctor: Sir John Hill' (*Journal of American
Medical Association*, vol.212, 1970,
pp. 103–8). G.S. Rousseau: 'John Hill,
universal genius manqué: remarks on his
life and times with a checklist of his works'.
In J.A.L. Lemay and G.S. Rousseau: *The
Renaissance Man in the Eighteenth Century*,
1978, pp. 49–129. G.S. Rousseau: *Letters
and Papers of Sir John Hill*, 1982.
C.E. Dolman Essay review: 'That
impudent fellow Hill' (*Annals of Science*,
vol.40, 1983, pp. 281–8). K.J. Fraser:
'John Hill and Royal Society in eighteenth
century' (*Notes and Records of R.S.*, vol.48,
1994, pp. 43–67.)

HILLIER, John Masters
(*c.* 1861–1930)

b. Teddington, Middx *c.* 1861 *d.* Osterley,
Middx 5 Oct. 1930

Assistant, Kew Museums, 1879; Keeper,
1901–26.

References: *Kew Bulletin*, 1926, pp. 220–1;
1930, p. 495.

HOLTTUM, Richard Eric
(1895–1990)

b. Linton, Cambs. 20 July 1895 *d.* London
18 Sept. 1990

FLS 1927. VMH 1972. Assistant
Director, Botanic Garden, 1922–5;
Director, 1925–49. Professor of Botany,
University of Malaya, 1949–54. Authority
on ferns and orchids. Holttum spent more
than 30 years of his retirement using the
Kew Herbarium in his researches. He is a
good example of those visiting botanists who
make intensive use of Kew's resources.

References: *Kew Bulletin*, 1986,
pp. 485–9.

HOWES, Frank Norman
(1901–1973)

b. Natal 2 Aug. 1901 *d.* London 26
Feb. 1973

BA Durban 1922. DSc 1935. FLS 1925.
On staff on Botanical Survey of S. Africa,
1922–4. Economic botanist, Department
of Agriculture, Gold Coast, 1925–6.
Assistant, Kew Museums, 1926; Keeper,
1948–66.

References: *Journal of Kew Guild*, 1974,
p. 340.

HUBBARD, Charles Edward
(1900–1980)

b. Appleton, Norfolk 23 May 1900
d. 8 May 1980

FLS 1935. VMM 1970. Gardener,
Sandringham, 1916–20; Kew, 1920.
Assistant in Herbarium, 1922; Keeper,
1957. Deputy Director, 1959–65.
Authority on grasses.

References: *Journal of Kew Guild*, 1956,
pp. 343–4; 1980, pp. 940–1. *Kew
Bulletin*, 1981, pp. 437–40.

HUTCHINSON, John
(1884–1972)

b. Wark, Northumberland 7 April 1884
d. Kew 2 Sept. 1972

FRS 1947. FLS 1918. VMM 1944.
VMM 1945. Hutchinson, W.B. Hemsley,
C.E. Hubbard and S.A. Skan are
outstanding examples of Kew gardeners who
subsequently established themselves as
botanists. Hutchinson transferred from the
gardens to the Herbarium in 1905. Keeper
of Kew Museums, 1936–48. Studied plant
classification and evolution.

References: *D.N.B. Kew Bulletin*, 1974,
pp. 1–14. *Biographical Memoirs of Fellows
of R.S.*, 1975, pp. 344–57.

JACKSON, John Reader
(1837–1920)

b. Chelsea 26 May 1837 *d.* Lympstone,
Devon 28 Oct. 1920

ALS 1868. Keeper, Kew Museums,
1858–1901.

References: *Journal of Kew Guild*, 1921,
pp. 37–8. *Kew Bulletin*, 1920, 368–9.

MASSEE, George Edward
(1850–1917)

b. Scampton, Yorks. 20 Dec. 1850
d. Sevenoaks, Kent 17 Feb. 1917

FLS 1895–1915. ALS 1916. VMH
1902. Principal assistant, Kew Herbarium,
1893–1915. First President of British
Mycological Society, 1896–8. Wrote a
number of books on fungi.

References: *Kew Bulletin*, 1915,

pp. 118–20, 1917, pp. 84–5; 1922,
pp. 335–48.

MELVILLE, Ronald (1903–1985)

b. Bristol 12 March 1903 *d.* Kew 6 Aug.
1985

PhD Imperial College 1934. FLS 1938.
Assistant, Kew Museum, 1934; transferred
to Herbarium, 1950. Represented Kew on
Vegetable Drugs Committee during last
war. Postulated a "new theory of the
angiosperm flower". Compiled the first Red
Data Book of threatened plant species in
1970.

References: *Journal of Kew Guild*, 1985,
pp. 488–9. *Kew Bulletin*, 1986,
pp. 761–8.

METCALFE, Charles Russell
(1904–1991)

b. Whiteparish, Wilts 11 Sept. 1904
d. 16 June 1991

PhD Cambridge 1930. Assistant Keeper,
Jodrell Laboratory, 1930 (regraded
Keeper); retired 1969. Authority on
anatomy of monocotyledons and
dicotyledons.

References: *Taxon*, 1970, pp. 85–7. *Kew
Bulletin*, 1991, pp. 221–2.

MORRIS, Sir Daniel (1844–1933)

b. Loughor, Glam. 26 May 1844
d. Boscombe, Hants 9 Feb. 1933

MA Dublin 1876. DSc Dublin 1894.
KCMG 1893. FLS 1903. VMH 1897.
Director, Public Gardens, Jamaica,
1879–86. Assistant Director, Kew,
1886–98. Imperial Commissioner of
Agriculture, West Indies, 1898–1908.

References: *Journal of Kew Guild*, 1933,
pp. 277–8. *Kew Bulletin*, 1933,
pp. 110–1. *Proceedings of L.S.*,
1932–3, pp. 197–9.

OLIVER, Daniel (1830–1916)

b. Newcastle upon Tyne 6 Feb. 1830
d. Kew 21 Dec. 1916

LLD Aberdeen 1891. FRS 1863. FLS
1853. Assistant, Kew Herbarium, 1858;
Keeper, 1864–90. Professor of Botany,
University College London, 1861–88.
Wrote several editions of the official guides

to the Gardens and Museums. Gave lectures to gardeners from 1859. "He used to walk very solemnly from his house [55 The Green] to the Herbarium clad in a frock coat and top hat; and last thing at night, on winter evenings, he used to see his staff off the premises, carrying a little lamp in front of him." (R. Henrey: *The King of Brentford*, 1946, p. 137). *Curtis's Botanical Magazine* dedicated to him in 1867.

References: *Kew Bulletin*, 1893, pp. 188–9; 1917, pp. 31–6. *Proceedings of R.S. part B.*, 1917, pp.xi–xv.

SCOTT, Dukinfield Henry
(1854–1934)

b. London 28 Nov. 1854 *d.* Oakley, Hants. 29 Jan. 1934

BA Oxon. PhD Würzburg 1881. FRS 1894. FLS 1880. Lecturer, Royal College of Science, London, 1884–92. Honorary Keeper, Jodrell Laboratory, 1892–1906. Palaeobotanist. W.C. Worsdell, G.T. Williams and L.A. Boodle were employed by him to help in the Laboratory.

References: *D.N.B.* F.W. Oliver, editor: *Makers of British Botany*, 1913, pp. 243–60. *Journal of Kew Guild*, 1934, pp. 376–7. *Kew Bulletin*, 1934, pp. 128–33. *Obituary Notices of Fellows of R.S.*, 1934, pp. 205–27.

SAMPSON, Hugh Charles
(1878–1953)

b. Simla 2 May 1878 *d.* 29 Nov. 1953

Indian Agricultural Service, 1906–23. Kew's first economic botanist, 1927–38.

References: *Journal of Kew Guild*, 1952–3, p. 133.

SANDWITH, Noel Yuri
(1901–1965)

b. Harworth, Notts. 8 Sept. 1901 *d.* Kew 7 May 1965

MA Oxon. FLS 1936. Kew Herbarium., 1924–65. Authority on Bignoniaceae.

References: *Journal of Kew Guild*, 1966, pp. 710–1.

SHAW, Herbert Kenneth Airy
(1902–1985)

b. Woodbridge, Suffolk 7 April 1902 *d.* Kew 19 Aug. 1985

BA Cantab. Kew Herbarium, 1925–52. Authority on Euphorbiaceae.

References: *Kew Bulletin*, 1987, pp. 3–21.

SMITH, Alexander (1832–1865)

b. Kew 17 Dec. 1832 *d.* Kew 15 May 1865

Son of the first Curator, John Smith. He was appointed the first Curator of the Kew Museums, 1856–8.

References: *Journal of Kew Guild*, 1895, pp. 30–1. *Kew Bulletin*, 1914, pp. 85–7.

SPRAGUE, Thomas Archibald
(1877–1958)

b. Edinburgh 7 Oct. 1877 *d.* Cheltenham, Glos. 22 Oct. 1958

BSc Edinburgh 1898. FLS 1903. Assistant, Kew Herbarium, 1900; Deputy Keeper, 1930–45. Authority on plant nomenclature and herbals.

References: *Journal of Kew Guild*, 1958, p. 594. *Proceedings of L.S.*, 1959–60, pp. 134–5.

STAPF, Otto (1857–1933)

b. Ischl, Austria 23 March 1857 *d.* Innsbruck 3 Aug. 1933

PhD Vienna. FRS 1908. FLS 1908. VMH 1928. VMM 1931. Assistant to Professor Kerner von Marilaun, 1882–92. Assistant, Kew Herbarium, 1891; Keeper, 1909–22. Edited *Index Londinensis*, 1929–31.

References: *Journal of Kew Guild*, 1933, pp. 283–4. *Kew Bulletin*, 1933, p. 366, pp. 369–90. *Journal of R.H.S.*, 1934, pp. 127–30. *Obituary Notices of Fellows of R.S.*, 1933, pp. 115–8.

SUMMERHAYES, Victor Samuel
(1897–1974)

b. Street, Somerset 21 Feb. 1897 *d.* 27 Dec. 1974

BSc London 1920. Orchidologist in Kew Herbarium, 1924–64. Wrote *Wild Orchids of Britain*, 1951; 2nd ed., 1968.

References: *Journal of Kew Guild*, 1976,

pp. 515–6. *Orchid Review*, 1934, pp. 194–5; 1975, pp. 77–8.

TURRILL, William Bertram
(1890–1961)

b. Woodstock, Oxfordshire 14 June 1890 *d.* Kew 15 Dec. 1961

BSc London 1915. DSc 1928. FRS 1958. FLS 1925. VMH 1956. VMM 1953. Assistant, Kew Herbarium, 1909; Keeper, 1946–57. Studied flora of Balkans and phytogeography. Author of *Royal Botanic Gardens, Kew*, 1959 and *Joseph Dalton Hooker*, 1963.

References: *Kew Bulletin*, 1960, pp. 218–9. *Journal of Kew Guild*, 1948, pp. 639–40; 1962, pp. 203–4. *Biographical Memoirs of Fellows of R.S.*, 1971, pp. 689–712.

WAKEFIELD, Elsie Maud
(1886–1972)

b. Birmingham 3 July 1886 *d.* Richmond 17 June 1972

MA Oxon. FLS 1911. Assistant, Kew Herbarium; Deputy Keeper, 1945–51. Mycologist.

References: *Journal of Kew Guild*, 1944, pp. 343–4; 1972, pp. 163–4. *Transactions of British Mycological Society*, 1973, pp. 167–74.

GARDENERS

Head Gardeners and Curators:

DILLMAN, John (–c. 1760)

A German by birth, Dillman was gardener to Lady Elizabeth St André (née Capel) at Kew House when Frederick Prince of Wales leased the property. He retained Dillman, eventually making him responsible for the gardens of his other properties. Princess Augusta also employed Dillman whose responsibilities at Kew were confined to the kitchen garden, melon yard and orangery in 1753. He resigned in 1756. His salary which had been £50 in 1730 was £700 by 1752; he owned land in Kew.

GREENING, Robert (-1758)

d. March 1758

His father, Thomas Greening, had been in charge of Richmond Gardens from 1738. Robert Greening was appointed head gardener at Kew in 1753. He continued Dillman's creation of a lake and a mound – part of the late Prince of Wales's 'Plan' for the garden – and laid out the 'Great Lawn'.

HAVERFIELD, John
(*c.* 1694–1784)

d. Kew 21 Nov. 1784

Surveyor, Twickenham. Head gardener to Princess Augusta at Kew, 1758. He and his son, also named John, were in charge of Richmond Gardens as well with an annual allowance of £1,513 from which they were to provide equipment and seeds.

AITON, William (1731–1793)

b. Boghall, Lanarkshire 1731 *d.* Kew 2 Feb. 1793

Apprentice at Chelsea Physic Garden. In charge of small physic garden at Kew, 1759. Succeeded John Haverfield as head gardener, 1784. Putative author of *Hortus Kewensis*, 1789.

References: D.N.B. *Kew Bulletin*, 1891, pp. 298–9. C.C. Gillispie: *Dictionary of Scientific Biography*, vol.1, 1970, pp. 88–9.

AITON, William Townsend
(1766–1849)

b. Kew, 2 Feb. 1766 *d.* London 9 Oct. 1849

Son of W. Aiton whom he succeeded as head gardener at Kew in 1793. He also took charge of Richmond Gardens when John Haverfield junior resigned in 1795. FLS 1797. Gardens at Kensington and St James's Palaces came under his control in 1804. George IV made him Director-General of all the royal gardens. Relinquished control of Kew's small Botanic Garden in 1841, and of the Pleasure Grounds in 1845. Involved in the production of *Hortus Kewensis*, 1810–13.

References: D.N.B. *Journal of Kew Guild*, 1966, pp. 688–93. C.C. Gillispie: *Dictionary of Scientific Biography*, vol.1, 1970, pp. 89–90.

SMITH, John (1798–1888)

b. Aberdour, Fife 5 Oct. 1798 *d.* Kew 14 Feb. 1888

ALS 1837. Gardener in Scotland including Botanic Garden in Edinburgh. Kew, 1822; foreman in propagating department, 1823; Curator, 1841–64. He resented being overruled by both Hookers in gardening matters. His publications include *Cultivated Ferns*, 1857, *Ferns; British and Foreign*, 1866, *Historia Filicum* 1875, *Bible Plants* 1878 and *Records of Royal Botanic Gardens, Kew*, 1880. From 1849 he provided cultivation notes to *Curtis's Botanical Magazine*.

References: *Cottage Gardener*, vol.16, 1856, pp. 15–17. R. Desmond: 'John Smith, Kew's first Curator' (*Journal of Kew Guild*, 1965, pp. 576–87). R.E. Holttum: 'John Smith of Kew' (*British Fern Gazette*, vol.9, 1967, pp. 330–67.)

SMITH, John (1821–1888)

b. Kelso, Roxburghshire 1821 *d.* Kew 11 May 1888

Gardener to Duke of Northumberland. Curator, Kew, 1864–86.

References: *Gardeners' Chronicle*, 1873, pp. 1701–2. *Journal of Kew Guild*, 1897, pp. 32–3.

NICHOLSON, George
(1847–1908)

b. Ripon, Yorks. 7 Dec. 1847 *d.* Richmond 20 Sept. 1908

ALS 1886. FLS 1898. VMM 1894. VMH 1897. Kew, 1873; Curator, 1886–1901. Made great improvements to the Arboretum about which he wrote some 20 articles in the *Gardeners' Chronicle*, 1881–3. Compiled *Hand-list of Trees and Shrubs grown at Kew*, 1894–6. His major publication, however, was *Dictionary of Gardening*, 1885–89 which was the precursor of the R.H.S. *Dictionary of Gardening*. *Curtis's Botanical Magazine* was dedicated to him in 1903.

References: D.N.B. *Journal of Botany*, 1908, pp. 337–9. *Kew Bulletin*, 1908, pp. 422–7.

WATSON, William (1858–1925)

b. Garston, Liverpool 14 March 1858 *d.* St Albans, Herts. 30 Jan. 1925

ALS 1904. VMM 1891. VMH 1916. Kew, 1879; Assistant Curator, 1886; Curator, 1901–22. Revised part of R. Thompson's *Gardener's Assistant*, 1902, 1905. Wrote books on orchids, cacti, rhododendrons and climbing plants. *Curtis's Botanical Magazine* dedicated to him in 1904.

References: *Gardeners' Chronicle*, 1892, pp. 812–3. *Journal of Kew Guild*, 1926, pp. 422–5. *Kew Bulletin*, 1925, pp. 94–6.

BEAN, William Jackson
(1863–1947)

b. Malton, Yorks. 26 May 1863 *d.* Kew 19 April 1947

VMH 1917. VMM 1922. Kew, 1883; Arboretum foreman, 1892; Assistant Curator, 1900; Curator, 1922–9. Regular correspondent to *Garden* and *Gardeners' Chronicle*. Author of numerous books of which the two most important are *Trees and Shrubs hardy in the British Isles* 1914, 1933 and *Royal Botanic Gardens, Kew*, 1908. *Curtis's Botanical Magazine* dedicated to him in 1938.

References: *Journal of Kew Guild*, 1928, pp. 575–6; 1930, pp. 771–2; 1946–7, pp. 598–600. *Kew Bulletin*, 1929, pp. 139–40. *Quarterly Journal of Forestry*, 1947, pp. 65–6.

TAYLOR, Thomas William
(1878–1932)

b. Glos. June 1878 *d.* London 4 March 1932

Kew, 1902; Assistant Gardener, 1922; Curator, 1929–32.

References: *Journal of Kew Guild*, 1932, pp. 191–2. *Kew Bulletin*, 1932, p. 112, pp. 156–7.

COUTTS, John (1872–1952)

b. Lochnagar, Aberdeenshire 17 Jan. 1872 *d.* Woking, Surrey 21 Dec. 1952

VMH 1933. VMM 1937. Kew, 1896–1900. Gardener at Killerton, Devon, 1900–9. Foreman, Kew, 1909; Deputy Curator, 1929; Curator, 1932–7. Co-author of *Complete Book of Gardening*, 1930 and of *Lilies* (with H.D. Woodcock), 1935.

References: *Gardeners' Chronicle*, 1953, p. 10. *Journal of Kew Guild*, 1930, p. 749; 1938, pp. 757–8; 1952–3, p. 132. *Kew Bulletin*, 1937, p. 396. *Lily Yearbook*, 1953, pp. 9–10.

CAMPBELL, William Macdonald (1900–1964)

b. May 1900 *d.* 3 Oct. 1964

VMH 1957. Kew, 1922–4. Superintendent, Parks Department, Southend-on-Sea, 1932. Curator, Kew, 1937–60.

References: *Gardeners' Chronicle*, 1932, p. 214; 1937, vol.2, p. 264. *Journal of Kew Guild*, 1964, pp. 456–7.

STENNING, Lewis (1901–1965)

b. Sorn, Ayrshire 17 Nov. 1901 *d.* London 4 March 1965

VMH 1964. Kew, 1925; Assistant Curator, 1929; Curator, 1960–5.

References: *Gardeners' Chronicle*, 1939, p. 226; 1965, p. 295. *Journal of Kew Guild*, 1965, pp. 601–2.

SHAW, Richard L.

Curator of Younger Botanic Garden, Benmore. Assistant Curator, Wakehurst Place, 1965; Curator at Kew, 1966–72; returned to Edinburgh Botanic Garden as Curator.

SIMMONS, John B.E.

VMH, 1987. Student gardener, Kew, 1958; Supervisor, Tropical Department, 1961; Assistant Curator, 1964; Deputy Curator, 1968; Curator, 1972.

Some other gardeners who worked at Kew:

ALDRIDGE, John (*c.*1787)

Gardener, Kensington Palace. In charge of Royal Kitchen and Forcing Garden, Kew, 1834.

BROWN, George Ernest (1917–1980)

b. Jinja, Uganda 19 March 1917 *d.* Devon 20 July 1980

Kew, 1946. Lecturer, Swanley College, 1947. Assistant Curator, Kew Arboretum, 1956–77. Author of *Pruning of Trees, Shrubs and Conifers*, 1972 and *Shade Plants for Garden and Woodland*, 1980.

References: *Journal of Kew Guild*, 1975, p. 295; 1977, p. 581; 1980, pp. 934–5.

BRUTY, Herbert John (1910–1983)

b. Stoke by Clare, Suffolk 6 April 1910 *d.* 29 Dec. 1983

Gardener, Stoke College. At Kew in 1946 where he was in charge of ferns until his retirement in 1970.

References: *British Pteridological Society Bulletin*, vol.3, 1984, pp. 31–2.

DALLIMORE, William (1871–1959)

b. Tardebigge, Worcs. 31 March 1871 *d.* Tonbridge, Kent 7 Nov. 1959

VMM 1924. VMH 1931. Kew, 1891; foreman in Arboretum, 1901–8; Assistant in Museums, 1908; Keeper of Museums, 1926–36. Laid out National Pinetum at Bedgebury from 1925. Author of *Holly, Yew and Box*, 1908 and *Handbook of Coniferae* (with A.B. Jackson), 1923.

References: *Journal of Kew Guild*, 1926, pp. 365–6; 1959, pp. 702–3. *Empire Forestry Review*, 1960, pp. 1–2. *Quarterly Journal of Forestry*, 1960, pp. 81–2.

DEWAR, Daniel (*c.* 1860–1905)

b. Perthshire *c.* 1860 *d.* New York 7 May 1905

Kew, 1880–93. Curator, Glasgow Botanic Garden, 1893–1902. Wrote *Synonymic list . . . of Genus Primula*, 1886.

References: *Journal of Kew Guild*, 1905, pp. 266–7.

DONN, James (1758–1813)

b. Monivaird, Perthshire 1758 *d.* Cambridge 14 June 1813

ALS 1795. FLS 1812. Foreman gardener at Kew under William Aiton. Sir Joseph

Banks's influence obtained him the post of Curator at Cambridge Botanic Garden, 1794–1813. Author of *Hortus Cantabrigiensis*, 1796.

References: *D.N.B. Gentleman's Magazine*, 1813, p. 663.

GOULD, John (1804–1881)

It is sometimes stated that John Gould, the well-known ornithologist, worked at Kew but this is an error. His father worked in the gardens at Windsor Castle where John Townsend Aiton, the son of William Aiton, was head gardener. "Between the ages of fourteen and twenty he [i.e. John Gould] spent most of his time under the care of the late J.T. Aiton, at the Royal Gardens, Windsor, where he soon acquired a taste for botany and horticulture." (*Gardeners' Chronicle*, 1881, p. 217). Gould left Windsor in 1823.

GREEN, Thomas (*fl.* 1770s)

Foreman, flower garden, Kew. Invented fumigating bellows for use in glasshouses. On 2 Nov. his bellows were tested at Kew and it was reported that the instrument was "of the greatest use of anything yet tried for killing all sorts of insects, etc. either in hot-house, cherry, melon or peach houses . . ." (R. Dostie: *Memoirs of Agriculture and other Oeconomical Arts*, vol.3, 1782, p. 349).

HOULSTON, John (*fl.* 1840s–1850s)

At Birmingham Botanic Garden before coming to Kew, 1848–55 where he was foreman in orchid and fern houses. Contributed papers on cultivated ferns to *Gardener's Magazine of Botany . . .* 1851 and *Garden Companion and Florists' Guide*, 1852.

References: *Journal of Kew Guild*, 1894, pp. 40–1.

IRVING, Walter (1867–1934)

b. Wickham Market, Suffolk 3 Aug. 1867 *d.* Lightwater, Surrey 23 April 1934

Kew, 1890; Assistant Curator, 1922. Author of *Everyman's Book of the Greenhouse*, 1907 and *Rock Gardening*, 1925.

References: *Gardener's Magazine*, 1912, pp. 791–2. *Gardeners' Chronicle*, 1934, p. 337. *Journal of Kew Guild*, 1928, p. 555; 1934, p. 374.

NIVEN, James Craig (1828–1881)

b. Dublin 1828 *d*. Hull 16 Oct. 1881

Gardener, Belfast Botanic Garden, 1843. Kew, 1846 where he compiled *Catalogue of the Hardy Herbaceous Plants in the Royal Gardens at Kew*, 1853. Curator, Hull Botanic Garden, 1853.

References: *Gardeners' Chronicle*, 1881, vol.2, pp. 541–2, 589. *Journal of Kew Guild*, 1898, pp. 37–8.

OSBORN, Arthur (1878–1964)

b. Sonning, Berks 16 Dec. 1878 *d*. Kew 24 Feb. 1964

Kew, 1899; Deputy Curator, 1932. Author of *Shrubs and Trees for the Garden*, 1933.

References: *Journal of Kew Guild*, 1934, pp. 313–4.

PEARCE, Sydney Albert (1906–1972)

b. near Romsey, Hants. 16 June 1906 *d*. Rowde, Wilts. 28 March 1972

Kew, 1928; Assistant Curator, 1937; Deputy Curator, 1967–8. Author of *Flowering Shrubs*, 1953, *Climbing and Trailing Plants*, 1957, *Ornamental Trees for Garden and Roadside Planting*, 1961, and *Flowering Trees and Shrubs*, 1965.

References: *Journal of Kew Guild*, 1964, pp. 386–7. *Arboricultural Association Journal*, 1972, pp. 48–51.

PURDOM, William (1880–1921)

b. Heversham, Westmorland 10 April 1880 *d*. Peking 7 Nov. 1921

At Kew, 1902–8, when he was spokesman in the student gardeners' campaign for better pay and working conditions. Plant collector for Messrs Veitch and Arnold Arboretum in China, 1909–12. Travelled in Kansu region with Reginald Farrer. Forestry adviser to Chinese Government, 1915–21.

References: *Journal of Kew Guild*, 1922, pp. 115–6. *Garden*, 1980, pp. 397–9. *Gardeners' Chronicle*, 1980, pp. 37–8.

WILLIAMSON, Alexander (1819–1870)

b. Dunkeld, Perthshire 1819 *d*. Kingston, Surrey 1870

Kew, 1843; foreman of herbaceous department; Curator of Pleasure Grounds, 1848–66. When he retired in 1866 the post of Curator of Pleasure Grounds was abolished and his duties transferred to the Curator of the Botanic Garden.

WILSON, Ernest Henry (1876–1930)

b. Chipping Campden, Glos. 15 Feb. 1876 *d*. Worcester, Mass., USA 15 Oct. 1930.

VMH 1912. VMM 1906. A student gardener at Kew in 1897–9, he was recommended by the Director for the post of plant collector in China for Messrs Veitch in 1899. He introduced many new plants to British and American gardens.

References: D.J. Foley: *Flowering World of 'Chinese' Wilson*, 1969.

Kew gardeners overseas:

A letter from Sir Joseph Hooker (10 October 1870) informed his friend, George Maw, that he had "just returned from two days in Gloucestershire to a pile of letters, including a request from the Colonial Office to find a good botanic gardener for Jamaica, and from the India Office for 6 gardeners for cotton experiments – this is sharp and responsible work." Kew's Annual Reports frequently note the selection of gardeners for overseas posts, usually in the colonies: "Six more active and intelligent young gardeners have been sent to the cotton, etc. plantations in India, in which country there are upwards of 30 former Kew employees engaged in various departments of horticulture and arboriculture" (1870 report). "Under instructions from the Secretary of State for the Colonies a skilled propagator has been sent from Kew to superintend the extension of forest plantations in the island of Mauritius. Others have been selected for tea and coffee plantations in India" (1872 report). "At the present time there are about 50 men in parts of India, in government and private employ, engaged in cultivating tea, cinchona and cotton" (*Journal of Kew Guild*, 1893). In 1910 it was estimated that about 160 Kew gardeners were serving in Asia, Africa, America and Australia, the majority as curators and superintendents. These former student gardeners usually kept in touch with Kew, seeking advice and not infrequently sending plants and seeds.

PLANT COLLECTORS

The early collectors who worked for Kew were recruited from the staff or were known to Sir Joseph Banks who also occasionally employed them himself – e.g. George Caley. The despatch of collectors ceased shortly after the appointment of Sir William Hooker, who obtained his plants and seeds mainly through exchange and purchase.

BARCLAY, George

b. Huntly, Aberdeenshire *d*. Buenos Aires, Argentina

Kew gardener. Collected on the survey of Pacific coast of S. America by HMS *Sulphur*, 1836–41. Kew only got some seeds, the dried plants were appropriated by Robert Brown at British Museum.

References: *Gardeners' Chronicle*, 1882, pp. 305–6. *Aliso*, vol.5, 1964, pp. 469–74. *Phytologia*, vol.43, 1979, pp. 281–6. Correspondence at Kew ('Kew collectors' vol.4); MS of 'Journal of Voyage round the World' at Natural History Museum, London.

BOWIE, James (c. 1789–1869)

b. London *c*. 1789 *d*. Claremont, Cape Town 2 July 1869

Kew gardener, 1810. Collected plants with Allan Cunningham in Brazil, 1814–16. In Cape 1816–23; returned there in a private capacity in 1827. *Clivia nobilis* was one of his introductions.

References: *D.N.B. London Journal of Botany*, 1842, 232–8. M. Gunn and L.E. Codd: *Botanical Exploration of S. Africa*, 1981, pp. 101–2. Correspondence at Kew ('Kew collectors' vol. 2); Cape journal at Natural History Museum, London.

BURKE, Joseph (1812–1873)

b. Bristol 12 June 1812 *d*. Harrisonville, USA 23 Jan. 1873

Gardener and collector of live animals for Lord Derby. Sir William Hooker shared the expenses with Lord Derby of his plant collecting trip to Canada and USA, 1843 7 but his sponsors were disappointed with his collections.

References: *London Journal of Botany*, 1843, pp. 163–5; 1845, pp. 643–53; 1846, pp. 14–22, pp. 430–5. S.D. McKelvey: *Botanical Exploration of Trans-Mississippi West 1790–1850*, 1955, pp. 792–817. J. and N. Ewan: *Biographical Dictionary of Rocky Mountain Naturalists*, 1981, p. 32. Correspondence and journal at Kew ('Kew collectors', vol. 2).

CALEY, George (1770–1829)

b. Craven, Yorks. 10 June 1770 *d*. London 23 May 1829

Sir Joseph Banks engaged him to collect in Australia, 1800–10. Allan Cunningham described him as "a most accurate, intelligent and diligent botanist". One of his introductions was the Stag's Horn Fern, *Platycerium bifurcatum*.

References: J.E.B. Currey: *Reflections on Colony of New South Wales by G. Caley*, 1967. A.E.J. Andrews: *Devil's Wilderness: George Caley's Journal to Mount Banks, 1804*, 1984.

CUNNINGHAM, Allan (1791–1839)

b. Wimbledon, Surrey 13 July 1791 *d*. Sydney 27 June 1839

Collected plants in Brazil with James Bowie, 1814–6. In Australia, 1816; visited New Zealand, 1826, Norfolk Island, 1830. Returned to Kew in 1831. Succeeded his brother, Richard, as Superintendent of Botanic Garden, Sydney, 1837. One of the early explorers of Australia.

References: *D.N.B.* W.G. McMinn: *Allan Cunningham, Botanist and Explorer*, 1970. Correspondence and journal at Kew ('Kew collectors', vol. 1).

GOOD, Peter (–1803)

d. Sydney 11 June 1803

Assisted Christopher Smith in escorting plants from Kew to Calcutta Botanic Garden in 1795; returned with large consignment of plants from India and East Indies for Kew. Foreman at Kew until he was appointed an assistant to Robert Brown on Flinders's exploration of Australian coast, 1801–3, on HMS *Investigator*. Sent seeds to Kew.

References: *Journal of Kew Guild*, 1946–7, pp. 561–3. Journal of Peter Good (*Bulletin British Museum (Natural History)*), *Historical Series*, vol. 9, 1981.

HOOKER, Sir Joseph Dalton (1817–1911)

See also entry under 'Directors'. Collected plants in Sikkim, Nepal and Bengal, 1848–51. Introduced many new species of rhododendrons and alpines such as *Primula sikkimensis* and *Meconopsis napaulensis*. Collected with John Ball in Morocco, 1871, and with Asa Gray in the Rocky Mountains and the California Sierra Nevada, 1877. Wrote *Himalayan Journals*, 1854 and *Journal of Tour in Marocco and Great Atlas* (with J. Ball), 1878.

HOVE, Anton Pantaleon

Polish born gardener at Kew. Collector on HMS *Nautilus* off coast of West Africa, 1785. Collected in India, 1787–8.

References: A.M. Coats: *Quest for Plants*, 1969, p. 57, pp. 145–8.

KERR, William (–1814)

d. Ceylon 1814

Foreman gardener, Kew. Collected Chinese plants, based on Canton and Macao, 1803–12. His introductions included *Kerria japonica*, *Nandina domestica*, *Pieris japonica*. Superintendent of first botanic garden in Ceylon, 1812.

References: K. Lemmon: *Golden Age of Plant Hunters*, pp. 1968, 109–19.

LOCKHART, David (–1846)

d. Trinidad 1846

Kew gardener. Assistant to Christen Smith on Captain Tuckey's Congo expedition,

1816. Superintendent of Botanic Garden, Trinidad, 1818–46, whence he sent plants, especially orchids, to Kew.

References: *D.N.B. Kew Bulletin*, 1891, pp. 310–1.

LUNT, William (1871–1904)

b. Ashton-under-Lyme, Lancs. 16 Dec. 1871 *d*. St Kitts, West Indies 3 Jan. 1904

Kew gardener. Collected plants for Kew on Theodore Bent's expedition to Hadramaut Valley, S. Arabia, 1893–4. Assistant Superintendent, Trinidad Botanic Garden, 1894. Curator, Botanic Station, St Kitts-Nevis, 1898.

References: *Journal of Kew Guild*, 1904, pp. 208–9. *Kew Bulletin*, 1893, p. 366; 1894, pp. 328–43.

MANN, Gustav (1836–1916)

b. Hanover 20 Jan. 1836 *d*. Munich 22 June 1916

Kew gardener. "one of our well educated Kew gardeners as botanist to join the present Niger expedition under Dr Baikie, as successor to the late Mr Barter" (Kew Report 1859). During 1859–62, Mann collected some 3,000 species and varieties for the Herbarium. Indian Forest Service, 1863–91. *Curtis's Botanical Magazine* dedicated to him in 1896.

References: *Kew Bulletin*, 1916, p. 237. *Journal of Kew Guild*, 1917, pp. 373–4.

MASSON, Francis (1741–1805)

b. Aberdeen Aug. 1741 *d*. Montreal 23 Dec. 1805

FLS 1796. Kew gardener and the Royal Gardens' first plant collector. Cape of Good Hope, 1772–74. Canaries and Azores, 1776–8. West Indies, 1779–81. Spain, Portugal and Tangier, 1783–5. Cape, 1786–95. New York and Montreal, 1798 Wrote and illustrated *Stapeliae Novae*, 1796–7.

References: *D.N.B.* M.C. Karsten: 'Francis Masson, a gardener-botanist who collected at the Cape' (*Journal of South African Botany*, 1958, pp. 203–18; 1959, pp. 167–88; 1960, pp. 9–15; 1961, pp. 15–45).

MENZIES, Archibald (1754–1842)

b. Aberfeldy, Perthshire March 1754
d. London 15 Feb. 1842

FLS 1790. Assistant surgeon, Royal Navy, 1782. Surgeon and naturalist on HMS *Discovery* under Captain Vancouver, 1791–4. Collected for Banks. Introduced Monkey Puzzle tree, *Araucaria araucana*, a specimen of which Banks presented to Kew.

References: *D.N.B. Archives of Natural History*, vol.14, 1987, pp. 3–43.

MILNE, William Grant (–1866)

d. Creek Town, Old Calabar 3 May 1866

Gardener, Edinburgh Botanic Garden. Assistant naturalist on HMS *Herald*'s voyage to Fiji, 1852–6. Sir William Hooker paid him to collect plants for Kew but the results were disappointing.

References: *Transactions of Botanical Society of Edinburgh*, vol.8, pp. 71–3, 485–6.

NELSON, David (–1789)

d. Koepang, Timor 20 July 1789

Recommended as a collector to Sir Joseph Banks by the nurseryman, James Lee. Joined Captain Cook's third voyage, 1776–80; brought back over 200 packets of seeds for Banks. Remained at Kew as a gardener until 1787 when he sailed on HMS *Bounty* to collect and care for breadfruit trees. He had also collected plants for Kew at the time of the celebrated mutiny.

References: K. Lemmon: *Golden Age of Plant Hunters*, 1968, pp. 79–106.

OLDHAM, Richard (1837–1864)

d. Amoy 13 Nov. 1864

Kew gardener. Kew's last official collector left for Japan to join a survey ship in 1861. Sent over 13,000 herbarium specimens and introduced, among other plants, *Elaeagnus multiflora*, *Rhodotypos kerriodes* and *Styrax japonica*.

References: *Journal of L.S.*, 1865, pp. 163–70.

PURDIE, William (*c*. 1817–1857)

d. Trinidad 14 Oct. 1857

Gardener at Edinburgh Botanic Garden. Hired by Sir William Hooker and Duke of Northumberland to collect in Jamaica and Colombia, 1843–6. Made many despatches to Kew. Superintendent, Botanic Garden, Port of Spain, Trinidad, 1846–57.

References: A.M. Coats: *Quest for Plants*, 1969, p. 348. Correspondence at Kew ('Kew collectors', vol.5).

SEEMANN, Berthold Carl (1825–1871)

b. Hanover 25 Feb. 1825 *d*. Nicaragua 10 Oct. 1871

FLS 1852. Trained to be a plant collector at Kew, 1844–6. Naturalist on HMS *Herald* on voyage in Pacific, 1847–51. Sent plants to Kew. *Narrative of Voyage of HMS Herald*, 1853; *Botany of Voyage of HMS Herald*, 1852–7.

References: *D.N.B. Journal of Botany*, 1872, pp. 1–7. *Journal of Kew Guild*, 1895, pp. 31–2. Correspondence at Kew ('Kew collectors' vol.7).

WILFORD, Charles (–1893)

d. Wimbledon, Surrey 1893

Assistant, Kew Herbarium, *c*.1854–7. Chosen by Sir William Hooker to collect plants in Japan and China, 1857. Collected in Hong Kong.

References: E.H.M. Cox: *Plant Hunting in China*, 1945, pp. 94–5. Correspondence at Kew ('Kew collectors', vol.9).

FLOWER PAINTERS

ATKINSON, Gerald (1893–1971)

b. Hull 26 Aug. 1893 *d*. Cornwall 1 July 1971

Studied at Hull College of Art. ALS 1955. Botanical artist and photographer, Kew, 1922–59. Restored panorama in Sherman Hoyt Cactus House, 1951. Contributed plates to *Curtis's Botanical Magazine*, 1923–49.

References: *Kew Bulletin*, 1922, p. 349. *Journal of Kew Guild*, 1972, pp. 59–60. Hunt Library: *Artists from Royal Botanic Gardens, Kew*, 1974, pp. 8–9.

BAUER, Franz (Francis) Andreas (1758–1840)

b. Feldsberg, Austria 4 Oct. 1758 *d*. Kew 11 Dec. 1840

He and his brother, Ferdinand, were briefly botanical artists at the Imperial Gardens in Vienna. From 1790 Sir Joseph Banks engaged him to paint plants at Kew, especially new introductions. FRS 1821. FLS 1804. An outstanding artist in the golden age of flower painting. Examples of his published work can be seen in *Delineations of Exotick Plants cultivated in Royal Gardens at Kew*, 1796–1803, *Strelitzia depicta*, 1818 and J. Lindley's *Illustrations of Orchidaceous Plants*, 1830–38. His drawings of Kew's plants are in Natural History Museum, London (see Appendix 11).

References: W. Blunt and W.T. Stearn: *Art of Botanical Illustration*, 1994, pp. 223–5. W.T. Stearn: 'Franz and Ferdinand Bauer: masters of botanical illustration' (*Endeavour*, 1960, pp. 27–35).

BOND, George (*c*. 1806–1892)

d. Lydbury, Shropshire 6 Sept. 1892

Kew gardener. During 1826–35 he drew about 1,700 plants at Kew for W.T. Aiton who wanted them as a record for a new edition he was preparing of *Hortus Kewensis*. Gardener to Earl of Powis at Walcot, Shropshire, 1835–80.

References: *Garden*, 1880, p. 75; 1892, p. 709. *Gardeners' Chronicle*, 1892, vol.2, p. 381. *Kew Bulletin*, 1916, p. 164. MS list of his drawings at Kew.

DUNCANSON, Thomas

Gardener, Edinburgh Botanic Garden; Kew, 1822–6, where he drew about 300 plants for W.T. Aiton's revision of *Hortus Kewensis*.

References: *Garden*, 1880, p. 75. *Journal of Botany*, 1912, supplement 3, pp. 14–15. *Kew Bulletin*, 1916, p. 165. MS list of his drawings at Kew.

FITCH, Walter Hood (1817–1892)

b. Glasgow 28 Feb. 1817 *d.* Kew 14 Jan. 1892

FLS 1857. Employed by Sir William Hooker in 1830s primarily to illustrate *Curtis's Botanical Magazine* which he did for over 40 years. Became a freelance artist in 1860. Produced over 10,000 published drawings including those in H.J. Elwes's *Monograph of Genus Lilium*, 1877–80, J.D. Hooker's *Rhododendrons of Sikkim-Himalaya*, 1849–51 and *Illustrations of Himalayan Plants*, 1855.

References: Kew Bulletin, 1915, pp. 277–84, 392. *Kew Magazine*, 1987, pp. 3–14, pp. 51–6. R. Desmond: *Celebration of flowers*, 1987, *passim*. J. Lewis: *Walter Hood Fitch*, 1992. W. Blunt and W.T. Stearn: *Art of Botanical Illustration*, 1994, pp. 261–8.

HOOKER, Sir William and Sir Joseph

Both father and son were talented artists. Sir William contributed over 600 plates to *Curtis's Botanical Magazine*. When Sir Joseph was in northern India he made about 1,000 drawings of the landscape and vegetation which are now at Kew.

MEEN, Margaret

Came from Bungay in Suffolk to London to teach flower and insect painting. Exhibited at Royal Academy, 1775–85. Founded and illustrated a periodical which lasted only two issues: *Exotic Plants from the Royal Gardens at Kew*, 1790.

References: Kew Bulletin, 1893, p. 147; 1925, appendix 2, p. 65; 1933, pp. 2–3. *Gardeners' Chronicle*, 1894, pp. 197–8.

NORTH, Marianne (1830–1890)

b. Hastings, Sussex 24 Oct. 1830 *d.* Alderley, Glos. 30 Aug. 1890

Pupil of Valentine Bartholomew. Travelled around the world painting plants in their native habitats. Financed the building of a gallery at Kew to house her paintings. *Recollections of a Happy Life*, 1892; *Some further Recollections of a Happy Life*, 1893.

References: D.N.B. Anon. *Visions of Eden: Life and Work of M. North*, 1980. L. Ponsonby: *Marianne North at Kew Gardens*, 1990.

SMITH, Matilda (1854–1926)

b. Bombay 30 July 1854 *d.* Kew 29 Dec. 1926

ALS 1921. A distant cousin of Sir Joseph Hooker who recruited her as an artist for *Curtis's Botanical Magazine*. She contributed over 2,300 plates to the magazine between 1878 and 1923. From April 1898 she was appointed a part-time artist in the Herbarium – the first official botanical artist on the staff; her predecessors had received their salaries from other sources.

References: Journal of Kew Guild, 1927, pp. 527–8. *Kew Bulletin*, 1921, pp. 317–8; 1927, pp. 135–9. R. Desmond: *Celebration of Flowers*, 1987, *passim*.

TAYLOR, Simon (1742–*c.* 1796)

Attended William Shipley's Drawing School. When he was about 17 Lord Bute engaged him to paint flowers. He also received commissions from John Fothergill and John Ellis. "I suppose you know Ehret is dead! We have nobody to supply his place in point of elegance. We have a young man, one Taylor, who draws all the rare plants of Kew Garden for Lord Bute; he does it tolerably well, I shall employ him very soon." (Ellis to Linnaeus 28 Dec. 1770). A number of his drawings are in the Kew Library.

References: D.N.B. Kew Bulletin, 1971, pp. 167–9.

Many of the drawings commissioned by *Curtis's Botanical Magazine* were, and still are, of plants grown at Kew. They include more than 700 plates by Lilian Snelling; amongst contemporary artists are Ann Farrer, Mark Fothergill, Victoria Goaman, Mary Grierson, Christabel King, Joanna Langhorne, Valerie Price, Rodella Purves, Stella Ross-Craig, Pandora Sellars, Margaret Stones and Ann Webster.

ARCHITECTS AND GARDEN DESIGNERS

BRIDGEMAN, Charles (–1738)

d. London 19 July 1738

Trained as a designer in Brompton Park Nursery, Kensington under G. London and

H. Wise. Succeeded Wise as Royal Gardener to George II and Queen Caroline, 1728–38. He was engaged in landscaping Richmond Gardens from about mid-1720s.

References: P. Willis: 'Charles Bridgeman; the royal gardens' (*Furor Hortensis* 1974, pp. 41–7). P. Willis: *Charles Bridgeman and the English Landscape Garden*, 1978.

BROWN, Lancelot ('Capability') (1716–1783)

b. Kirkdale, Northumberland 1716 *d.* London 6 Feb. 1783

At Stowe, Bucks in early 1740s, eventually becoming head gardener. In 1750s set up his own practice as landscape designer. Master gardener at Hampton Court, 1764. Engaged by George III to relandscape Richmond Gardens, thereby destroying the work of Charles Bridgeman and William Kent.

References: D. Stroud: *Capability Brown*, 2nd edition, 1975. D. Jacques: *Georgian Gardens*, 1985, *passim*. R. Turner: *Capability Brown and 18th Century Landscape*, 1985.

BURTON, Decimus (1800–1881)

b. 30 Sept. 1800 *d.* 14 Dec. 1881

Superintending architect for the great conservatories at Chatsworth, 1836–41, and Royal Botanical Society, Regent's Park, 1845–6. With the engineer Richard Turner, he designed the Palm House at Kew. Also designed the Temperate House, the Museum, the Main Gates and restored Temple of Eolus. Supervised the work of the landscape designer, W.A. Nesfield.

References: *D.N.B.* P. Miller: *Decimus Burton*, 1981. E.J. Diestelkamp: 'Design and building of the Palm House, Royal Botanic Gardens, Kew' (*Journal of Garden History*, vol. 2, part 3, 1982, pp. 233–72). E.J. Diestelkamp: 'Conservatories of Decimus Burton' (*Country Life*, 19 May 1983, pp. 1342–6). G. Williams: *Augustus Pugin versus Decimus Burton*, 1990.

CHAMBERS, Sir William (1723–1796)

b. Gothenburg, Sweden 23 Feb. 1723 *d.* London 8 March 1796

FRS 1776. Educated in England. In the service of Swedish East India Company he visited Bengal, 1740–2, and Canton, 1743–5, 1748–9, where he studied and sketched Chinese architecture. In private practice an architect in London, 1755. In 1757 became architectural tutor to the Prince of Wales (future George III) and was engaged as architect by Princess Augusta. Architect to Office of Works, 1761; Comptroller of Works, 1769; Surveyor-General and Comptroller, 1782. His *Plans, Elevations, Sections and Perspective Views of Gardens and Buildings at Kew* 1763 describe and illustrate the buildings he designed there.

References: J.W. Chase: 'William Mason and Sir William Chambers' Dissertation on Oriental Gardening' (*Journal of English and German Philology*, vol.35, 1936, pp. 517–29). N. Pevsner: 'The other Chambers' (*Architectural Review*, vol.101, 1947, pp. 195–8). R.C. Bald: 'Sir William Chambers and the Chinese garden' (*Journal of History of Ideas*, vol.11, 1950, pp. 287–320). J. Harris: *Sir William Chambers, Knight of the Polar Star*, 1970.

FERGUSSON, James (1808–1886)

b. Ayr 20 Jan 1808 *d.* 9 Jan. 1886

Architectural historian. Owned indigo factory in India where he studied indigenous architecture. He tested his theories on the lighting of Greek and Roman temples in his design for the Marianne North Gallery, 1882.

References: *D.N.B.*

KENT, William (*c.* 1685–1748)

b. Bridlington, Yorks. *c.* 1685 *d.* London 12 April 1748

Architect, furniture designer, painter and garden designer. His patron, Lord Burlington, advanced his career as interior decorator and architect. The garden buildings he designed for Richmond Gardens included the Hermitage and Merlin's Cave.

References: M. Jourdain: *Work of William Kent*, 1748. M. Wilson: *William Kent*, 1984. J.D. Hunt: 'A silent and solitary Hermitage' (*York Georgian Society Annual*

Report, 1970, pp. 47–60). J.D. Hunt: *William Kent, Landscape Garden Designer*, 1987.

KIRBY, John Joshua (1716–1774)

b. Wickham Market, Suffolk 1716 *d.* Kew 20 June 1774

Coach and house painter at Ipswich where he met Thomas Gainsborough. Through the influence of Lord Bute he was appointed teacher of perspective to the Prince of Wales who, as George III, made him and his son joint Clerks of Works at Richmond Lodge and Kew Palace. Kirby drew three of the views of Kew Garden which were engraved in W. Chambers's 1763 book. He monitored the work of 'Capability' Brown in Richmond Gardens (Works 4/13 P.R.O.) He lived in Ferryhouse near the riverside steps to the Brentford Ferry. Thomas Gainsborough expressed a wish to be buried next to him in St Anne's churchyard, Kew.

NESFIELD, William Andrews (1793–1881)

b. Chester-le-Street, Durham 19 Feb. 1793 *d.* London 2 March 1881

Watercolour artist and garden designer. Employed at Kew, 1844–8, to landscape the grounds: the parterre around the Palm House, the Pond and vistas in the new Arboretum.

References: B. Elliot: 'Master of the Geometric Art' (*Garden*, 1981, pp. 488–91). S. Evans: 'Master Designer' (*Antique Collector*, Oct. 1992, pp. 52–5).

NESFIELD, William Eden (1835–1888)

Son of W.A. Nesfield. Trained as architect under A. Salvin. Set up his own office in 1859; in partnership with Norman Shaw, 1862 – both involved in Queen Anne style of architecture. Designed Temperate House Lodge, 1867.

References: C. Aslet: 'Country Houses of W.E. Nesfield' (*Country Life*, 16 March 1978, pp. 678–81; 23 March 1978, pp. 766–9).

TURNER, Richard (*c.* 1798–1881)

b. Dublin *c.* 1798 *d.* Dublin 31 Oct. 1881

Iron manufacturer and engineer, Hammersmith Ironworks, Dublin, 1833. Specialised in wrought iron conservatories. Collaborated with Decimus Burton in the building of the conservatory of the Royal Botanical Society, Regents Park, 1845–6, and Palm House at Kew, 1844–8.

References: *D.N.B. Missing Persons*, 1993, pp. 684–5. E.J. Diestelkamp: 'Richard Turner and his glasshouses' (*Glasra* no. 5, 1981, pp. 51–3). E.J. Diestelkamp: 'Design and building of the Palm House, Royal Botanic Gardens, Kew' (*Journal of Garden History*, vol.2, part 3, 1982, pp. 233–72). E.C. Nelson: 'Richard Turner: an introductory portrait' (*Moorea*, vol.9, 1990, pp. 2–5).

WYATT, James (1746–1813)

b. Burton Constable, Staffs. 3 Aug. 1746 *d.* near Marlborough, Wilts. 4 Sept. 1813

Surveyor-General and Comptroller of Works, 1796. Designed Castellated Palace at Kew, 1801–11.

References: *D.N.B.* A. Dale: *James Wyatt*, 1956. H. Colvin: *Biographical Dictionary of British Architects*, 1978, pp. 940–52.

WYATVILLE, Sir Jeffry (1766–1840)

b. Burton-on-Trent, Staffs. 3 Aug. 1766 *d.* London 10 Feb. 1840

Architect. Nephew of James Wyatt. Undertook transformation of Windsor Castle in 1824. In 1834 designed a new palm house for Kew and a wing to Kew Palace (both unexecuted). Adapted conservatory transferred from Buckingham Palace to Kew, 1836. Designed King William's Temple, 1837. Enlarged Kew church, 1837–8.

References: *D.N.B.* D. Linstrum: *Sir Jeffry Wyatville*, 1972. H. Colvin: *Biographical Dictionary of British Architects*, 1978, pp. 959–64.

PRINTED AND ARCHIVAL SOURCES
FOR THE HISTORY OF KEW

The sources of Kew's history, both printed and archival, are copious and rich. A series of four engraved maps of Richmond Gardens by John Rocque between 1734 and 1754 is the first printed record we have. Edmund Curll's *Rarities of Richmond: being exact Descriptions of the Royal Hermitage and Merlin's Cave* (1736) provides the first of numerous accounts of William Kent's extraordinary buildings. After he had completed his commission for Princess Augusta, William Chambers wasted no time in publishing his *Plans, Elevations, Sections, and Perspective Views of the Gardens and Buildings at Kew in Surry* (1763). Undoubtedly an act of self-promotion, but it has left us with one of the best illustrated records of any eighteenth century English garden. In addition to Chambers's architectural drawings, it includes a series of views by William Marlow, Thomas Sandby and Joshua Kirby, all superbly engraved. They were frequently copied without acknowledgement, one of the earliest being the anonymous *Description of the Gardens and Buildings at Kew in Surry* about the mid-1760s. The most unscrupulous plagiarist was George-Louis le Rouge whose *Jardins Anglo-Chinois à la Mode* (1776–*c.* 1787) is a compendium of engravings taken unacknowledged from other works, including pictures of Kew.

Contemporary periodicals such as the *Craftsman, London Magazine* and *Royal Magazine* reported developments in the Royal Gardens, and topographical guides to London and its neighbouring counties described its buildings and other features. R. and J. Dodsley's *London and its Environs Described* vol.3 (1761), John Gwynn's *London and Westminster Improved* (1766), and *Les Delices des Chateaux royaux: or, a Pocket Companion to the Royal Palaces of Windsor, Kensington, Kew and Hampton Court* (1785) are examples of this genre. A word of warning which I have repeated elsewhere: some guides, especially those of the nineteenth century, describe garden buildings at Kew as still standing when in fact they had been demolished. Obviously their compilers had slavishly copied earlier guides without checking whether any changes had occurred. General accounts of the Thames usually include Kew and Richmond Gardens, the best being J. and J. Boydell's *History of the River Thames* (1796).

Poets celebrated royal patronage of these gardens with ingratiating verses: George Ritso in *Kew Gardens – a poem* (1763) and Henry Jones in *Kew Gardens: a poem in two cantos* (1767), for example. Some quotable lines can usually be found in this literary trivia.

Earnest tourists, assiduously visiting country houses and their gardens, usually included Richmond and Kew in their itinerary. John Macky saw Ormonde Lodge before the Prince and Princess of Wales acquired it. Daniel Defoe dismissed Richmond Gardens with a cursory nod of approval, but this was before Bridgeman and Kent got to work there. As the reputation of the two Royal Gardens grew, visitors gave them more attention in their journals, diaries and letters. Count Frederick Kielmansegge's *Diary of a Journey to England in the Years 1761–1762* (1902), Arthur Young's *Farmer's Tour through the East of England* (1771) and Faujas de Saint Fond's *Journey through England and Scotland to the Hebrides in 1784* (1907) offer perceptive impressions of the Royal Gardens.

A pioneering attempt at a history of Kew Gardens was written by Frederick Scheer, a City merchant and a resident on Kew Green, where he had a garden with one of the best collections of cacti in the country. His slender octavo, *Kew and its Gardens* (1840), hastily written to alert the public to its threatened dismemberment, is a vigorous defence of a place he clearly loved. Sir William Hooker, in the midst of all his changes and improvements, found time to write a brief history of Kew in his *Companion to the Botanical Magazine* in 1845. Edwin Simpson, evidently pleased with a lecture he gave on the history of Kew in 1849, duly published it. He made the interesting observation that the House of Confucius, which had been scheduled for demolition in 1843/4, was "now in a meadow near Richmond Bridge". But since he also referred to the "remains of the far famed Merlin's Cave" which had, in fact, been pulled down by 'Capability' Brown, one has some doubts about his reliability. Edward Brayley contradicts Simpson by asserting in his *Topographical History of Surrey* (1850) that the Chinese summerhouse was still at Kew. P.H. Gosse's *Wanderings through the Conservatories at Kew* (1856) is, as its title proclaims, an account of its greenhouses, and its wood engravings are the only known illustrations of some of these structures.

Sir William Thiselton-Dyer, who had wanted to stage a permanent exhibition on Kew Gardens in Kew Palace, brought together a collection of comments and extracts on his institution's personalities and buildings prior to 1841 in an article in the *Kew Bulletin* in 1891. Sir William encouraged his Assistant Curator, W.J. Bean, to write the first full-scale history of the Gardens. *The Royal Botanic Gardens, Kew: Historical and Descriptive* (1908) is still a useful work, especially as a commentary on the landscaping and living collections at the time when Bean was a senior member of staff. W.B. Turrill's *The Royal Botanic Gardens, Kew: Past and Present*, published in 1959 to coincide with the Gardens' bicentenary, describes in detail the plants in cultivation in his day. Wilfrid Blunt's *In for a Penny. A Prospect of Kew Gardens: their Flora, Fauna and Falballas* (1978) is an anecdotal account, urbanely written, a pleasure to read but rather frivolous in its treatment. *Royal Kew* (1985), by Ronald King, a former Secretary of the Gardens, is the first to make use of archives but only those housed at Kew. Lucile H. Brockway who turned her PhD thesis at the City University in New York into a book in 1979 started a trend by assessing Kew's involvement in imperial affairs. *Science and Colonial Expansion: the Role of the British Royal Botanic Gardens* examines the links between science and politics, the exploitation of poorer nations by the major colonial powers, but fails to substantiate her polemic with archival evidence.

Amongst relevant biographies, Olive Hedley's *Queen Charlotte* (1975) and Harold Carter's *Sir Joseph Banks, 1743–1820* (1988) are pre-eminent. Mea Allan's *The Hookers of Kew, 1785–1911* (1967) never discloses the sources of her facts and annotations and is rather tendentious. Much more scholarly is Leonard Huxley's *Life and Letters of Sir Joseph Dalton Hooker* (1918). Mrs Papendiek, Assistant Keeper of the Wardrobe, wrote her *Court and Private Life in the Time of Queen Charlotte* (1887) when she was very old and her memory faulty; nevertheless it gives intimate glimpses of the Royal Family in the seclusion of Kew.

Biographies and accounts of plant collectors tell us of the flowers they introduced into cultivation. Plants introduced to Kew are listed in the two editions of the *Hortus Kewensis* (1789 and 1810–13). John Smith's *Records of the Royal Botanic Gardens, Kew* (1880) concentrates on plants grown at Kew between the 1820s and 1860s. When Sir William Hooker took over the editorship of *Curtis's Botanical Magazine* in 1827 he increased the number of Kew plants covered in its pages; after he became Director in 1841 he featured many more of his institution's plants, an editorial policy observed to the present day.

Periodicals are manifestly a rich repository of contemporary comment: the *Gardener's Magazine* (1826–44), *Gardeners' Chronicle* from 1841, and the *Cottage Gardener* (later changed to *Journal of Horticulture*) (1848–1915) kept a close eye on Kew, sometimes highly critical of its management. The *Journal of the Kew Guild*, an annual publication since 1893, chronicles the careers of its members. Kew's annual reports are a primary source of information but it must be remembered that there is a certain amount of 'window dressing' in their presentation. They cover the institution's activities from 1841, initially rather sketchily, but more expansively in the second half of the nineteenth century. Unfortunately they ceased in 1882 but were resurrected for some years in the *Kew Bulletin*, which essentially is a periodical devoted to taxonomic research.

These periodicals and much other printed material can be consulted in the Kew Library, which also has a large collection of letters and reports. When W.T. Aiton retired he took with him all the correspondence and papers accumulated during his administration from 1793 to the 1840s. These were destroyed on his death, the only items to escape being a series of ledgers recording the receipt and despatch of plants starting in 1793. This catastrophe was reported by John Smith, the Gardens' first Curator. Smith spent his own retirement dictating – he was almost blind – his recollections of Kew, where he had served for more than 40 years. "No one knows more about the matter of the Gardens for the last 50 years than I do," he told the Botany Department of the Natural History Museum in 1872. Smith's disappointment when Sir William Hooker was chosen over him as director when the Royal Gardens were transferred to state control, and his resentment towards both Hookers, whom he believed never valued his professional advice, distort his interpretation of events. Prejudiced though his manuscript account is, as well as sometimes marred by inaccuracy, it is nevertheless valuable for being an insider's record of what was happening at Kew after the death of Sir Joseph Banks in 1820. Another unpublished memoir is William Dallimore's reminiscences of Kew from 1891 to 1936, largely anecdotal and particularly informative about Sir William Thiselton-Dyer's administration, and the creation of the Bedgebury Pinetum.

The core of Kew's archives is the correspondence of Sir William and Sir Joseph Hooker. Anxious that Kew should be the permanent repository of all his collections, Sir William extolled their worth in a memorandum to his superiors only months before his death. His correspondence with more than 4,000 government officials, botanists, businessmen and travellers over nearly 60 years amount to some 30,000 letters:

> I know of no person who has conducted a foreign correspondence upon scientific subjects so extensively as myself . . . I further believe that it [i.e. the collection] contains more material for the history of botanical science & its progress during the present century, than any other business whatever.

The most important collections outside Kew are the Royal Archives at Windsor; the household accounts of Frederick and Augusta at the Duchy of Cornwall; Sir Joseph Banks's correspondence at the Natural History Museum, London; the Map Library and Department of Manuscripts at the British Library; the Public Record Office (the papers of the Lord Steward's Department, the Office of Woods and Forests, the Office of Public Works, the Ministry of Agriculture, and the Ministry of Works); and the Local Studies Library, Richmond.

BIBLIOGRAPHY

Unless otherwise stated, the place of publication of books is London.

Anon. 1740s. *A Description of the Royal Gardens at Richmond in Surry, the Village, and Places Adjacent.*

Anon. 1760s. *A Description of the Gardens and Buildings at Kew in Surry: with the Engravings belonging thereto in Perspective. To which is added a short account of the principal seats and gardens in and about Richmond and Kew.*

Anon. 1785. *Les Delices des Chateaux royaux: or, a Pocket Companion to the Royal Palaces of Windsor, Kensington, Kew and Hampton Court.*

Aiton, W. 1789. *Hortus Kewensis.* 3 vols.

Aiton, W.T. 1810–13. *Hortus Kewensis.*

Allan, M. 1967. The Hookers of Kew, 1785–1911.

Baker, J.G. 1886. *On Kew Gardens and some of the Botanical Statistics of the British Possessions.*

Bartholomew, J. 1988. *The Magic of Kew.*

Bailey, M. and others. 1994. *Arcadian Thames: the River Landscape from Hampton to Kew.* Barn Elms.

Bean, W.J. 1908. *The Royal Botanic Gardens, Kew: Historical and Descriptive.*

Bickham, G. and Norbury, P. c. 1771. *Description of the Gardens and Buildings at Kew.* Richmond.

Blomfield, D. 1994. *Kew Past.*

Blunt, W. 1978. *In for a Penny. A Prospect of Kew Gardens: their Flora, Fauna and Falballas.*

Boulger, G.S. 1915. 'The History of Kew Gardens: the Connection of Kew with the History of Botany'. *Proceedings and Transactions South East Union of Scientific Societies,* vol.20.

Bowden, J.K. 1989. *John Lightfoot, his Work and Travels; with a Biographical Introduction and a Catalogue of the Lightfoot Herbarium.* Pittsburgh.

Brett-James, N.G. 1926. *Life of Peter Collinson.*

Brockway, L.H. 1979. *Science and Colonial Expansion: the Role of the British Royal Botanic Gardens.*

Brockway, L.H. 1979. 'Science and Colonial Expansion: the role of the British Botanic Gardens'. *American Ethnologist,* vol.6, part 3, pp. 449–65.

Chambers, W. 1763. *Plans, Elevations, Sections and Perspective Views of the Gardens and Buildings at Kew in Surry.*

Charlton, J. 1983. *Kew Palace.*

Cloake, J. 1991. *Richmond Past: a Visual History of Richmond, Kew, Petersham and Ham.*

Colton, J. 1974. 'Kent's Hermitage for Queen Caroline at Richmond'. *Architectura,* no. 2, pp. 181–91.

Colton, J. 1976. 'Merlin's Cave and Queen Caroline: garden art as political propaganda'. *Eighteenth Century Studies,* vol.10, Fall, pp. 1–20.

Colvin, H. 1968. *Royal Buildings.*

Colvin, H., editor. 1973–6. *The History of the King's Works. vol.5: 1660–1782.* 1976; *vol.6: 1782–1851.* 1973.

Conner, P. 1979. *Oriental Architecture in the West*.

Conner, P. 1979. 'China and the landscape garden: reports, engravings and misconceptions'. *Art History*, vol.2, part 4, pp. 429–40.

Cooper, D. 1948. 'Richard Wilson's views of Kew'. *Burlington Magazine*, vol.90, pp. 346.

Cotton, A.D. 1943. 'The Cambridge Cottage garden'. *Journal of Kew Guild*, supplement, 23p.; 1943, pp. 291–2.

[Curll, E.] 1736. The Rarities of Richmond: being exact Descriptions of the Royal Hermitage and Merlin's Cave.

Dallimore, W. 'A Gardener's Reminiscences'. MS in Kew archives; extracts published in *Journal of Kew Guild*, 1933–58.

Daniels, G.S. 1974. *Artists from the Royal Botanic Gardens, Kew*.

Davis, R.M. 1926. *Stephen Duck, the Thresher-poet*. Orono, Maine.

Desmond, R. 1969. 'Kew Herbarium and Library'. *Journal of Kew Guild*, pp. 1071–87.

Desmond, R. 1972. 'Who designed the Palm House at Kew?' *Kew Bulletin*, pp. 295–303.

Desmond, R. 1987. *A Celebration of Flowers*.

Desmond, R. 1992. *The European Discovery of the Indian Flora*. Oxford.

Desmond, R. 1994. *Dictionary of British and Irish Botanists and Horticulturists*. Basingstoke.

Desmond, R. and Hepper, F.N. 1993. *A Century of Kew Plantsmen: a Celebration of the Kew Guild*. Kew.

Diestelkamp, E.J. 1982. 'The design and building of the Palm House, Royal Botanic Gardens, Kew'. *Journal of Garden History*, vol.2, part 3, pp. 233–72.

Elliott, B. 1986. *Victorian Gardens*.

Erdberg, E. van. 1936. *Chinese Influence on European Garden Structures*.

Evans, Sir G. 1942. 'Economic botany at Kew in wartime'. *Journal of Kew Guild*, pp. 143–7.

Ewan, J. 1979. 'John Cree, collector for Kew in Carolina'. *Taxon*, vol.28, part 1, pp. 41–3.

Fisher, R. and Johnston, H. 1979. *Captain James Cook and his Times*.

Gilbert, L. 1986. *The Royal Botanic Gardens, Sydney: a History, 1816–1985*. Melbourne.

Goldney, S. 1892. *Kew: our Village and its Associations*. Richmond.

Goode, P. and Lancaster, M. 1986. *Oxford Companion to Gardens*. Oxford.

Goodison, I.C. 1936. 'History and associations of the Royal Palace at Kew'. *Journal of London Society*, no. 216, pp. 28–32.

Gosse, P.H. [1856]. *Wanderings through the Conservatories at Kew*.

Grosley, P.J. 1772. *A Tour of London; or, New Observations on England and its Inhabitants*. 2 vols.

Gunn, S. 1991. 'Banking for the future [Kew's Seed Bank]'. *Kew (Friends of Kew Magazine)*, Spring, pp. 16–21.

Guthrie, J.L. and others. 1988. 'Royal Botanic Gardens, Kew: restoration of the Temperate House'. *Proceedings of Institution of Civil Engineers*, vol.84, part 1, pp. 109–43.

Halliwell, B. 1981. 'A new Alpine House at Kew'. *Quarterly Bulletin of Alpine Garden Society*, vol.49, pp. 23–6.

Halliwell, B. 1983. 'A history of the Kew Rock Garden'. *Quarterly Bulletin of Alpine Garden Society*, vol.51, pp. 27–35.

Halsband, R. 1973. *Lord Hervey*.

Harris, J. 1959. 'Fate of the royal buildings'. *Country Life*, pp. 1182–4.

Harris, J. 1959. 'Two lost palaces'. *Country Life*, pp. 916–8.

Harris, J. 1963. 'Exoticism at Kew'. *Apollo*, vol.78, pp. 103–8.

Harris, J. 1970. *Sir William Chambers, Knight of the Polar star*.

Harvey, J.H. 1981. 'A Scottish botanist in London'. *Garden History*, vol.9, pp. 40–75.

Haverfield, T.T. 1862. 'Notes on Kew and Kew Gardens'. *Leisure Hour*, no. 570, 29 November, pp. 767–8.

Hedley, O. 1975. *Queen Charlotte*.

Henrey, B. 1975. *British Botanical and Horticultural Literature before 1800*. 3 vols. Oxford.

Hepper, F.N., editor. 1982. *Royal Botanic Gardens, Kew: Gardens for Science and Pleasure*.

Hepper, F.N., editor. 1989. *Plant Hunting for Kew*.

Hill, A.W. 1923. 'The work of the Royal Botanic Gardens, Kew'. *Journal of Royal Society of Arts*, 28 December, pp. 87–102.

Honour, H. 1959. 'Pagodas for the park'. *Country Life*, pp. 192–4.

Honour, H. 1961. *Chinoiserie: the Vision of Cathay*.

House of Commons. 1840. *Copy of the Report made to the Committee appointed by . . . the Treasury in January 1838 to enquire into the management, etc. of the Royal Botanic Gardens, by Dr Lindley . . . who . . . made an actual survey of the botanical garden at Kew . . .*

House of Commons. 1859. *A copy of all communications made by the officers and architect of the British Museum to the Trustees respecting the want of space for exhibiting the collections in that institution, as well as respecting the enlargement of its buildings . . .*

House of Commons. 1872. *Copies of papers relating to changes introduced into the administration of the Office of Works affecting the direction and management of the gardens at Kew. And of the correspondence between the Treasury and Dr Hooker on the same subject.*

House of Commons. 1901. *Botanical Work Committee return . . . reports . . . minutes of evidence, appendices and index.*

House of Lords. 1991. *Select Committee on Science and Technology (Sub-committee II: Systematic Biology Research).*

Howard, R.A. and Powell, D.A. 1965. 'Indian Botanic Garden, Calcutta and the gardens of the West Indies'. *Bulletin of Botanical Survey of India*, vol.7, pp. 1–7.

Howes, F.N. 1941. 'Kew's part in colonial agricultural development'. *Crown Colonist*, June, pp. 253–5.

Howie, W. 1982. 'Building study: restoration of the Temperate House, Royal Botanic Gardens, Kew, Surrey'. *Architects' Journal*, 16 June, pp. 47–64.

Jackson, B.D. 1924. 'History of the compilation of the 'Index Kewensis''. *Journal of Royal Horticultural Society*, pp. 224–9.

Jacques, D. 1983. *Georgian Gardens: the Reign of Nature*.

Jacques, D. 1990. 'On the supposed Chineseness of the English landscape'. *Garden History*, vol.18, pp. 180–91.

Johnson, M.S. 1948. 'Kew Green'. *Journal of London Society*, no. 298, pp. 45–8.

Johnstone, J. 1938. *History of the Lodge of Harmony, no. 255, 1785–1937*.

Kielmansegge, Count F. 1902. *Diary of a Journey to England in the Years 1761–1762*.

King, R.W. 1970. 'The Queen's Garden at Kew'. *Journal of Royal Horticultural Society*, pp. 335–48.

King, R.W. 1976. *The World of Kew*.

King, R.W. 1985. *Royal Kew*.

Lewis, G. 1989. *Postcards from Kew*.

Mabberley, D.J. 1985. *Jupiter Botanicus: Robert Brown of the British Museum*. Braunschweig.

Mabey, R. 1988. *The Flowering of Kew: 200 Years of Flower Paintings from the Royal Botanic Gardens*.

Mackay, D. 1985. *In the Wake of Cook: Exploration, Science & Empire, 1780–1801*.

Mackenzie, B. 1987–8. 'The Cree nursery at Addlestone'. *Surrey History*, vol.3, part 4, pp. 165–75.

Macleod, R.M. 1974. 'The Ayrton incident: a commentary on the relations of science and government in England, 1870–1873'. *In* A. Thackray and E. Mendelsohn: *Science and Values*, pp. 45–78.

Macnaghten, A. 1970. 'Stephen Duck: the thresher poet'. *Country Life Annual*, pp. 82–3.

Meikle, R.D. 1971. 'History of the Index Kewensis'. *Biological Journal of Linnean Society*, vol.3, pp. 295–9.

Metcalfe, C.R. 1976. 'History of the Jodrell Laboratory as a centre for systematic anatomy'. *In* P. Baas and others, editors: *Wood Structure in Biological and Technological Research*, pp. 1–19.

Meynell, G. 1980. 'The Royal Botanic Society's garden, Regent's Park'. *London Journal*, vol.6, part 2, pp. 35–46.

Meynell, G. 1982. 'Kew and the Royal Gardens Committee of 1838'. *Archives of Natural History*, vol.10, pp. 469–77.

Minter, S. 1990. *The Greatest Glasshouse: the Rainforests Recreated.*

Minter, S. 1990. "To show the rainforests of the world': replanting the Palm House at Kew, 1989–90'. *Kew Magazine*, vol.7, part 1, pp. 22–31.

Moncrieff, A.R.H. 1908. *Kew Gardens.*

Morris, D. 1898. 'Report on the economic resources of the West Indies'. *Kew Bulletin. Additional series* no. 1.

Pain, S. 1990. 'Sir Joseph Banks's buried treasure [Sir Joseph Banks Building]'. *New Scientist*, 17 March 1990, pp. 57–61.

Pain, S. 1991. 'Saving orchids with Sainsbury's'. *Kew (Friends of Kew Magazine)*, Winter, pp. 23–5.

Papendiek, C. 1887. *Court and Private Life in the Time of Queen Charlotte, being the Journals of Mrs Papendiek, Assistant Keeper of the Wardrobe and Reader to Her Majesty.* 2 vols.

Papendiek, G. E. *c.* 1820. *Kew Gardens: a Series of Twenty-Four Drawings on Stone.*

Port, M.H. 1984. 'A contrast in styles at the Office of Works. Layard and Ayrton: aesthete and economist'. *Historical Journal*, vol.27, part 1, pp. 151–76.

Powell, D. 1977. 'The voyage of the plant nursery, HMS *Providence*, 1791–1793'. *Economic Botany*, vol.37, pp. 387–431.

Preston, G.H. 1961. 'A decade at Kew'. *Quarterly Bulletin of Alpine Garden Society*, vol.29, part 2, pp. 176–82.

Rauschenberg, R.A. 1968. 'Daniel Carl Solander, naturalist on the "Endeavour".' *Transactions of American Philosophical Society*, vol.58, part 8, pp. 1–66.

Report of the Commission appointed to inquire into the constitution and government of the British Museum with minutes of evidence. 1850.

Riddell, J. and Stearn, W.T. 1994. *By Underground to Kew: 90 years of Poster Art advertising the Delights of Kew Gardens.*

Rorschach, K. 1983. *The Early Georgian Landscape Garden.* New Haven.

Rorschach, K. 1991. 'Frederick Prince of Wales: taste, politics and power' *Apollo*, vol. 34, pp. 239–45.

Rowan, A. 1968. *Garden Buildings.*

Royal Commission on scientific instruction and the advancement of science. 1874. 2 vols.

Royal Commission on national museums and galleries, 1927–1929. 1928–9.

Royal Botanic Gardens, Kew. 1987. *The Princess of Wales Conservatory.* Kew.

Royal Botanic Gardens, Kew. 1991. *The Marine Display.* Kew.

Rupke, N.A. 1994. *Richard Owen: Victorian Naturalist.*

Russell, T.A. 1962. 'Palms at Kew'. *Principes*, vol.6, part 2, pp. 53–63.

Ratton, W.L. 1905. 'The royal residences of Kew'. *Home Counties Magazine*, vol.7, pp. 1–13, 85–98, 157–70.

Salisbury, E.J. 1946. 'The work of Kew in the interests of the overseas empire'. *Journal of Royal Society of Arts*, 26 April, pp. 336–40.

Scheer, F. 1840. *Kew and its Gardens.*

Scott, D. 1984. 'Kew Gardens and the public, 1841–1914'. *Richmond History*, no. 5, pp. 30–45.

Scott-Elliot, A.H. 1956. 'The statues by Francavilla in the Royal collection'. *Burlington Magazine*, March, pp. 77–84.

Simmons, J.B.E. 1981. 'The new Alpine House at Kew'. *Garden*, pp. 149–52.

Simmons, J.B.E. 1981. 'The history and development of the Arboretum at the Royal Botanic Gardens, Kew'. *Arboricultural Journal*, vol.5, pp. 173–88.

Simmons, J.B.E. 1982. 'The Temperate House at Kew'. *Garden*, pp. 173–80.

Simmons, J.B.E. 1986. 'Conservation and the living collections: Royal Botanic Gardens, Kew'. *Kew Magazine* No. 3, part 1, pp. 39–48.

Simmons, J.B.E. 1987. 'The new conservatory at Kew [Princess of Wales Conservatory]'. *Garden*, pp. 257–66.

Simmons, J.B.E. 1989. 'Grand restoration [of the Palm House]'. *Garden*, pp. 162–9.

Simmons, J.B.E. and Clamp, H. 1986. 'A living showcase: [Sir Joseph Banks Building]'. *Landscape Design*, December, pp. 54–7.

Simpson, E. 1849. *The History of Kew in the County of Surrey; being a Lecture delivered in January 1849 at the Queen's School, Kew.*

Siren, O. 1950. *China and Gardens of Europe of the Eighteenth Century.* New York.

Smith, J. 1880. *Records of the Royal Botanic Gardens, Kew.*

Stacey, M. 1986. 'Glass conservatory at Kew'. *Architects' Journal*, 28 May, pp. 55–60.

Stafleu, F.A. 1966. 'The Index Kewensis'. *Taxon*, vol.15, pp. 270–4.

Stearn, W.T. 1981. *The Natural History Museum at South Kensington: a History of the British Museum (Natural History), 1753–1980.*

Stearn, W.T. 1990. *Flower Artists of Kew: Botanical Paintings of Contemporary Artists.*

Steward, J., editor. 1992. *Orchids at Kew.*

Stone, M. and P. 1984. 'The new alpine display house at the Royal Botanic Gardens, Kew'. *Rock Garden*, vol.19, no. 74, pp. 54–63.

Strong, R. 1992. *Royal Gardens.*

Symes, M., Hodges, A. and Harvey, J. 1986. 'The plantings at Whitton'. *Garden History*, vol.14, part 2, pp. 138–72.

Thiselton-Dyer, Sir. W. 1880. 'The botanical enterprise of the Empire.' Paper read at the Colonial Institute, May 11.

Thiselton-Dyer, Sir W. 1891. 'Historical account of Kew to 1841'. *Kew Bulletin*, pp. 279–327.

Turrill, W.B. 1959. *The Royal Botanic Gardens, Kew: Past and Present.*

Turrill, W.B. 1959. 'Royal Botanic Gardens, Kew'. *Journal of Royal Horticultural Society*, pp. 256–64.

Uggla, A.H. 1943–4. 'Jonas Dryander (1748–1810)'. *Proceedings of Linnean Society*, pp. 99–102.

Victoria History of the Counties of England. A History of Surrey. 1911. vol.3.

Watkins, J. 1985. 'Reference collection and exhibition buildings, Kew [Sir Joseph Banks Building]'. *Construction*, Autumn, pp. 72–5.

Webb, M.I. 1954. *Michael Rysbrack.*

Willis, G.M. and Howes, F.N. 1950. 'Notes on early Kew and the King of Hanover'. *Kew Bulletin*, pp. 299–318.

Worboys, M. 1990. 'The Imperial Institute: the state and the development of the natural resources of the colonial empire, 1887–1923'. *In* J.M. Mackenzie, editor: *Imperialism and the Natural World*, pp. 164–86.

NOTES & SOURCES

ABBREVIATIONS

B.L.   British Library
D.C.O.   Duchy of Cornwall Office
D.T.C.   Dawson Turner copies of the letters of Sir Joseph Banks in the Botany Library of the Natural History
          Museum, London
I.O.L.R. India Office Library and Records
L.S.   Linnean Society, London
P.R.O.   Public Record Office
R.A.   Royal Archives
R.S.   Royal Society

CHAPTER 1

1 Site of the house is indicated on 25 inch Ordnance Survey map, 1894–96; i.e. about 400 yards south of the Queen's Cottage.

2 J. Macky: *Tour through England* vol.1, 1724, p.66. The house and southernmost reach of the Terrace with the Duke's summer house are shown in an oil painting by Jan Griffier *c*.1710 (J. Harris: *The Artist and the Country House*, 1979, fig. 131).

3 G. Sherburn: *Correspondence of Alexander Pope*, vol.2, 1956, p.14.

4 *Ibid.* p.372.

5 C. de Saussure: *A Foreign view of England in the Reigns of George I & George II*, 1902, p.143.

6 Revd John Laurence: *A New System of Agriculture, being a Complete Body of Husbandry and Gardening*. London, 1726.

7 1st Earl of Egmont Diary. Historical MSS Commission, vol.2, 1923, p.138.

8 MR 696. P.R.O.

9 Rocque's first published plan of any English garden was that of Richmond Lodge in 1734 when he described himself as "*dessinateur de jardins*". Subsequent versions appeared in 1736, 1748 and 1754 showing relatively minor additions to the estate; the most obvious differences are the inset illustrations of buildings and garden features.

10 1st Earl of Egmont Diary. Historical MSS Commission, vol.1, 1920, p.101.

11 *Gentleman's Magazine*, August 1733, p.436. This still survives but rather overgrown in the south west corner of the Queen's Cottage grounds.

12 Introduced, according to Sir John Clerk of Penicuik near Edinburgh, for rearing pheasants and partridges. GD/18/2110 15 and 19 May 1733. Scottish Record Office.

13 i.e. in the neighbourhood of the present Azalea Garden and the Beech Clump.

14 William Kent designed a similar building at Stowe a few years later, even incorporating the same solitary turret.

15 No. 480, 13 September 1735, p.107.

16 Now in the Royal Collection, they are on display at Kensington Palace.

17 Letter of 17 March 1879 inserted in Kew's copy of E. Curll's *Rarities of Richmond*, 1736.

18 *Gentleman's Magazine*, 1735, pp.331, 498, 532–5, 715–6.

19 Works 6/16, f.90v. P.R.O.

20 J. Colton (1976), pp.12–20.

21 No. 480, 13 September 1735.

22 R. Sedgwick, *editor: Lord Hervey's Memoirs*, 1963, p.109.

23 Works 4/13. Minutes and Proceedings, 10 October 1766. P.R.O.

24 *Archaeologia*, vol.7, 1785, p.127.

25 Treasury Minute Book, vol.28, pp.93–4. P.R.O.

26 Works 6/8, ff.57–61. P.R.O.

CHAPTER 2

1 First Earl of Egmont Diary, vol.2, 1923, p.355.

2 From various fleeting references, it is fairly certain that Frederick was in residence in 1730 after Lady Elizabeth had moved out.

3 *Archaeologia* ed. 2, vol.12, 1809, p.185.

4 William IV had a sundial, originally one of a pair at Hampton Court, erected on the lawn at Kew to commemorate the event.

5 Add. MS.24397, f.36. B.L.

6 James Bradley: *Miscellaneous Works and Correspondence*, 1832, p. xv.

7 *The Craftsman*, 6 September 1735.

8 23 December 1734. *Manuscripts of Earl of Carlisle*. Historical MSS Commission, vol.6, 1897, pp.143–4.

9 Details of acreage in 1730s and later are to be found in Sir George Lee's papers in the Buckingham Record Office.

10 Lease of White House 553/1. Surrey Record Office.

11 Duchy of Cornwall: *Household Accounts*, vol.5, ff.254–8.

12 Smith to Sir Joseph Hooker, 17 March 1879, letter inserted in Kew's copy of E. Curll's *Rarities of Richmond*, 1736.

13 Works 16/921. P.R.O.

14 Sir George Lee's papers. Buckingham Record Office.

15 24 April 1759. Bute Archives, Mount Stuart, Rothesay, Isle of Bute.

16 *Walpole Society*, vol.16, 1927–28, p.23.

17 G. B. Dodington: *Diary . . . 1749 . . . 1761*, ed. 3, 1785, p.66.

18 *Walpole Society*, vol.30, 1955, p.153.

19 *Ibid.* vol.18, 1929–30, p.14.

20 Lord Scarborough: *Ledgers*. Victoria and Albert Museum.

21 Walpole Society, vol.18, 1929–30, p.13.

22 Add. MS.19027, f.80. B.L.

23 R.A. 55240.

24 J. Harris: *Sir William Chambers*, 1970, pp.33–5.

25 R.A. 55240.

26 It is mentioned in Princess Augusta's household accounts for 1754: the "Chinese Arch in our Garden at Kew".

27 *Extracts from the . . . Correspondence of Richard Richardson*, 1835.

28 A. H. Scott-Elliot: 'The Statues by Francavilla in the Royal collection', *Burlington Magazine*, March 1956, pp.77–84.

29 W. Darlington: *Memorials of John Bartram and Humphry Marshall*, 1849, pp.184, 367.

CHAPTER 3

1 Sir George Lee's papers. Buckingham Record Office.

2 R. Pulteney to Sir J. E. Smith, 4 December 1791, Smith letters vol.24, f.167. L.S.

3 Reproduced in O. Hedley (1975), p.138, R.A.

4 *Walpole Society*, vol.16, 1927–28, p.23.

5 *Plans . . . of the Gardens and Buildings at Kew*, 1763, p.2.

6 John Rocque's 1748 plan of Richmond Gardens is dedicated to Thomas Greening and his son.

7 G. Jackson-Stop: *An English Arcadia, 1600–1990*, 1992, p.44.

8 This Lake, subsequently known as George III's Lake, was made, according to John Smith, from a number of 'lagoons and swamps' which stretched from the river opposite Syon House to the village of Mortlake.

9 *Letters to Gilbert White of Selborne from the Revd John Mulso*, 1907, p.86.

10 North MS b15, ff.17–21v. Bodleian Library. Could this be the wilderness just south of the Pagoda?

11 North MS d9, f.160. Bodleian Library.

12 Sir George Lee's papers. Buckingham Record Office.

13 *Leisure Hour*, 1862, p.767; letters from Haverfield family in Kew archives.

14 *Ibid.*

15 Bute archives. Cardiff Public Library.

16 Walpole to H. Zouch. 5 January 1761. (W. S. Lewis: *H. Walpole's Correspondence*, vol.16, 1952, p.42).

17 Lady Hill: *An Address to the Public . . . setting forth the Consequences of the Late Sir John Hill's Acquaintance with the Earl of Bute*, 1788, pp.7–8.

18 J. Hill: *The Construction of Timber from its early Growth*, 1770, p.33.

19 C. Knight (1785), p.34.

20 *Walpole Society*, vol.16, 1927–28, p.23.

21 Acc. 1132. Surrey Record Office.

22 F. Kielmansegge (1902), p.98.

23 Richardson letters. MS Rad. Tr.c12, f.45. Bodleian Library.

24 Hales to John Ellis. 21 November 1758. (J. E. Smith: *Selection of the Correspondence of Linnaeus*, vol.2, 1821, pp.40–2).

25 Identified by College of Arms.

26 J. Smith: *Records of the Royal Botanic Gardens, Kew*, 1880, p.242.

27 Letter to Lord Hillsborough. Ellis MSS. L.S.

28 Letter to Governor Tryon. 2 January 1771. Ellis MSS. L.S.

29 Letter to Lord Hillsborough. 16 November 1769. Ellis MSS. L.S.

30 Letter to William Stonehouse. 27 February 1763. B. Henrey (1986), p.257.

31 *Transactions of Linnean Society*, vol.10, 1811, p.275.

32 J. H. Harvey: 'A Scottish Botanist in London.' (*Garden History*, vol.9(1), 1981, pp.40–75).

33 Letter to Sir James Wright. 21 September 1767.

34 Collinson to John Bartram. 21 August 1766. (W. Darlington: *Memorials of John Bartram and Humphry Marshall*, 1849, p.282).

35 Volumes 12 (1767), 13 (1768) and 16 (1770).

CHAPTER 4

1 22 August 1757. CHA/1/5. Royal Academy.

2 Hardwick (1825), p.4.

3 *Walpole Society*, vol.16, 1927–28, p.23.

4 Royal Academy.

5 Hardwick (1825), p.5.

6 19 March 1844. Works 1/26, f.494. P.R.O.

7 Chambers (1763), p.4.

8 C. Knight (1785), p.35.

9 5 July 1761. W. S. Lewis: *Walpole's Correspondence*, vol.35, 1973, p.308.

10 *Designs of Chinese Buildings* (1757), plate 5.

11 11 August 1761. Translation of a letter to Linnaeus. 'Linnaeus letters', vol.14. L.S.

12 Chambers (1763), p.5.

13 Works 1/13, f.336. P.R.O.

14 *Kew Bulletin*, 1915, p.412.

15 Hooker to Dawson Turner. 9 July 1845. Kew archives.

16 Hooker to Commissioners of Woods and Forests. 28 June 1845. Works 16/590. P.R.O. The canard that the Prince Regent had sold the dragons to pay some of his debts is without foundation.

17 *Kew Bulletin*, 1915, p.411.

18 Burton to Commissioners of Woods and Forests. 10 July 1845. Kew archives and Kewensia pK120.

19 c.61.c.13, p.21. B.L. The title page is annotated in ink: "with

additions & corrections – H.W. Earl of Orford".

20 Chambers (1763), p.6.

21 *Entwurft einer historische Architectur*, 1725; English versions in 1730 and 1737.

22 A more classical version is in the drawings collection in the Royal Library.

23 Kew Annual Report 1864, p.4.

24 *Journal of Kew Guild*, 1933, p.240.

25 Michael McCarthy: *The Origins of the Gothic Revival*, 1987.

26 Works 5/96: Kew House, Michaelmas quarter 1807. P.R.O.

27 *Walpole Society*, vol.16, 1927–28, p.23.

28 Kielmansegge (1902), pp.77–8.

29 See Appendix for approximate dates when these garden buildings disappeared.

30 Chambers (1763), p.2.

31 Sir George Lee's papers. Buckingham Record Office.

32 Chambers (1772), p.53.

33 Draft letters 29 October 1761. Peter Collinson: small commonplace book, ff.180–83. L.S.

34 *Walpole Society*, vol.16, 1927–28, p.23.

35 Chambers: *Designs of Chinese Buildings*, 1757, preface.

36 *Plans, Elevations, Sections and Perspective Views of the Gardens and Buildings at Kew in Surry, the Seat of Her Royal Highness, The Princess Dowager of Wales*, 1763.

37 Dec. 1763. Add.MS.5726 Cf.126. B.L. The original drawings are now in the Metropolitan Museum of Art, New York.

38 MS. ENG. Misc.e 522. Bodleian Library.

39 Parnell: *Journal of a tour thro' England and Wales 1769.* MS MA.11 Folger Shakespeare Library, Washington.

40 P. J. Groseley (1772), vol.2, p.117.

41 *Détails de nouveaux jardins à la mode*, 1776–87, 21 parts.

CHAPTER 5

1 Plan dated 'Dec. 10th 1764' in Kew archives; another in Royal Archives at Windsor.

2 Minutes and Proceedings of Board of Works 11 October 1765. Works 4/13. P.R.O.

3 *Ibid.* 10 October 1766.

4 W. B. Cooke: *The Thames*, vol.2, 1811.

5 *Middlesex Journal*, 17 July 1773.

6 Brown introduced the ha-ha in 1767.

7 Arthur Young: *The Farmer's Tour through the East of England*, vol.2, 1771, p.247.

8 *London Magazine*, August 1774, p.360.

9 *Ibid.*

10 *Middlesex Journal*, 17 July 1773.

11 J. J. Boydell: *History of the Principal Rivers of Great Britain*, vol.2, 1796, p.36.

12 R. and J. Dodsley: *London and its Environs Described*, vol.5, 1761, pp.255–61.

13 It can now be seen in Kew Palace.

14 Collinson to William Bartram. 16 February 1768. (W. Darlington: *Memorials of John Bartram and Humphry Marshall*, 1849, p.296).

15 Annotation in A. Robertson: *Topographical Survey of the Great Road from London to Bath and Bristol*, 1792, 21. c.61.c.13. B.L.

16 Count Karl von Zinzendorf in 1768. *Journal of Garden History*, vol.9(2), 1989, p.42.

17 Chambers (1763), p.3.

18 An estimate for its demolition dated 3 July 1772, is in R.A. 15981.

19 An amateur botanist and artist who gave drawing lessons to Princess Elizabeth. Lord Bute presented the Bishop's wife (Mrs Jane Barrington) with one of the 12 copies of his *Botanical Tables*.

20 Crest 2/1245. P.R.O.

21 *Gentleman's Magazine:* vol.71, 1801, p.567.

22 *The Craftsman*, 29 May 1731.

23 Works 6, vol.16, f.109. P.R.O.

24 30 December 1793. Add. MS.33979, f.226. B.L.

25 *London Magazine*, August 1774, p.361.

26 ". . . the furniture is all English prints of elegance and humour." Lord Ailesbury 8 September 1786: "There are neat rooms in house here with Hogarth's prints." Historical

MSS Commission Ailesbury MSS.270.

27 *Journal of Kew Guild*, vol.7, 1955, p.319.

28 *Home Counties Magazine*, vol.7, 1905, p.166.

29 LS10/4, f.64. P.R.O.

30 25 March 1791. (A. Aspinall: *Later Correspondence of George III*, vol.1, 1962, p.525).

31 Banks to Sir William Hamilton. 4 July 1794. Add. MS.2641, ff.151–2. B.L.

32 Banks to A. L. de Jussieu. 29 June 1788. D.T.C. vol.6, f.40.

33 19 March 1788. Bute Archives, Mount Stuart, Rothesay, Isle of Bute.

34 18 November 1792. Smith letters, vol.18, f.129. L.S.

35 *The Times*, 28 September 1803; F. Scheer: *Kew and its Gardens*, 1840, p.17; D. Lysons: *Environs of London* ed. 2, vol.1(1), 1811.

36 The site is clearly demarcated on Thomas Chawner's plan of Kew and Richmond Gardens, July 19, 1839, Surrey County Record Office; copy in Kew archives.

37 26 September 1803. R.A. 16734. Extract quoted by gracious permission of Her Majesty The Queen.

38 Plan of Castellated Palace at Victoria and Albert Museum. E 1896 B–D–1948.

39 Joseph Farington's diary, 14 December 1807.

40 During its brief existence it was also known as the Castellated Mansion, New Palace and Kew Palace.

41 24 and 30 August 1827. Works 1/15, ff.384, 389–90. P.R.O.

42 John Charlton learned about the Gothic window in Aiton's residence (now Descanso House) from Sir Sydney Cockerell. H.M. Colvin: *King's Works*, vol.6, 1973, p.357; A. Dale: *James Wyatt*, 1956, p.189.

43 *Diaries of Colonel the Hon. Robert Fulke Greville*, 1936, p.219.

44 *Ibid.*, 20 February 1789.

45 After Robert Brown's death, it is believed that the herbarium was purchased by G. S. Gibson who bequeathed it to Saffron Walden Museum in Essex. In 1880 part of

the herbarium which had been acquired by Revd Samuel Goodenough came to Kew, to be joined by the Saffron Walden Museum Collection in 1921. (D. H. Kent and D. E. Allen: *British and Irish Herbaria*, 1984, p.188).

CHAPTER 6

1 19 August 1768. (J. E. Smith: *Correspondence of Linnaeus*, vol.1, 1821, p.231).

2 2 August 1806. Misc. letters E/1/ 114, f.36. I.O.L.R.

3 *Observations made during a Voyage round the World*, 1778, p.203.

4 January 1777. Bute Archives, Mount Stuart, Rothesay, Isle of Bute.

5 To Bernando Campo d'Alange 10 April 1796. D.T.C. vol.10(1), ff.38–43.

6 21 February 1789. (F. M. Bladon: *Diaries of . . . Robert Fulke Greville*, 1930, p.239).

7 23 February 1789. *Gentleman's Magazine*, August 1820, p.99.

8 10 August 1787. E. Smith (1911), p.102.

9 Vol. 66, 1776, pp.268–317.

10 Banks to Clarke Abel, 10 February 1816. D.T.C. vol.19, f.240.

11 Banks to Henry Shirley, Kingston, Jamaica. 2 April 1795. D.T.C. vol.9, ff.205–07.

12 Banks to Alexander Garden. 2 January 1771. (J. E. Smith: *Correspondence of Linnaeus*, vol.1, 1821, p.583).

13 These are listed in a MS 'Catalogue of trees & shrubs in his Majesty's botanic collection at Kew, planted in the year 1773'. Kew archives.

14 Banks to Robert Brudenell. 29 November 1782. D.T.C. vol.2, f.221.

15 Sir A. Geikie, *ed.: A Journey through England and Scotland to the Hebrides in 1784 by B. Faujas Saint Fond*, 1907, pp.77–84.

16 Add.MS 8097, ff.162–3. B.L.

17 East to Banks. 27 April 1789. 'Banks letters,' vol.1, f.342. Kew archives.

18 29 August 1785. Hyde collection, Somerville, New Jersey.

19 Banks to Sir Thomas Gery Cullum. 15 November 1789. E2/21/5. Suffolk Record Office, Bury St Edmunds Branch.

20 Sir James Everard Home. 'Robert Brown letters,' vol.1, f.214. Natural History Museum, London.

21 19 March 1790. Stockholm University Library.

22 *Leisure Hour*, 1862, p.768. There is no supporting evidence for this statement and we know the lake was still intact in 1796. Another version is that the lake gradually silted up leaving only the Pond now in front of the Palm House.

23 Menzies to Banks. 1 March 1791. 'Banks letters,' vol.2, f.34. Kew archives.

24 Banks to Olaf Swartz. 16 April 1792. MS.24783. Stockholm University Library.

25 LS 10/4, f.12. P.R.O.

26 10 December 1794. Cambridge Botanic Garden archives.

27 Banks to Sir J. E. Smith. 10 August 1804, 'Smith letters,' vol.1, f.127. L.S.

28 30 March 1797. (F. M. Bladon: *Historical Records of New South Wales*, vol.3, 1895, p.202).

29 Michaelmas 1801. LS10/5. P.R.O.

30 Banks to B. Campo d'Alange. 10 April 1796. D.T.C. vol.10(1), ff.38–43.

31 Banks to Lord Hawksbury. 13 September 1802. D.T.C. vol.13, ff.252–54.

32 18 April 1803. D.T.C. vol.4, ff.61–8.

33 8 April 1803. Add.MS 32439, f.95. B.L.

34 10 July 1804. D.T.C. vol.15, ff.1–2.

35 LS10/5, LS10/6. P.R.O.

36 LS10/4, f.90. P.R.O.

37 Banks to N.T. Host. 25 February 1812. D.T.C. vol.18, ff.146–7.

38 7 June 1814. D.T.C. vol.19, ff.40–1.

39 13 February 1817 'Kew collectors', vol.7A, f.18. Kew archives.

40 To Cunningham. 7 August 1818. Sutro Library, California.

41 26 August 1818. MS A300. Mitchell Library, Sydney.

CHAPTER 7

1 'Small as the book appears, the composition of it has cost him [i.e. W. Aiton] a large portion of the leisure allowed by the daily duties of his station, during more than sixteen years.' Aiton in Dedication of *Hortus Kewensis*, 1789.

2 15 August 1787. 'Sir J. E. Smith letters', vol.1, ff.87–88. L.S.

3 Sir J. E. Smith: *Selection of Correspondence of Linnaeus*, vol.2, 1821, p.3.

4 Journal of Botany 1912, supplement, 16.

5 25 April 1803. 'J. Dryander letters', f.136a. Natural History Museum, London.

6 There were 2 variant issues of the *Epitome (Journal of Botany*, 1912, supplement, 13–14).

7 Banks to Bowie. 26 March 1817. Sutro Library, California.

8 Dates of plant introductions are manifestly difficult to confirm and those cited in the *Hortus Kewensis* are not always correct.

9 'Robert Brown letters', vol.1, f.214. Natural History Museum, London.

10 24 May 1791. Add. MS. 56301. B.L.

11 B. Henrey (1975), vol.2, 253.

12 90 copies of the first part printed, of which 10 were spoilt in colouring and hot pressing; 80 copies in second part; 50 in the third and final part.

13 Bauer to J. von Jacquin. 19 August 1814. Russel Collection MS. Add. 254. Univ. College London.

14 *Annual Register*, vol.85, 1843, pp.224–5.

15 Barbara Hofland: *A Descriptive Account of the Mansion and Gardens of White-knights, a Seat of His Grace the Duke of Marlborough*, 1816.

16 Memorandum. 25 January 1806. *Historical Records of New South Wales*, vol.6, 1898, pp.16–9.

17 See Appendices for an account of the migration of Bauer's drawings to Natural History Museum, London and to Göttingen.

CHAPTER 8

1 Banks to W. Aiton. 29 August 1785. Hyde collection, Somerville, New Jersey.
2 Charles Blagden to Banks. 15 September 1785. 'Banks letters', vol.1, f.204. Kew archives.
3 16 July 1798. D.T.C., vol.11, ff.15–16.
4 18 September 1814. 'Kew collectors', vol.1, f.15. Kew archives.
5 Banks to Governor P. G. King. 22 June 1801. Mitchell Library, Sydney.
6 Banks to G. Harrison. 1 September 1814. D.T.C., vol.19, ff.56–63.
7 Banks to Governor P. G. King. 29 August 1804. D.T.C., vol.15, ff.73–8.
8 18 December 1817. 'Kew collectors', vol.7, f.33. Kew archives.
9 18 April 1803. D.T.C., vol.14, f.67.
10 Transactions of Horticultural Society of London, vol.3, 1822, f.424.
11 18 September 1814. 'Kew collectors', vol.1, f.15. Kew archives.
12 7 January 1787. D.T.C., vol.5, f.122.
13 18 September 1814. 'Kew collectors', vol.1, f.15.
14 Banks to F. Masson. 3 June 1787. D.T.C., vol.5, ff.73–4.
15 10 June 1815. 'Kew collectors', vol.1, f.22.
16 Banks to Viscount Sydney. August 1787. D.T.C., vol.5, ff.210–6.
17 Banks to D. Nelson. 1787. D.T.C., vol.5, ff.217–25.
18 A. Menzies to Banks. January 1793. D.T.C., vol.8, ff.142–55.
19 A. Menzies to Banks. 14 September 1795. 'Banks letters', vol.2, f.127. Kew archives.
20 Maggs Bros. Catalogue 686, item 444.
21 Banks to A. Menzies. 22 February 1790. 'Kew collectors', vol.11, f.282.
22 Banks to John King. 13 December 1798. D.T.C., vol.11, ff.122–4.
23 W. Bligh: A Voyage to the South Seas, 1792, p.5.
24 Bulletin of Botanical Survey of India, vol.7, 1965, p.1.
25 16 October 1798. MSS.6, f.65. R.S.
26 Banks to W. Price. 4 August 1796. D.T.C., vol.10(1), f.34.
27 Transactions of Horticultural Society of London, vol.1, 1812, p.276.
28 Banks to H. Dundas. 15 June 1787. D.T.C., vol.5, ff.184–91.
29 Aiton to Sir Henry Strachey. 18 November 1807. LS10/5, ff.136–7. P.R.O.
30 10 February 1816. D.T.C., vol. 19, f.240.

CHAPTER 9

1 W. T. Aiton to Sir William Knighton. 28 October 1827. R.A. 24208.
2 Works 1/12,f.170; Works 32/123. P.R.O.
3 Its position is roughly the path to the left just beyond the present ticket office at the Main Gate on Kew Green.
4 Works 5/109. P.R.O.
5 Curtis's Botanical Magazine, 1827, plate 2710.
6 J. Bowie to J. Smith. 5 July 1850. Inserted at rear of 'Kew collectors. Bowie 1814–21', vol.2. Kew archives.
7 Gardener's Magazine, 1828, p.333.
8 J. Smith (1880), p.262.
9 Philosophical Magazine, vol.4, 1824, pp.365–6; Botanical Miscellany, vol.1, 1830, p.65.
10 Gardener's Magazine, 1829, pp.379–84.
11 Voyage agronomique en Angleterre fait en 1829, Paris, 1830.
12 W. Herbert: Amaryllidaceae, 1837, pp.247–8.
13 F. Scheer: Kew and its Gardens, 1840, p.60.
14 4 June 1793. D.T.C., vol.8, f.210.
15 Plate 119, 1790.
16 20 December 1790. Add MS 8097, ff.400–1. B.L.
17 1 December 1797. D.T.C., vol. 10(1), ff.210–3.
18 4 December 1797. D.T.C., vol.10(1), ff.214–6.
19 Sutro Library, California.
20 Gardener's Gazette, 21 October 1837.
21 W. T. Aiton to Office of Works 2 March 1824. Works 1/12,f.375. P.R.O.
22 5 February 1825. 'Kew collectors', vol.7, f.134. Kew archives.
23 Gardener's Magazine, vol.1, 1826, p.54.
24 J. P. Burnard: 'Remarks on the policy pursued in the management of the King's Botanic Garden at Kew.' Gardener's Magazine, vol.2, 1827, pp.313–5.
25 Penny Cyclopaedia, vol.11, 1838, p.74: anonymous but probably by J. Lindley.
26 W. Herbert: Amaryllidaceae, 1837, p.410.
27 J. Smith: Unpublished memoirs of Kew Gardens, p.486. Kew archives. Probably (Sir) Robert William Gardiner, later General-Commandant of the Royal Horse Artillery. Lived at Claremont, Surrey.
28 1830, p.731.
29 J. Smith: Records (1880), pp.vi–vii.
30 Kew Bulletin, no.3, 1950, p.309.
31 Kew file 284, passim. Kew archives.
32 'Kew gates 1847–1913' f.213 for sketch map showing these alterations. Kew archives.
33 J. Smith: Unpublished memoirs of Kew Gardens, ff.26–7. Kew archives. Works 16/589. P.R.O.
34 L. J. Jennings, ed.: Croker Papers, vol.2, 1885, p.101.
35 M. Charlot: Victoria, the Young Queen, 1991, p.63.
36 13 May 1836. Works 16/29/4. P.R.O.
37 J. Lindley et al.: 'Report on Royal Gardens', 1838, f.144. Kew archives.
38 'Kew collectors', vol.4. Kew archives.
39 Gardener's Gazette, 7 October 1837.
40 The Times, 31 October 1837.
41 Morning Chronicle, 14? October 1837.
42 Horticultural Journal, 1 November 1837, p.97.
43 J. C. Loudon: 'Hints for a national garden'. SP.716. L.S.
44 Sutro Library, California.
45 Vol. 7, 1831, p.96.

46 *Gardener's Magazine*, vol.14, 1838, pp.68–71.

47 T90/189. P.R.O.

48 *Botanical Garden (Kew): Return to an order of the Honourable The House of Commons dated 11 May 1840.*

49 T90/190. P.R.O.

50 Letter bound in *Copy of a Letter addressed to Dawson Turner* 1840. Kew archives.

51 'Creation of Royal Botanic Gardens in 1840', f.6. Kew archives.

52 J. Smith (1880), pp. xii–xiii.

53 A. B. Lambert to J. Smith. 9 March 1840. 'A. B. Lambert letters', Kew archives.

CHAPTER 10

1 19 June 1813. 'Banks letters', vol.1, f.39. Kew archives.

2 14 August 1824. 'Extracts from letters from Sir W. J. Hooker to Sir J. Richardson, 1819–1843', f.27. Kew archives.

3 8 March 1838. 'Sir W. Hooker letters', vol.46, f.12. Kew archives.

4 3 March 1838. 'Lindley letters', f.473. Kew archives.

5 18 January 1839. 'Extracts from letters . . . to Dawson Turner, 1834–50', ff.8–9. Kew archives.

6 *Copy of a Letter addressed to Dawson Turner . . . on the Occasion of the Death of the late Duke of Bedford particularly in reference to the Services rendered by His Grace to Botany and Horticulture*, 1840.

7 3 April 1840. Letter bound in Kew's copy of *Copy of a Letter addressed to Dawson Turner*, 1840.

8 11 August 1840. 'English letters 1840, I–Z', f.37. Kew archives.

9 Not 15 acres as stated in Lindley's 1838 report, although Hooker later contradicted himself and quoted 14 acres. (*Companion to Botanical Magazine*, 1845, p.7).

10 17 May 1845. 'Extracts from letters . . . to Dawson Turner, 1834–50', f.56. Kew archives.

11 24 March 1841. 30/22/4A, ff.155–6. P.R.O.

12 4 October 1842. 'Extracts from letters . . . to Dawson Turner, 1834–50', f.48. Kew archives.

13 13 July 1843. 'Extracts from letters . . . to Dawson Turner, 1834–50', f.50. Kew archives.

14 It included Methold's Garden purchased by George III in 1794.

15 The site is now occupied by the present Herbaceous Ground or Order Beds but the arrangement is still that of Bentham and Hooker. The earlier systematic arrangement adopted the prevailing Linnean classification.

16 *Catalogue of the Hardy Herbaceous Plants in the Royal Gardens of Kew*, 1853.

17 Sketch plan dated October 1843. Kew archives.

18 Hooker to Lord Lincoln. 1 February 1844. Nec 8969/1. Nottingham University.

19 Sketch plan by John Smith, 1844. Kew archives.

20 Burton to Woods and Forests. 7 March 1844. Works 16 29/8. P.R.O.

21 Burton to Woods and Forests. 23 August 1844. Works 16 29/8. P.R.O.

22 Turner to Charles Gore. 9 May 1844. Works 16/29/8 ff.13–14. P.R.O.

23 Turner to Woods and Forests. November 1848. Works 16/29/8 ff.178–9. P.R.O.

24 J. Smith: Unpublished memoirs of Kew Gardens, p.76. Kew archives.

25 *Gardeners' Chronicle*, 24 July 1847, p.486.

CHAPTER 11

1 Sir John Hill: *Hortus Kewensis*, 1768.

2 J. Smith (1880), p.260.

3 Board of Green Cloth to Treasury. 19 November 1831. T1/3963: bundle of letters on Kew and Deer Park, 1825–40. P.R.O.

4 J. C. Loudon: *Arboretum et Fruticetum Britannicum*, vol.1, 1838, p.75.

5 *Florist's Journal*, 1840, p.100.

6 Smith (1880), p.269.

7 Hooker to Lord Lincoln. 1 February 1844. Newcastle MS 8969. Nottingham University Library.

8 Nesfield to Hooker. 15 February 1844. 'Sir W. Hooker letters', vol.22, f.146. Kew archives.

9 Hooker to Lord Lincoln. 20 February 1844. Newcastle MS 8971. Nottingham University Library.

10 Nesfield to Hooker. [2 March] 1844. 'Sir W. Hooker letters', vol.22, f.148. 'Sketch for arboretum at Kew', not dated. Kew archives.

11 Nesfield to Hooker. 25 April 1844. 'Sir W. Hooker letters', vol.22, f.153. Kew archives.

12 Gore to Hooker. 13 January 1845. 'Kew Pleasure Grounds, 1845–1911', f.218. Kew archives.

13 Burton to Hooker. 18 February 1845. 'Sir W. Hooker letters', vol.23, f.109. Kew archives.

14 Burton to Hooker. 19 February 1845. 'Sir W. Hooker letters', vol.23, f.110. Kew archives.

15 20 February 1845. Kew 207. Kew archives.

16 Burton to Board of Woods and Forests. 21 February 1845. Kew 207. Kew archives.

17 Hooker to Dawson Turner. 17 May 1845. 'Extracts from letters . . . to Dawson Turner, 1834–50', f.50. Kew archives.

18 'Observations on the proposed arboretum in conjunction with the Royal Botanic Gardens at Kew'. 'Kew Pleasure Grounds, 1845–1911', ff.216–7. Kew archives.

19 Burton to Hooker. 28 May 1845. 'Sir W. Hooker letters', vol.23, f.92. Kew archives.

20 Burton to Hooker. 4 June 1845. 'Sir W. Hooker letters', vol.23, f.113. Kew archives.

21 'Sir W. Hooker letters', 1845–1851', f.56. Kew archives.

22 Copies of Nesfield's report are filed in Kew 207, Kew 258 and 'Kew Pleasure Grounds, 1845–1911', ff.214–5. Kew archives.

23 Hooker to Board of Woods and Forests. 18 August 1849. Kew 39. Kew archives.

24 *Cottage Gardener*, 18 September 1860, p.369.

25 Works 32/114 and 115. P.R.O.

26 Nesfield to Hooker. 19 November 1848. 'Sir W. Hooker letters', vol.26, f.406.

27 Nesfield to Hooker. 28 December 1848. 'Sir W. Hooker letters', vol.26, f.407.

28 Works 1/31, f.206; Works 1/33, ff.11-2. P.R.O.

29 Burton to Board of Woods and Forests. 21 February 1845. Kew 207. Kew archives.

30 18 July 1844. Kew 88. Kew archives.

31 24 August 1844. Kew 88. Kew archives.

32 Hooker to Dawson Turner. 5 August 1843. Kew 88. Kew archives.

33 6 September 1843. Kew 88. Kew archives.

34 Upton to Lord Lincoln. 20 September 1843. Kew 88. Kew archives.

35 28 and 29 April 1845. Kew 88. Kew archives.

36 Hooker to Board of Woods and Forests. 2 May 1845. Kew 88. Kew archives.

37 8 May 1845. 'Extracts from letters . . . to Dawson Turner, 1834-1851', f.35.

38 19 August 1851. 'Kew Pleasure Grounds, 1845-1911', ff.10-1. Kew archives.

39 4 December 1853. 'Sir W. Hooker letters to Mr Bentham, 1842-1862', f.743. Kew archives.

40 19 July 1856. 'Sir W. Hooker letters', vol.36, f.201. Kew archives.

41 7 October 1856. Ibid. f.208.

42 10 October 1856. House of Commons. Copies of papers relating to changes . . . affecting . . . the Gardens at Kew, 1872, p.76.

43 J. Smith (1880), p.339.

44 Gardeners' Chronicle, 1866, p.880.

45 Hooker to Office of Works, 8 December 1855. Works 16 31/5. P.R.O.

46 Cottage Gardener, 18 September 1860, p.369.

47 Board of Woods and Forests to D. Burton. 29 August 1848. Works 1/32, f.449. P.R.O.

48 Cottage Gardener, 23 October 1855, p.52.

49 Ibid. 16 October 1855, p.35.

50 21 December 1856. 'J. D. Hooker letters to W. Munro'. Kew archives.

51 Cottage Gardener, 22 September 1857, p.388.

52 Ibid. 6 October 1857, p.2.

53 Ibid. 20 September 1859, p.363.

54 Ibid. 21 August 1860, p.310.

55 Now known as Victoria amazonica.

56 Curtis's Botanical Magazine, 1847, plates 4275-8.

57 15 June 1858. House of Commons. Copies of papers relating to changes . . . affecting . . . the Gardens at Kew, 1872, pp.83-4.

58 Gardeners' Chronicle, 5 June 1858, p.456.

59 Burton to Hooker. 28 March 1859. 'Sir W. Hooker letters', vol.39, f.142. Kew archives.

60 Kew Annual Report, 1871. p.5.

61 August 1863; repeated on 27 November 1863. pK 96. Kew archives.

CHAPTER 12

1 19 July 1893. Add MS 33981, ff.117-8. B.L.

2 27 December 1841. 'Wallich letters'. Kew archives.

3 Garden, 24 January 1880, p.75; J. Smith (1880), vi-vii; letter to J. Britten, 11 March 1883, filed with F. Masson drawings at Natural History Museum; J. Smith's unpublished memoirs of Kew Gardens, ff.310-1.

4 Annals of Botany, vol.16, 1902, p. lxv.

5 Hooker to D. Turner. 22 November 1845. 'Extracts from letters . . . to Dawson Turner, 1834-50', f.60. Kew archives.

6 J. Hooker to G. Bentham. 20 April 1851. 'Sir. J. D. Hooker letters', f.49. Kew archives.

7 10 and 21 December 1851. 'Sir W. Hooker's letters to Mr Bentham, 1842-62', ff.650, 651. Kew archives.

8 J. Hooker to W. H. Harvey. 15 January 1852, 'Letters from J. D. Hooker', vol.5, f.248. Kew archives.

9 Sir B. Hall to Col. Phipps. 30 August 1855. R.A. Add MS Qf.224.

10 18 July 1859. Hansard House of Commons Debates, p.1434.

11 Minutes of evidence, 1838, para. 107. T190. P.R.O.

12 J. Smith: Unpublished memoirs of Kew Gardens, pp.249,294. Kew archives.

13 Kew 207. Kew archives.

14 14 September 1846. Works 1/30, f.194. P.R.O.

15 3 October 1846. 'Extracts from letters . . . to Dawson Turner', 1834-50', f.66. Kew archives.

16 18 October 1848. Ibid. f.73. Kew archives.

17 J. D. Hooker to W. J. Hooker. 21 June 1850. 'Letters from J. D. Hooker', vol.9, f.208. Kew archives.

18 18 August 1850. 'Letters from J. D. Hooker', vol.9, f.225. Kew archives.

19 27 September 1852. Works 16 31/4. P.R.O.

20 Receipt for sale (£18,250) 3 July 1823. Kew 276. Kew archives; R. A. Geo 34784.

21 Kew 276. Kew archives.

22 3 February 1857. 'J. D. Hooker letters'. Kew archives.

23 Catalogue of the plants distributed at the Royal Gardens, Kew (under the sanction of the Secretary of State for India) from the herbaria of Griffith, Falconer and Helfer, 1865.

24 4 August 1842. 'W. Hooker letters, 1833-1844', f.424. Kew archives.

25 D. J. Mabberley (1895), p.392.

26 Report of the Commission (1850), para. 2681.

27 Ibid. para 3468.

28 J. Hooker to Sir W. Hooker. 24 July 1848. 'Letters from J. D. Hooker', vol.8, f.223. Kew archives.

29 15 January 1852. 'Letters from J. D. Hooker', vol.5, f.248. Kew archives.

30 Hooker to T. W. Philipps. 16 February 1854. 'Kew Presentations and Purchases, 1846-1898'. Kew archives.

31 L. Huxley (1918), vol.1, p.378.

32 7 July 1855, p.452.

33 18 June 1858. 'Letters from J. D. Hooker to Huxley, 1851-94', f.37. Kew archives.

34 14 June 1858. 'Royal Botanic Gardens, Kew and the British Museum'. Kew archives.

35 *Copy of all communications made by the officers and architect of the British Museum to the Trustees respecting the want of space for exhibiting the collections in that institution* . . . 1859.

36 18 June 1858. 'Letters from J. D. Hooker to Huxley, 1851–94', f.40. Kew archives.

37 *Correspondence of Charles Darwin* vol.7, 1991, pp.522–30.

38 Kew Annual Report 1875, p.2.

39 Geo. 34766–7. R.A.

40 Geo. 34779. R.A.

41 *Athenaeum*, 26 December 1840, pp.1025–6.

42 'Robert Brown letters', vol.1, f.214. Natural History Museum. This statement also appears in *Athenaeum* 1840, pp.1025–6 and *Annals of Natural History*, vol.7, 1841, p.78.

43 1 March 1865. 'Kew Presentations and Purchases, 1846–1898'. Kew archives.

44 12 March 1899. 'J. D. Hooker letters to W. T. Dyer', f.276. Kew archives.

CHAPTER 13

1 Kew Annual Report 1844, p.3.

2 G. MacNab to R. Graham. 7 December 1843. 'Kew collectors. W. Purdie', f.13. Kew archives.

3 15 September 1847. 'Kew collectors, 1791–1865', f.9. Kew archives.

4 31 March 1851. 'Sir W. Hooker's letters to Mr Bentham, 1842–62', f.629. Kew archives.

5 Sir W. Hooker to Wilford. 4 January 1859. 'Kew collectors', vol.8, f.74. Kew archives.

6 3 July 1861. 'Kew collectors', vol. 9, f.18. Kew archives.

7 Oldham to John Smith. 2 October 1862. *Ibid*, f.1. Kew archives.

8 Oldham to Hooker. 8 June 1864. *Ibid*, f.44. Kew archives.

9 Oldham to Hooker. 13 February 1864. *Ibid*, f.28. Kew archives.

10 Oldham to Hooker. 13 February 1864. *Ibid*, f.25. Kew archives.

11 *Report on the present condition of the Botanical Garden at Kew, with recommendations for its future administration*, 1840, p.5.

12 Kew Annual Report 1844.

13 18 November 1844. 'Sir W. Hooker's letters, 1833–1844', f.545. Kew archives.

14 8 July 1849. 'Sir W. Hooker's letters to Mr Bentham, 1842–1862', f.548. Kew archives.

15 29 October 1859. 'India Economic Products. Cinchona', f.19. Kew archives.

16 14 May 1857. *Kew Bulletin*, no.2, 1905, p.12.

17 *Ibid*, p.13.

18 Duke of Newcastle to Sir W. Denison. 28 June 1868. *Ibid*, p.14.

19 *Flora Indica*.

20 *Colonial Floras*, 1863. Kew archives.

21 Hooker to Dawson Turner. 23 October 1848. 'Sir W. Hooker letters 1845–1851', f.319. Kew archives.

22 *Gardener's Magazine*, vol.1, 1826, p.12.

23 W. B. Hemsley, a Kew employee since 1860. *Journal of Kew Guild*, 1895, p.28.

24 *Annals of Botany*, vol.16, 1902, p. xlvii.

25 Kew Annual Report 1849.

26 Sir J. Hooker to Asa Gray. 12 October 1865. Archives, Library of the Gray Herbarium, Harvard University.

CHAPTER 14

1 J. Hooker to Sir W. Hooker, 15 April 1849. 'Letters from J. D. Hooker', vol.9, f.53. Kew archives.

2 Memorial to Lord Seymour. May 1851. The eight petitioners who signed this memorial included Robert Brown, John Lindley and George Bentham.

3 24 March 1854. 'Letters from J. D. Hooker', vol.5, f.16. Kew archives.

4 J. Hooker to I. Palgrave. 26 April 1900. L. Huxley (1918), vol.2, p.381.

5 J. Hooker to C. Darwin. October? 1863. *Ibid*, p.68.

6 November 1865. *Ibid*, pp.48–9.

7 28 September 1866. *Ibid*, p.80.

8 J. Hooker to C. Darwin. February 1868. *Ibid*, p.80.

9 10 March 1870. 'Letters to J. D. Hooker', vol.1, f.182. Kew archives.

10 Hooker to R. Lingen. 19 February 1872. 'Ayrton controversy', vol.1, f.105. Kew archives.

11 *Curtis's Botanical Magazine*, 1867, pl.5621.

12 17 August 1872, p.150.

13 W. Robinson: *Parks and Gardens of Paris*, 1869, p.72.

14 *Garden*, 16 March 1872, p.379.

15 24 June 1845. 'Extracts from letters . . . to Dawson Turner, 1834–50', f.53. Kew archives.

16 *Gardeners' Chronicle*, 12 September 1874, p.333.

17 1875. 'Kew Gardens. Forenoon opening, 1866–1889', f.144. Kew archives.

18 *Richmond and Twickenham Times*, 15 September 1877.

19 Kew Annual Report 1877, p.7.

20 Memorandum. 29 October 1877. 'Admission of the public, 1853–1925'. Kew archives.

21 *Observer*. 8 September 1878.

22 Hansard House of Commons Debates.

CHAPTER 15

1 R. J. Callender to Hooker. 17 August 1871. *Copies of papers relating to changes . . . affecting . . . the Gardens at Kew*, 1872, p.107.

2 Hooker to Ayrton. 19 August 1871. *Ibid*, p.107.

3 Hooker to Sir Charles Lyell. 12 January 1871. "He [Ayrton] is an incarnation of low cunning and falsehood, and as ignorant and dull. If he perseveres in his efforts to remove Smith, my Curator, from Kew, or add to his work here, I will go to Mr Gladstone and expose him, and demand an investigation." 'Letters from J. D. Hooker', vol.10, f.312. Kew archives.

4 Hooker to Gladstone. 19 August 1871. 'Papers relating to Kew, 1867–72. Ayrton controversy', f.18. Kew archives.

5 19 August 1871. 'Letters from

Hooker to Huxley, 1851–94', f.109. Kew archives.

6 Hooker to Ayrton. 31 August 1871. *Copies of papers relating to changes . . . affecting . . . the Gardens at Kew*, 1872, pp.108–10.

7 Ayrton to Gladstone. 26 September 1871. *Ibid.*, pp.111–2.

8 4 October 1871. 'Papers relating to Kew, 1867–72. Ayrton controversy', f.45. Kew archives.

9 10 October 1871. *Ibid.*, f.48. Kew archives.

10 16 and 23 October 1871. *Ibid.*, f.59. Kew archives.

11 30 October 1871. *Ibid.*, ff.72–7. Kew archives.

12 1 November 1871. *Ibid.*, f.83. Kew archives.

13 14 January 1872. Gray Herbarium archives.

14 'Papers relating to Kew, 1867–72. Ayrton controversy', f.128. Kew archives.

15 *Ibid.*, f.159. Kew archives.

16 Hooker to Asa Gray. 22 May 1872. Archives, Library of Gray Herbarium, Harvard University.

17 Tyndall to Hooker. 18 May 1872. 'Papers relating to Kew, 1867–72. Ayrton controversy', f.210. Kew archives.

18 *Ibid.*, ff.41–9. Kew archives.

19 *Copies of papers relating to changes . . . affecting . . . the Gardens at Kew* 1872, pp.51–64.

20 31 December 1868. 'British Museum (Natural History), 1858–1901', ff.201–09. Kew archives.

21 Kew 73. Kew archives.

22 17 August 1872. 'Papers relating to Kew, 1867–72. Ayrton controversy', f.457. Kew archives.

23 6 September 1872. Kew 73. Kew archives.

24 21 September 1872. Kew 73. Kew archives.

25 31 October 1872. Archives, Library of Gray Herbarium, Harvard University.

26 10 October 1872. 'Letters from J. D. Hooker', vol.2, f.284. Kew archives.

27 3 January 1873. *Memorial addressed to Right Hon. W. E. Gladstone on the subject of the national herbaria.* Reprinted in

Gardeners' Chronicle, 18 January 1873, pp.72–3.

28 *Memoranda of the Fellows of the Royal Society*, 1873, p.1. 1991e, 128(15). Bodleian Library.

29 A. West: *Recollections, 1832 to 1886*, vol.2, 1899, p.14.

30 Lord Henry Gordon-Lennox to H. Ponsonby. 29 September 1875. Add. MSS Q913. R.A.

31 Biddulph to Lord Henry Gordon-Lennox. 16 October 1875. Add. MSS Q916. R.A.

32 Hooker to Frank Darwin. 18 April 1887. 'Letters from J. D. Hooker, BOL-DAR', f.160. Kew archives.

33 Hooker to Frank Darwin. 13 March 1893. *Ibid.*, f.208. Kew archives.

34 Royal Commission. 4th report, 1874, p.10.

35 Jodrell to Hooker. 11 November 1874. Kew 25. Kew archives.

36 21 December 1874. *Ibid.*

CHAPTER 16

1 *Memorial addressed to Right Hon. W. E. Gladstone on the subject of the national herbaria*, 1873, p.3.

2 J. D. Hooker to Secretary of Office of Works. 27 October 1874. 'Appointment of Assistant Directors, 1874–98', f.2. Kew archives.

3 Treasury to Lord Henry Gordon-Lennox. 24 March 1875. *Ibid.*, f.21. Kew archives.

4 26 May 1875. 'Letters from J. D. Hooker', vol.11, f.260. Kew archives.

5 *Journal of Royal Society of Arts.* 7 April 1876, p.476.

6 'India Office. Caoutchoac', vol.1, f.2. Kew archives.

7 H. A. Wickham: *On the Plantation, Cultivation and Curing of Para India Rubber*, 1908.

8 W. Dean: *Brazil and the Struggle for Rubber*, 1987.

9 *Kew Bulletin*, 1912, p.65.

10 Thiselton-Dyer to India Office. 12 August 1876. 'India Office. Caoutchoac', vol.1, f.32. Kew archives.

11 Thiselton-Dyer to India Office. 17 April 1878. *Ibid.*, vol.2, f.463.

12 Government of India to India

Office. 21 August 1876. 'India Office. Caoutchoac', vol.1, ff.60–4. Kew archives.

13 4 April 1851. N. B. Ward: *On the Growth of Plants in loosely glazed Cases*, 1852, pp.131–2.

14 27 May 1905. 'India Office. Caoutchoac', vol.2, f.303. Kew archives.

15 9 September 1872. 'Letters from J. D. Hooker', vol.1, f.196. Kew archives.

16 Kew Annual Report, 1877, p.8.

17 *Botanical Enterprise of the Empire*, 1880, p.6.

18 'Kew administration, 1864–1925', f.61. Kew archives.

19 Works 16 29/5. P.R.O.

20 This metal shed, known as the 'Iron Room', subsequently served as a lecture hall for student gardeners until as recently as 1962.

21 11 August 1879. Kew 249. Kew archives.

22 Thiselton-Dyer to David Prain. 3 August 1907. 'Letters from Thiselton-Dyer', vol.1, f.151. Kew archives.

23 'Kew administration, 1864–1925', f.36. Kew archives.

24 25 November 1912. 'Letters from Thiselton-Dyer', vol.2, f.150. Kew archives.

25 J. D. Hooker to Asa Gray. 4 October 1885. 'Letters from J. D. Hooker', vol.5, f.149. Kew archives.

26 'J. D. Hooker letters to Thiselton-Dyer', f.196. Kew archives.

27 *Proceedings of Linnean Society*, 1897–98, p.33.

CHAPTER 17

1 Sir A. Geikie, ed.: *A Journey through England and Scotland to the Hebrides in 1784 by B. Faujas Saint Fond*, vol.1, 1907, p.83.

2 *Penny Magazine*, 1841, p.267.

3 *Gardener's Magazine*, 1843, p.455.

4 *Garden*, 16 March 1872, p.379.

5 *Ibid.*, 22 August 1874, p.174.

6 *Ibid.*, 10 February 1906, pp.81–2.

7 *Ibid.*, 6 January 1906, pp.1–2.

8 Minute to Office of Works. 24 May 1891. Kew 207. Kew archives.

9 28 March 1896, p.323.

10 W. Dallimore: 'A Gardener's Reminiscences', p.465. Kew archives.

11 Thiselton-Dyer to Office of Works. 31 January 1898.

12 'Household papers', vol.14, p.1346. R.A.

13 21 August 1883. 'Kew administration, 1864–1925', f.75. Kew archives.

14 8 December 1896. Works 19/229. P.R.O.

15 Works 19/229. P.R.O.

16 *Ibid.*, 16 April 1898.

17 Sir J. D. Hooker to H. Bolus. 20 November 1883. L. Huxley (1918), vol.2, pp.246–7.

18 Thiselton-Dyer to Office of Works. 24 November 1894. Kew 93. Kew archives.

19 Sarah A. M. Brown writing to Miss Symonds on behalf of Sir Joseph Hooker. 20 March 1902. 'Letters from J. D. Hooker', vol.13, f.261. Kew archives.

20 *Kew Bulletin*, 1892, p.42.

21 M.A.F. 46/60. P.R.O.

22 Thiselton-Dyer to Secretary of Board of Agriculture. 14 April 1905. M.A.F. 46/59. P.R.O.

23 L. Linder: *editor: Journal of Beatrix Potter from 1881 to 1897*, 1966, p.414.

24 Turrill (1959), p.36.

25 Departmental Committee on Botanical Work and Collections at the British Museum and at Kew. Minutes of evidence . . . 1901.

26 Thiselton-Dyer to Office of Works. 7 January 1899. *Ibid.*, f.lv.

27 *Nature*, 24 November 1910, p.104.

28 Board of Agriculture to Treasury. 10 April 1902. 'Kew. Transfer from Office of Works to Board of Agriculture'. Kew archives.

29 Thiselton-Dyer to D. Prain. 28 March 1913. 'Letters from Thiselton-Dyer', vol.2, f.167. Kew archives.

CHAPTER 18

1 Sir C. Bruce: *The Broad Stone of Empire*, vol.2, 1910, p.117.

2 Thiselton-Dyer to Sir Mountstuart E. Grant Duff. March 1885. *Notes from a Diary kept chiefly in Southern India, 1881–1886 by Sir M.E.G. Duff*, vol.2, 1899, p.21.

3 Jamaica. Annual Report 1892, p.6.

4 16 June 1890. 'Kew. Royal Niger Company', f.34. Kew archives.

5 See Appendices for comments on Kewites in senior appointments in the Empire.

6 *India-rubber and Gutta-percha*, 9 June 1890.

7 *Kew Bulletin*, 1888, p.150.

8 9 March 1890. 'Kew. Royal Niger Company', f.64. Kew archives.

9 T.V. Lister to Thiselton-Dyer. 2 March 1891. *Kew Bulletin*, 1894. p.18.

10 H. Shirley to Sir J. Banks. 20 December 1794. D.T.C. vol.9, ff.141–3.

11 16 October 1789. Misc. MSS 6, f.65. R.S.

12 19 July 1803. Add MS. 33981, ff.117–8. B.L.

13 Sir J. Banks to Sir G. Yonge. 15 May 1787. D.T.C. vol.5, ff.159–66.

14 Report on the gardens in Jamaica, 1869.

15 20 July 1872. 'Papers relating to Kew, 1867–72. Ayrton controversy', f.313. Kew archives.

16 Kew Annual Report 1878, p.25.

17 *Notes on Liberian Coffee, its History and Cultivation*, 1881.

18 4 December 1883. 'West Indies botanic stations, 1884–97', f.6. Kew archives.

19 Thiselton-Dyer to Colonial Office, 15 December 1893. 'Grenada Botanical Garden, 1885–1900', f.100. Kew archives.

20 Thiselton-Dyer to Colonial Office. 2 June 1890. *Kew Bulletin*, 1891, p.106.

21 Reprinted in *Kew Bulletin. Additional series I*, 1898.

22 *Kew Bulletin*, 1895, 205–8.

CHAPTER 19

1 Thiselton-Dyer to Office of Works. 1 January 1885. 'Kew Bulletin correspondence, 1882–1919', f.2. Kew archives.

2 *Journal of Botany*, 1904, p.160.

3 6 July 1905. 'Letters from J. D. Hooker', vol.13, f.114. Kew archives.

4 Was it just a slip when William Dallimore in his unpublished memoirs said Sir William had resigned, suggesting that he had been humiliated by having to reinstate the two gardeners he had just dismissed?

5 *Proceedings of the Royal Society. B.*, vol.106, 1930, p.xxvii.

6 L. Linder, *editor: Journal of Beatrix Potter from 1881 to 1897*, 1966, p.426.

7 Minute to Office of Works. 6 April 1886. Kew 261. Kew archives.

8 J. D. Hooker to Sir I. Palgrave. 6 July 1905. 'Letters from J. D. Hooker', vol.13, f.114. Kew archives.

9 Introduction to W. J. Bean's *Royal Botanic Gardens, Kew*, 1908, p. xviii.

10 *Journal of Kew Guild*, 1907, p.354.

11 MAF 46/36. P.R.O.

12 Audrey le Lièvre: *Miss Willmott of Warley Place: her Life and her Gardens*, 1980, p.188.

13 House of Lords Debates, 10 July 1917.

14 'A list of economic plants native or suitable for cultivation in the British Empire', *Kew Bulletin*, 1917, pp.241–96.

15 MAF 46/34. P.R.O.

16 *Obituary notices of Fellows of the Royal Society*, vol.4, 1942, p.88.

17 Cultivated crop plants of the British Empire and the Anglo-Egyptian Sudan, *Kew Bulletin. Additional series*, vol.12, 1936.

18 *Interim Report*, 1928.

19 *Final Report*, part 2, 1930, p.22.

20 MAF 46/21. P.R.O.

CHAPTER 20

1 *Biographical Memoirs of the Royal Society*, vol.26, 1980, p.511.

2 Sir G. Taylor's autobiography. Acc. 9533/321. National Library of Scotland.

3 Court Roll for Richmond, 1647.

4 Sir G. Taylor's autobiography. Acc. 9533/321. National Library of Scotland.

5 Office of Woods and Forests to Treasury. 12 March 1838. 'Report

on Royal Garden at Kew 1838',
ff.204–10. Kew archives.
6 See Appendices for a full statement.
7 Memorandum submitted by Kew to
Select Committee on Science and
Technology (Sub-committee II:
Systematic Biology Research).
House of Lord's Sessional Paper,
1991, 92.
8 Kew's floras and monographs are
listed in the chronological
appendix.

9 Annual Report. Peradeniya Botanic
Garden, 1844.
10 Daniel Corneille to Banks.
17 March 1787. 'Banks letters',
vol.1, f.262. Kew archives.
11 Instructions to J. Wiles about to
join *Providence*. 25 June 1791.
D.T.C. vol.7, f.224.
12 7 May 1791. 'Banks letters', vol.2,
f.45. Kew archives.

13 His herbarium and drawings of
St Helena plants are now at Kew.
14 Unpublished journal quoted by
L. C. Brown in *Flora and Fauna
of St Helena*, 1982, p.5.
15 'St Helena. Cinchona, 1868–98',
f.48. Kew archives.
16 Royal Botanic Gardens, Kew.
*Corporate Strategic Plan. 1994/
95–1998/99*, p.2.

INDEX

Page numbers in bold refer to illustrations